History of

American Technology

JOHN W. OLIVER

PROFESSOR EMERITUS IN HISTORY
UNIVERSITY OF PITTSBURGH

THE RONALD PRESS COMPANY ~ NEW YORK

3

Library of Congress Catalog Card Number: 56-6269

PRINTED IN THE UNITED STATES OF AMERICA

FOREWORD

The distinction of this volume is that it represents the first comprehensive historical account of American technology and invention as a basic contribution to the nation's culture. The oft-told story of Puritanism and politics, of laws and constitutional struggle, of wars and the beckoning frontier here give way to the wonderful record of technological achievements. Its archives begin with the pioneer's struggles against nature and carry the reader to the age of automation and the atom. The American colonist in a virgin land brought to his tasks the old technologies, the newer political ideas, and often the insurgent religious faiths of the Europe he had left behind. He blended these with physical stamina into an equipment that was ideal for his times and his tasks.

It is here that Professor Oliver takes up the long-neglected story of how the crude technologies of colonial days were developed by ingenuity and genius into a dominant characteristic of American culture. The pace is slow at first, but the application of science swells the stream to a tumultuous torrent that challenges man's social inventiveness to direct and control it. Can we dedicate this new physical world to the task of social and spiritual triumphs? This is a challenge that I put in print, in other words, some twenty-five years ago. The question is even more compelling today, which explains my deep interest in this most useful book.

I know it to be a book which is the fruitful result of twenty years of teaching and study of its theme. It is a book which should stimulate courses in the subject in liberal arts colleges and in every school of science, engineering, and technology in the land. Such a course will give students of the sciences new pride in their profession and a larger sense of their social responsibility in the new world created by science and technology.

Guy Stanton Ford

Washington, D. C.

PREFACE

This book is a historical survey of the development of American science and technology and of their effect on the growth of America's culture. Americans have developed many tools and techniques, inventions, and devices to provide themselves with food, clothing, shelter, transportation, and communication in abundance never before seen on the earth. With these applications of technology, Americans have conquered a continent and have made great strides toward harnessing the forces of nature and putting them to man's use.

History is commonly thought of as a record of struggles for power between dynasties, nations, or economic groups. The United States has had its share of such struggles but it has something besides that has made its history unique as it has made America's culture and its place in world affairs unique. This something is what is frequently called American know-how, inventiveness, or Yankee ingenuity, all of which are terms that refer to the characteristic development of science and technology that has taken place in our country since its beginning and has given us today a standard of living unequalled in the world—plus a position of world leadership and responsibility.

The names of such men as Franklin, Evans, Morse, Whitney, Edison and a host of others loom larger in the pages of American history than do their counterparts in the history of other nations. These men, who worked not for personal power but for the common welfare of mankind, made contributions that exemplify the American belief in the value of technology. This belief so characterizes our culture that no history of our country is complete without a review of American scientific and technological achievements.

American technology had its beginnings in the efforts of the first settlers to carve homes and farms from the wilderness and to provide the necessities of life for themselves and their families. The landmarks of our history since that time have been the men and the inventions that have made it possible for us today to enjoy the benefits of labor-saving machinery, mass production, and automatic manufacturing.

One point to be borne in mind throughout this survey of American technology is that we have not yet reached our highest development nor have we exhausted all the possibilities of science. In spite of its long history, our technology is still, in a sense, in its infancy;

the advent of atomic power and automation are merely indications of what lies ahead. It can safely be assumed that among the readers of this book are some whose names will have important places in subsequent chapters of the history of our struggle to control nature and adapt her forces to the ends of man.

It is impossible to list the names of all those who, through the years, have contributed to the formation and writing of this book. Students, colleagues, and research scientists in all fields have been a constant inspiration. Their enthusiasm for this reappraisal of American civilization was so contagious that a great many people joined the venture.

Among those to whom I am most indebted are Alexander J. Allen, William J. Bining, William G. Buchanan, Robert J. Corbett, James Coul, George L. Davis, Myrl I. Eakin, G. R. Fitterer, John Abram Foster, John Geise, Anna M. Grau, Robert C. Gorham, William A. Gruse, Charles A. Harrison, John W. Harpster, William A. Hamor, William Hummel, William A. James, Webster N. Jones, Nora Lynch Kearns, William F. Keller, John W. Khouri, Alice S. Linton, George H. McCune, E. H. McClelland, Charles S. Marshall, John W. Mochnick, John W. Oliver, Jr., Andrew J. Sherockman, Carolyn D. Patterson, Victor C. Showers, William S. Tacey, and Stanley P. Wagner.

To Dr. E. Ward Tillotson, who read a number of chapters and saved me from many mistakes in the use of technical terms, I am especially indebted.

To J. Steel Gow, former assistant to the Chancellor of the University of Pittsburgh, who made possible the establishment of a Science and Technology Library, I express genuine appreciation.

To Stanton C. Crawford, Dean of the College, University of Pittsburgh, who made it possible for me to have a research assistant, I am also greatly indebted.

Most of all, though, I am indebted to Jessamine DeHaven Lewis, my able research assistant. She rendered invaluable services over the months and years of research and went beyond her regular duties to help with the editing and writing. Whatever errors or shortcomings the book may have, however, are not hers but mine.

Finally, to my wife Helen I owe everything. She not only typed the manuscript but has been my constant companion in exploring the miraculous achievements of American scientists and technologists. It has been pleasant pioneering all the way.

John W. Oliver

Pittsburgh
February, 1956

CONTENTS

PART I. YANKEE INGENUITY: from Jamestown and Plymouth to the American Revolution

PART II. LAYING THE FOUNDATIONS: from the American Revolution to the Civil War

PART III. RISE OF THE UNITED STATES AS A WORLD POWER: Civil War to 1900

PART IV. WORLD LEADERSHIP: from 1900 to the Present

History of
American Technology

PART I

YANKEE INGENUITY: from Jamestown and Plymouth to the American Revolution

When, in 1620, the Pilgrim Fathers stepped from the "Mayflower" onto Plymouth Rock, the most precious physical possession they unloaded was the collection of one hundred and two tools, implements, and utensils with which they were to start life anew in the New World.

The food they had brought with them was soon exhausted. Their clothing wore out. It was upon the use of their tools and implements, and their skill in handling them to provide food and shelter and to make their own clothes that their hope of survival depended. The Jamestown Colony in Virginia had at first met with failure because of its lack of tools, and it was only after its settlers realized the need for tools that the colony survived.

As later shiploads of settlers arrived and other colonies were founded, their needs became more urgent. It was then that the ingenuity of the early settlers was taxed in meeting these needs from the materials at hand. New tools were designed. New methods of planting, fertilizing, and growing crops were adopted. New ways to tan leather and to process wool and flax for clothing were discovered. American technology was born.

Migrations from the Old World to the New continued, and by the 1730's, settlements extended all the way from Maine to Georgia. Transportation and communication developed, bringing the colonies closer together. Exchange of ideas and practices in raising livestock, in building homes, and in manufacturing goods followed. Iron works, grist mills, and sawmills were established; shipbuilding became an important business.

By the time of the American Revolution, the colonists had built up an economy of their own. They were producing iron, leather, glass, and other basic commodities in sufficient quantities to cause alarm to the mother country. Conscious of their ability to "go it alone," they declared their independence, political and economic, and condemned the King, "For cutting off our Trade with all parts of the World." During the seven years of war that followed, their economic and creative ingenuity was put to a test that was just as trying as was their struggle to establish political freedom.

Chapter 1

TRANSPLANTING OF TOOLS, SKILLS, AND TECHNOLOGIES

THE JAMESTOWN SETTLEMENT, 1607

"I entreat you send me thirty carpenters, husbandmen, gardeners, blacksmiths, masons, and diggers-up of trees' roots rather than a thousand such as we have," wrote Captain John Smith in 1609, in a desperate effort to save the settlement at Jamestown, Virginia. The London Company, which had planted the settlement two years earlier, had originally sent over "gentlemen," soldiers of fortune, traders, goldsmiths, and silversmiths, with instructions to search for gold, to gather herbs for drugs, and to discover other valuable products that would bring fabulous prices in the London market.

Colonization Problems. But the plan had not worked. There was no gold to be discovered. Such herbs as were found had little or no medicinal value. The settlers, instead of planting crops and raising their own food, depended on the Indians. Or, if the Indians failed to provide provisions, the settlers waited for the arrival of new supplies from England. Sickness and starvation took a heavy toll during the first two years.

But New World technology was not to be founded on such settlers as these. The London Company heeded Captain Smith's request. When Captain Christopher Newport arrived on his second voyage, he brought a shipload of new settlers, including, in addition to a few "gentlemen," six carpenters, some blacksmiths, one mason, and twelve laborers. Of equal importance were the tools, implements, and utensils that the craftsmen brought with them, such as axes, hatchets, chisels, mattocks, pickaxes, knives, and pike heads. They also brought a supply of shot and powder, and, of great importance, livestock consisting of six mares and a horse, a hundred swine, a number of chickens, and some goats and sheep. Also, a quantity of wheat was brought.

Among later arrivals was a long list of workers. There were husbandmen, gardeners, brewers, bakers, sawyers, carpenters, joiners, boatwrights, shipwrights, coopers, weavers, tanners, potters, fowlers, fishhook makers, leather dressers, blacksmiths, iron workers, hemp and flax dressers, and miners. From Devonshire, came one hundred men experienced in husbandry. From Warwickshire and Staffordshire came one hundred and ten men, many of whom were craftsmen, and from Sussex came more than forty men skilled in iron working.

The third supply of emigrants that arrived in 1610, according to Tyler, "were chiefly artisans of all sorts." [1] In August, 1611, Sir Thomas Gates arrived with two hundred men, "nearly all artisans and mechanics." Some "Dutchmen" were sent over to erect sawmills, Poles were sent to work in naval stores, Italians came to set up glass works. By May, 1620, one hundred and fifty persons had been sent to set up three iron works; and directions had been given for making cordage, as well as hemp and flax, from a native grass which grew in great abundance and which was found to make "the best cordage in the world." In 1623 the ship "Bonny Bess" brought twenty-four tradesmen and only "ten gentlemen." [2]

Reorganization of the Colony. The London Company now advised the settlers to abandon their search for gold and turn to the making of pitch, tar, pot- and soap-ashes. Orders were also given for cutting timber for shipping, for making masts and boards. This marked a change in the original policy. The establishment of a self-sustaining colony became the primary objective. The company no longer wanted the adventurer or the cavalier. Neither did it want the peasant who had been driven to labor, nor the refugee who had escaped bondage. Rather, it wanted the worker skilled in the use of tools, trained to an honest craft, one who was willing to stay and make this his home. Within a few years the Colony's survival was assured. Production of a few clapboards, pitch, tar, pearlash, and glass was undertaken in a small way. The early settlers began to live here much as they had lived in England.

The settlers soon found new adaptations for their skills. A new plant, tobacco, indigenous to the New World, was discovered to be profitable. Captain John Rolfe early produced a successful crop, and within a few years tobacco became the staple crop of Virginia. It proved to be the gold which the settlers had originally sought. Tobacco production, requiring as it did a number of workers, led to the importation of slaves and the establishment of a plantation economy in Virginia.

The Pilgrims

The Pilgrims, who came in 1620, unlike the first settlers at Jamestown, were determined to establish themselves permanently in America. Although the "Saints," who migrated from Scrooby, England, to Holland, and thence to Plymouth, were in the minority on the "Mayflower," they—as well as the "adventurers"—were skilled husbandmen and artisans. They brought with them a variety of arts, tools, and skills. During their twelve years' stay in Holland, the Pilgrims had lived in the midst of the most industrious people in Europe. Holland was the leading nation of the world in scientific, engineering, and technological developments. Leyden, a city of some 80,000 population, was the center of numerous handcraft industries. Improvements in windmills, in sawmills, in shipbuilding, in textile technology, in papermaking, and in printing were being developed during the very period that the Pilgrims were residing in that country.

Among the skilled workers on the "Mayflower" were William Bradford who had learned the craft of the fustian maker (corduroy or moleskin) while living in Amsterdam, William Brewster who had become a ribbon maker while in Leyden, William White and Francis Cook, wool combers, Isaac Allerton and James Chilton, tailors, John Alden, a cooper, and Samuel Eaton, a ship's carpenter. Still others had been employed as silk workers, feltmakers, button makers, drapers, tailors, hatters, glovers, leather dressers, cobblers, metal workers, carpenters, cutlers, brewery workers, stone and brick masons, hod carriers, printers, brickmakers, pipemakers, pumpmakers, and in numerous other crafts. Also listed were fifteen artisans and mechanics, two carpenters, two silk workers, one printer, one fustian maker and silk dyer, a cooper, a blacksmith, a tailor and a hatter. A goodly proportion of the others were familiar with "the prior art of English husbandry." [3]

Also aboard the "Mayflower" was a valuable cargo of tools, implements, and utensils. These included all the hand tools and tools of husbandry then in use in both England and Holland. There were hoes, spades, shovels, sickles, mattocks, garden rakes, and hay forks. Also, there was a meager supply of grain and garden seed. There were tools needed to erect adequate shelter, such as saws, hammers, augers, chisels, hatchets, nails, vises, and an iron jackscrew. There were blacksmith tools, cooper's tools, and a supply of bar iron and some steel. Equipped with the latest tools and techniques from both

England and Holland, and with Old Testament determination, the Pilgrims came to stay. They came equipped to lay the foundation for a New World technology.

Later settlers were likewise skilled in trades. "The Fortune," which arrived at Cape Cod in November, 1621, with thirty-five passengers, had a number of skilled craftsmen. On board were William Bassett of Middlesex, listed as an iron worker, Clement Briggs of Surrey, a fellmonger, Robert Cushman of Kent, a wool carder, Stephan Deane, a miller, and a Mr. Ford, a leather dresser of Southwark, William Hilton of Chester, a vintner, William Palmer of London, a nailer, and William Pitt of London, an armorer.

Growth of Massachusetts Bay Colony. Other New England settlements attracted the same class of artisans. The organizers of the Massachusetts Bay Colony brought a number of skilled craftsmen and husbandmen. Included among the original founders of Salem were two experienced surveyors and six shipwrights. The latter arrived in 1629 and brought with them a supply of pitch, tar, cordage, and sailcloth. Within two years a 30-ton sloop, the "Blessing of The Bay," was built and launched for Governor Winthrop. The sloop was used for coastwise trade with the Dutch at New Amsterdam.

Among the ninety-two grantees of Newbury (1635) were two clergymen and eight "gentlemen," with thirty-seven skilled tradesmen. In this same year, five other ships arrived in Massachusetts bearing a number of husbandmen and craftsmen. In the summer of 1636 six ships arrived in Massachusetts bringing workers equipped with many tools and utensils. The next year the ship "Confidence" brought eight husbandmen, two shoemakers, a tanner, and a saddler.

The need for other artisans increased. When horses were brought over, harness makers and saddlers were induced to come. Tanners, glovers, and cordwainers were also induced to settle in all the New England towns. The demand for new homes brought joiners, glaziers, locksmiths, and sawyers. Experienced tallow candlemakers began to relieve the housewife of this time-consuming chore. Ropemakers were in demand for the fishing and maritime enterprises.

Special inducements were offered to skilled workers to settle in a community and ply their trade. There was always a demand for shoemakers and blacksmiths. In 1635 Newbury (town) invited a blacksmith to come and settle, and allotted him 20 acres of land. In 1658 Haverhill gave to John Johnson, a blacksmith, a house and land costing 20 pounds. The sum was raised by subscriptions from twenty persons who stipulated that Johnson was to follow his trade for seven years and could not work for any nonsubscribers. Other crafts necessary to the development of the town were likewise encouraged. At Windsor, Connecticut, an expert currier was furnished a house and shop for "him and his heirs," if he would live and ply his trade there until he died.

Needs of Other Colonies

Other colonies wanted the same kind of settlers. The New Netherlands settlement issued a pamphlet in 1624 inviting craftsmen and tool workers to come and settle there, promising them free passage to America. In 1628 Reverend Jonas Michaelius of New Netherlands, writing about conditions there, stated that the newly arrived workers "have cut much timber here to carry to the Fatherland. . . . They are making a windmill to saw lumber, and we also have a grist mill. The harvest, God be praised, is larger than ever." [4]

Tools and implements necessary to the trades of the colonists were found on many ship's lists. The following farm tools were listed in a leaflet issued by the Secretary of the Dutch India Company in 1660:

Tools Needed

12	plough shares with coulters
1	first class wheel plow with pulleys
12	two-prong hay and grain forks
13	three-prong hay and grain forks
100	iron teeth to make harrows
24	best scythes
24	good reaping hooks
50	steeled axes
20	winnowing fans
30	spades
30	shovels
30	hoes
20	iron rakes
12	hay knives

The transit of skilled settlers to New Sweden on the Delaware was similar to that of other colonies. Governor Rising of New Sweden, writing in 1654, asked for pottery makers, brickmakers, lime burners, cabinet makers, shoemakers, tanners, and a French hatmaker.

When William Penn advertised for emigrants to follow him to the New World, he sought "industrious husbandmen, carpenters, masons, weavers, shoemakers, and other mechanics, and industrious spirits that are oppressed about a livelihood. . . ." He wanted only those "whom Providence seems to have best fitted for our plantations, farmers and skilled craftsmen." In 1685, Germantown, Pennsylvania, was laid out by twelve German families, nearly all mechanics and weavers. Besides their trade and tools, they also brought some iron stoves to the colonies.

Even in the South where a somewhat different economy developed, due to the staple crops and the early plantation system, care was exercised in selecting settlers. In 1665 Governor William Berkeley of Virginia requested the home government to send a number of experienced flaxmen. Virginia filed a strong objection to the number of felons and other desperate villains who had been sent from the prisons of England. Governor Berkeley asked for skilled workers from Naples and Sicily to teach his settlers "the art of growing mulberry trees and working in silk," believing that he could produce more than enough to meet England's total needs. In a letter to a ship captain in 1670, he promised to give "skilled workers or tradesmen who will come and live in our midst large privileges." A few years later a ship arrived with eight combers, five serge weavers, four clothiers, fifteen plowmen, two carpenters, two shoemakers, two masons, a tanner, a soap boiler, and a ribbon weaver, but no silk growers.

Governor Spotswood of Virginia desired, above all others, mechanics, flax growers, and iron workers. He was anxious to develop the iron deposits near the falls of the James River, and in 1714 persuaded a small colony of Germans to settle on the Rapidan River "to mine iron ore and other ores." These settlers were provided with free houses for five years. This marked the beginning of that rather famous, although not very productive, mining settlement of Germans in Spotssylvania County.

In a letter from a North Carolina settler, dated July 7, 1711, to the Society for Propagating the Gospel, the writer announced that "workmen are dear and scarce." He described the population as consisting mainly of carpenters, joiners, wheelwrights, coopers, tanners, shoemakers, tallow chandlers and what not." Women workers, soapmakers, starchmakers, and dyers were also mentioned. "He or she that can not do these things will have a bad time of it." [5]

These examples of the transplanting of tools, skills, and technologies of England and Western Europe to America could be multiplied a hundredfold. By the time of the American Revolution the skill or tool not known in America was indeed a rare one in Europe. The European techniques imported to America quickly underwent changes and differed from their European counterparts. The alteration was almost never one of mistake but of necessity. These alterations resulted from a new environment and were the factors that led to the introduction of later technical improvements.[6]

New World Native Techniques

But the technology of the New World was not merely a replica of the Old World. The first European settlers learned much from the North American Indians. The Indians taught them how to grow maize, how to make hominy, succotash, samp, pemmican, and other dishes. They taught the white man the value of wild fruits, nuts, and other products of the forest. Berries were plentiful. Wild blackberries, raspberries, and cranberries were all indigenous to North America, and all found their way to the tables of the early settlers through the aid of the Indians. DeVries, writing in 1639, described the lands around the New Netherlands settlement as "covered with chestnut, plum, and walnut trees, and hazel-nut bushes, and the level lands were covered with large crops of strawberries." Along the rivers were vines bearing two kinds of grapes, "good blue grapes . . . , the other kind, like the grapes that grow in France on trellises . . . , and as large as the joints of the fingers."

Corn (maize) was the Indian's basic crop. The early settlers soon discovered that it was also to be their chief food supply. It was worth more than all their European crops. It yielded more per acre, grew faster, ripened earlier, and was fully matured before the first killing frosts came.

In 1609 Captain John Smith captured two Indians as a retaliatory measure and made them "teach us how to order and plant our fields." Forty acres were planted in corn with the assistance and under the supervision of those Indian prisoners. Within five years, the Jamestown Colony was planting 200 acres in corn.

The Pilgrims were aided in their early agricultural technology by the friendly Indian, Chief Squanto. He and other Indians taught the settlers their first lesson in plant fertilization. Dead fish, dropped in a hill of corn at planting time, hastened the plant growth and greatly increased its yield. Governor Edward Winslow wrote in December, 1621, that during the past season they had planted 20 acres of Indian corn, "according to the manner of the Indians." They had manured their ground with herring and shad, "which we have in great abundance and take with ease at our doors." They used from a ton to a ton and one-half of dead fish per acre. This transfer of maize technology from the Indians to the first English settlers best exemplifies the early blending of European and native American agricultural techniques.

The Indian, while learning to live in the shadow of a virgin forest, also had learned how to girdle trees and bleed them to death in order to clear land for planting. This method of clearing the forests was quickly adopted by the new settlers, especially since it required a minimum of labor.

Food, cleared lands, and agricultural "know-how," transmitted through friendly intercourse, barter, and conflict were the chief contributions made by the Indians. These contributions were responsible in a large degree for the successful establishment along the Atlantic coast of those early outposts of European civilization, a civilization whose inexorable expansion was to dilute, and later to destroy, the Indian culture.

A New Society Emerges

The early settlers brought with them more than skills and tools. They came with a devotion to the land. While most of them had been deprived of land ownership at home, yet they had that hard-to-define feeling, that given a piece of land as their very own, they would become self-sustaining. They had grown up in the manorial agricultural period in England. On the English manor the husbandmen made their own household utensils, and all the wooden parts for the scythes, rakes, hoes, flails, and other implements for tillage. Their wives spun, wove, dyed, and cut out linen and woolen cloth, brewed ale, and made soap. These skills had been handed down from one generation to another and were carried to America. The New England villages with their common pastures, cowherds, and family production units were in many respects merely a transfer of English manorial technology to the New World. Since a portion of the Atlantic coast region possessed physical characteristics somewhat similar to those in western Europe and England, many of the Old World crops were grown successfully and because of the virgin soil produced more abundantly in the New World. These experiments together with the arts learned from the Indians represent the first attempts of agricultural technology in America.

While the role of adventurers and religious refugees has been much emphasized in Colonial histories, the fact is that the early settlements in America were made permanent not alone because of the religious zeal of the first inhabitants, nor by their search for treasure, but also by their skilled manual labor in tilling the soil, building homes, making their own clothing, and establishing their own industries.

Notes for Chapter 1

1. Lyon Gardiner Tyler, *England in America, 1580–1652* (1904), p. 63.
2. Richard B. Morris, *Government and Labor in Early America* (1926), pp. 25-26.
3. For a description of the cargo and passengers on the "Mayflower," see Azel Ames, *The Mayflower And Her Log* (1901).
4. David Peterson DeVries, *New Netherlands, Old South Leaflets,* No. 168.
5. John R. Commons, *Documentary History of American Society,* I (1910–1911), p. 166.
6. For a recent study bearing on this general subject, see Wallace Notestein, *The English People On The Eve Of Colonization: 1603–1630,* the "New American Nation Series" (1954).

Chapter 2

ADVANCES IN AGRICULTURAL TECHNOLOGY

Clearing the Land

The forests presented the greatest physical obstacle to Colonial agriculture. The first settlers at Jamestown and Plymouth did little clearing of the land. They discovered a considerable amount of rich and fertile land that had been cleared either by some natural cause, such as fire started by lightning, or by the Indians before the new settlers had arrived. The Pilgrims also profited in the latter instance when they found a few cleared corn fields deserted by some Indian tribe. While this search for cleared land in the immediate area of the settlements was the easiest method of obtaining land for cultivation, it was not one that would supply a growing population. As the population grew, the forests had to go.

The first woodcutting technique of the Colonial farmer was almost as crude as that used by the Egyptians four thousand years earlier. The tools were ill adapted for chopping down large trees. This necessitated either the technical improvement of ancient tools or the development of new skills in using them.

The early settler brought with him the tools which he had used in England. Chief of these were the axe and the mattock. The axe was so useful that it became the traditional symbol of America's frontier. The dense forests could never have been mastered without it. The European axe, however, was ill adapted for use in a thickly wooded country. It was a heavy, awkward tool with a long bit and a light poll and was designed for limited use.

The efficient use of the American axe depended to a large degree upon its handle. Each community had at least one craftsman specialized in making axe handles. He became an expert in the use of the drawing knife, and was in great demand. The axe handlemaker took a special pride in his work. To him, an axe handle was not just another handle.[1] It was specially designed, perfectly balanced, and fitted to the grip of the individual user. No one other than the individual for whom the handle was made could use it with the same unfailing accuracy. There was a special feel about the axe handle—its grip, and its balance—similar to that which exists today between a champion baseball player and his special bat, or the professional golfer and his matched clubs.

Benjamin H. Latrobe, in describing the American chopping axe, pointed out how it differed from the Old World axes. The handle was set forward, and it acted upon the tree "more from its wedge-like form, its own weight, and skillful swing, which gives it impetus, than from any great exertion of strength on the part of the woodsman." Skill, more than strength, was employed in its use, a skill which excited the admiration of the Europeans. Latrobe declared that, "the American axe may well rank with maize and steam as one of three things which have conquered the western world." [2] The aesthetic grace of the swinging, gleaming axe, and the rhythmic feel of the smooth axe handle were intimate details known only to a breed of men who were building a new civilization.

The axe blade underwent a change also. Skilled axmen developed a special technique for hewing. For this craft they had their axe blade made thinner, broader, and sharper. This improved implement became known as the broadaxe, and with it the woodsman could hew out a piece of timber as smooth as if it had been sawed. Walt Whitman, in his "Song of The Broad-axe," written in the mid-nineteenth century, paid tribute to this simple, efficient tool and the products it helped to create.

> The axe leaps!
> The solid forest gives fluid utterances,
> They tumble forth, they rise and form,
> Hut, tent, landing, survey,
> Flail, plough, pick, crowbar, spade,
> Shingle, rail prop, wainscot, jamb, lath, panel, gable,
> Citadel, ceiling, saloon, academy, organ, exhibition-house, library,
> Cornice, trellis, pilaster, balcony, window, turret, porch,
> Hoe, rake, pitchfork, pencil, wagon, staff, saw, jack-plane, mallet, wedge, rounce,
> Chair, tub, hoop, table, wicket, vane, sash, floor,
> Work-box, chest, string'd instrument, boat, frame, and what not . . . .

Less glamorous, but equally important to the early settler for clearing his field, was the mattock, sometimes called the grubbing hoe. This implement had two stout iron blades, a digging hoe blade and a smaller cutting blade. With these two tools, the axe and the mattock, the first Americans cleared their ground for plowing and planting.

Tilling the Land

Once the land had been cleared of trees and brush the next task was to loosen and turn under the top soil, a most laborious and difficult task in a virgin land. The early Colonial farmers used wooden spades, clumsy wooden forks, and now and then a wooden plow. The Pilgrims imported a plow in 1632 from England. Five years later the Massachusetts Bay Colony reported thirty-five plows in the entire Colony. In 1682 the Dutch West India Company was asked to send to the Dutch settlers on the Delaware, "12 plowshares with coulters, and one first class wheel plow with pulleys. . . ." [3]

In a few of the northern colonies some early improvements were made in plow construction. Since craftsmen and blacksmiths fashioned plows to suit the desires of the individual owners, no effort was made toward standardization. This custom was merely another expression of the individuality of the early settlers, and this characteristic of Colonial plowmaking became so imbedded in Yankee tradition that as late as the 1850's large plow manufacturers were still making individual plows to order in their own blacksmith shops.

However, several new developments were introduced. Wrought iron began to replace wood for shares and points. Strips of iron were fashioned to the wooden moldboards. This was one of the most revolutionary developments in Colonial plow manufacturing, for it made the plow run more smoothly and made it more effective in plowing under the grass and weeds. It marked a distinct advance in the history of Colonial agriculture. By the time of the American Revolution this type of plow, often referred to as the Carey plow, was widely used, and it was commercially advertised as a standard plow. On the Carey plow, the landside was made of wood, but the wooden moldboard was plated roughly with pieces of old saw plate, or sheet iron. The share was made of wrought iron; the wooden handles were held in place by two wooden pins.

Another type of the re-enforced plow was known as the "Bull Plow." It had an upright cutting bar (forerunner of the coulter) and a wooden moldboard. Although it did not scour easily, it did tear up the ground. Crude as were these early plows, they marked an improvement over the shovel plows. They tore up the ground to a depth of 4 to 6 inches. With two men to operate them, and teams of two to four oxen to pull them, they could plow an acre of ground a day—a decided improvement.

After the ground had been plowed, the harrow was used to break up the clods and level down the rough surface for planting. The early Colonial harrow, like the plow, was a crude, simple implement. At first a branch of a tree, usually forked, and bearing numerous small limbs or twigs served the purpose. Soon, however, the early settlers designed an improved harrow of their own. A square wooden frame, with teeth of wood or iron inserted at intervals of 4 to 5 inches was devised. The square-shaped harrow was ill suited, however, to newly cleared ground where the stumps had not been removed.

Some ingenious farmer came up with another design. He took the crotch of a tree of desired size, inserted one tooth in front at the point, and four or five others in each side of the frame. This was called the crotch harrow and was widely used. Later, some one hewed out two pieces of seasoned timber of even size, fastened two of the ends together, with the opposite ends spread apart in the form of a letter "A." Then he bored or burned holes in each side, and inserted the teeth. This harrow, known as the "A" harrow, and still later as the "double A" harrow, marked a great improvement in agricultural technology.

Selection of Crops

Corn. The planting and cultivation of maize (corn) which the settlers first learned from the Indians, proved to be a significant development in Colonial agricultural technology. Maize was the most valuable crop, more valuable than all other grain crops combined. The early settlers' method differed little from that of the Indian's, except that better hoes were used and men as well as women helped with the planting. In preparing the soil for corn planting, strips of ground of a hoe's width were loosened 4 to 5 inches deep, and grains of corn were dropped into the soil. An old jingle of the Colonial period explains the reason for the number of seeds planted.

> One for the blackbird,
> One for the crow,
> One for the cutworm,
> And three to grow.

The practice of fertilizing the ground by depositing dead fish was borrowed from the Indians. Corn was usually planted late in April, or early in May, when the oak leaf was as "big as a squirrel's ear." John Winthrop, Jr., delighted in telling his English friends of the

British Royal Society of the success of this new grain. He described its variety of colors as "red, yellow, white, blue, olive, greenish black, speckled and striped, with large ears, defended from the cold and storms by strong thick husks." He said the stalks grew from 6 to 8 feet high, with joints like cane stalks, and were full of sweet juice like the sugar cane, and a syrup as sweet as sugar could be made from the stalks.

Tobacco. Another crop of great importance to the early Colonial farmer was tobacco. When John Rolfe, sometimes called the first agricultural scientist in Virginia, persuaded the London Company to let him grow and cure tobacco rather than continue the futile search for gold and silver, he profoundly affected the Colony's future development, and indeed, the whole future of American history. Tobacco was cultivated in all parts of eastern North America, but in Maryland, Virginia, and North Carolina it was planted, cultivated, and sold as an export crop. In the cultivation of tobacco a topping system was introduced: by cutting off the top of the plant, a heavier middle leaf was produced. This technique greatly increased the quantity and quality of tobacco production.

Tobacco was raised from seed sown in the spring season in a bed of rich mold. When it grew to the height of 4 or 5 inches, the planter waited for a rainy day and transferred the plants to specially prepared soil of a rich black mold. Fresh woodlands were considered best. Rolfe, the pioneer tobacco grower in Virginia, experimented with a number of varieties, and boasted of growing as much as 500 pounds per acre.

Rice. Governor William Berkeley of Virginia was probably the first to introduce rice culture in the colonies in 1647. The experiment did not prove too successful, however, but in 1693 another colonial governor, Landgrave Thomas Smith of South Carolina, made his colony the first center of rice culture in the New World. While traveling in Madagascar, Smith obtained some rice seeds, which he planted in his garden, and grew a luxurious crop. He distributed samples of the seed to his planter friends, and within a few years rice became one of the major crops.[4]

Rice culture developed rapidly. The seed was planted in early spring in furrows 18 to 20 inches apart and about 3 inches deep. The seeds were covered with a brush or harrow, or sometimes with a plow. A peck of seed was sufficient for 1 acre of land. The cultivation of rice required a considerable amount of time and labor. After the grain was planted, the land was submerged in water. The object was to keep the crops clear of weeds and grass; grass was the worst enemy. Slaves were employed almost exclusively to plant and harvest the rice.

Grain. Throughout the Colonial period there was some experimenting in various regions to ascertain what crops were best suited for particular areas. Certain plants were attacked by the ravages of mildew or "blast" as it was called. In Connecticut the blast was first noticed by that keen observer, John Winthrop, Jr., in 1660. It attacked wheat in all the three New England colonies.

> The corn (wheat) flourished well till it came to be eared, and the ears would at first appeare faire, and as if full, but no corne in them. There have been thousands of acres in that manner every year. What the cause was, whether naturall, or a blasting from heaven we know not. Our old husbandmen of England thought mildew, others that the original defect is in the roote; the peas, barly, rye, and India corne were not touched with it, as was generally observed.

According to Winthrop's report the causes of the "blast" were shrouded in mystery, but through yearly observations it was discovered that the areas near barberry bushes were most affected. Farmers began to destroy the bush, and Connecticut enacted legislation to aid in removing this pest. Township officials were authorized to eradicate barberry bushes wherever found. In 1772 Rhode Island imposed fines on farmers who failed to destroy barberry bushes on their lands.

Harvesting Crops

The system of harvesting and threshing was carried on by the early Colonial farmers as it had been done for centuries in the Old World. Small grain such as wheat, oats, rye, and barley, was cut with the sickle, later supplemented by the scythe. Since a farmer could seldom cut more than an acre a day unless he could count on some extra help at harvest time, it was useless to sow more than 5 or 6 acres of grain.

A new harvesting implement called the "cradle" was introduced some years before the American Revolution. The handle and the cutting blade of the cradle resembled the scythe, but in addition, it had four or five wooden prongs called fingers which were attached to a light frame, shaped like the blade of the scythe, and arranged parallel with the blade. Considerable skill and strength were required to swing a cradle 10 or 12 hours a day, but with the use of this new implement, a good, strong cradler could cut 3 to 4 acres of grain a day.

The introduction of this implement meant that a farmer could double the amount of acreage sown to grain.

Threshing the grain was accomplished by methods and tools as old as the sickle. In the New England settlements, the flail was commonly used; in the middle and southern colonies, treading by oxen and horses was generally employed. One man, using the age-old flail, could beat out 6 to 7 bushels of grain a day. The amount that could be threshed by treading depended upon the size of the treading floor and the number of animals used. The treading floor was a smooth, bare plot of ground, but occasionally a large wooden platform, laid out like a circus ring, was erected and surrounded by a fence. A few horses or cattle could trample out 200 to 300 bushels of grain a day.

Winnowing, separating the chaff and dirt from the grain, at first was accomplished by the simple method of tossing the threshed grain from large wicker baskets into the air on a windy day. The chaff was blown away while the grain fell to the ground in a pile. A sheet was often used as a fan to produce a breeze if there was no wind blowing. The wheat was thrown into the air with shovels, and as it passed before the briefly flopping cloth, the chaff was blown away. An improvement was introduced a few years before the Revolution when some hand-operated "Jams" were imported from England. Thus winnowing, the simplest of all the threshing operations, was the first hand labor to become mechanized.

The grinding of grain was an operation performed with two round buhrstones, held in a strong solid frame. The circular stones were from 3 to 4 feet in diameter, and 6 to 8 inches thick. The upper stone rotated on the "nether," or lower stone, which remained stationery. Small grooves cut in each stone allowed the meal or flour to sift out over the edge of the lower stone. Grain was fed into the center of the two stones from a hopper, a funnel-shaped box, fixed above the buhrstones. Horse power, and later water power, were the main sources of energy for operating these early gristmills.

In some communities, one mill operating one day a week could supply the entire neighborhood, and that day came to be known as "milling day." In many communities the gristmills became sizeable commercial enterprises during the late Colonial period. American flour was exported in considerable quantities, proof that the gristmills were as good if not superior to the mills in other countries.

Animal Husbandry

The raising of livestock, a major development in the American agricultural economy, was carried on by the colonists not only for their meat supply, but also for the purpose of supplying their leather and woolen products. A surplus of animals was being produced within a half-century after the first New England settlements were established. By the close of the seventeenth century a rough pattern of livestock industry had been established. The severe New England winters led the more progressive farmers to provide shelter for their livestock for a few months of the year. The thrifty Germans in Pennsylvania began to build warm, weatherstripped barns for their animals. In the warmer climate of the South, livestock was allowed to roam at large throughout the year.

Cattle Forage. The problem of providing grass and hay for livestock demanded early attention. The indigenous grasses and native forage plants proved disappointing. Captain John Smith observed: "There was grasses plenty, though very long and thickly stalked . . . , which although they be good and sweet in summer, they will deceive your cattal in the winter."

The lack of nutritive grasses, the periodic sessions of drought, and the increasing number of livestock led the colonists at an early date to cultivate forage grasses, the most important being timothy. This grass proved to be valuable in some of the New England colonies and was later introduced by Timothy Hanson into other colonies, notably Virginia and North Carolina, where the native grass was too tough for forage. In some sections it was called "Herd's grass." Timothy became one of the major hay crops of America. An English observer, author of "Animal Husbandry," who visited this country in 1793–1794 as a representative of the English Board of Agriculture, wrote that timothy grass was extensively cultivated in the New England and middle states and yielded abundantly.

Cowpeas and red clover, both used for feed for livestock, were grown in New England in the late seventeenth century and were introduced into North and South Carolina before 1700. Red clover was being grown in New York before the American Revolution. The rapid yield of forage crops was a great boon to those who raised livestock. The most progressive farmers began early to utilize portions of their grain crop for fodder and feed. The straw of various small grains was used for roughage, while the cornstalks and blades supplemented the basic feed of hay.

To offset the dearth of natural pastures, the industrious Germans in Pennsylvania early devised an irrigation system of rerouting small streams through their natural meadows. Peter Kalm noted, in his travels in 1749–1750, that many farmers laid out their fields in regions where springs abounded, and dug small channels to direct the water out over their fields to irrigate the crops. In some cases they built dams near the springs, stored up the water, and later drained it off as it was needed. Kalm was amazed at the abundant crops that were produced in the irrigated areas where, he said, farmers grew two and sometimes three crops per year.

Breeding Livestock. The colonists began at an early date to try to improve their livestock. In many New England towns animals selected for breeding purposes were not allowed to graze in the common pastures. In 1668 Massachusetts Bay Colony passed an act which prohibited stallions under fourteen hands high from running at large. Similar legislation was enacted in Connecticut six years later.

William Penn, a lover of horses, brought three blooded mares and five white horses with him on his first trip to America in 1682. On his second visit, 1699, he brought the renowned colt "Tamerlane," of the celebrated Godolphin Barb, to whom (says Fletcher) "some of the best horses in England, even today, trace their pedigree." In Virginia, some years later, Thomas Jefferson and other plantation owners imported a number of blooded horses and mares.[5]

Two famous breeds of horses were developed during the mid-eighteenth century. In the Conestoga Valley in eastern Pennsylvania, a powerful draft horse was bred, known in American history as the Conestoga horse. This was a type rather than a breed. These horses were powerful and efficient draft horses. They stood from fifteen to sixteen hands high, had large legs, arched necks, and weighed from 1,600 to 1,800 pounds. They possessed great pulling power, and when hitched to the huge, lumbering Conestoga wagons, became the symbol of America's inland commerce until the coming of the railroads. The second breed of horses was the celebrated Narragansett pacers of Rhode Island. Noted for their gait and speed they were in great demand during the latter part of the eighteenth century.

The colonists also made early attempts to improve the breed of their sheep. Sheep were first raised for their wool, not for meat. For many years the yield of wool from American sheep was small and of inferior quality. But in spite of English legislation, which forbade the exporting of live animals, a few carefully selected rams were brought to this country at fancy prices and used to improve the breed of American sheep. George Washington took a special interest in improving his flocks. By selective breeding with choice rams, he developed sheep that compared favorably with the English sheep in producing wool. His sheep produced from 5 to 5½ pounds per fleece, compared with 2 and 3 pounds from average sheep.

Fencing. Fencing was another problem that confronted the early progressive farmer. He had to protect his crops from the animals that roamed at large, and if he had superior livestock of his own he had to keep them fenced in. Fencing was essentially a labor problem. Wood was plentiful, as was stone in many sections. But the simplest fence required many days of strenuous toil to build. In New England a co-operative system was practiced, whereby each man cared for his particular section of the common fence. When some farmers decided to expand their holdings, the common fence began to disappear along with the common fields. Like the farmers in the middle and southern colonies, each farmer became individually responsible for the protection of his own fields.

The type of fence varied in different sections. In New England stone fences were common. Large stones and boulders which cluttered up the land were dragged to the edge of the cleared field, and carefully arranged to form a solid stone fence.

The post-and-rail fence was another type found in New England and in some of the middle colonies. The rails were fitted into holes that were cut in the posts, or the end of one rail rested upon the end of the other with two posts on each side to hold them in place. The worm fence, built of split rails, was the most prevalent in the middle colonies, but required much more timber and used more ground than the post-and-rail fence, since it was not built in a straight line.

Good fences were essential to good farming. Special acts relating to the building of fences and their upkeep were passed by various colonial assemblies. If a farmer in Pennsylvania in the early 1700's surrounded his field with a fence 5 feet high, of sufficient rails or logs, and closed at the bottom, then, "the owner of any horse, kine, sheep, hogs or goats, or any kind of cattle [that] shall break into any man's enclosure, the fence being of the height and efficiently prescribed," was liable for all damages.[6]

Agricultural Progress and Experiments

Throughout the Colonial period the more advanced British farmers were attempting to aid their American cousins in introducing new crops. John Ellis published a book in 1770 in which he devoted several pages to the proper method of packing seeds so as to prevent spoilage on the long journey overseas. Among other methods, he especially recommended packing seeds in beeswax. China, according to Ellis, was the source of many valuable seeds, and he urged East India sea captains and missionaries there to help in obtaining olive plants, black pepper, mangoes, spices, and other species, with the hope that they could be successfully grown in North America. Upland rice from Canton was imported into South Carolina.[7]

An early attempt to improve crop cultivation in the New World occurred on the banks of the Ashley River in South Carolina in 1699. The Lord Proprietors established a farm in order to determine the adaptability of certain agricultural crops. Attempts were made to grow grapes for wine, mulberry trees for silk worms, flax, indigo, tobacco, and hemp, for export. After a few years of experiments, it was discovered that sugar cane and cotton (perennial) were not able to withstand the South Carolina winters. Later, however, the annual variety of cotton became a profitable crop for that state.

Another interesting advance in agricultural and horticultural technology occurred in 1733 in Georgia, when an experimental garden was established. The promoters imported grains, plants, and herbs indigenous to southern Europe, with the aim of cultivating them in our southern climate. A 10-acre site on a bluff outside of Savannah was selected on which were planted orange trees, mulberry trees, and subtropical plants such as coffee, cocoanut palm, and bamboo. Botanists were employed to scour the world for plants which they believed suitable to this experimental garden. The project was supported by contributions from English friends, among them Hans Sloane, who was interested in developing a new enterprise in America. The experiment was not financially successful, but it may be regarded as the forerunner of the agricultural experimental stations of the following century.[8]

In Philadelphia, in 1773, a group of "gentlemen farmers," businessmen, merchants, doctors, lawyers, judges and educators, working through the American Philosophical Society and the Society for Promoting Agriculture, took steps to encourage better farming. They called upon the farmers of Pennsylvania to adopt the European practice of enriching their soil with "Marle and other composts." Gold medals were offered for the best method of recovering worn-out lands, and for the best yield of clover from a 10-acre field. Such rewards proved to be a great incentive to good farming.

While the Colonial farmer may have been less concerned about scientific agriculture than his European counterpart, nevertheless he took advantage of other opportunities at hand. In Europe where labor was cheap and land poor and scarce, careful farming was not only a virtue but also a necessity. In America, land was plentiful and fertile, and labor was scarce and expensive. Here the farmer found it was more advantageous to clear a new field than to rebuild an old worn-out farm. Although he was not concerned with soil conservation, he was most eager to adopt all new labor-saving devices. He improved the axe, the plow, and the harrow. He substituted the cradle for the age-old hand scythe for cutting grain. He introduced selective breeding of livestock and developed a breed of horses that became the pride of the American farmer for almost two hundred years.

While performing all these tasks, the Colonial farmer was thinking all the while of some new tool that would lighten his labor. Throughout most of the Colonial period the spectre of failure was the figure of famine. The farmer's capacity to produce during the first hundred years was so limited that he was unable to store up a sufficient quantity to carry him comfortably over a season of crop failure. As a result of this constant, continuing toil, there developed a close-mouthed, hardheaded, inventive, self-reliant people—the kind of people that would soon be designing new methods, and inventing new tools that would enable them to establish an economy of their own. The step from economic independence to political indepedence was a short one.

Notes for Chapter 2

1. The curved axe (Helve) handle did not come into general use until the mid-nineteenth century. See Henry C. Mercer, *Ancient Carpenter Tools* (1929).

2. *Pittsburgh Gazette* (Sept. 8, 1835).

3. Stevenson Whitcomb Fletcher, *Pennsylvania Agriculture and Country Life, 1640–1840* (1950), p. 89.

4. Mildred Kelley Ginn, "A History of Rice Production in Louisiana to 1806," *Louisiana Historical Quarterly,* XXIII (1940), No. 2, 544-586.

5. See *Thomas Jefferson's Farm Book, With Commentaries and Relevant Extracts From Other Writings,* edited by Edwin Morris Betts (1953), p. 87.

6. Fletcher, *op. cit.,* p. 85.

7. For an excellent, brief account of early importation of seeds, see Nelson Klose, *America's Crop Heritage* (1949).

8. James W. Holland, "The Beginnings of Public Agricultural Experimentation in America," *Agricultural History,* XII (1938), 271 *ff.*

Standard works on the early history of American agriculture include: Percy W. Bidwell and John I. Falconer, *History of Agriculture in Northern United States, 1620–1860* (1925); Lewis C. Gray and E. K. Thompson, *History of Agriculture in Southern United States to 1860,* 2 vols. (1933); and files of *Agricultural History,* 1926 continuing; also Everett Eugene Edwards, *United States Department of Agriculture,* XXXIII (1938); Philip E. Bruce, *Economic History of Virginia,* II (1898); Carl Raymond Woodward, *Ploughs and Politics: Charles Read of New Jersey and His Notes On Agriculture, 1715–1774* (1941).

Chapter 3

ADVANCES IN BUILDING TECHNOLOGY TO THE REVOLUTION

It was necessary for the settler to clear the forests in order to have fields for agriculture, but it was just as necessary for him to have the wood to build his home. The trees that were a hindrance to agriculture were, at the same time, a blessing in providing material for shelter. The climate of the country from Maine to Virginia made it necessary for the settlers to provide shelter or die from exposure; therefore, they built homes as rapidly as they could. They wanted a roof over their heads to keep out the rain and snow, and walls around them to protect them from wild animals and Indians.

First Colonial Shelters

The phrase *building technology* can hardly be applied to the construction of the first homes in the New World. In a few of the colonies (Plymouth, for example) the settlers enjoyed the modest luxury of living temporarily on board their ships. Later arrivals also remained on board their vessels as long as they could. But such respite was not always possible, nor was it permanent. The ship's captain carried only a limited supply of rations for his crew, and he had to set out on his return trip before they were exhausted. Left on their own, the early settlers had to erect hastily some kind of primitive shelters. Many of them literally burrowed into the ground for a period. They dug holes in the side of a hill, laid small poles across the top, and filled in the crevices with bark and sod. These primitive shelters kept out some of the cold, wind, and rain, but long rains and melting snows seeped through to the great discomfort of the occupants.

Surprisingly enough, while the first settlers in America readily imitated the Indian in many of his agricultural methods, there is no evidence that they tried to do so in building. Once the settler became certain that he would stay here, his primary aim was to imitate the buildings of his homeland rather than the Indian wigwam. While the settlers planted their corn and tobacco after the Indian method, they soon discovered that the red men's homes were greatly inferior to those they had known in Europe.

Building Materials. Building technology involves the use of materials, tools, and skills, all of which the settlers had learned in Europe. The materials first used for erecting shelters were those at hand. Wood was plentiful and easily obtained. Many of the settlers, such as Colonial divine Francis Higginson, observed that:

> For wood there is no better in the world I think; here being foure sorts of Oke, differing both in Leafe, Timber, and Colour, all excellent wood. There is also good Ash, Elme, Willow, Birch, Beech, Sasafras, Juniper, Cypress, Cedar, Spruce, Pines, and Firs that will yield abundance of Turpentine, Pitch, Tarre, Masts and other material for building. . . .[1]

With wood, they used thatch and clay to fill in the cracks, materials with which the early colonists were familiar in the ordinary Elizabethan homes. Dried grass made a quick, light covering for the roof. Daubs of mud or clay were used for the walls, and often for inner insulation in the early wooden chimneys.

The wood, thatch, and clay, while satisfactory in preventing the early settler from being constantly ill from exposure, needed frequent repairing. These materials did not, however, provide the comforts that the settlers had known in Europe. The danger of fire, because of the wooden chimneys (even though insulated with clay), was always imminent. The loss from fires was tremendous. The Pilgrims reported in 1624 that out of thirty-two dwellings, seven were lost in one year due to fire.[2]

During the period from about 1700 to the start of the Revolution, brick and stone began to replace wood in a number of settlements. The skill to build such homes was already here, and the only thing that was lacking was the time needed to build them. Most of the bricks were made on the building site, although some were brought over as ballast in ships. By 1750, Boston had about 3,000 houses, of which one-third were brick and the other two-thirds were of wood. Williamsburg, the capital of Virginia, had buildings of both brick and wood. Norfolk's buildings were chiefly of wood. Fredericksburg, at the head of the Rappahannock, had buildings constructed of wood and brick, and its houses were roofed with shingles painted to resemble slate. In Philadelphia, the largest of all the cities, most of the buildings were built of brick. The use of brick was the most noticeable change in building technology that occurred along the

Atlantic coast before the Revolution. The change, however, took place only along the coast. The frontier continued to build wooden structures.

Buildings of stone, although erected as early as those of brick, were not as common. In the southern colonies few stone buildings were erected, due to the scarcity of stone. When greater strength was needed, as for forts, a mixture of mortar (concrete) and sea shells was used. In the northern colonies, on the other hand, field stone was easily and cheaply obtained. Henry Whitfield erected a stone house in Guilford, Connecticut, as early as 1639. Winchester in 1755 was described as, "a town built of limestone and covered with slate with which the hills abound." The Dutch in New Amsterdam showed a fondness for stone.

The vital material for both stone and brick building was lime for mortar. Some sections of New England had little limestone. Oyster shells provided a high quality of lime but could be used only where there were abundant oyster beds. The mortar made from such lime contained shell fragments which provided an excellent binder. On Staten Island, oyster shell lime was used in constructing large stone homes. Large oyster beds were found in Jamestown, Virginia, as early as 1623. Clay was used in masonry in less expensive dwellings.

Other important material such as glass and nails were at first shipped over from England, but by the middle of the eighteenth century these were being made almost entirely at home.

Building Tools and Machinery

The axe was the most vital tool used in preparing building materials. After the trees had been felled, they were cut into logs. Usually the bark was stripped off immediately after the tree was cut down. Then the logs were moved, either by rolling, or by dragging with teams of oxen, to the building site where they were split, hewn, and shaped according to plan. Here again, the American type of axe was used for hewing and smoothing the logs. It was also employed to prepare the sills, corner posts, beams, and main supports. Other tools were perfected gradually, and new skills rapidly developed in order to accomplish the work quickly and with a minimum amount of labor.

The handsaw, hammer, mallet and froe, which were brought over in the early days of colonization, were all used by the early settlers. With a wooden mallet (maul, sometimes called "beetle") and an iron or wooden wedge, the builder split the logs into halves, quarters, or any smaller size. One of the earliest materials prepared with these tools and, likewise, one of America's technological contributions to building was the clapboard.[3]

The earliest method of splitting the log cuts for clapboards was the same method the settlers had used back home; namely, by using a froe.[4] With the mallet and the froe, the logs were split again and again until thin pieces, wedgelike in shape, with one edge as thin as a feather and the other from 3/8 to 1/2 of an inch thick, were produced. This technique, however, proved too slow and cumbersome for the Americans who had urgent need of quantities of clapboards.

A speedy method, known as pit sawing, was developed for providing boards. A pit was dug in the earth and a log placed across it. One man stood in the pit while another stood on the surface, as they pulled a whipsaw in an up-and-down position. The whipsaw soon gave way to the more efficient "framed pit saw," in which the saw was fastened within an oblong wooden frame. This permitted a thinner blade to be used.

The pit sawing method was never satisfactory, and as soon as the settlers discovered the facilities needed to construct sawmills, they discontinued the pit sawing method. The sawmill produced boards of more even surface, and with the improved materials, the primitive Colonial architecture was improved. Hovels were transformed into the kind of houses the settlers had known back home. In addition, the sawmills also furnished valuable articles for export by the infant commerce of the colonies.

Waterpower, supplied by fast flowing streams, was employed early. In many instances, windmills were used to drive the sawmills. The motion of the saw in either case was up-and-down. The saw was fastened to one end of a wooden beam that joined a crank on a water wheel, while the other end was held taut by attaching it to the end of a spring pole. The log, spiked on a rough timber carriage, moved against the saw.

Sawmills were erected in a number of the colonies. The second shipload of settlers at Jamestown included a number of Poles and Dutch who had been sent over for the purpose of erecting sawmills.[5] Massachusetts had a sawmill in 1635, and in the same year another one was constructed on Manhattan Island by the Dutch. The mills built by the Dutch could saw more lumber in one hour than ten men could saw in a day. By 1700, the use of sawmills was rather general throughout New England. On one small stream in Massachusetts there were six sawmills within a distance of a quarter of a mile.[6] The later colonies, like Pennsylvania, had sawmills from the beginning of their settlements.

Some new technological improvements occurred in the operation of the sawmills in an effort to increase production. Instead of moving the logs by spikes manipulated by hand, a carriage driven by a cogwheel was designed to carry the log forward and backward, and instead of a single saw, two or more were added. In one of the mills in New York State, twelve saws worked in a gang. By such improvements the production was increased fourfold.

Coastal House Design and Construction

As soon as the Americans decided to build more permanent homes, the brick kiln became an important factor in the preparation of building material. There is some confusion as to when the first brick kiln was constructed in America, but regardless of the exact year, the surprising fact is that the first generation of settlers began to manufacture brick. Higginson, writing in 1629, said, "It is thought here is good clay to make Bricks and Tyles and Earthen pots as need be. At this instant we are setting a Bricke-Kill on works to make Brickes and Tyles for the building of our homes." [7] The technique of using clay and straw was the same as that known in Europe for centuries. An early contribution to brickmaking technology in America occurred when a law was passed by a Massachusetts court in 1667 regulating the size of bricks manufactured. Alexander Hamilton in his "Report on Manufactures" in 1790 reported that the manufacture of brick, coarse tile, and potters ware had grown to be most considerable. The production of bricks kept pace with the demand, and by 1790 the American brickmakers began to export brick in small quantities.

The construction of the wooden house was carried on according to the building methods known for many centuries in England and on the continent. The foundation was built of stone, brick, or wooden posts. The next step was to lay the sills upon the foundation. The sturdy sills were usually 5x8 inches, and rested upon the cornerstones of the foundations. Into the sills were mortised the tenoned thick vertical posts, which in turn supported the girder or summer beams. The girders, which carried the second floor, if there was to be one, corresponded to the sill beams on the foundation; the summer beams ran across the room, parallel to two of the girders, and were notched into two of the upright posts.

Between the rather massive posts and between the posts and the girts were smaller pieces which served as studs and diagonal braces. The summer beam, the posts and sills, and even the joists were all exposed in the interior of the early frame houses, and often presented an appearance of picturesque strength. Later in the Colonial period most of these construction features were hidden as a gesture of architectural refinement, only to appear again in the twentieth century as a decorative feature of "Colonial houses."

The sturdy framing of these early houses has been universally acclaimed by students of Colonial architecture. The excellent craftsmanship of the builders belies the scarcity of skilled carpenters. Many of these houses remain today and stand as a fitting memorial to the painstaking skill and labor which literally carved and hewed the sturdy structures out of the forests.

In filling in the frame, the Colonial builders departed somewhat from the procedure common in the seventeenth century English practice. The half-timbered houses of England contained a filling between the framing timbers, which was left exposed, of "cot and clay" or "wattle and daub." Conditions in America, however, caused the colonists to add a layer of clapboard siding over the exterior walls. Wood was abundant, lime was relatively scarce. This led to the lavish use of boards. The use of shingles for siding did not become common until the nineteenth century.

The same boards that were used for siding were used for roofing, and often laid two or three thick. Shingles were a luxury in the early period, even for roofs, and therefore many of the early houses had thatched roofs. By 1700, however, the abundance of timber and the introduction of sawmills put shingles within the reach of most people. Many of the early roofs were of the gambrel design. This type of roof, with its double pitch, provided more height and more space on the top floor. While probably of Dutch origin, its greatest development was in England. This explains its use in America.

An outstanding feature of the Colonial house was the chimney. The cold New England winters demanded that the center of the house should be, even more than in England, the hearth. The house very often looked as though it had been planned around the chimney. In erecting new homes or enlarging old ones, the chimney (or chimneys) largely determined the arrangement of the rooms. The most common plan in New England was to group the rooms about a central chimney which served two or more fireplaces in the house. Some of the more expensive homes had two chimneys, one at each end.

Another feature in early building technology was the construction of additions to the original dwelling. As the families increased in size and wealth, so they expanded their homes. Often they built additions on every side. The one-story house was enlarged into a two-story dwelling, and an entry hall was added. Later, more rooms were provided by building an addition of the same style and shape as the original building. Still another method used to increase the size of the original building was to add a lean-to. This was usually a one-story supplement attached to the rear of the house. The roof of the lean-to continued the slope of the original roof and dropped down to within 6 or 8 feet of the ground. These lean-to additions served as tool sheds, bedrooms, or at times were cut up into smaller rooms.

Inland House Construction and Design

Inland from the towns and villages of the coast a totally different type of structure was introduced. It was neither the frame house nor one of brick or stone, but a cabin made of logs, a structure unknown to the English. The log cabin was introduced into America by the Swedes as early as 1638. Its construction received a fresh impetus by the influx of the Germans and Scotch-Irish who came into Pennsylvania early in the eighteenth century. The log cabin was admirably suited to the wilderness regions, and it spread out into central Pennsylvania, and down through the Shenandoah Valley. As later generations of frontier Americans crossed the Appalachian Mountains and spread out into the Ohio Valley and the Kentucky and Tennessee regions, they carried the log cabin plans with them. Owing to its manifest advantages the log cabin had, by the time of the American Revolution, become the typical frontier dwelling.[8]

The first log cabins were built by using small, round, unbarked logs, 12 to 14 feet in length. The ends were crudely notched and the logs were laid one above the other to a height of 6 to 7 feet. Small logs were chosen since they were more abundant and could be raised into position by one man with the aid of his wife. Cracks between the logs were filled with small pieces of timber, twigs, and moss. Sometimes clay mud was used. Anyone who could swing an axe could throw together this crude type of dwelling within a few days.

These one-story, one-room dwellings were topped with roofs of bark shingles or rough-hewn clapboard shingles. An improvement was made when clapboards were later used to cover over the cracks in the walls. The floor, if not bare ground, consisted of split logs. These were called "puncheon" floors. The early cabins built of round, unbarked logs were never meant to be permanent. The bottom logs rotted quickly or sank into the earth, and then the building toppled down.

The first crude cabins were followed years later by hewn-log houses. The two-story hewn-log house became a mark of distinction. Special care was used in selecting the materials and in constructing the buildings. Long straight oaks, firs, and pines were selected. They were cut by woodsmen who knew how to fell trees properly and how to notch logs.

Before the walls were built, a foundation of flat stones was laid, on which the sills rested. On these sills, logs properly hewn were placed one above the other. As the walls approached the second floor a special technique of using hand spikes and skid poles was devised. A man with an axe stood on each corner of the building and carefully fitted the logs into place. Crevices or cracks were filled with clay, sand, and flat stones. In many cases the logs were so evenly dressed and the notches cut so accurately that chinking was unnecessary.

The roof was formed by making the end logs shorter, laying one on top of the other. A single log formed the ridge of the roof. As the end logs were made shorter they were held in place by poles which extended across the cabin from end to end. These same poles served as rafters upon which the clapboards were placed. Straight-grained oak or ash trees, 3 to 4 feet in diameter, were selected for making clapboards. The clapboards were laid side by side over the rafters, with a pole laid lengthwise to hold them in place.

Heating Facilities. The interior of the Colonial home was dominated by the fireplace. The entire family activity revolved around it, winter and summer. The fireplace had functional values; it often served as the only source of heat in the entire house and was also used for cooking and baking.

These early fireplaces involved a technological achievement worth noting. The amount of masonry used in the construction was enormous; foundations often measured 10 and 12 feet square. The upper part of the foundation was corbeled out to provide support for the hearth stones above and for the timbers extending from the front to the rear sills, into which the first floor joists were framed. Sturdy firedogs, andirons, or cobirons held the smaller logs. The necessary cooking implements and utensils were arranged around the fireplace. Many designs of an artistic and technological value evolved in the equipping of these central heating plants. Hooks, cranes, gridirons, tripods, spits, ladles, pots, and skillets all challenged the imagination and ingenuity of the designer.

Stone was generally used for the early fireplaces, but stone could not withstand the intense heat as well as brick. Consequently, as the supply of brick continued to increase throughout the eighteenth century, the brick fireplace became more common. The rear of the fireplace was protected by a large iron plate, called a fireback or fire-plate.

These large cavernous fireplaces consumed huge quantities of wood, from 30 to 60 cords a year. The problem of heating the early homes, especially those in the Northern colonies, led to a number of new experiments. The first settlers followed the pattern of heating with the old English fireplaces, but they found the climate more severe than in England. Provisions often froze during the winter when stored only a few feet from the fireplace. As the years passed and new settlements developed, the forests that were believed to have been inexhaustible began to recede. Wood was still free, but it became more expensive to transport from distant points. In an effort to overcome these discomforts many experiments were attempted by the settlers.

While more wood could be piled into the fireplaces with straight walls, some of the settlers began to observe that fireplaces thus constructed did not necessarily produce the most heat. They discovered that fireplaces with splayed jambs, built in a V shape rather than a U shape, required less wood, and funneled out more heat.[9] Another improvement was introduced some years later by immigrants from the Rhine Valley who brought iron fireplaces which were set inside the larger fireplace. These "stoves," as they were called, gave off more heat and used very little wood. The English colonists seldom used stoves. They knew only the warming pan before Benjamin Franklin introduced his "Pennsylvania Fireplace."

This Pennsylvania fireplace, devised in 1742 by Franklin, proved to be his most popular invention, and marked the most significant advance in heating technology for years to come. Franklin's aim was to utilize the heating effects of circulating hot, fresh air. His first experiments involved the use of an iron, boxlike affair, with an open front, which was built into the fireplace. Inside this main box was a small air box that extended to the sides, but not quite to the top of the stove. The smoke from the fire passed over this air box, down between the air box and the back of the wall, around the false back of the fireplace, and up the flue. The air in the air box was thus warmed, and escaped from the sides of the stove through shutters.

In writing of the purpose of this invention, Franklin commented on the inconvenience of the Colonial homes, and declared that by using his invention, "People need not crowd so close around the

Fire, but may sit near the Window, and have the Benefit of the Light for Reading, Writing, and Needlework, etc. They may sit with comfort in any Part of the Room, which is a very considerable Advantage in a large family." The Pennsylvania fireplace, as later developed, came to be known as the Franklin stove.

Lighting. For lighting, candles were used. During the earliest years the colonists used the fat of deer and bear for making candles. When beef and mutton tallow became more abundant, candles were made by dipping flax or cotton threads into molten tallow. Few household duties were of greater importance than the preparation of the precious supply of candles for the winter. It was with a feeling of relief and contentment that the labor of making candles was completed, and the family was assured of having light for the long winter months. However, only a few families ever succeeded in storing up a sufficient supply to last through the season. This meant that some member of the household would again have to take up the smelly task of preparing extra candles at times when other chores needed attention. In later years the traveling candlemaker made his annual visitation, dispensed the news, and prepared candles from the bits of tallow saved for him by the housewives.

Before the end of the seventeenth century, candles were made from the bayberry of the myrtle bush. It was found that these candles burned more slowly, gave more light, and had less odor. The myrtle plant grew abundantly in the marshes and swamps, and since the berries were gathered in great quantities and converted into wax by boiling, both the affluent and the poor provided themselves with this wax. The candles made of myrtle were often used in public meetings.

In the mid-1700's the daring American whalers began to bring back quantities of spermaceti from which candles were made. These candles afforded about three times more light and four or five times the flame of an ordinary tallow candle. A factory for the manufacture of spermaceti candles was opened in Boston in 1750 by Benjamin Crabb, an Englishman, who obtained the exclusive right to manufacture them for fourteen years in Massachusetts. His factory was burned within a year, and another was begun by Obadiah Brown in 1753. By 1761 there were eight such factories in New England and one in Philadelphia.[10]

The factory manufacture of candles began on a rather large scale in the late 1700's. The factory technique of making candles was to suspend wicks, made of hemp, tow, cotton, or the "silk down" of the milkweed, from wooden poles or crossbars and dip them into melted tallow or wax. Many drippings were required to fill out the candle to the desired thickness.

Another method was to pour the tallow or wax around the wick, which was suspended in metal candle molds. The molds were usually built-in units, some containing as many as twenty-four tubes. The tallow candle industry was one of the first enterprises in the early history of American technology to take on a degree of specialization. Candlemaking was one of the first family tasks to be removed from the list of household chores.

An improvement in lighting was introduced when reflectors were installed in sconces on the walls; and later myriads of candles were set in sparkling cut glass chandeliers which greatly increased the power of the light. The designing of these glass chandeliers, as well as those made of metal, was a test of the maker's ingenuity, artistic skill, and workmanship.

Glass was one of the earliest products manufactured in the colonies, but not on a large scale. The early settlers continued to depend upon Europe for most of their supply. The first glass windows were of the hinged, casement type, with small panes cut in oblong or diamond shapes and framed in heavy lead. The casement windows swung inward on hinges in contrast to the sliding vertical movement of the doublehung sash type. This latter type came into more general use during the eighteenth century. In these windows the lower sash was raised or lowered by weights suspended inside a box frame, the same as used today.

Furnishing and Decoration. Although some furniture was imported, the cabinet maker assumed an early enviable position in Colonial history. There was a noticeable development in making chairs, tables, and bed-steads. The principal skill involved in the construction of these until as late as 1700 was that of "joinery." This involved notching the wood and making joints with straight pieces of wood always meeting at right angles.

An important development occurred in furniture making with the introduction of mahogany wood around 1700. The strength of this wood made possible a new method of carving, delicate and lace-like, which reached its perfection in some of the Chippendale models. The early American cabinet makers were not far behind the English in introducing the use of mahogany for furniture, and this new wood gave rise to new skills. It was too expensive and required too much labor to gain wide use in frontier communities. In its place inexpensive substitutes for mahogany, such as cherry, maple, walnut, and pine were used in the inland districts.

The walls and floors of the Colonial houses were as simple as the furniture. In the earliest settlements the colonists were compelled to forego all interior ornamentation which they had known in Europe. Few of the homes were painted. The housewife preferred a well-scoured and sanded floor.

The interior walls were generally covered with a plaster of clay or mud, plus a rough coating of lime plaster. The use of paint for walls was discouraged as a useless luxury. Bishop cites a case where the Reverend Thomas Allen, of Charlestown, was called to account in 1639 for having paint about his dwelling, but was discharged upon his showing that the painting was done before his time and was disapproved by him. The first church erected in Boston was never painted within or without while it stood. Indeed, a list of artisans made out by the General Court of Massachusetts in 1670 does not contain the name of a painter.

The Colonial houses of the eighteenth century, inspired by the renaissance in English architecture, took on the new motif which flattened the roofs, raised the ceilings, emphasized symmetry, decorated the door frames with classical pediments and pilasters, and added other embellishments.

The importation and dissemination of the new styles were effected chiefly by skilled craftsmen and by knowledge gained from books on architecture. One of these building guides appeared as early as 1624, and a number of others were printed before the close of the century. Professional architects were few, even in England. It became a fashion for gentlemen to assume an interest in architecture and it often became the intellectual occupation of amateurs. All this, coupled with the skill of the craftsmen and the use of popular handbooks, resulted in fashioning many fine Georgian houses in America that compared favorably with those in England.

Andrews, in summing up the developments in building technology during the Colonial period, states that:

> Taking the colonial area through its entire length and breadth, we find individual houses of almost every description, from the superb mansion of the Carters in Virginia to the small wooden frame buildings, forty by twenty or thereabouts, with a shed on the backside and a porch on the front, and the simple houses of the country districts or the western frontier, hundreds of which were small, of one story, unpainted, covered with rough hewn or sawn flat boards, weather-stained, with few windows and panes of glass, and without adornment of architectural taste.[11]

Notes for Chapter 3

1. Martin Shaw Briggs, *The Homes Of The Pilgrim Fathers In England and America* (1932), p. 127. Quoted by permission of the Oxford University Press.
2. Briggs, *op. cit.,* p. 123.
3. The difference between a clapboard and a shingle is that the clapboard is longer and usually narrower.
4. An essential early tool now almost extinct. It had a blade with a wooden handle at a right angle to it. It was used by the early settlers for splitting logs. It was steered by the handle and was pounded on the opposite end of its blade with a wooden implement called a "fore club," or mallet.
5. J. Leander Bishop, *A History of Manufacturing,* I (1864), p. 93.
6. Bishop, *op. cit.,* p. 100.
7. Briggs, *op. cit.,* p. 136.
8. For an excellent history of this type of building, see Harold Robert Shurtliff, *The Log Cabin Myth* (1939).
9. William Babcock Weeden, *Economic and Social History of New England,* II (1890), p. 654.
10. Bishop, *op. cit.,* I, p. 208.
11. Charles McLean Andrews, *Colonial Folkways* (1921), p. 55.

See also: Luke V. Lockwood, *Colonial Furniture in America* (1926); Nancy Vincent McClement, *Furnishing The Colonial And Federal Houses* (1947); Sidney Fiske Kimball, *Domestic Architecture Of The American Colonies.* These are only a few of the dozens of excellent works about Colonial architecture.

Chapter 4

CLOTHING AND LEATHER TECHNOLOGY

Clothing Imports

The first English settlers in America depended more upon the mother country for clothing than they did for any other necessity. They could fish and hunt and trap, and thus secure food immediately. They could throw together a hovel in a day or two that could withstand the blasts of winter, but the making of clothing required time and skill. The first settlers brought some clothing with them—articles made of wool, linen, leather, and even silk, which lasted a year or two but not much longer. Chopping down trees, grubbing stumps, plowing and tilling the rocky soil were all hard on clothes. Providing suitable materials and fashioning them into clothing made early demands upon the colonists.

Lists of the type of clothing and the amount needed were drawn up by many of the early settlers and sent back to prospective emigrants. Captain John Smith, who resided in Jamestown long enough to form some idea of the nature of the climate, advised prospective settlers to bring "a monmouth cap, three falling bands, three shirts, one waistcoat, three pairs of Irish stockings, a pair of garters, four pairs of shoes, and a dozen pairs of pants." His selections were made with the idea of discouraging extravagance and finery. The articles were named because of their simplicity, serviceability, and cheapness. The entire list above mentioned could be purchased, wrote Smith, for 59 shillings.

Not every one followed instructions, however. Many of those who came during the next fifty years brought along the finest of clothing. While not serviceable, this finery proved a constant reminder of the best, and thus became an incentive to produce better clothing. Clothes such as those of Thomas Warnet, a prominent merchant of Jamestown, represented some of the finest tailoring known in Europe. His will, dated 1629, listed "a coif, a cross-cloth of wrought gold, a pair of silk stockings, a pair of black hose, a pair of red slippers, and silk green scarf with gold edges, six dozen buttons of silk and thread, a felt hat, a black beaver hat, a Polish fur cap, a doublet of black camlet, etc." [1] These articles would have been envied even in the London of that day.

Need for Raw Materials

The raw materials which the colonists sought for processing into clothing were wool, leather, and linen. In the eyes of the colonists the most important raw material for clothing was wool. Relative to cotton, wool was used more than at present. The North American climate made wool the staple material for heavy clothing, both in the northern and middle colonies. Gradually its production and manufacture shifted from England to the colonies.

Raising Sheep. Sheep were introduced into Jamestown in 1609, into Massachusetts in 1624, into New York in 1625. The early Dutch and Swedish settlers on the Delaware brought sheep from Holland and Sweden. Governor Printz was commanded by the Swedish Crown . . . "above other things he shall direct his attention to sheep, to obtain them of good kind . . . , so that presently a considerable supply of wool of good quality may be sent over to this country." [2]

The importation of sheep was officially encouraged in other colonies. The general court of Massachusetts passed an order in 1645 requesting families, "to bring as many sheepe as convenient." [3] Three years later another act was passed relating to the pasturing of sheep on the commons; and a still later act provided bounties for the killing of wolves. With such incentives Massachusetts Colony was able to report in 1650 that three thousand sheep were being raised there. An equal number was reported from Virginia. One clipping of this number of sheep meant an immediate supply of 12,000 pounds of wool. By 1661, the Council for Foreign Plantations in England estimated that there were ten thousand head of sheep in New England. In 1678, according to a report of sixty-three families paying taxes in Southampton, Long Island, fifty were sheep owners.[4]

A Parliamentary Act of 1699 disclosed that sheep raising had become a major industry in the American colonies. Thirty years later the Board of Trade reported that in New England, New York, Pennsylvania, and in the county of Somerset, Maryland, the citizens had undertaken the manufacture of woolen cloth from the fleece of their own sheep.

Wolves were a constant threat to early sheep growers, and this danger led the more alert farmers to adopt a new protective measure.

When they discovered that the islands and peninsulas could easily be cleared of wolves, they transferred the raising of sheep to such coastal islands as Nantucket, where fifteen thousand sheep were pastured in 1770; Martha's Vineyard, peninsulas like Hempstead Heath, and Long Island became favorite sheep growing areas, chiefly because of the safety from wolves.

Leather Tanning. Leather, and the clothing products made from leather, were equal to wool in importance during the early Colonial period. In some respects leather was even more closely identified with Colonial America than wool, for while the colonists always depended upon England for much of their wool supply, they soon became independent in their production of leather. Although leather was adapted to many uses, its greatest importance was for boots and shoes.

It has long been said that civilization marches on foot, and the well-shod races have always overcome the unshod. Whether or not the early settlers were aware of this, they were continually calling for tanners, shoemakers, and a greater supply of leather. To satisfy these demands for leather, hides were sought from all sources. Doe and buck skins were bought from friendly Indians or were secured from trappers; domestic stock was drawn upon; and hides were imported from the Spanish colonies in the West Indies.

Some English cattle were landed in Virginia in 1608, and a herd consisting of three heifers and a bull was introduced into the Plymouth Colony by Edward Winslow as early as 1624. The same year the Dutch brought cattle to New Amsterdam. An additional herd of twelve cows arrived at Cape Ann, Massachusetts, in 1626, and thirty more arrived in 1629. In October, 1683, the "Concord" (the "Mayflower of the Germans") arrived at Philadelphia with a score of cattle on board.[5] Cattle, sheep, and goats were kept in a common herd in the early settlements and guarded by an overseer.

Linen Production. While wool and leather offered the best materials for clothing needed to withstand the vigorous American winters, yet they were not always the most comfortable. Undergarments made from either cotton or linen were much preferred, but cotton had not yet become king. The cotton supply was limited, and the cost of processing it into cloth was prohibitive. Only the wealthiest could afford cotton calicoes.

Linen made from flax and hemp served the general purposes for which cotton, nylon, rayon, orlon, and dacron are now employed. Consequently, early attention was given to the cultivation of flax and hemp. In the early years, bounties were frequently offered to encourage farmers to sow fields of hemp and flax.

A Massachusetts Act of 1640 directed the magistrates of towns to determine the number of men and women "skilled in breaking, spinning, and weaving and to determine what provisions should be made for teaching boys and girls the art of spinning yarn." A bounty of three pence was offered for every shilling's worth of linen, woolen, and cotton cloth produced. Similar acts were passed in later years, aimed to stimulate the increased production of linen and cotton cloth. Other colonies followed the same practice.

In Virginia in 1673, an act was passed alloting to "each tythable person" one quart both of flax and hemp seed, with the provision that each such person must produce one pound of "flax drest" and one pound of "hemp drest," or two pounds of either. Failure to do so resulted in a fine of fifty pounds of tobacco for each pound of flax or hemp not produced.

In 1682 another Virginia act was passed for the encouragement of the manufacture of both linen and woolen cloth. Twenty-four pounds of tobacco were offered to the person who produced one peck of flaxseed; twenty pounds of tobacco for each peck of hemp seed; six pounds of tobacco for each ell of linen cloth three-fourths of a yard wide, and six pounds of tobacco for each yard of woolen cloth or linsey-woolsey.

The Provincial Council of Pennsylvania in 1686 reminded the people of "ye Governor's promise to him who shall make the first and finest piece of linen cloth." The council encouraged the further production of hemp "by bringing in persons of skill to lead the way." A premium of five pence per pound for hemp, and eight pence per pound for flax was offered.

The offering of such awards, prizes, and premiums paid off. By the end of the seventeenth century, a sufficient amount of flax was being produced to permit exporting a considerable quantity of flax seed to Ireland and Scotland. In 1730 William Byrd, Esq., of Virginia, was paid five pounds, four shillings, for raising two thousand six hundred pounds of hemp. Awards in lesser amounts were paid to other growers.

Manufacturing Processes

Three Technological Principles. After the raw materials such as wool, flax (linen), and leather had been supplied, it was then necessary to process these materials into finished products. A number of basic features characterized textile technology during the Colonial period. First, the clothing material was homespun, a factor that distinguished it from the workshop products of contemporary Europe. Each Colonial family supplied its own clothing needs, but the task did not demand the entire time of any one person; neither did it require the continuous pursuit by the family, collectively.

Second, the only labor employed was hand labor. Third, the techniques employed were those the settlers had inherited from Europe. The technological changes in the preparation of materials therefore did not take place until after the Industrial Revolution in England, and following the American Revolution in this country.

The processes involved in the preparation of wool represented the stage of development then current in England. The crude implements used were the hand tools which the settlers brought with them: spinning wheels, hand cards, brushes, and combs that were in common use in the mother country. Those who came from the regions of Yorkshire brought tools a little more refined, designed to handle different grades of wool.

Woolen Technology

The year 1638 is an important date in the history of American woolen technology. That year five professional weavers settled in Rowley, Massachusetts, where they manufactured the first piece of cloth in the colonies of which we have record. This first piece of cloth measured 83½ yards in length. In 1643, twenty more families of weavers came directly from Yorkshire, England, and also settled in Rowley. This settlement is noteworthy because here was introduced the most advanced English textile technology in the manufacture of cloth.

The steps involved in the process of woolen manufacture, from clipping the wool from the sheep to scouring, carding, combing, spinning, weaving, and napping the woven cloth, were all simple hand operations. Washing or *scouring* was done in the family wash tub with homemade soap and weak lye. Repeated washing, dipping, and rinsing removed most of the dirt and grease.

Carding prepared the wool for spinning. A hand card consisted of a small rectangular piece of leather, fitted with small pieces of wire with bent ends, and provided with a handle. A handful of wool was brushed between two of these cards until the fibers of the wool were straightened. Strength was given to the wool by intermixing fibers of different lengths. Carding was an operation usually performed by the women or the older children of the family.

The *combing* operation required greater strength and skill than hand-carding. Frequent references are made to the wool combers of the early Colonial period. The comb, like the card, was also a hand tool, but was made of metal. It had fewer but longer points than the card. Combs were used to draw out further the long fibers in parallel position, and remove the shorter ones. This operation was carried on by the use of two combs with which the wool combers, in wide dexterous sweeps, drew the wool into long fibers. This action was repeated until only the long fibers remained in the top comb. The loose wool was then removed and set aside for the spinner.

Spinning was done on a hand spinning wheel by the women and girls of the family. The American women abandoned rather early the old distaff and spindle and substituted the handwheel. The wool that had been carded and combed was held in the left hand and fastened to the spindle. As the large wheel turned, an outward movement of the spinner's left hand elongated the yarn to the desired length; and by giving a twist to the yarn at the same time it was wound on the spindle. This gave it added strength.

The *weaving* equipment was much more complex. The looms built by the colonists were large and heavy, and considerable strength was required to operate them. Usually, the looms were set up in the loft or in a shed, or in a special lean-to attached to the house. When complete with beams, heddles, batten, and shuttle, men of skill and strength were assigned to this particular task. Early in Colonial history the work of a weaver became a specialized occupation.

Some of the handcraft weavers maintained shops to which the yarn was brought; while others moved from place to place, and were called the itinerant weavers. The latter traveled about the country, offered their services to any who desired them, and wove enough cloth for the settlers of entire communities. These weavers played an important part in early colonial manufacturing: in certain areas their services were indispensable. They were always welcome because they were purveyors of news.

After the wool was woven into cloth, the next process involved was called *fulling,* or *felting* the fibers. Prior to the general introduction of the fulling mill, about the middle of the seventeenth century, fulling was accomplished in a primitive manner. The most common method was to soak the fabric thoroughly in warm soapy water, then spread it out on a flat surface, and beat it with flat sticks. The early settlers often made a game of this arduous task by inviting the neighbors for a "fulling bee." The soaped cloth was placed on the floor and the guests, after removing their boots, shoes, and stockings, kicked it around with their bare feet.

With the erection of fulling mills, however, special attention was given to the proper treatment of the cloth. The fulling mill consisted of two rollers fixed above a trough filled with soapy water. A piece of cloth was formed to resemble the common roller towel of a later day. The cloth passed through the water and was compressed by the rollers located at the top. The rollers were driven by a waterwheel.

A new technical improvement was made when wooden beaters were added to the action of the rollers. The rollers raised the cloth gradually, and as they did so, the water was squeezed from the cloth. The squeezing action forced the fibers of the cloth to knit closer together, and the action of the wooden beaters on the damp cloth aided in the process of intermeshing the warp and woof of the cloth. The fuller's art lay in the knowledge gained from experience of the right mixture of soap and water, and (later) the amount of fuller's earth to be added (a highly absorbent clay that soaked up the oil and helped mat the fibers). The skilled fuller learned how to direct these processes so that the cloth would be prepared evenly and thoroughly without damage to the fiber.

Fulling mills were introduced early. One was erected in Roxbury in 1657, one in Dorchester in 1659, and one in Watertown in 1662. During the 1670's, mills were set up at Andover, Ipswich, and Salem, and by 1700 some eighteen fulling mills had been built in New England. There was one in Virginia in 1692, and one in Pennsylvania in 1698. After 1700 they spread rapidly throughout the colonies and almost every village had at least one.

Other phases in woolen technology included those of *shearing* the cloth and *dyeing*. Dyeing was commonly carried on in the household. This was especially true of cloth that was "dyed in the wool." In this procedure the wool was dyed while it was still in the form of fleece. Early dyestuffs were rare and difficult to obtain. The colonists bought them from itinerant peddlers, or made them in a crude fashion from the barks of trees of the native wood. The art of dyeing cloth demanded considerable skill that was acquired only after long years of practice.

The process of shearing the cloth was likewise a difficult one. In this procedure the purpose was to clip any protruding ends of the fiber that had not been pressed into the cloth during the fulling process. When a product like broadcloth was desired, shearing was preceded by napping. The napping process was accomplished by scratching the surface of the cloth with hand cards. This action raised the nap. Ordinary shears were then used for the clipping process. The main idea was to clip evenly and produce a surface as smooth as possible. Shearing was seldom carried on in the household process of making cloth, due to the long time required.

Cloth making was carried on throughout most of the colonies by 1700. In a report to the British Board of Trade in 1705, Governor Cornbury, Governor of the Province of New York, declared that once they began to manufacture serge they would soon be making both coarse and fine cloth. He pronounced the cloth that was being made good enough for any man to wear, and predicted that his colony would soon be producing enough clothing to supply all local needs. By 1710, three-fourths of the linen and woolen goods in New York were home produced. Governor Spotswood, reporting to the Council of Trade for Virginia in the same year, declared that in his colony they were making a mixture of cotton and wool for clothing for slaves, and in one county 40,000 yards were made in one year.

Linen Technology

Linen, the second material in importance in making clothes, required an entirely different type of textile technology. Linen cloth was made from flax which was grown throughout the colonies. This hollow-stemmed plant, with its long fibers running the entire length of the stalk, presented a difficult problem. Since that part of the plant that surrounds the fibers has no textile value, it had to be removed before the fiber could be used for spinning. The difficult and distasteful task of removing this waste material of the stalk was accomplished by a "retting" process; that is, by exposing the stalk to water.

Retting Processes. Two different processes of retting were developed. In one, the stalk was immersed in a stream of running water where it was allowed to remain from ten to twenty days, depending upon the temperature of the water and the maturity of the plant. This process was known as "water retting." The second method was known as "dew retting." In this case, the stalks were laid out on the grass in thin layers and remained there until they were partially rotted. This process required a longer period of time, but it was the only method that a farmer could use if he had no access to a running stream. When the right stage was reached in either process, the retted flax was gathered up and thoroughly dried. Drying was accomplished by exposing it to the sun on clear days, and by placing it in the large bake ovens of the kitchen fireplace during the nights.

When the flax was thoroughly dried it was broken, a handful at a time, on a wooden beam called the "flaxbrake." This device consisted of a sawhorse arrangement, on top of which was a wooden beam. This beam was fastened by a hinge to a lower beam. The flax was placed between the two, and the upper beam was brought down sharply against the flax.

The next step was to beat the flax with a wooden knife. In this manner, most of the broken waste that had been loosened by pounding was removed. Then came the "hetcheling." The flax processed in this manner was drawn, a handful at a time, through a comblike device to remove the remaining waste. The product was then called dressed flax.

Most of the well-ordered homes in New England during the Colonial period had a flax wheel. Daniel Denton wrote in 1670 that, "everyone makes cloth for their own wearing, also woolen cloth and linsey-woolsey." Around 1700, wheels for spinning linen from flax were introduced widely throughout most of the Mid-Atlantic and New England colonies. From that time on production of linen cloth was common.

It should be noted that there was almost no resemblance between the flax wheel and the wheel for spinning wool. The technological operations of the two were wholly different. The flax spinner sat while at work, and with skillful fingers separated and fed to a foot reel the long fiber filaments drawn from the heavy strand of dressed flax that was thrown over the distaff. This was an art that required great skill. Men obviously were not suited for it. The linen thread from the spinner was woven into cloth (linen) on the handloom. This linen cloth was then washed and bleached in the sun, and when dry, was ready to be cut to measure and made into clothing.

As the Colonial settlements expanded, the household manufacture of cloth, both woolen and linen, became more widely extended. The net result was a dominant family manufacture of cloth for more than one hundred and fifty years. With the production in the home of woolen and linen textiles, and leather for shoes, the economy in the making of clothing was complete. Between the earlier and later Colonial period there were no major technological changes.

Leather Technology

In preparing leather, the early colonists first borrowed a method used by the Indians. The Indians were the first tanners on the North American Continent. Before the coming of the first settlers the Indians had to depend entirely on leather for their hunting shirts, breeches, moccasins, and even their shelter.

Tanning. While the Indians knew nothing of the bark method of tanning, they had a method of dressing skins, called the "buckskin tan," which the colonists eagerly adopted. With a stone knife, or sharpened stone, the hide of the animal was removed, and thoroughly cleaned. The skin was then buried in the ground for a few days. When it was taken up the hair was easily scraped off. After the hide was thoroughly cleaned on both sides, it was rubbed with the dust of dead trees. The small amount of tannin in the dust served to tan the skins.

Next, the Indians rubbed fat or suet into the leather until it was soft and pliable. Some of the tribes prepared a mixture from the brain of the animal which they used in place of fat. Occasionally, to remove the surplus hair and flesh from the hide, they made a mixture of water and ashes. With these simple processes of a chemical nature, but not so named, the Indians produced a soft pliable leather, impervious to all sorts of weather. The early settlers soon learned and adopted this process from the Indians.

But the early colonists soon began to improve on the Indian method of tanning. Every shipload of early settlers brought skilled tanners. In the list of workmen that arrived in Virginia in 1620, tanners and leather dressers were mentioned. In 1623, Experience Mitchell, a tanner, came to Plymouth on the "good ship Anne." In the Salem Colony in 1628, one Francis Ingalls, a tanner, was granted leave to go where he would, and in 1632 he built a tanyard in what is now Lynn, Massachusetts, a town that was later to become the center of the boot and shoe industry of America. There were three tanners among the first settlers of Boston; also one shoemaker.

The Colonial assemblies tried hard to stimulate the tanning industry by offering bounties and prizes. Virginia considered the tanning of leather of sufficient importance to pass an act in 1661 requiring every county in the state to have at least one tanner, currier, and shoemaker. The following year another act was passed providing for the erection of a tanning house in each county, to be kept up by the county. Any county that failed to comply could be fined to the amount of 500 pounds of tobacco. At the same time each inhabitant of the county was allowed 2 pounds of tobacco for each pound of animal skin or hide brought to the tannery.

In New York expert tanners were in such demand that a monopoly was granted in 1676 to two skilled leather workers as an inducement to have them come and settle there. A number of towns in Massachusetts offered inducements for skilled tanners to settle within their borders. The town of Newbury granted Ebenezer Knowlton nine rods of land for setting up a tannery in that town. George Barnhall, reported as the first tanner in Maine, carried on a successful business between 1680 and 1700. With such inducements as these, advances in tanning technology were inevitable.

As the demand for leather continued, stringent regulations were passed to prevent waste and to preserve hides. In 1638 the General Court of Massachusetts prohibited the export of all skins and hides, and appointed two searchers and sealers to enforce the law. In 1640 Connecticut passed an act penalizing anyone who did not preserve all animal skins and hides for tanners. The scarcity of hides was so acute in Massachusetts that an act was passed in 1642 prescribing the process to be used in tanning. This was done so that no hides would be lost by using improper methods. In New Haven, one Goodman Osborne, who had spoiled several hides in tanning because of want of experience, was ordered to replace all the hides he had spoiled. By 1650 there were fifty-one tanners listed in the Massachusetts Bay Colony. In 1662 Virginia passed an act forbidding the exportation of hides under penalty of 100 pounds of tobacco; and a few years later, the legislature prohibited the exportation of skins of all kinds.

The technology of tanning which these early craftsmen adopted could hardly be called a scientific process. Rather, it was a procedure by rule-of-thumb. The hides were first thoroughly cleaned, with each tanner using a secret process, usually employing lye obtained from leaching wood ashes as the cleaning agent. Oak bark, crushed with a club or stone, was used for tanning; a crude box made of rough plank and sunk in the ground served as a vat. Alternate layers of oak bark and hides were spread in the vat. Water was poured in until the vat was filled.

The hides were kept in this liquor for six or eight months, depending on the use for which the leather was intended. Periodically, it was necessary to renew the liquor; the hides were taken out, and a new supply of oak bark and water added. Hides that were to be used for sole leather or harness were kept in the vats for a longer period. After the hides had gone through this tanning process, they were removed from the vat and thoroughly washed. They were then hung in a barnlike room until they dried. During the drying period they were occasionally taken down, rubbed, and rolled to remove the stiffness. The finished product was then ready for use, and it was sold to dressmakers, shoemakers, harness makers, and saddle makers. Most of the leather produced went into shoes worn by the early colonists. The first shoes were rather crude and ill-fitting, and were built "straight," with neither a right nor left. They could be worn on either foot, and were interchanged regularly so as to insure uniform wear of heels and soles. All shoes were handmade throughout.

Shoemaking was learned by apprentices from experienced craftsmen cobblers. It was common practice for shoemakers to travel from village to village, carrying their kit of tools and a supply of leather. They repaired old shoes and made new ones for every member of the family. A Massachusetts Act of 1633 allowed shoemakers twelve pence a day for work done at their customer's home. Gradually some of the most skilled and progressive shoemakers managed to accumulate small savings, and began to hire other skilled craftsmen to do special piece work.

In parts of New England and in Pennsylvania shoemaking became a family task in many Colonial homes. During the long winter evenings, the entire family gathered around the huge fireplace, and each member was assigned a special task. The father and the grown-up boys cut and trimmed the heavy sole leather. The mother bound the "uppers," as they were called, or sewed the heavy leather of the tops by hand. The younger boys cut the wooden pegs which fastened the sole to the upper. The finished shoe, while not remembered for beauty or style, nevertheless was known for its long, useful wear.

In 1750, John Adams Dagyr, often called "the Father of American shoemaking," arrived from Wales and settled in Lynn, Massachusetts. He opened a shoe shop, began to employ skilled workers, and assigned each a special operation. Here was instituted the beginnings of a small factory system.

Notes for Chapter 4

1. Philip E. Bruce, *Economic History of Virginia,* II (1898), p. 187.
2. Stevenson Whitcomb Fletcher, *Pennsylvania Agriculture and Country Life, 1640–1840* (1950), p. 191.
3. J. Leander Bishop, *A History of Manufacturing,* I (1864), p. 310-11.
4. Victor S. Clark, *History of Manufactures in the United States,* I (1928), p. 78-79.
5. Fletcher, *op. cit.,* p. 174.

William Babcock Weeden, *Economic and Social History of New England, 1620–1879* (1890), and J. Leander Bishop, *History of American Manufactures* still remain excellent references for the early history of clothing and textiles. Also of help are such books as: William S. Murphy, *The Textile Industries* (1912); Arthur Harrison Cole, *American Wool Manufacture* (1926); Mary Schenck Woolman and E. A. (Beers) McGowan, *Textiles* (1913). For a discussion of the hemp industry, see James F. Hopkins, *A History Of The Hemp Industry In Kentucky* (1951).

Chapter 5

EARLY TRANSPORTATION AND COMMUNICATION TECHNOLOGY

METHODS OF COAST TRAVEL

The first developments in transportation technology occurred when the early settlers began to navigate the rivers, bays, and inlets along the coast. Not since the days of the Phoenicians had people anywhere in the world depended so completely on water transportation as those who came to the New World. In the first place, they had crossed 3,000 miles of ocean before they even reached a new place of settlement. After arriving they built their homes along the water's edge, either on the sea coast or on the river banks. Here, the first colonists spent their entire lives; and here their children grew up inured to the sea. They had to master every form of water transportation, ranging from the light canoe to the heavily rigged sailing vessel.

Early Small Watercraft. The construction of small watercraft began in the Virginia Colony about 1611. This activity sprang from absolute necessity, since the plantations were located on rivers and creeks and could be reached only by waterways. Within a few years numerous barks, pinnaces, and rowboats, both large and small, were being built. No craft described as ships, however, were constructed in Virginia in the early days.

The New England settlers, somewhat more ambitious, built small craft, such as rowboats and shallops used in 1622 to explore the shores of Massachusetts Bay. In 1624 a ship carpenter came to the Colony with the technical heritage of English shipbuilding, and "quickly built two very good and strong shallops, and a great strong lighter." Unfortunately for the colonists, this skilled artisan died early of fever.

First Shipbuilding. In 1629 shipbuilding was firmly launched in New England when six shipwrights reached Salem, bringing pitch, tar, cordage, and sailcloth. After building several small vessels, they undertook what they called their masterpiece, a 30-ton sloop for Governor John Winthrop, called "The Blessing of The Bay." The ship was launched July 4, 1631, and in Weeden's words, "She carried the future of this continent in her lading."[1] With this undertaking, the art of shipbuilding was marked by a number of technological developments. Skill, labor, material, and money were assembled; and the expanding, sea-minded society of New England was favorable to the industry.

Following the construction of this first vessel, a number of others were built. In 1640 a vessel was built at Wethersfield, and in 1641 Richard Hollingsworth of Salem built a "prodigious ship of 300 tons," while another vessel called the "Snow" was built the same year. A 60-ton vessel, the "Rebecca" built in Medford in 1663, was used for trade on the Connecticut River.

By the middle of the seventeenth century a new type of vessel called a "ketch" was being built. It was used for fishing until the innovation of the sloops, and was also employed in coastwise trade between the settlements. A few ketch-built boats ventured down to the West Indies, and even to England. Their breadth of beam made them good sea-going barks, safe and fast.

Colonial Shipbuilding

The demand for ships was always greater than the supply. By 1665 some two hundred American-built vessels, ranging from 20 to 600 tons had slid down the ways in New England shipyards. By the end of the seventeenth century more than one thousand vessels had been built in the New England colonies; Salem was fast becoming a center of the shipbuilding industry and the port for maritime adventure.

Shipbuilding in Pennsylvania began with the establishment of the Colony. Stately oaks, measuring 60 and 70 feet in height, were brought in from nearby forests. The first sails and rigging were brought from England, although by the time of the Revolution these were being made in Philadelphia. Various types of boats were designed, including the pink or galley and a special draft type of ship for transporting timber that carried upward of 500 tons. These boats, like the rivercraft, were designed to be broken up on the completion of their first trip. By the 1770's Philadelphia had become a major shipbuilding city and almost half of the ships registered in the colonies were built there.

This rapid development of shipbuilding in the American colonies caused some alarm among the shipbuilders in England. In 1720 the master shipwrights in London filed a complaint with the King, declaring that their trade was being injured, and that workmen were deserting them because of Colonial competition. They had good reason for registering such a complaint because the American colonists were building ships faster than any other people in the world. Moreover, in 1713 the Colonial shipbuilders had added to their models a new type of ship known as the "schooner," a Yankee invention which was largely the work of Andrew Robinson. This small vessel rode the surface of the water and had the speed and maneuverability that later characterized the clipper ships.

Early Shipbuilding Technology. A description of the building of one of the early ships, the "Mary," gives a fair picture of the technological advances in shipbuilding in the mid-eighteenth century. The "Mary" measured 58 feet in length at the keel, 21½ feet across the beam, and 10½ feet deep in the hold. She was not an unusually large ship, for British merchants emphasized stability rather than tonnage. The builders laid great importance on seasoning of all timbers. When the frame of the vessel was completed, all work stopped and the vessel stayed on the shipways for months before the housing-in, planking, decking, finishing, and launching were undertaken. The purpose of the delay was to allow the keel, stem, and stern posts, floors, transoms, fashion pieces, and other framework to season by natural processes.

Wooden nails made of locust and hickory wood steeped in tar or tallow were used throughout the vessel in pinning the timbers. Once they swelled in place with the aid of tar, they held fast. No iron was used in fastening the ship together where it would be exposed to the action of salt water. A few iron and copper nails were used inside and on the top structure of the vessel where there was little danger of corrosion. The blacksmith for the "Mary" forged the chains, chain plates, eyebolts, dump bolts, and rings to drive them through for clinching. He also fashioned the saucer and collar-headed bolts, ring and hook bolts, keys, hatch bars, ratchets, and spawls for the windlass; he also forged the small hardware such as hinges, screws, locks, and nails.

After six to ten months of seasoning, the sheathing was applied, the finishing touches were added to the interior, and the "Mary" was ready for launching. This was accompanied by an elaborate program including songs and speeches planned for the ceremony. When "to quarters" was sounded, all hands—shipwrights, blacksmiths, joiners, blockmakers, carvers, ship chandlers—and their families, together with spectators, stood at attention and listened to the long roll of drums and the singing of "God Save The King." Then two men, at the direction of the shipwright, knocked down the dog-shores and the ship slid down the ways by force of her own weight, accompanied by the shouts and cheers of the spectators. The anchor was released and the colors were run up, while a lone gun in the shipyard boomed out a salute. The "Mary" representing another triumph of the technical skill of Philadelphia shipwrights, moved out to sea.

Development of New Designs. Other types beside the "Mary" continued to slide down the ways in ever increasing numbers, always incorporating some improvements. The superstructure with which foreign ships were overloaded was scaled down, permitting the ships to carry the sail more stiffly, to sail faster, and to keep cargo and passengers drier than the tall-topped vessels of Europe. From the beginning, attention was given to speed and efficiency.

Another important advance in shipbuilding technology occurred when the colonists decided to build their ships in the woods, some miles from shore. This was done because they found it cheaper and easier to transport the completed ship to the water's edge than to carry or drag the supplies from the woods. Moving the finished ship from the forests to the water's edge was accomplished by an army of men who bodily lifted the vessel and placed it on a cradle arrangement, equipped with side wheels. With several yokes of oxen hitched to the conveyance, and followed by cheering onlookers, the ship was moved slowly down to the shore.

The reputation of the Colonial shipbuilders for skillful workmanship caused many British shipowners to engage them to build their vessels. They could be built cheaper in America. A merchantman of 250 tons built in England or Scotland in 1740 cost five or seven pounds per ton; whereas the American shipbuilders built the same ship for as low as four pounds per ton. Ships of superior craftsmanship were built in America for twenty-four pounds a ton, compared with a cost of sixty pounds in England. By 1760 some two hundred ships a year were being built in New Hampshire alone for the Royal Navy, and by 1775 one-third of England's merchant marine had gone to sea from the shipyards of Colonial America.

Seamanship

Equal in importance to the technique followed in developing a world merchant marine was the psychological development of a people used to the sea, which produced daring and skillful seamen. Boys from earliest youth looked to the sea as an exciting and challenging highway made especially for them to conquer. As they grew up, they learned to love the sea and were willing to spend their lives on it. At home, the womenfolk had for generations learned to stay behind and wait for their menfolk who were being trained on the whaler, the merchantman, and the slaver. This training developed generations of incomparable seamen, "whose Captains felt at home in Surinim and the Canaries, who trimmed their yards in the reaches of the Mediterranean and the North Sea, or bargained thriftly in the Levant." [2] From their ranks came characters who were likened to Captain Kidd and Blackbeard. Blackbeard is described as the most flamboyant of all Colonial pirates who filled the stage with swaggering success, "chewing wine glasses in his cabin, burning sulphur to make his ship seem more like hell, and industriously scourging the whole Atlantic coast." [3]

A number of factors accounted for the daring ventures of America's early seamen. The fact that they lived along the coast had something to do with it, especially in New England where the soil was often too rocky or too thin for agriculture, and where men turned to the sea for livelihood. Their English heritage was also a factor. But of far more consequence was the community venture of shipbuilding and ship investments. The blacksmiths, the riggers, the caulkers took their pay in shares. They became part owners, as did the merchants who supplied stores and material. When the ship was afloat, even the seamen were allowed cargo space for articles which they might buy and sell to their own advantage. Thus early did a large number of New England seamen learn to bargain shrewdly, for every voyage directly concerned all the seamen and all the neighborhood they left behind.

The early American seamen were a daring group. Edmund Burke, member of the British Parliament, reminded his fellow members in 1775 that there was no sea that was not being vexed by the American seamen, "whose daring exploits have carried them from the Arctic Circle to the Antipodes . . . , from Africa's equinoxial heat to the winter climes of the poles."

Inland Water Travel

The development of transportation technology on the inland waters, while less spectacular than that on the ocean, was nevertheless of considerable importance. The rivers were America's first highways.

First Rivercraft. In constructing their first coastal- and rivercraft the settlers followed somewhat the technique employed back in England, with some additional skills which they had learned from the Indians. Peter Kalm, an early traveler, observed that the people around Albany, New York, had discarded the fragile birch bark canoe of the Indian, and were making improved canoes of white pine. They hollowed out the white pine log, and by sharpening the ends made a craft faster and more easily maneuvered than those employed by the Indians. By tarring the outside of the canoe to prevent rotting—something which the Indians did not know how to do—they made it durable.

The canoe built from white pine and from the tulip tree was light enough to be carried across portages from one stream to another. Canoes of this type carried successive waves of frontier settlers up the inland streams and, once they crossed the Appalachian Mountain ranges, on westward with the westerly course of the rivers pouring into the Ohio and the Mississippi Rivers.

Flatboats and Ferries. On one of these rivers, Jacob Yoder, a German, around 1750 launched an awkward boxlike craft which drew only a few feet of water, and which he called a "flatboat." It was constructed by building the sides and roof of a cabin upon a flat bottom. The flatboat became the principal cargo vessel on the western rivers before the days of the steamboats. Driven by the current and guided by one or more long sweeps and manned by a crew of five or six men, it carried grain, flour, furs, and all kinds of produce downstream to different destinations. Flatboats ranged from 10 to 20 feet wide, and from 40 to 60 feet long. When the flatboats reached their destination they were broken up and sold for lumber.

During the eighteenth century the ferry came into extensive use. Where a road reached a broad, deep stream, a nearby settler established a ferry. Frequently the ferry boat was little more than a raft, and the traveler often took his life in his hands when operating it during the season of spring floods. The ferries were slow, but in the absence of bridges, provided the only means of crossing the many streams.

Land Travel

Roads that could be used for vehicle travel, linked by bridges, were scarcely in existence in North America during the Colonial period. Wild animals and Indians had made paths between water holes and to clearings, but these were not roads. The first improvement in the Indian paths and trails made by the early settlers was to widen them by cutting back bushes and trees a few feet to allow packhorses and ox carts to pass uninterrupted. In a few of the colonies, Acts were passed early, aimed at improving the roads.

Road Building. In 1630 the General Court of Massachusetts enacted a law declaring that: "in common grounds, or where the soil is wet or miry, they [road supervisors] shall lay out the ways wider, six or eight, or ten rods or more. . . ." In 1643 an Act was passed requiring all able-bodied men from eighteen to sixty to give one day's labor a year, with their teams, to improving the roads. New York in 1664 passed an Act requiring that roads be made clear of "standing trees and lying trees," and that the road be 10 feet broad, with all stumps and shrubs cut to the ground. Bridges were ordered to be built and kept in repair over all marshes, swamps, and "difficult dirty places," and, "what ever else shall be thought more necessary." A law of Maryland in 1704 specified that one, two, or three notches on the trees were to designate the various routes.

Improved, all-season roads developed slowly, however. By mid-eighteenth century a few advances had been made by the methodical Dutch in New Amsterdam. They adopted the technique of filling the holes with loose stones, to be crushed into the bed by passing vehicles. When they laid out roads along the shore, they dumped oyster shells in the low places to promote drainage. They blasted with gunpowder the large boulders that impeded the way and threw the pieces into the low sections.

Bridge Building. Most of the roads required bridges of some kind. The first ones used were the footbridges that the Indians had known, a log thrown across the stream. The colonists improved this, however, by throwing two or more logs across and binding them together. Another improvement was made by hewing off the tops of the logs to secure a smoother surface, and by adding a handrail as a safety device.

These answered the purpose for narrow streams, but when the width of the stream exceeded the length of the timber available the bridge builder had to devise something else. He did so with a multiple structure built on a support in midstream. The type of structure and the rigidity of the supports varied, depending on the flood stage of the stream and the service the bridge was designed to render. If the bridge was to be permanent, a timber crib filled with stone was constructed in the middle of the stream. To prevent undermining, stone was packed around the outside of the crib.[4]

Bridge Engineering Advances. About the middle of the eighteenth century Major Samuel Sewall made some technological advances in a bridge which he built across the York River in Maine. The bridge was built in thirteen sections, each supported by four pilings. The pilings were saplings of varying lengths, floated into position, guyed, and driven into the bed of the stream by hand-operated mauls and by an ingenious pile-driving device, an oak beam fastened on shore. Pulleys were used to raise the beam, and when it was released it fell with sufficient impetus to drive the piling into the mud. The sections in the center of the bridge were arranged as a drawbridge to permit the passage of vessels up or down stream.

As the bridge builders came to understand the principles of the arch and the truss, the long-span bridge developed rapidly. The famous Leffingwell Bridge over the Shetucket River was built shortly before the Revolution. This bridge, 120 feet in length and 28 feet above water, was said to be the first application of the truss principle in American bridge building. Nothing was placed between the abutments to support the bridge. The truss work was arranged in diagonals, suggesting triangles. This feature caused the bridge to be known as the "Geometry Bridge." The truss principle was widely employed in a later period.

Vehicular Modes of Travel

The vehicles that creaked and rumbled over the early Colonial roads were certainly not conveyances of comfort, although the claim was often made that overland travel facilities in the American colonies were superior to those in England of that day. Nevertheless, as late as the Revolution, the quickest, safest, and most comfortable means of travel in the colonies was still by horseback.

Early Vehicles. In winter, sleds (sleighs) were used extensively, but for summer travel something had to be found that could plow through the mudholes and ruts in the roads. The simplest conveyance was the two-wheeled cart, with solid wood wheels as tall as a man and fitted on wooden axles. The tires were made wide so that they would not sink into the mud. Rimming the tires with iron was the local blacksmith's task. Always alert for faster means of transportation, the colonists developed two other vehicles shortly after 1700. One was the American sedan chair, a two-wheeled vehicle, capable of carrying two passengers. The other vehicle was the gig, a light one-seat, open cart, drawn by one horse.

During the early part of the eighteenth century the Conestoga wagon was developed, so called because it originated in the Conestoga Valley, Lancaster County, Pennsylvania. Probably inspired by the English covered wagon, it improved upon that flat-top type in that it carried more goods (2 to 4 tons) with greater safety. One characteristic of the American model was the decided curve toward the bottom of the bed. This was designed to keep the freight from rolling and shifting while traveling over the Pennsylvania hills. Wooden bows, arched over the wagon bed, were highest at the two ends and slightly lower in the middle. Over these bows a rough covering was drawn and fastened to the rings along the side of the wagon. As noted earlier, a special breed of horses, known as Conestoga horses, was developed to pull these heavy freighters. The horses were noted for their size, endurance, and pulling power.

Development of Public Conveyance. The early wagons and coaches were constructed for utility and few refinements were considered other than occasional leather seat paddings and wrought iron or leather suspension springs. Brakes were crude "lazy boards," or a chain attached to an iron handle, which could be lowered to drag on the road, or merely a heavy log dragged behind the wagon. Tar carried in all-leather buckets served as grease to keep wheels running smoothly. Tallow heated with hot stones served as a lubricant for the wood-on-wood axles and rendered them as hard as iron. Speed was of small consequence, but the forerunner of the speedometer appeared when Benjamin Franklin devised an instrument to measure the distance traveled.

Carts, wagons, and small coaches were the property of private citizens for their personal use. At first travelers were entirely dependent upon their own resources to get to their destination. Soon after 1700, however, public transportation was provided in awkward, clumsy wagons drawn by four or six horses. Lines of freight traffic employing such wagons were in regular service between New York and Philadelphia soon after 1700, making trips once a fortnight. By 1750 the Boston–New York stagecoach line began operating on an irregular schedule. Regular trips were announced between New York and Philadelphia. While the crude covered wagons required three days to travel the 90 miles between New York and Philadelphia, by 1766 the "flying machines" were covering the distance in two days in fair weather.

Robert G. Albion, in describing the time required to spread the news of the Battle of Lexington, states that when the first shots were fired at Concord and Lexington on Wednesday, April 19, 1775, a messenger was immediately dispatched with the news to points south. Four days later at noon on Sunday, citizens of New York at a distance of 225 miles received the news. Farther south poor roads, deep inlets from the sea, and swampy sections slowed the rate. Five days were required to carry the news across Virginia, a distance of more than 200 miles. Six more days were needed to reach Charleston, South Carolina, a similar distance.[5]

Postal System

Early action was taken by the British government to provide a postal system in the colonies, when, in 1688, an order in Council authorized a "letter office to be settled in His Majesty's plantations." Five years later a patent was granted to a Mr. Neale for establishing a post office in America, and he instituted a system of uniform charges which made it mandatory that each letter be stamped with the receiving date. In 1707 the postal system in America was made a branch of the British post office, and remained such until 1775, when the Provincial Congress established an independent postal service.

When Benjamin Franklin became postmaster general in 1753, his first task was to improve the service so that "settlers in one colony may have their letters speedily dispatched to settlers living in another colony." In order to speed up the service from Boston to New York, and thence to Philadelphia, he required post riders to continue day and night, and thus make the round trip from Boston to Philadelphia in six days instead of taking three weeks. By 1775 a complete chain of post offices from Portsmouth, New Hampshire, to Williamsburg, Virginia, was established with semi-weekly service between New York and Philadelphia. As bags of mail became too heavy to carry on horseback, stagecoaches were employed. In 1773 a stagecoach driver was appointed a special carrier, and this marked the inauguration of the stagecoach as an official mail carrier.

Publishing and Printing

The printing press came to America almost as soon as the first settlers. The first publication in the United States, the *Freeman's Oath,* came off the press at Cambridge, Massachusetts, at the hands of Stephen Day in the same year Harvard College was founded, 1636. Press, paper, type, ink, and skill, all were brought from England—another illustration of the transit of tools, skills, and technology. The second product of the press was *The Almanac;* and then came the *Bay Psalm Book;* and after further importations of type, John Eliot's translation of the Bible for the Indians.

First Colonial Newspapers. Before Addison began writing the *De Coverley Papers,* forerunner of the British newspaper, the idea of having a newspaper was implanted in the American Colonies. In 1690 a small four-page sheet, 7x11 inches in size appeared, called *Public Occurrences.* It was published by Benjamin Harris, and printed by Richard Pierce of Boston. It was suppressed by the Massachusetts authorities and did not survive the first issue. The first successful newspaper was the *Boston News Letter,* which appeared April 24, 1704, edited by the postmaster, John Campbell. This marked the beginning of a long line of newspapers published by postmasters and introduced the practice of postmasters' printing newspapers. Another early newspaper, *The American Weekly,* was established by William Bradford in Philadelphia, 1719.

Printing Facilities. Though the printing business became firmly established in America in Colonial times, there were few technological improvements. Benjamin Franklin learned the printing trade on a press little improved over the one Johannes Gutenberg contrived when he adapted a wine press to operate his newly discovered movable type. The press in Colonial times was a clumsy, wooden affair, worked with the exertion of terrific strength on each "pull," and required nine different hand operations on each run. Most of the presses were made in England, except those contrived by a Pennsylvania printer, Christopher Sower. The printing operation consisted of lowering a heavy platen, which was suspended from two uprights over a flat bed of type. This was first done by adjusting a screw to a moving bar; later a lever device was instituted. Both types of press were in operation in the colonies.

The type used was imported foundry type and was tediously set, letter by letter in a stick or line, column width. The type was laboriously distributed again after the printing was completed. Type was frequently broken or lost and printers were constantly harassed by shortages of s's or m's. In 1728 Franklin contrived molds in which to cast some missing letters. By 1770 Christopher Sower was able to cast a complete set of type from matrices he imported. By the time of the American Revolution some printing was being done from type cast in the colonies.

The first supply of ink, applied by a dauber on the flat bed of type before the platen was lowered, was imported. Soon printers began making their own ink of linseed oil, boiled with resin and colored with lampblack. The quality of ink varied; much of it was poor.

Paper Manufacture. All paper was imported until William Rittenhouse established a paper mill in the colonies in 1690 at Germantown, near Philadelphia. By the outbreak of the American

Revolution there were a number of paper mills in operation in Pennsylvania, Delaware, New Jersey, Massachusetts, and Rhode Island which produced 100,000 pounds of paper annually.

Paper was made from rags. In paper manufacture the rags were first fermented and then beaten to a pulpy mass, which was then mixed with water and stirred to interlace the fibers. This mixture was drained on a wire sieve and again shaken to secure an even texture. The pulp was spread evenly between layers of felt and squeezed on a hand-screw press into sheets ready to be hung up to dry. The "Hollander," a beating machine, was introduced from Holland about the time of its invention, 1750. It tore and shredded the rags, thus eliminating much of the hard work. These machines were operated by windmills and watermills.

There were few technological advances in the printing and publishing industry during the Colonial period. Moreover, there was a scarcity of practical printers. Prices of paper and printing material continued to be ruinously high, and in many cases the publisher's business was ill-managed. Franklin, in speaking of the only journal published in Philadelphia, said, "it was a paltry thing, wretchedly managed." The American Revolution was to bring about a marked change. It was to educate a generation of readers and bring forth a generation of skilled mechanics ready to supply them with printed matter.

Transportation and Communication Expansion

Before America had declared her political independence she had already developed her own system of transportation and communication technology. By means of her transportation technology she was able to dispose of her surplus products in many of the different ports of the world and to procure the products she needed. She had opened roads and trails into the interior, had pushed small boats up her inland waters, and had developed a means of contact and exchange of commodities and ideas within her own borders. In communication technology the newspapers, with the aid of the recently established postal system, disseminated the news that was considered of most importance; and post riders speeding from Massachusetts to Georgia carried personal letters and official tracts as fast as they could be dispatched in any country of the world.

Notes to Chapter 5

1. William Babcock Weeden, *Economic and Social History of New England,* I (1890), 128.

2. Ralph Paine, *The Old Merchant Marine* (1919), p. 14.

3. *Ibid.,* p. 20.

4. Llewellyn N. Edwards, *The Evolution of Early American Bridges: The Newcomen Society Transactions,* London, XII (1933), 95-116.

5. Robert G. Albion, "The Communication Revolution 1760–1933; *Mechanical Engineering,*" LV (September 1933), 531-539.

For early shipbuilding, see *The Shipbuilding Business in the United States,* edited by F. G. Fassett, Jr., Carnegie Institution of Washington; [Yearbook] *Society of Naval Architects and Marine Engineers* (1948); Edward Kable Chatterton, *The Marvels of the Ship and Sailing* (1921); Charles Lyman Chandler, *Early Shipbuilding in Pennsylvania, 1683–1812* (1932); Samuel Eliot Morison, *The Rope Makers of Plymouth, A History of the Plymouth Cordage Company* (1950).

Books on early printing include: Douglas McMurtrie, *Civilization Follows the Press* (1934); Harry Myron Konwiser, *Colonial and Revolutionary Posts* (1931); Pasko, *American Dictionary of Printing* (1894); Isaiah Thomas, *History of Printing in America* (1810); Robert Hoe, *Short History of Printing Press* (1902).

For the early history of the U. S. Post Office, see Marshall Cushing, *Story of Our Post Office* (1893); Clyde Kelly, *United States Postal Policy* (1931).

Chapter 6

EMERGENCE OF EARLY AMERICAN TECHNOLOGY

SPIRIT OF AMERICAN TECHNOLOGY

The transit of tools, skills, and technologies from Europe to the American colonies gradually declined as the colonists began to design implements of their own. During the first generation of settlers the spirit of technology had centered largely in Europe. When there was a need for a new tool or skill to meet the challenge of living in the New World, the practice was to look to the mother country. But this dependence on the homeland lessened as the years passed. It became evident long before the American Revolution that there was a spirit in American technology capable of surviving without further dependence upon Old World practices. While its full character did not become evident until the outbreak of the Revolution, its origin dated back to the very first settlements.

Development of Yankee Ingenuity. The colonists looked to men of skill in their own communities for their necessities, and they began to feel that such men by their own skill could create anything that was needed. This feeling of technical self-sufficiency developed slowly at first, but eventually it grew into an independent characteristic which came to be called "Yankee ingenuity," and which prevailed in all the colonies.

Self-sufficient Economy. In providing their first needs—food, shelter, and clothing—the colonists early improved the means for securing these necessities. To cure their fish they made their own salt. To construct their houses, they made bricks and lime and tile, and what paint they used. Their bottles of common glass, their soap, and their candles, they made for themselves. They provided the sugar for their own beverages, beer, rum, and cider. They made their own gunpowder for use in hunting for food and in defending their homes. They extracted glue from the horns of animals. They tanned their own leather. They made their own paper, and dyes, and lampblack. Potash and pearlash they produced in such quantities that there was a surplus for export. Naval stores, such as turpentine, pitch, and tar, became articles of trade. Pig iron and bar iron were sent to England. Shipbuilding became an industry that attracted the maritime interests of the homeland. In all, Yankee ingenuity devised a way to utilize the raw materials at hand without depending on England. Edmund Burke, an early friend of the American colonists, said, "they grew up in salutary neglect."

Financial Support for Industry

The production of these various commodities was at first a household industry, although in some instances companies were formed to provide capital for the establishment of an enterprise. The capital was usually subscribed in England.

Rewards and Bounties. As settlements increased in population, special Acts were passed by the various governing bodies aimed to encourage new inventions. Bounties, subsidies, and monopolies were offered to skilled workers, both as incentives to develop new improvements and as rewards for work well done. While these acts were designed primarily to promote greater production, they likewise encouraged new inventions and improvement of skills.

Industries Encouraged. All kinds of industries were encouraged. In 1638 one Richard Dummer of Newbury, Massachusetts, was granted a monopoly on milling rights, provided he built a mill, "fit to grind [corn] wheat." The mill proved fit, and succeeding generations of the Dummer family operated it for almost three hundred years. In 1639 the builder of a glass plant in Massachusetts was voted a generous grant of land as a subsidy, and two years later a money grant of thirty pounds was voted with the understanding that the amount would be paid back when the company was in position to do so.

As early as 1643 the Plymouth Company and the Massachusetts Bay Colony had granted monopolies to private individuals for setting up ironworks. Also in 1643, a "Company of Undertakers for Iron Works" was granted the exclusive privilege by the General Court of Massachusetts to manufacture iron for twelve years, if after two years they manufactured enough iron for the use of the citizens of the Colony. Further, they and their workers were exempt from all public charges and taxation, and their workmen were exempt from military duty.

In 1651 John Winthrop, Jr., was promised a generous monopoly of land, timber, and water sites for three miles around, if he would set up a plant to manufacture saltpeter, and erect other plants "for digging, washing, and melting other materials." In 1667 Massachusetts offered a monopoly to any person who would construct a dry dock; and three years later some Boston merchants were given the sole right to make and sell pitch, turpentine, and resin.

Land Grants and Other Inducements. Other colonies likewise offered special inducements for developing new enterprises. In 1676 New Jersey authorities granted Colonel Lewis Morris and his friends, who had just arrived from Barbados, extensive grants of land, and the exclusive right to erect an ironworks and to exploit the ores near Tinton Falls, Monmouth County. In 1712 Ebenezer Fitch was given sole right to erect a slitting mill to make nails. The following year Edward Hinman was granted a monopoly to make molasses with the understanding that he would make it from cornstalks, and "as good and cheap as the West Indies variety." In 1718 Rhode Island granted one of its citizens exclusive right "to set up a linseed oil mill and produce."

In 1719 Maryland passed an Act granting 100 acres of land, if said land was found suitable for the erection of an ironworks and a forge mill, to any person willing to set up an ironworks. The owner was to have free ingress and egress to and from his works over adjoining property. Workmen engaged in the production, not exceeding eighty in number, were exempt from tax levies for four years. Two years later, the Connecticut Assembly offered a prize to a Miss Woodhouse to encourage the plaiting of leghorn hats. Straw, spear grass, and red top grass were used for manufacturing the hats.

In 1727 the Virginia Assembly passed an Act "to encourage Adventurers in Iron Works." The adventurers were granted authority to cut down, take, or use any wood needed for repairing bridges or roads to their works, and the ironworkers, like those in the New England colonies, were exempt from military duty and from payment of public taxes while employed, but not from county or parish taxes.

Development of Patent Protection. The granting of special patents offered another incentive for new technological developments. In 1684 the legislature of Connecticut declared, "There shall be no monopolies granted or allowed among us, but of *such new inventions as are profitable to the country.*" [1] Although designed to prevent corruption in the granting of monopolies, the law proved to be a direct encouragement to new and profitable inventions. Early in her history, South Carolina passed an Act "for the better encouragement of the making of engines for the propagating of the staples [products] of this colony."[2] The British trade laws of the late seventeenth and early eighteenth centuries were often regarded by the colonists as an infringement upon their political and economic liberties, but, in a very real sense, they resulted in speeding up new technological developments. There was hardly an industry in the colonies that did not at one time or another receive encouragement either from British laws or from the colonizing companies that were planting settlements in America.

Spurred by the offer of such monopolies, special prizes, and exclusive grants, it is no wonder that the colonists made phenomenal advances in inventing new methods, new devices, and new techniques in their effort to conquer their new environment. Nowhere else in the world and at no time in history had people lived under such favorable conditions for the promotion of individual skills. "Yankee ingenuity" developed at a remarkable pace.

Advantages of Natural Resources. Among the incentives that contributed to the early technological developments, especially in the New England colonies, was the topography of the region. Geologically, that part of North America differs greatly from the Old World. Here was an environment different from the homeland. The short crop season, a stony soil, and the shortage of labor all meant that an agricultural society modeled on Old World plans would not thrive. With an abundant source of water power, a reasonable supply of ore immediately at hand, and a virgin forest at their back, it was inevitable that both a diversified and a somewhat specialized culture would develop. The fact that ironworks, gristmills, and sawmills were set up so early in the New England colonies attests the speed with which the first settlers responded to the challenge of their environment.

Chemical Technology. Scientific enterprises developed also. There appeared at an early date a number of ingenious persons who fostered the spirit of scientific investigation. Among these was the aristocrat, John Winthrop, Jr. Equipped with scientific knowledge and organizing ability, he succeeded in procuring a small amount of British capital. In 1635 he set up the first chemical industry in the New World. His curious establishment, labeled a "chymist's" plant, could hardly be called a plant, Haynes says, but rather a strange combination of druggist's shop, metallurgist's workroom, chemist's laboratory, and alchemist's den.[3] Nevertheless, within this crude laboratory Winthrop developed some experimental batches from which there radiated a series of primitive industrial enterprises designed to provide the colonists with many needed materials.

There was immediate demand for all the chemicals Winthrop could produce. Alum was needed in curing skins and furs for export, and saltpeter was used in making the much needed gunpowder. The manufacture of this product involved a careful chemical operation in which "sandiver," used by the alchemists in the manufacture of explosives and by physicians in the treatment of gallstones, was employed. There was need for lead, tin, and copper, which Winthrop learned to mine from the New England hills. He also established saltworks and glassworks. He developed potash, wood pitch, tar, indigo, and other natural dye stuffs. Winthrop became the pioneer industrial chemist in the New World.

Moral Support by the Clergy. But more than science was needed for the establishment of New England enterprises. Science required the endorsement of the clergy. It was none other than the Reverend John Cotton, "the mightiest man in Massachusetts," who gave the sanction of the clergy to the development of new enterprises. He pointed out that there was something of "the workmanship of God in man," and he declared that by the study of the nature of all things, man would "enlarge his heart to God" and his "skill to usefullness to ourselves and others." [4] The Reverend Cotton urged his parishioners to seek out many inventions and to make practical use of scientific knowledge. Thus, backed by scripture, Puritan New England early developed a utilitarian, capitalistic philosophy that was but one step from practical technology.

Early Iron Industry

The iron industry was one of the first to develop in New England, and again it was John Winthrop, Jr., who pioneered the new enterprise. When a company known as "The Undertakers for Iron Works In America" was launched by Winthrop and his friends, with a capital stock of one thousand pounds, Joseph Jenks, master mechanic, was engaged to build the works.

Jenks erected the first furnace at Saugus (Lynn) Massachusetts. He raked up some iron ore deposits from the nearby bogs, brought in some shells and corals from the seashore for flux, and with charcoal as fuel, produced samples of iron. Jenks and his associates were soon producing pots, pans, nails, hinges, horseshoes, and other needed tools.

First American Patents. The heavy demand for iron products led Jenks to petition the Massachusetts General Court to grant him a patent in 1646. He declared:

> Whereas the Lord hath been pleased to give me knowledge in making and erecting engines of mills to go by water for the speedy dispatch of much work with few men's labor in little time, it is my desire to improve this talent for the public good and benefit and service of this country; to which end my intention and purpose is (God permit) to build a mill for making of sithes: with a new invented Sawemill, and divers other engines for making of divers sorts of edge tools. . . .

The Court granted Jenks a monopoly for fourteen years, reported as the first patent issued in America.[5]

Jenks' reputation as a skilled metalworker spread throughout the Colony, and in 1652 he was commissioned to cast a die at his ironworks for the first money coined in America. The dies were cast for three-pence, six-pence, and a shilling, stamped with the word "Massachusetts." This famous pine-tree shilling showed a pine tree on one side and "New England, Anno 1652" stamped on the reverse side. In 1654 the selectmen of Boston contracted with Jenks to build an engine "to carry water in case of fire." This was the first fire engine in America.

The following year Jenks fashioned a distinctly new type of scythe "for the more speedy cutting of grass." He took the heavy English scythe, made the blade longer and thinner, and then strengthened it by welding a bar of iron along the back of the scythe. The permanence of this new technological improvement is attested by the fact that there has been practically no change in the shape or size of the scythe to this day. Jenks is also credited with introducing the wire-drawing industry into the colonies in 1667. The iron that he manufactured by the charcoal process was rolled into plates and then sheared into strips of small square rods, ready to be drawn into wire. This venture, however, was not successful.

Early Iron Technology. Examples of Yankee ingenuity were demonstrated in the early development of the iron industry. The first type of ironworks was the bloomery, patterned after those commonly used in England. A bloomery resembled somewhat the blacksmith's forge, but was wider and deeper. Requiring only a small investment, bloomeries could be set up anywhere. The ore was broken into small pieces and placed in the firepot of the forge with the charcoal that had previously been heated. A hot blast, made by a water-powered bellows, introduced at the side of the furnace, reduced the iron ore to a semimolten condition. The forgeman stirred or worked this mass with a long bar until it gathered into a lump, called a "loupe." While the lump (loupe) was still in a semimolten mass, he transferred it to the anvil where he fashioned it with a heavy hammer, into a bar of iron. Successive reheatings and successive hammerings produced bars for the blacksmith trade. The iron produced in this manner was commonly called wrought iron. A number of bloomeries were early established in New England.

Adaptation of the Blast Furnace. In Pennsylvania a different process was developed. Here the blast furnace appeared almost from the beginning. One was built at Colebrookdale, Berks County, in 1720. Thomas Rutter and Thomas Potts led the way in an enterprise that was to become a major contribution to the iron industry of Pennsylvania. By 1771 there were more than fifty iron furnaces and forges operating in Pennsylvania.

The blast furnaces that were operating in England had slowly evolved from the days of Henry VIII, and by 1700 had become standardized. When introduced in Pennsylvania, the furnaces took on larger proportions. Usually 25 to 30 feet high, they were built on the side of a hill. The stacks were square at the bottom, tapering at the top, with a bridge connecting the head of the stack with the bank. Creaking wagons drawn by teams of horses or oxen hauled the iron ore, limestone, and charcoal up the furnace road to the bridge where workers, known as "fillers," carried it in baskets across the bridge and dumped it into the stack, alternating the layers until the stack was filled.

After a few days of burning, the molten metal was run out into a bed of sand, forming a stream of metal which, because of its resemblance to a sow and suckling pigs, was called "pig iron." The molten metal was ladled into molds in the casting shed near the furnace, where pots, pans, kettles, and Dutch ovens were made. Pig iron to be made into bars was further refined in the smelting furnace and was pounded into bars by a heavy hammer on a forge. Bars cut into convenient lengths were sold to the blacksmiths, locksmiths, and other artisans to be worked up into tools and articles of hardware.

Furnace operation required the services of ten or twelve men. In addition to the ironmaster, there were founders, guttermen, two or three fillers, one ore roaster, a few laborers, and a potter to make hollow ware. The founders regulated the furnace, made the sand molds, and cast the iron. When the blast started, the men worked two twelve-hour shifts. Work was usually suspended during the coldest winter months, and sometimes during the summer if the heat became oppressive.

The blast of air required for the furnace and forge was supplied by a large double bellows made of leather, driven by a water wheel. Some of these water wheels were of enormous size, with a diameter of 28 to 30 feet. In the 1750's some extra-large wheels, 40 feet in diameter, were built for iron plantations in Pennsylvania. Dams were built upstream from the furnaces and the water was conveyed to the wheel by a small ditch or through hollow logs.

Increased size in the hammers marked a distinct advance in iron technology. The hammer was operated by water power. During the eighteenth century, iron was seldom rolled, although Henry Cort of England understood the rolling process as early as 1754. But rolling mills were not in general operation in the United States until the early part of the nineteenth century.

Iron Plantation System

The manufacture of iron was the most significant industry in Pennsylvania during the middle of the eighteenth century. Here the plantation system developed and reached its highest degree of efficiency. On these plantations, consisting of several thousands of acres of land, the most advanced technological methods were adopted. The plantation contained the mansion house, the furnace, the forges, the casting sheds, the ironmaster's store, the homes of the workers, a gristmill, a sawmill, and fields for tillage. Everyone about the iron plantation was taught some kind of skill. Some worked in the mines, others at the furnace or forge, still others in the fields.

Near the ironmaster's mansion stood the furnace, a huge stone pyramid, built on the side of a hill so that it could be easily charged at the top. The main casting shed was placed just below the furnace to facilitate running the molten metal from the hearth of the furnace to the molds of blackened sand. The forge, where the raw iron was refined, was adjacent to the furnace. On the stream below the furnace and the forge, the gristmills and sawmills were located. These were run by giant water wheels. The mills were built of heavy logs or stone.

Plantation Community Stores. At some central point on the plantation was located the community store. In many ways these stores resembled the old country stores of a generation ago, common to every farming and mining community in the nation. Here one could buy or barter for the grain and vegetables that had been grown on the plantation, could purchase axes, shovels, chains, hoes, and other implements made by the plantation blacksmith. Also available was sugar and molasses brought from the West Indies, rum from New England, and imported goods from Great Britain. Broadcloth, linen, flannel, shoes, deerskins, scythes, stoves, and castings were listed in the inventories of the iron plantations.[6]

The iron plantations of Peter Hasenclever of New York, and later of New Jersey, came to be known as the most grandiose iron business in Colonial America. Baron Hasenclever, as he was known, son of an iron and steel merchant in Prussia, arrived in New York in 1764. A company known as The American Iron Association, formed in 1762 to secure iron for Great Britain, employed Hasenclever as their superintendent. He operated on a magnificent scale, purchased 50,000 acres of land in New York and New Jersey, and imported five hundred skilled workmen from Germany. Among them were experienced miners, foundrymen, forgemen, colliers, carpenters, masons, and laborers. He erected six blast furnaces with a capacity of 60 to 70 tons, and seven forges with twelve hammers. He also erected numerous houses, barns, gristmills, sawmills, and blacksmith shops. He built canals, dams, bridges, and reservoirs, all a part of the plantation equipment.

New Process of Iron Production

Hasenclever and his associates introduced a new process in iron technology by adopting the practice of using iron recovered from the banks of discarded cinders. Hasenclever discovered that cinders mixed with other ores produced a tougher iron, and his was pronounced the best in America. Some of his products were sent to London and the English ironmasters declared it to be the best drawn iron sent from the colonies. Locally, cinder iron was pronounced superior for wagon tires, plowshares, and other farm tools.

This innovation in iron technology resulted in the erection of stamping mills, in practically all ironworks to crush the furnace cinders and slag in order to secure the metallic iron. The cinders were dumped into the furnace with the other ores, charcoal, and limestone, and the blast was applied as in the regular process of making iron. These technological advances of Hasenclever, effected by the skilled workers he employed, greatly stimulated the iron industry of the pre-Revolutionary period. There was a constant demand for skilled German ironworkers, colliers, and forgemen.

American vs. British Iron. The advances in American iron technology early attracted the attention of the home government. Hearings held in the House of Commons in 1740 on the competition that was developing between the iron manufacturers in the colonies and the English manufacturers revealed a falling off of one-third in the use of bar iron in England. English axes had not been shipped to New England, New York or Pennsylvania, and only a few to the Carolinas, since 1722.

William Carey, a noted ironmaster of England, testified that iron made in the colonies was equal to the Swedish iron, and he recommended that pig iron be admitted into England duty free, since the wood supply in England was rapidly disappearing. The British iron manufacturers objected and blamed the current depression upon their loss of trade to the colonies. They petitioned Parliament not to allow the colonists to make axe heads, shovels, nails, or other products from bar iron. They observed that garden tools, carpenters' axes, and nails were being manufactured in the colonies in such great quantities that there was no demand for English goods. The Iron Act of 1750 resulted, which imposed drastic restriction upon the erection of ironworks in the colonies.

Technological Improvements. This did not stop the march of the ironman, however. On the contrary, iron technology made rapid strides during the years preceding the outbreak of the American Revolution. In 1762, the London Society of Arts awarded a gold medal to the Reverend Jared Eliot for producing in a common bloomery, with the aid of "magnetic sand," a bar of iron weighing 50 pounds. Slitting mills and plating forges came into common use during the 1760's. Shears were introduced for cutting bar iron. Rollers were employed to draw the sheared iron, after it had been heated white hot, to a thickness suitable for nails. Cutters slit the bars into rods, varying in width. Water wheels were used to operate the rollers and the cutters. The flattened bars as they came from the rollers were used to make shovels, spades, parts for plows, and other tools.

Manufacture of Iron Implements and Products

Tool makers and artisans who could fashion tools from iron were always sure of employment. One of these, Hugh Orr, a mechanic who migrated to New England in 1740 from Scotland, stands out above others of his time. A master gunsmith and locksmith, he was employed by a Mr. Keith, a maker of scythes. Orr demonstrated his skill when he converted an old skillet handle into a keen-edged razor. He was widely known for his skill in producing sharp-edged tools. He devised and built an improved trip hammer, reported as the first one in the colonies. Mechanics of all kinds, carpenters, shipwrights, millwrights, and wheelwrights for miles around came to his shop. Orr showed them how to improve their tools or make new ones.

Joseph W. Roe considers him America's first toolmaker, one who devoted his technical genius and skill to the production of fine edged tools and to teaching the art to others.[7]

In New England iron bars were used chiefly for making nails. This was a semihousehold industry. The sheets of iron that came from the plating mills were cut into strips and sold directly to the farmers. Almost every farmer had his own anvil, and on this he hammered out nails, rivets, bolts, braces, and other tools. During the decade preceding the American Revolution, the making of nails and tacks became one of the principal products of home manufacture. The farmers plied their skill in nailmaking during the winter months. With a small furnace, hammers, and kneading bars, they turned out several hundred nails a day.

Most of the bar iron produced at the forges, however, was hammered into shape by blacksmiths, who played a distinctive role in early American technology. They were the highly skilled artisans and craftsmen of that day. The blacksmith needed only a rough idea of what was wanted, and he did the rest. Without a drafting board or precision instruments, he fashioned axes, shovels, chains, scythes, guns, bolts, ship irons, and dozens of other articles. After 1752 blacksmiths hammered out the lightning rods invented by Franklin.

During the American Revolution the blacksmith was regarded as the nonexpendable man of the community. The case of a blacksmith, one James Galloway of Westmoreland County, Pennsylvania, is often cited to prove how necessary blacksmiths were in helping to win the war. Westmoreland County was a hotbed of frontier patriotism. It was in that community that the famous "rattlesnake" flag, with the insignia "Don't Tread on Me," was first designed. If a man did not enlist for service he was branded a traitor and treated accordingly—but not Galloway. His neighbors demanded that he remain at home, and they took up a collection to hire a substitute. They paid the substitute twelve dollars, gave him a rifle and a butcher knife, and sent him off to war. Galloway remained at home and repaired guns, hammered out horseshoes, made wagons, and performed the necessary iron work for the entire community.

Steel and Steel Products

Steel was manufactured in a small way in several of the American colonies in "the German manner," or by the more clearly defined *cementation* process. The art of making steel was kept a profound secret and was shrouded in mystery. The quality of the steel was supposed to depend on the liquor in which it was quenched. Tench

Coxe reported in 1791 that about one-half the steel consumed in the United States was made here. The making of steel had not reached the proportions in Colonial America that the iron industry had attained.

Processes of Manufacture. In 1728 Samuel Higley of Rhode Island and Connecticut manufactured what is believed to be the first piece of steel made in the American colonies. A book written in 1720 by a French physicist, René Antoine Réaumur, explaining the technological process of making steel known as the cementation process, was probably the source of Higley's information. Steel made in this process was known as "blister steel" because the bars were covered with blisters when withdrawn from the furnace.

In making steel by the cementation process, bar iron was used, and the time element was essential for the iron to carbonize. The process was carried on in a steel furnace, in the hearth of which iron bars were laid, sprinkled with charcoal dust, horn shavings, and ashes. To this a blast was applied which continued until test bars revealed the "right" color. Then the fires were extinguished and the bars left to cool in the furnace, a process lasting from seven to ten days.

The Germans in Pennsylvania had a more direct method of making steel. In this process, pig iron or crude iron was used instead of the bar iron, and special qualities of wood employed. This particular technique was always referred to as "after the German manner."

Uses of Steel. Steel was used chiefly for swords, bayonets, and sharp-edged tools. Axes with steel edges were especially needed in opening up new clearings. At the outbreak of the Revolution, the Pennsylvania Provincial Congress recommended to the people of the Colony that they manufacture steel, iron, wire, and tin plate. The Continental Congress contracted with a Mr. Whitehead Humphreys to make steel weapons for the army. In 1776 William Hawxhurst, who was operating a half-dozen furnaces, advertised in the newspapers for men "to make pig metal into steel in the German way."

By the time of the American Revolution the colonies were operating nearly two hundred iron forges, but producing only a small amount of steel. The number of iron forges exceeded the total number in England and Wales. American-made iron ranked third among the iron producing countries of the world, led only by Russia and Sweden. America's total production equaled one-seventh of the world's output.[8]

Glassmaking

Glassmaking was the first industry to be undertaken in America. In 1608 eight Poles and Germans arrived in Virginia and set to work making pitch tar, glass, and soapashes. They brought their crucibles with them and within a few weeks had produced samples of window glass, glass bottles, lamps, tableware, and glass beads. A few years later six Italian glassmakers arrived in Jamestown and began making glass beads and bottles.

In 1639 a glassworks was set up in Salem, Massachusetts, and as an incentive to promote this infant industry, the General Court offered the glassworkers "several acres of land adjoining to their homes." The town of Salem was authorized to lend the workers thirty pounds in money from the town taxes. A small quantity of window glass was produced, but like most of the early glass, it was not satisfactory. The glass was coarse and of green or brown color, depending upon the sand used, and was not very transparent.

Materials Used in American Products. Yankee ingenuity was exerted early in discovering the kind of sand most suitable for glass, and in producing satisfactory glassware from the materials at hand. The first successful large glass plant in the colonies was set up in 1738 by Caspar Wistar, a German emigrant, on a tract of pine-wooded land in Salem County, West Jersey. Here he had access to an ample supply of fuel and silica, and to water transportation. He imported experienced glassworkers from Belgium, Germany, and Portugal, and made flint glass, later known as lead glass. He produced glass bottles, scent bottles, table crystal, snuff bottles, and some window glass.

The Wistar factory was successful in developing a number of glass products. In the 1750's the company was advertising window glass, sizes 6x8, 9x11, 10x10, and glass of "any uncommon sizes under 16x16" could be cut on short notice. Bottles of quart and half-gallon sizes were made, as were retorts, tubes, and globes of various sizes, and snuff bottles. Buyers were reminded that, "our glass is of American manufacture—and it is of interest to America to encourage her own manufactures, more especially those upon which duties have been imposed solely to raise revenue." [9]

Tools and Technology. The composition of the glass made at the Wistar plant was essentially the same as today, potash, sand, lime, and soda. These materials were fused in a firepot usually made from imported clay. The blowpipe was the principal implement of the glassblower. This was a tube, 4 to 6 feet in length, having a handle and a wooden mouthpiece, and at the other end an expanded lip.

The size of the lip depended upon the size of the gob of glass required. The next important tool was called the "pontil." This was a long solid tube that held the glass in place until the proper shape had been reached. Shears were used to cut the material. Molds were commonly used to produce articles that could not be shaped in any other way.

From 1609 to about 1740 there were some half-dozen or more attempts to produce glass in the colonies. This seems a little surprising, since glass was not one of the necessities. In fact, in those countries from which most of the early settlers came, glass was a luxury, manufactured mostly by nobles who held a monopoly.

Development of a Glass Industry. Henry William Stiegel, German emigrant who settled in eastern Pennsylvania and who became a pioneer iron manufacturer, also gained fame as a pioneer glass manufacturer. After making a fortune in the iron business, he ventured into the manufacture of glass. He went to England in 1763–1764, visited several glass plants, and returned with a number of skilled English and German glassworkers who had been trained in the British technique, and who were skilled in the latest enameling processes. He built the town of Manheim, in Lancaster County, Pennsylvania, and laid out a great plantation somewhat similar to his iron plantation. He erected a huge glasshouse, the largest in existence, and it was said that "a coach and four could turn around within the brick dome of his melting house."

Stiegel started producing flint glass tableware, and introduced blue, amethyst, and violet ware of clear and light variety. The colored glass was in rich shades of blue, purple, amethyst and occasionally emerald green. Stiegel glassware was later to become a collector's item.

The Results of Yankee Ingenuity

Yankee ingenuity had brought young America to a high level of self-dependence by the time of the American Revolution. The work of the farmer, blacksmith, tailor, tanner, shoemaker, miller, wheelwright, mason, gunsmith, printer, sailmaker, all combined to produce a practical, keen witted, self-reliant, dependable people. These people had given up any fanciful dreams they may have had about exploiting gold and silver mines, or finding precious gems. They turned instead to producing such commonplace articles as iron, salt, bricks, lumber, potash, leather goods, and clothing. By combining their various skills with men of science, they paved the way for a new culture, founded upon a number of diversified technologies. From these foundations there developed within the next two and one-half centuries those numerous inventions and techniques that revolutionized the feeding, clothing, housing, communication, transportation, and manufacturing systems of the New World, and the Old.

Notes for Chapter 6

1. Author's italics.
2. Victor S. Clark, *History of Manufactures in the United States,* I (1928), 48.
3. See William Haynes, *Chemical Pioneers* (1939).
4. See Theodore Hornberger, *Puritanism and Science, The Relationship Revealed in The Writing of John Cotton* (1937).
5. This 300-year-old ironworks has been restored to its original condition on the historic site of the first operating manufacturing plant in America.
6. An excellent description of early iron manufacture in the colonies is Arthur C. Bining, *Iron Plantations;* also his "Early Iron Masters of Pennsylvania," *Pennsylvania History,* V, XVIII (April, 1951) No. 2.
7. See Joseph W. Roe, *English and American Toolmakers* (1926).
8. Arthur C. Bining, *British Regulation of the Colonial Iron Industry* (1933), p. 4.
9. Richard Wistar, son of Caspar Wistar, is said to have been the first to use the slogan, "Buy American Manufactured Goods." This advertisement appeared in issues of Franklin's *Pennsylvania Gazette* in 1769.

For general reading: Warren C. Scoville, *Revolution in Glass Making* (1948); I. Bernard Cohen, *Some Early Tools of American Science* (1950); Dirk Jan Struik, *Yankee Science in the Making* (1948); Carl Bridenbaugh, *The Colonial Craftsman* (1950); Scott Graham Williamson, *The American Craftsman* (1940).

Material relating to the early history of glass may be found in the files of *The American Glass Review* (1882–on); *Directory of the Glass Trade* (1905); and in such works as J. Leander Bishop, *A History of American Manufactures,* 3 vols. (1860); Victor S. Clark, *History of Manufactures in the United States, 1607–1928,* 3 vols. (1929); George S. and Helen McKearin, *American Glass* (1941); C. J. Phillips, *Glass; The Miracle Maker. The History, Technology and Application* (1941).

Chapter 7

ROLE OF SCIENCE IN EARLY AMERICAN TECHNOLOGY

Just as the early American colonists brought with them the tools, skills, and techniques from the Old World, and altered and improvised them to meet the demands of the new country, so, too, they brought with them the scientific ideas and the current philosophies of their homeland. These also became modified and adjusted to the new environment.

Colonial Scientists

It was America's good fortune to be colonized during a period when a scientific awakening was stirring the western world. The fascination of syllogisms and abstract dogma that had dominated medieval learning had given way to a zeal for experimentation in the physical world. A knowledge of nature came to be regarded worthy of real philosophers. Elizabethan England was fired with a spirit to investigate the phenomena of the universe. Coupled with this was the desire to put science to practical use. Learning for learning's sake had been supplemented by a utilitarian objective. Francis Bacon was the exponent of this doctrine which declared that the true objective of all science was to relieve man from drudgery, to alleviate human misery, and to promote the comforts of life. It was this idea of scientific investigation and its application to practical needs that was transplanted to America, where it found a fertile field for growth and expansion without the limitations imposed in the Old World.

There were scientists among the early settlers as well as craftsmen and artisans. These men of learning were the equal of those found in any country on the globe. They brought with them the true scientific spirit of investigation and co-operation. They had pioneered in the method of collecting data and pooling findings with other scientific investigators. They carried on extensive correspondence with such men as Isaac Newton, Galileo, John Milton, Linnaeus, Dr. Sackville, and Sir Christopher Wren. In this manner they contributed to, and received much from, the common pool of scientific knowledge. Their activities led the British scientists to refer to them as that group which was building up "vast riches in the lap of England's principal daughter."

Old World Scientific Societies. In England the followers of the Baconian school of scientific inquiry had organized a "Society for the Promotion of Knowledge." The aim of the Society was to promote "Natural Knowledge and Experimental Philosophy." Because of the political turmoil existing during the reign of Charles I, this group of scholars met in secret at Gresham College at Oxford where some of their leaders were lecturing; notably, Robert Boyle, Sir Christopher Wren, Lawrence Rock, and others. It was during this period that one of their number, John Winthrop, Jr., (see Chapter 6) sought solace in America, and came with a band of followers to Massachusetts Bay Colony where his father was governor. Keeping up his contact with his fellow scientists, Winthrop urged his colleagues to migrate to America. His proposal met with favorable consideration until the advent of Charles II to the throne, when political order was restored. Later, royal patronage brought the scholarly group into the open, and in 1662 it became known as "The Royal Society of London."

For more than a century this society was the center of scholarly endeavor for both England and America. Americans who achieved distinction in science were elected Fellows of the Royal Society and their discoveries were published in the Society's Transactions.[1]

Collection of Scientific Data. One of the early projects of the Society was the collection of scientific data. The Society was designated as a repository for the accumulation of facts from which deductions and evaluations could later be formed. In this collecting area, America, because of the vast opportunity afforded here, became a most valuable asset. Collections were made of the flora and fauna of America; of the mineral specimens; of fossils, and of Indian artifacts. Astronomical calculations founded upon New World observations proved invaluable later in making deductions. The Americans interested in science began to prepare and classify their collections for the Society.

Leading Early American Scientists. Leading all the American scholars in making these collections was John Winthrop, one of those who met in the secret group, and who, after the Society was formed, was formally elected an "original" member. Winthrop arrived in the colonies in 1631, but by correspondence kept in touch with his friends in England. He began collecting numerous mineral and plant specimens found in the New World. He wrote a number of papers based on his observations and sent them to the Society. He described the best methods for growing maize, methods of making sugar from the cornstalks, the "stellar fish," and wrote an interesting article on the hummingbird. In still another article on "The Observations in New England" he described the dwarf oak and the cotton tree. On one of his trips back to England, Winthrop carried a mineral collection which he exhibited to his scientific friends. On the same trip he read a paper before the Society on the subject, "The Manner Of Making Tar and Potash." In all, Winthrop contributed ten articles to the Society's *Transactions,* and discussions of his contributions played a prominent part in stimulating an interest in science.

Others who made collections and told of their findings, and who were elected to the Royal Society in the early period, were Roger Williams, educator and distinguished linguist; William Byrd of Virginia, a close friend of Sir Hans Sloane, president of the Royal Society; and Cotton Mather, a Massachusetts divine. Byrd was primarily interested in the plant and animal life of Virginia, and described his meetings with Dr. John Tennant (who later became a member of the Society) and Mark Gatesby, both wandering botanists. Cotton Mather faithfully carried out his obligations to the Society to contribute collections of material and sent packets of bones, teeth, and other skeletal remains found in New England. His collections included sun-dried plants, data on the moose, the rattlesnake, and pigeons. He also made observations as to the Indian's use of herbs in the treatment of diseases. In all, Cotton Mather sent thirty-seven communications to the Society.

Another contributor to the accumulation of scientific data was Thomas Brattle, successful merchant and trustee of Harvard. With the little telescope at the college, he made careful and exact observations that added to the sum of available information. He compiled an almanac of celestial motions, observed two solar and three lunar eclipses, and made the earliest precise observations on the variations of the magnetic needle. Brattle's observations were helpful in the determination of the exact longitude of Boston, a contribution adding greatly to the Society's project of ascertaining the longitudes of various localities. Brattle's vast collection of data was added to the store of the Society after his death, though he, himself, was never formally a member of the Society. Others contributing meteorological data were Thomas Robie, a Boston physician, Governor William

Burnet of New York, from whose observations the longitude of New York was determined, and John Leverett, president of Harvard, who introduced courses in natural philosophy.

Two other American scientists of this early period not only accumulated data, but through their original investigations also made discoveries that enlisted the attention of the scientists of the world. Paul Dudley, a Massachusetts lawyer, was inveterate in his zeal for knowledge. He contributed twelve papers to the *Transactions* on weather, earthquakes, and natural phenomena. But his original findings were in the field of horticulture, where he anticipated Luther Burbank by two hundred years. Observing the growth of vegetables and fruits, the fertility of the soil, and their cultural habits, Dudley noted that maize was grown in a variety of colors, but when kept separate would produce according to the seed. When planted in the open, however, the seed produced ears of motley color. This information was a great boon to the horticulturists of the time who were just beginning to advance the theory that plants had sex.

A Boston physician, Zabdiel Boylston, made a name for himself in scientific circles before he was elected to the Royal Society. He had read in the *Transactions* of the accomplishments in Constantinople of inoculating against smallpox. Boylston developed his own method which proved effective in the great epidemic of that disease in Boston in 1721. Despite the edict existing in the colony against the practice, Boylston made many secret inoculations that were successful. Backed by the powerful Cotton Mather, who had his own family inoculated, Boylston openly engaged in the practice. His method, which did not vary in principle from that in effect today, was demonstrated with great acclaim before the Royal Society in 1726, where he was honored for his pioneer work.

Perhaps the most diligent of the collectors for the repository in the Royal Society was John Winthrop, Esq., grandson of John Winthrop, Jr., who carefully classified, labeled, and dispatched to the Royal Society more than eight hundred specimens of minerals, fossils, shells and stones found in America. The fortieth volume of the *Transactions* was dedicated to him in recognition of the work of both himself and his illustrious grandfather.[2]

Young America's Contributions to Science. America's chief contribution to the world of science during the late Colonial period, was the work of Benjamin Franklin. When Franklin, in 1743, received from Peter Collinson an electric tube with instructions for using it, the findings of the Royal Society proved most helpful to him. Spurred on by the desire to contribute something new, Franklin acquainted himself through the *Transactions* with all that was then known about electricity. When his own papers started to appear in 1752, his role was that of judge and counselor, rather than contributor. His opinions on scientific matters were solicited, and even before he was elected a member of the Society (1756), he was awarded the Copley Medal. His contributions to the Royal Society were made largely through the eminent British scientist, William Watson, who noted that Franklin was, "a very able and ingenious man who had a head to conceive and a hand to carry into execution."

The Royal Society aided a number of American scientists in the field of botany by providing them with previous findings, and afforded them an opportunity for the advancement of their own theories. Outstanding in this field were Dr. John Mitchell, Dr. John Tennant, Dr. Arthur Lee of Virginia, Dr. Alexander Garden of South Carolina, and (though not a member of the Royal Society) John Bartram, friend and botanist to the King and a contributor to the Society's *Transactions.*

Likewise associated in the field of medicine was Dr. John Morgan of Philadelphia, founder of the first medical school in America (later the University of Pennsylvania) who was elected to the Royal Society in 1765 and contributed three papers to the *Transactions.* Tennant worked on medicinal herbs. Mitchell was a forerunner for Benjamin Rush on his diagnosis of the cause of yellow fever. Lee employed the methods of a true scientist in argument by proof. He conducted twenty experiments with Peruvian bark, the plant from which quinine is extracted.

Mitchell's scientific methods contributed to his proficiency in another line. He made a map of the American colonies for the home government on the eve of the French and Indian war that had been described as "the most important map in American history."

Alexander Garden, together with Cadwallader Colden of New York, made important contributions to the work of Linnaeus in the classification of plants. Garden started the botanical gardens at Charleston, South Carolina.

John Bartram was termed by Linnaeus "the world's greatest natural botanist." It was due to Bartram that the transit of plants from America to Europe was instituted. Bartram traveled the entire length and breadth of the colonies, from Ontario to Florida, gathering specimens for English gardens, which he carefully labeled and packed with cultural instructions so that the eighteenth century saw American species thriving in Kew Gardens and on the private estates of English noblemen.

Perhaps the most profound scholar in Colonial America whose contributions lay in almost all fields of intellectual activity, and who had a vast influence on establishing the scientific trend in America, was John Winthrop IV. For forty years he served as professor of mathematics and astronomy at Harvard, where he instituted the study known as calculus and inaugurated the first laboratory of experimental physics.

Thus in Colonial America science played a role in instituting methods of technology, and science was early adapted to utilitarian practice. Old World scientists had helped to point the way on which the New World would develop.

In the later Colonial period, and extending over into the early days of the republic were James Bowdoin, Governor of Massachusetts, Benjamin Thompson (Count Rumford), David Rittenhouse of Pennsylvania, and Elihu Bancroft, all contributing to the scientific approach to the American way of doing things.[3]

American Scientific Groups

Following the close of the American Revolution the relationship of American scientists to the Royal Society changed. Americans began to be listed as Foreign Fellows. Meanwhile, American scientific societies organized somewhat on the pattern of the Royal Society of London, had been started in this country.

Boston Philosophical Society, 1683. What is believed to be the first scientific society organized in the American colonies was the Boston Philosophical Society, 1683. Its brief existence was recorded by Increase Mather, the leader of the group. He explained that it was a private philosophical (scientific) society, of agreeable gentlemen of Boston who met every fortnight for discussion upon improvements in philosophy, and for increasing the storehouse of natural history. The famous Dutch scholar, Wolferdus Senguerdius, made use of the findings of this society in his *Philosophia Naturalis.*[4]

The American Philosophical Society, 1743. The first permanent scientific organization was the American Philosophical Society. It was founded by Benjamin Franklin who, in 1743, published "A Proposal for Promoting Useful Knowledge Among the British Plantations in America." His plan met with immediate response and the society was founded the following year with Thomas Hopkinson as president and Franklin as Secretary. Interest lagged for a period, and in 1769 it united with another scientific society founded by Franklin, with the imposing name, "The American Society Held At Philadelphia for Promoting Useful Knowledge," an outgrowth of Franklin's Junto.

This combined organization elected Franklin president, a post he held until his death. With the exception of a brief period during the Revolutionary War when its members were engaged in war activities, the society has continued to enjoy a vigorous life to this day.

The Original Junto. The original plan for the American Society as outlined by Franklin was for an intercolonial Junto. In Philadelphia there was a representative from each of the seven branches of science—a physician, a botanist, a mathematician, a geographer, a chemist, a mechanician, and a natural philosopher. Monthly meetings were planned to discuss philosophical interests and to carry on a correspondence with like-minded citizens in the other colonies. Franklin had observed in 1743 that, "the first drudgery of settling the colonies, which confines the attention of people to mere necessities, is [now] pretty well over." He noted that there were many citizens in each of the colonies "enjoying circumstances that set them at ease and afforded leisure to cultivate the finer arts and improve the common stock of knowledge."

Aims of the Philosophical Society. By means of systematic correspondence the members discussed such subjects as newly discovered plants and herbs, together with methods of propagating them and making them useful, new methods of preventing and curing diseases, the newly discovered fossils and archaeological treasures, improvements in mathematics and discoveries in chemistry, map making and surveys, mechanical inventions for saving labor, improvements in the breeding of animals and in gardening, and in all other experiments that "let light into the nature of things and tend to increase the power of man over matter, and multiply the conveniences or pleasures of life."

In the newly combined philosophical society the seven scientific interests as originally outlined were retained, but a number of new members were elected so as to secure representatives from all branches of knowledge. The same basis for membership was adopted as existed in the Royal Society of London, i.e., resident, nonresident, and foreign. The election of foreign members was indicative of the society's aim to become a world organization. In 1789, the first woman was elected, the Princess Dashkov of Russia, lady-in-waiting to Empress Catherine II.

David Rittenhouse succeeded Franklin as president of the Society, and he in turn was followed by Thomas Jefferson. Among the early members were: George Washington; Francis Hopkinson, composer and jurist; John Dickenson; John Bartram, botanist; Cadwallader Colden; Dr. Benjamin Rush; Reverend Ebenezer Kinnersley; Albert

Gallatin; James Madison; Robert Morris; John Quincy Adams; Thomas Smith, provost of the College of Philadelphia; and his successor, John Ewing; John Lukens, Surveyor General of Pennsylvania and Delaware; Benjamin Latrobe, architect and engineer; Jacob Bigelow; Philip Synge Physick, father of American surgery; and Dr. Caspar Wistar, distinguished professor of medicine and surgery.

Fortune favored the society in its very first undertaking. A rare phenomenon, the transit of Venus occurred on June 3, 1769, the year Franklin took office. The astronomers of the society made preparations for observation. Franklin, while in London, sent a telescope with a Dollond micrometer, and the Honorable Thomas Penn provided another telescope. David Rittenhouse constructed a clock for the time observation, so accurately made that it keeps time today from its position in the hall of the society. Nature was kind to the new army of scientists in giving them a cloudless day. European observers, less fortunate, met with overcast skies. As a result, the first approximately accurate measurement of the phenomena was given to the world by amateur devotees of science in young America, an achievement that brought international recognition to the American Philosophical Society.

Astronomy. In the field of astronomy the work of David Rittenhouse eclipsed all others. Notable in this was his celebrated "orrery" which illustrated the solar phenomena over a period of five thousand years. More than one-half of Volume I of the Society's *Transactions* is taken up with papers on astronomy. As an encouragement to this branch of science, a precedent was established for subsidizing research when the Pennsylvania House of Representatives granted the Society one hundred pounds to aid in erecting and maintaining an observatory on the State House lawn. A grant of land was voted, "forever free from taxation," for the site of a hall, a library, and a museum.

Agriculture. The American Philosophical Society was the first organized body in America to regard agriculture as a science. The early minutes of the Society are replete with suggestions on such items as the value of new plants, the prevention of plant diseases, and the cultivation and conservation of the soil. Beginning with Jared Eliot's "Essays on Field Husbandry," (1759) the Philosophical Society continued year after year to make available to the American farmer the latest in all scientific developments. A study of the Hessian fly, which was destroying the wheat crops, was of special importance since England had banned American wheat that was believed to be infested by the fly.

The Society also became interested in the culture of the mulberry tree and in silk production. Jefferson's account of his mold-board plow was first presented to this Society. Thomas Pinckney of South Carolina presented a plan for reclaiming marsh lands. These, and other papers, set the trend toward the recognition of agriculture as a science.

Entomology. The damage that the Hessian fly caused the early American farmers was responsible for the first scientific study in entomology in the United States. The Philosophical Society appointed a special committee in 1792, with Thomas Jefferson as a member, to collect material relating to the history of the Hessian fly. An award of sixty dollars was also made to John Ellis for his study on methods of preventing peach tree decay.

Botany. While the science of botany, so called, can be traced back to Aristotle, and while a number of efforts were made to classify plants during the sixteenth and seventeenth centuries, it was not until the Linnaeus system of classification was devised in the eighteenth century that a scientific system was established for collecting and classifying various species. In this field John Bartram (mentioned earlier), known for the several articles he had contributed to the *Transactions,* did pioneer work. His botanical gardens were frequented by a host of scientists.[5]

Jacob Bigelow, another noted botanist, is best remembered for his labors in preparing the American pharmacopoeia, and for his botanical lectures at Harvard. André Michaux, a Frenchman, spent ten years botanizing in America. Thomas Nuttall, English born, lived in America from 1808 to 1842 as curator of the botanical gardens of Harvard. He traveled to the Pacific coast and on to Hawaii to secure botanical specimens. The eccentric, yet gifted Constantine Rafinesque, who compiled a collection of medical flora of the United States and carried the science of botany to Transylvania University at Lexington, Kentucky, was an early member.

Engineering. Also listed among the members of the Society were America's first engineers who rendered valuable service to the young republic. The list included David Bushnell, of submarine fame, John Lukens, John Fitch, and Robert Fulton of steamboat fame, and Benjamin H. Latrobe, architect, civil engineer, and designer of the government's early federal buildings.

Medicine. Also included in the Society's membership were a number of the nation's distinguished medical men. To the names of Philip Syng Physick and Caspar Wistar, mentioned earlier, should be added two others, Dr. John Warren and Dr. Benjamin Waterhouse. The Society joined with Benjamin Franklin in sponsoring the first hospital and the first medical school in America. Every member of the first faculty of the College of Philadelphia (1765), now the University of Pennsylvania, was a member of the Philosophical Society. Many eminent physicians born and educated in Europe, upon coming to America became members of the Philosophical Society.

American Academy of Arts and Sciences. Another society somewhat similar to the American Philosophical Society was the American Academy of Arts and Sciences, founded in Boston in 1780. John Adams was the prime mover in its organization. James Bowdoin was its first president, and Dr. Samuel Cooper, pastor of Brattle Square Church in Boston and member of the corporation of Harvard College, was the first vice-president. Its membership, like the membership of the Philadelphia Society, included the most learned and distinguished citizens of Boston and Cambridge—Joseph Willard, president of Harvard, James Bowdoin, governor of Massachusetts, Samuel and John Adams, Thomas Cushing, John Lowell, John Hancock, Elbridge Gerry, and others. A few out-of-state members were elected, notably Franklin, Washington, and Owen Biddle of Philadelphia. Also a few European members were admitted, among them Joseph Priestley of England, and General Lafayette of France.

Purpose of the Academy. The charter of the Academy declared the arts and sciences to be "the foundation and support of agriculture, commerce, and manufacturing." The purpose of the Academy was to encourage the study of natural history, to determine the uses to which the nation's natural resources could be applied, and "to cultivate every art and science which tends to advance the honor, dignity, and happiness of a free, independent virtuous people."

Papers read at the early meetings of the Academy related to such subjects as lightning and electrical fluid, the strata of the earth, the culture of wheat, raising Indian corn on poor land, grafting of fruit trees, fire pumps, fire engines, distemper among cattle, and formulas for making good steel.

Expansion of Scientific Groups

During the last half of the eighteenth century young America became science conscious. Specialized societies were organized to stimulate the interest in particular fields.

Medical Science. In 1765 the Philadelphia medical society was organized under Dr. John Morgan of the medical school of the University of Pennsylvania. The Massachusetts Medical Society was founded in 1781, and the College of Physicians of Philadelphia in 1787.

Marine Societies. Even sea captains from New England became interested in scientific collections and established a repository to house their unique collections. They pooled their observations of phenomena discovered on the high seas. The Salem Marine Society (1766) recorded accurate data about the waters and coasts that proved a priceless legacy to the American fishing and sailing interests. The East India Marine Society (1799) became widely known and its priceless collections were placed in the Peabody Academy of Sciences at Salem, where they are regarded today as rare acquisitions.

State Academies of Science. In the early days of the republic several states toyed with the idea of establishing state-sponsored philosophical societies. One was proposed in Maryland in 1797, but to Connecticut goes the credit for the first establishment of a State Academy in 1799. It was called the Connecticut Academy of Arts and Sciences, and President Ezra Stiles of Yale was the moving spirit.

America's Practical, Utilitarian Sciences. From the very foundation of our nation men of learning organized themselves into societies intent on promoting the practical and useful arts and sciences. Through them the scientific spirit became more aggressive, and many of the inhabitants of young America came to look upon science as a way of life. They came to realize that scientific methods could be applied to a wide variety of human pursuits. The aspersion often cast on America that our developments in applied science far exceed our achievement in pure science—if it is rightly an aspersion—may be traced back to the very beginnings of our scientific societies. Those societies were organized for the distinct purpose of promoting useful knowledge. The members believed that men of learning should turn their attention to the utilization of knowledge. The young nation was facing up to a gigantic task: a vast continent had to be conquered, its natural resources were to be tapped and put to man's use. But there was a shortage of men. The inventive genius of the people was challenged, and it was inevitable that the practical, the applied sciences should be the first to develop. Pure science, theoretical science, would come later.

Notes for Chapter 7

1. See Royal Society of London, *Philosophical Transactions* (London, 1665), and XL (1737–1738), Preface, dedicated to John Winthrop III; see also Austin H. Clark, "Background and Origin of the American Association for the Advancement of Science," *Summarized Proceedings*, American Association for the Advancement of Science, June

1929–January 1934 (1934), pp. 15-20; Frederick E. Brasch, "The Royal Society and Its Influence on Scientific Thought in the American Colonies," *Scientific Monthly,* XXXIII (1931), 337 *ff*; Margaret Denny, "The Royal Society and American Scholars." *The Scientific Monthly,* LXV (1947), No. 5, p. 415; Dorothy Stimson, *Scientists and Amateurs: A History of the Royal Society* (New York 1948).

2. The dedication reads in part: "No sooner were the sciences revived at the beginning of the last century, and natural knowledge began to be thought a study worthy of a real philosopher, but the ingenious Dr. Winthrop, Esq., your grandfather, distinguished himself in the highest rank of learned men. . . . You have imitated the example of your worthy ancestor . . . , you have increased the riches of the repository (of the Society) with above 600 curious specimens . . . accompanied with a list containing an accurate account of each particular, thereby showing your great skill in natural philosophy and at the same time intimating to England the vast riches which lie hidden in the lap of her principal daughter." *Transactions of Royal Society,* XL, Preface.

3. Others elected to the Royal Society included: Henry Ellis, 1749; John Hyde, 1752; Francis Nicholson, 1706; James Oglethorpe, 1749; William Penn, 1681; Thomas Pownall, 1772; Elihu Yale, 1717; and Thomas Rutherford, 1742.

4. See Ralph S. Bates, *Scientific Societies in the United States* (1945).

5. The activity of the Society's botanists is evidenced by the list of such genera as *Bartramia, Marshallis, Jeffersonia, Bartonia, Washingtonia, Collinsia, Michauxia, Clintonia, Schweinitzia, Elliottia,* and others.

For further reading, see: Sir Henry George Lyons, *The Royal Society, 1660–1940* (1944); J. C. Wild, *History of the Royal Society* (1948); Royal Society of London, *Philosophical Transactions* (1665 to date); Charles Hutton, George Shaw, Richard Pearson, *Philosophical Transactions Abridged,* (1665–1800).

Chapter 8

AMERICAN TECHNOLOGY DURING THE REVOLUTION

Throughout our nation's history wars have had a pronounced effect on technological development, and the application of scientific discoveries has been a determining factor in all of our armed conflicts. This was as true in the Revolutionary War as it was in later wars. The fateful decision that brought forth the Declaration of Independence did not come solely from such dramatics as Patrick Henry's fiery oration, "Give me liberty or give me death," or from James Otis' "Sons of Liberty." Nor did the decision stem entirely from an adherence to the political ideals regarding man's inalienable rights. The final decision resulted largely from the belief that the colonists could go it alone. They had faith in their ability to provide for themselves. They knew they could invent any article or tool they needed and could provide the food they needed in a self-sustained economy. Demands made by the war spurred the colonists on to greater technological achievements—achievements which otherwise might have been postponed or never have occurred.

Severing Economic Dependence

Before the colonists could free themselves from the mother country politically, it was essential that they develop an economy that would free them from dependence on England. "A people who are dependent upon foreigners for food or clothes must always be subject to them," declared Benjamin Rush in 1775. The Revolutionary period was as much a test of the ability of the colonists to invent and improvise new weapons and to carry on an independent way of life, as it was a test of their ability to set up an independent government.

While the colonies had made considerable advances in technological development, they had not entirely supplanted the commodities imported from England. The mercantilistic legislation imposed by Parliament had caused the colonies to abstain from the use of the products on which the obnoxious tax was imposed, rather than supplant them with their own production. The hope of the colonists had been for the removal of the tax rather than for substituting their own commodities. The vast trade that England carried on with the colonies was proof of America's dependence on the mother country economically. To become independent of England it became necessary for America to prove that she could produce the commodities she needed and that she could get along on what she could produce.

Moreover, the technology needed to fight a war had not yet been developed. In all the Colonial wars thus far, England had directed the securing of all war materials. So, when the colonists decided to sever the ties with the mother country, a dual problem arose. It was their task to provide and equip an army to fight a war, and also to satisfy the demands of civilians who had been accustomed to rely on each shipload from England for the commodities they wanted.

Burke Evaluates the Colonies. When Edmund Burke made his eloquent plea before the British Parliament for conciliating the American colonies, he pointed out that they were worth holding at any cost. He noted that the colonies had developed with a vigor and rapidity unexampled in history, and that while other communities had scarcely had time to take root, the American colonies "had shot forth exuberant branches." Their commerce was out of all proportion to the number of people. It had become so great that England's trade with the colonies in 1772 was equal to her trade with the entire world in 1704. The colonies had made such advances in agriculture, that instead of importing corn, they were exporting it; "the Old World was being fed by the New." As to the wealth that the colonies had drawn from the sea, Burke declared: "that there is no sea but what is vexed by their fisheries; no climate that is not witness to their toils." They had become so powerful, he observed, that they were defying the united forces of that nation, which, "but a century ago, called them into being."

Thomas Paine's Views. Burke was not alone in his estimate of the economic and technical resources of the American colonies. Thomas Paine boasted early in the war that there was no country in the world so internally capable of raising a fleet as America. He pointed out that tar, timber, iron, and cordage were our natural products, and he declared that the colonists need not go abroad for anything since they possessed every needed article of defense. He called attention to our flourishing crops of hemp for cordage, to the supply of small arms, equal to that of any country; to our ability "to cast cannon at pleasure"; and he noted that each day saw an increase in our supply of gunpowder.

Lack of Technical Equipment. Despite these claims, the outbreak of the Revolution found the colonists poorly equipped for fighting a prolonged war. Their industrial resources were inadequate. Chemical technology was at a low level. Practically no gunpowder was being made in America, and little was known of the manufacture of saltpeter, an essential in the manufacture of gunpowder. The young army had no medical corps. There was no Peruvian bark from which quinine was obtained. The country lacked trained engineers for constructing defense bulwarks and encampments, and for transporting troops and planning maneuvers. Moreover, the civilian needs of six million dollars worth of commodities annually imported from England had to be provided. It was as necessary for American science and technology to find a way to meet these needs as it was for American statecraft to organize a new form of government.

Demands of War

To meet these emergencies the newly constituted Continental Congress, though possessing little power other than to resolve and recommend, set out to develop the industries of the country. To fight a war, ships had to be built, guns had to be manufactured, and there were camp kettles and army shoes and army clothes to be provided. Gunpowder and shot, and the saltpeter needed to make the gunpowder, had to be secured. Duck and sail cloth and rope had to be made, and there was need for paper for cartridges and for paper money. In addition there were other articles which the colonies had not heretofore supplied, or had supplied only in limited quantities, that had to be secured on an unprecedented scale. The forces of government and technology had to be united.

Support of the Continental Congress. When the Continental Congress passed the nonimportation agreements in 1774, they proclaimed their confidence in the ability of the colonists to provide their own food, shelter, and clothing, without aid from the mother country.[1] They adopted a resolution that called upon the people to abstain from the consumption of British merchandise and manufactures, stating: "we will endeavor to live without trade, and to recur for subsistence to the fertility and bounty of our native soil, which will afford us all the necessaries and some of the conveniences of life." [2] To implement this, they resolved "to encourage the arts and manufactures by all the means in our power."

Encouraged Production. Colonial legislatures were requested to take similar steps. Special inducements in the form of awards, prizes, and exemptions from prescribed duties were offered to spur new developments. Rewards were offered for the production of iron, leather, woolen goods, linen, potash, and firearms. Congress recommended that the makers of arms be encouraged, "to make good substantial muskets, with barrels 3½ feet in length to carry an ounce ball, and fitted with a good steel ram rod." Congress called upon the assemblies to keep their gunsmiths at work until a sufficient quantity could be manufactured. Gunsmiths and cannon makers were exempt from military duty. It was recommended that the furnace at Salisbury, Virginia, be enlarged for the casting of heavy cannon, and that supplies of brass and copper be commandeered.

Congress was particularly concerned with the necessity of increasing the supply of shot and securing adequate quantities of lead. Those engaged in shot-casting were exempt from military duty. Lead mines in Virginia were investigated. David Rittenhouse, to help offset the shortage, substituted iron weights for lead in the clocks he was making. Paper for cartridges and a special grade of paper for Continental currency had to be secured. The shortage of salt was also a matter to be considered.

One of the first concerns of Congress was the small supply of wool. They resolved, "to endeavor to improve the breed of sheep, and increase their number . . . , and to that end to kill them as seldom as may be." [3] To offset the shortage in wool, wearing leather waistcoats and leather breeches was recommended, and members of Congress agreed to set the example. Leather waistcoats and breeches were to be provided for the soldiers. Congress recommended that the several assemblies procure for each soldier a suit of clothes, a blanket, a felt hat, two shirts, two pairs of shoes, and two pairs of socks.

In some details, the response to the demands of Congress was surprisingly adequate. Virginia was able at one time to furnish ten thousand barrels of flour.[4] One blacksmith, a Thomas Bates, in two months' time produced two thousand camp kettles.[5] New England in 1776 furnished ten thousand pairs of shoes for General Washington's army.[6]

War Supply Industries

Iron Manufacture. Notable advances were made in iron technology. On the eve of the Revolution, the colonists possessed a line of forges extending from New Hampshire to South Carolina, and, in 1775, 30,000 tons of iron had been produced. Much of this iron, prior to the Revolution, had been exported as bar iron or pig iron, but with the outbreak of hostilities the iron manufacturers of Pennsylvania and New Jersey received large contracts for manufacturing war supplies. The hollowware furnaces of New England were converted into industries for casting shells for the army. By the close of the Revolution the manufacture of iron had been introduced into southwestern Virginia, Tennessee, and North Carolina.

Cannon were cast at the Durham Iron Works in Pennsylvania. The Cornwallis furnace, Lancaster County, Pennsylvania, cast shot and shells for the Continental Army and at a nearby forge round iron was drawn under the hammer for musket barrels. At the boring mills, located in secluded spots for protection, barrels were bored out. The bore of the gun was formed by a core of the exact diameter, placed vertically in the center of the mold. Boring and turning mills were not yet known.

In the manufacture of cannon, America was not so well provided as in making small arms, but again the Pennsylvania Germans proved to be the experienced craftsmen. At the Durham Iron Works, small brass swivel cannon were made, and at the nearby Warwick furnace cannon and cannon balls were produced. In 1776, sixty cannon of 12- to 18-pound caliber were cast. The famous Cornwall furnace began casting cannon soon after the outbreak of hostilities. In making cannon, the shortage of brass and copper was acute. One, William Downing, succeeded in casting cannon from wrought iron.

Ordnance Plants. When the Revolutionary War came, the need for the establishment of ordnance plants became apparent. General Henry Knox of Washington's staff as early as 1776 recommended that one or more plants be erected at some distance from the seat of war for the production of large quantities of ordnance stores, and that a sufficient number of able artificers be employed to make carriages for cannon, ammunition, tumbrils, harness, and other weapons of war. Following this recommendation, a depot was established at Carlisle, Pennsylvania. Since this was too remote from the New England hostilities, Springfield, Massachusetts, was selected as a site for another. By 1782 ordnance plants at West Point, New York, Yellow Springs, Pennsylvania, and New London, Virginia, had become depositories for magazines and arms, and continued to be used as repair and construction centers. The Springfield and West Point locations became permanent.

In the manufacture of war weapons, the colonies were better prepared than in many other lines. Small arms had always been a daily necessity, and by the time of the Revolution the Americans had become the greatest weapon-using people in the world. Some of the Colonial assemblies had passed Acts specifically ordering adult males to carry guns. In 1766, Georgia enacted a law that provided that if any male person should attend church without carrying a gun or pistol in good order and fit for service, with at least six charges of powder and ball, he should be fined ten shillings.

Development of Gun Designs. Many types of guns were in use. Some were made in state armories. Some were privately made by local gunsmiths who fashioned their weapons to suit the individual, and some were imported. Production rate reached surprising proportions. In 1770 a Massachusetts gunsmith, with the aid of helpers, produced five hundred muskets. In 1774 a gunsmith in Maryland produced over two thousand muskets. Pennsylvania led all the colonies and produced one hundred thousand guns in 1774.[7] Some establishments consisted of three and four barrel forges, a lockshop with twenty forges and benches, water mills for grinding and polishing, benches for forty workmen engaged in filing the castings and locks, benches for gunstock makers, a brass foundry for mountings, and numerous fitting benches.

More important than the quantity of the gunsmith's trade was the quality. The Colonial gunsmiths developed a high degree of technological skill. Without previous knowledge of ballistics, and unacquainted with the laws governing velocity, they patiently experimented. A new type of weapon called "the Pennsylvania rifle," designed by German and Swiss gunsmiths, appeared about 1750. This new gun became the standard weapon of the frontier and later came to be known as the Kentucky rifle, or the squirrel gun, because it was effective in hitting a small target. It had a long barrel and a rifled bore, which produced a weapon of greater muzzle velocity and greater accuracy than the old type of musket. In the hands of the skilled frontiersman it was a deadly weapon, greatly feared by the "Red Coats." The Pennsylvania rifle played a major part in winning the Revolution.

One of the important features of this rifle was the facility with which it could be loaded. This was due to an improvement instituted by the Pennsylvania gunsmiths, called the "greased patch method of loading." It was a perfect example of applied, practical technology. Formerly a heavy ramrod of iron and a small mallet were carried by every hunter to drive the bullet down the gun barrel. The greased patch obviated this. With the new method, the bullet was wrapped in a small piece of greased cloth and inserted in the barrel. With an easy shove from a wooden ramrod, it slid down to the butt end of the barrel. Because of this improvement, frontier history gave us the names of Indian fighters like Christopher Gist, Daniel Boone, and others who "loaded their rifles on the run."

Colonial Rifle Experts. Young George Washington, in his expeditions into western Pennsylvania during the 1750's, had become familiar with this new weapon. Later, when he took command of the Continental troops outside Cambridge, he sent for a battalion of riflemen from Pennsylvania. A picked company of these men, under Captain William Hendricks, made history. All volunteers, they set out on foot from Carlisle, Pennsylvania, in mid-July, 1775, and completed the cross-country march to Cambridge (430 miles) by August 9. They were dressed in white frocks (deerskins) and round hats (coonskin caps), and were remarkable for their accuracy of aim.

When this company was called on by Washington for review, General Gage stationed spies to witness their performance and issued orders that some of these men were to be taken as prisoners. There is no record of any being captured on this occasion, but when the company moved up to the plains of Quebec, a few Americans were captured and sent to England where they were made to exhibit their marksmanship. The British idea was that this demonstration would fire the people to a more ardent support of the war. But, instead, it produced the opposite effect and created among those who witnessed their performance marked admiration for the American riflemen.[8]

Defense Measures

The most striking illustration of iron for defense purposes was the construction of a large iron chain to block the Hudson River. The strategic importance of the Hudson was recognized early, and this chain was used to keep the British ships from sailing up the river. It was forged by Peter Townsend and his aides at the Sterling Iron Works, New York. It measured 500 yards long and was made of links varying in length from 2 to 3½ feet and from rods of iron 2 inches to 3½ inches in diameter. When the links had been forged they were carried two or three at a time on mule back, or in ox carts, over the hills to the banks of the river. The links were then welded into one solid chain at a furnace near West Point. A swivel was inserted at every 100 feet, and a clevis every 1000 feet.

In April, 1778, this novel technological device was stretched across the Hudson River and held in place by heavy anchors suspended from it. It was never broken and no enemy ship ever ascended the Hudson. Other chains lighter than this one were used to block entrances to rivers in other sections of the country.

The iron manufacturers were early confronted with a labor shortage. To overcome this shortage, Hessian prisoners, some of whom had been ironworkers, were leased from the Continental Army and assigned to the iron trade. Many of them had come from the iron regions of Germany and possessed special skills in the trade. Some remained here after the war.

Gunpowder for Offense and Defense

The shortage of gunpowder proved a matter of great concern to General Washington and his aides, and this shortage placed the Americans in a precarious position. Every means was employed to conserve the supply on hand, while steps were taken to increase the production. Because sources of saltpeter were lacking in this country at the outset of hostilities, practically no gunpowder had been made. When the war broke, all available supplies were seized. Importation was urged and manufacture was encouraged.

When Washington took command of the poorly equipped troops at Cambridge, Massachusetts, following the battle of Lexington, John Adams in Philadelphia wrote to Elbridge Gerry suggesting that General Washington and his officers be accorded a special welcome upon their arrival, with all pride and pomp, but under no circumstances *"should any powder be burned."* The Continental Congress called upon the state governments to offer prizes as inducements to spur production of gunpowder. Pennsylvania, in response, offered a prize of twenty pounds for every hundredweight of good merchantable saltpeter manufactured within the province and delivered to the Committee of Safety within three months. Similar resolutions were passed by other assemblies.

Powder Mills. Powder mills were erected in the neighborhood of Philadelphia and Lancaster, Pennsylvania. Robert Harris's mill, equipped with mortar and stampers, mixed 1 ton of saltpeter and 5 hundredweight of sulfur, in a four-to-one proportion, to produce 1 ton of powder. A much larger mill on a nearby creek in eastern Pennsylvania began producing 2½ tons of powder a week.

There was scarcely any available information concerning the manufacture of saltpeter. Congress published a pamphlet urging all citizens to save nitrogenous materials and gave directions on how to extract "nitrous salts." Dr. Benjamin Rush discovered an old European process which he published so that others might use it. The Pennsylvania Germans again demonstrated their "know-how." Although they were ignorant of the chemistry involved, many of their people knew from experience the method of extracting saltpeter. The saltpeter used in Europe came from Persia or the East Indies. France alone had evolved a method of extracting niter and was thus independent of importation. The French chemists had also discovered that niter was an "effluvia of animal bodies, and could be obtained from pigeon houses, stables, and barns."

It was this recipe that Dr. Rush had found, and the secrets of this same method had been handed down to the Pennsylvania Germans. A committee from York County petitioned the Committee of Safety in 1775, saying, "We have so many people who have wrought many years at making saltpeter in Germany, and understand it as any of our old women do in making soft soap. . . . They are but mechanics and do not understand theory, but let them make sufficient for our present wants, and let the theories improve and amend their defects at leisure." [9] How typical of early American technology! It was results, not theories, that counted!

After the disaster resulting from the shortage of powder at Bunker Hill occurred, New England was desirous of erecting powder mills of her own. Accordingly, Paul Revere, the handy man of the Revolution, was sent to Philadelphia to "obtain the exact plan of making powder, powder mills, the quantity of powder that may be made in one day in said mill, and the expense of the powder mill." [10] Revere visited the plant, but the manager of the plant, apparently fearing competition, hurried Revere through the factory so that he was unable to obtain any secrets. He failed to reckon with the ingenuity of Paul Revere, however, whose quick eye and mechanical mind grasped the technique. Revere returned to Massachusetts where, with the aid of Samuel Adams, he was able to advise the authorities on the erection of a powder mill at Canton. This mill supplied most of the powder used by the Massachusetts troops throughout the first three years of the war.

Despite all such efforts, the shortage of gunpowder continued. The total production from native saltpeter during the first two and a half years of the Revolution was slightly over 100,000 pounds. The amount of powder manufactured with imported saltpeter, mainly from France, was increased to almost 700,000 pounds.[11] French aid to the cause of American independence included more than General Lafayette's inspiring leadership; it also included the valuable contribution of her scientists.

American Military Engineers

The role of the military engineer in warfare is of first importance. His skill is needed to construct entrenchments, to provide roads, to move troops, as well as to carry out military maneuvers. Early in the war, Washington deplored the shortage of engineers, and asked the Continental Congress to provide immediate facilities for training a corps of engineers. He was disappointed with those he had because they were untrained and were lacking in an understanding of the duties and skills required for building fortifications.

French Engineers. A committee of Congress appealed to Franklin, who was then in France, and he secured four of France's finest military engineers. They came to America and were placed on Washington's staff where they served throughout the war. They rendered invaluable service in the several campaigns and were of special assistance in constructing the fortifications at Valley Forge. They later rendered valuable service at Yorktown. Another engineer, Kosciusko, a Polish patriot who espoused the American cause, planned the original fortifications at West Point.

Rittenhouse. One of the few native American engineers was David Rittenhouse, who was appointed to the Committee of Public Safety in Pennsylvania in 1775. Rittenhouse supervised the casting of cannon and the manufacture of powder. He experimented on the rifling of cannon and musket balls, and served as a one-man advisory board. His many wartime contributions caused Jefferson to say of him in 1778, "The world has but one Rittenhouse, and it never had one before."

Another engineer, though less versatile, was Louis Nicola, French Huguenot, who had spent twenty-six years as a military officer before arriving in Pennsylvania in 1766. Nicola published three *Military Manuals* for American use, and proposed a plan for the construction of vessels to defend the Delaware.

West Point Academy, 1794, 1802. Out of the need for engineers in the American Revolution grew the idea of establishing a military academy for the United States. Henry Knox of Washington's staff (in May, 1776) proposed such a training school, and in October of that year the Continental Congress passed a resolution to appoint a committee to draw up plans. A Philadelphia Corps of Invalids was formed the following year. In 1781 it was transferred to West Point where three buildings had already been erected for housing a military school. After Washington was elected president in 1789, he continued to urge the establishment of such a school, and in 1794 under his direction a school for artillerists, engineers, and cadets was founded. The work continued for two years when the buildings were destroyed by fire. It was not until 1802, under Jefferson's administration, that the school was formally opened, and it was not until ten years later that its permancence was assured. But the idea of a military academy had emanated from the Revolutionary War years and Washington's insistence on trained engineers.

Naval Operations

The confidence of the American people in their ability to construct a navy and to contend with the most formidable maritime nation of the world is evidence of their faith in themselves during the Revolution. Congress authorized the construction of a navy at the outset of the war. Shortly after hostilities started Washington wrote to Franklin, who was in Paris soliciting the aid of the French fleet, declaring that, "naval superiority was the pivot on which everything turned." In many respects the navy, authorized during the first months of the war, set the pattern for subsequent developments. A nation consisting of thousands of hardy maritime people, inured to the hardships of the sea, "whose whaling exploits had taken them to the Antipodes," and whose trading ventures encompassed the globe, might be expected to undertake naval warfare.

Organizing for Sea Battle. On May 10, 1775, the Continental Congress authorized the appointment of a provisional marine and naval committee. Included in its members were Robert Morris, John Hancock, Philip Livingston, Benjamin Harrison, Joseph Hewes, and Nicholas Van Dyke. The kind of navy to be built was yet to be determined. Young America had not constructed any naval vessels, "ships of the line." It had no fleet and no experience in conducting engagements in which a fleet of vessels was involved. America's prowess on the seas had been achieved with small, swift vessels, venturing alone after the prize they sought, winning as often by their ability to run away from danger as by being able to overcome it.

John Paul Jones, Naval Strategist. John Paul Jones is usually given credit for planning the first American Navy. Scottish born as John Paul, he adopted America as his country and took the surname of Jones. After achieving success in British shipping, he came to America and lived for a time in obscurity and poverty. Fired with enthusiasm for his adopted country, he drafted a plan for a new navy and submitted it to the recently appointed marine committee in July 1775. In this plan he specified the types of armed vessels needed and proposed plans for outfitting them. He advised against building "ships of the line," because they were too large and too costly. He recommended small vessels to harass the enemy, and advocated the construction of frigates for long voyages.[12]

America's First Fighting Ships. The first vessels of the new navy were authorized in December, 1775. The high standard of design exhibited in these first frigates indicates the widespread knowledge of naval technology in Colonial America. This knowledge was not confined to a single locality but prevailed along the entire coast. Shipbuilding plants had men who were acquainted with the technique of warship design and construction, since many of the early ship builders had served apprenticeships under admiralty trained shipwrights.

The naval skirmishes of the Revolution inspired a number of developments in naval construction. The ship builders produced designs of their own that led to preeminence of Americans as frigate builders. These frigates were built for great speed under sail and were equipped with heavy guns. They operated as single vessels rather than as a fleet or squadron. America became the "frigate nation." Small, fast frigates that could shoot and run were preferred to the large cumbersome fleets of boats characteristic of the English and continental naval strategy.[13]

First Submarine. Another American invention, an undersea boat, the forerunner of the deadly submarine, was first introduced in the early months of the American Revolution. David Bushnell, Yale graduate, is given credit for this invention. He constructed a wooden vessel, called the "Turtle" which, when weighted with lead, would submerge. It was driven by a hand-operated propeller, at a speed of 3 miles an hour. It had a small conning tower big enough for the operator's head and shoulders, and was equipped with glass windows for visibility. The vessel held enough air to enable a man to live inside it for 30 minutes, and it was provided with two ventilating tubes that closed automatically when the vessel submerged and gave access to fresh air when it appeared on the surface. The vessel submerged when a foot valve admitted water to the hull. It ascended when force pumps discharged the water. It carried 200 pounds of lead ballast which could be dropped in case rapid ascent to the surface became necessary.[14]

The egg-shaped magazine that carried from 100 to 125 pounds of powder was as ingenious in its operation as the submarine. The magazine was attached near the stern by means of a screw, one end of which protruded into the magazine, and acted as a stop on the movement of the clock to prevent premature explosion. A gunlock was attached for exploding the charge, and a clock was set to fire the gunlock at a predetermined time.

In August of 1776 the "Turtle," piloted by Sergeant Ezra Lee of the Connecticut Continental Regiment, attempted to blow up in New York harbor the sixty-four gun British warship "Eagle," commanded by Lord Howe. Lee succeeded in maneuvering the "Turtle" against the side of the man-of-war, but discovered that the iron screw would not penetrate the copper sheathing of the hull. When Lee was forced to surface he was discovered by the British, and the mine floated harmlessly away to explode in the East River.

The copper sheathing of the warship saved it from being sent to the bottom. The attempt itself failed, but the principle of the submarine had been successfully conceived. However, Bushnell was unable to win government support for his invention and submarine warfare missed becoming a reality in 1776.

Results of Wartime Technology

The American Revolution called forth a number of new technological developments, not the least of which were those connected with supplying new materials for civilian clothing and uniforms for the soldiers.

Wool Processing. Among these advances was the improved manufacture of hand cards used in processing woolen cloth. Before the Revolution these cards had been imported, but following the outbreak of hostilities the Americans were forced to supply their own. Financial aid was offered for the manufacture of wire suitable for the teeth, but the bending and installing of this wire was a tedious hand process. Oliver Evans came up with a machine that cut and bent the teeth and inserted them into a leather frame, a great improvement in hand carding.

Spinning. Christopher Tulley improved a spinning machine that would spin twenty-four threads of cotton or wool at one time and which was operated by one person. One Nathaniel Norgrove, who had recently come from Kensington, England, opened a manufactory for linen printing in eastern Pennsylvania, where all types of linens, cottons, calicoes, and velvets could be printed "in fast colors." To meet the demand for a strong durable fabric, a new cloth of part linen and part wool, called "linsey-woolsey," was devised. In weaving this cloth the cross (woof) threads were of wool and the longitudinal (warp) ones were of linen. This produced a closely knit, sturdy fabric.

Textile Manufacturers. Special encouragement was given to American textile manufacturers. Besides the premiums offered by the state legislatures and the stimulus granted by the Continental Congress, a number of societies offered rewards for both quantity and quality in home manufactured materials. In Philadelphia, subscriptions were taken for a fund for the establishment of a textile manufactory in which such leaders as Dr. Benjamin Rush participated. He told the people that "woolens and linens of all kinds made among us may be made and bought as cheap as those imported from Britain, and will wear twice as well. . . ." The wearing of homespun was encouraged on all sides. "Bounty Coats" were offered in the army, and the wearing of homespun became a symbol of loyalty.

Tanning and Leather Production. Tanning technology was encouraged, and local manufacturers were urged to supply leather uniforms and leather equipment for the army. All dealers in skins were urged to sell their products at the usual price, and were asked not to take advantage of any additional demand for their articles. A surplus supply of skins came from Georgia, enough to supply several regiments of men. Colonel William Smallwood's regiments from Maryland were fitted out in leather uniforms, and by the close of the war a number of other regiments were wearing uniforms of leather.

To stimulate greater leather production, a special commissary officer was appointed in the summer of 1777, "to receive all raw hides belonging to the United States and to arrange for immediate tanning of the leather at the custom rate of exchange." In case suitable terms could not be arranged, an officer of the supply unit was authorized to set up tanyards himself, engage workmen, and do the tanning. Also, he was authorized to import such quantities of dry hides from the Spanish West Indies as he deemed necessary to supply leather for the Army.

Scientific Advances. Public interest in the potential benefits of applied science and technology increased throughout the years of the American Revolution. It was in the midst of the war, three years before the treaty of peace was signed, that the American Academy of Arts and Sciences was founded at Boston (1780) with John Adams as the leading spirit. Adams proclaimed the thesis that "the promotion of scientific information was the best and most lasting protection for popular government," and he was later responsible for inserting in the Massachusetts state constitution in 1780 the provision that imposed on all future legislatures and magistrates the duty of promoting science. All schools and seminaries, including Harvard College, were required to promote the study of agriculture, arts, science, commerce, trade, and manufactures.[15]

Emergence of Cultural Inquiry

With the defeat of Cornwallis by General Washington at Yorktown in October, 1781, America's military and political independence was established. The young nation now faced many important decisions, mostly political. But, however great the strain from 1781 to 1787 may have been politically, the Americans had faith in their economic, industrial, and scientific capabilities. They set about zealously to free the young nation from ignorance, backwardness, and further dependence upon antiquated agriculture and handcraft industry. They believed with Jedidiah Morse, who wrote in 1785, that, "in this young country, aided by all the improvements of former ages, there will develop the highest degree of arts and sciences of any civilization."

Notes for Chapter 8

1. *Journals of the Continental Congress, 1774–1789* (Sept. 27, 1774).
2. *Ibid.* (Oct. 21, 1774).
3. *Ibid.* (Oct. 6, 1774).
4. *Ibid.* (Dec. 26, 1776).
5. *Ibid.* (Aug. 22, 1776).
6. *Ibid.* (Nov. 27, 1776).
7. William Babcock Weeden, *Economic and Social History of New England,* II, (1890), 793.
8. *Pennsylvania Archives,* Second Series, XV No. 1, 5 *ff*. Reprint of article printed in Glasgow, 1776.
9. *Pennsylvania Colonial Records,* X, 668.
10. Esther Forbes, *Paul Revere, and the World He Lived In,* (1942), p. 302.
11. For an interesting review of the efforts made by the colonists to develop new tools and techniques, see I. Bernard Cohen, "Science and the Revolution," *Technology Review,* XXVII (April 6, 1945).
12. See Lincoln Lorenz, *John Paul Jones* (1943).
13. For a detailed treatment of this idea, see Howard I. Chapelle, *History of the American Sailing Navy* (1949).
14. For a description of this early experiment, see *Transactions,* American Philosophical Society, IV, No. 37, 303-312, for "Letter from David Bushnell to Thomas Jefferson on the General Principles and Construction of a Submarine Vessel"; Lieut. Col. Henry L. Abbott, *Beginnings of Modern Submarine Warfare* (1881); Simon Lake, *The Evolution Of The Submarine* (1918).
15. Francis Newton Thorpe, *Federal and State Constitutions,* III (1909), 1907.

Chapter 9

OUR FOUNDING FATHERS IN SCIENCE AND TECHNOLOGY

Many of the "Founding Fathers" who fought to win political freedom were also the pathfinders in ushering in a new scientific and technological era. They believed that advances in science and technology would help to free the people of young America from economic subjection to Europe. They encouraged the development of useful arts and inventions, and boldly pioneered into new fields of scientific research and investigation. They believed political freedom could be assured only as long as the common welfare of the people could be assured. To understand the full history of the founding of our republic, we must understand the accomplishments of Franklin, Washington, Jefferson, Rittenhouse, and a galaxy of others in the field of science and technology, as well as in the political realm.

Benjamin Franklin, 1706–1790

The greatest of the "Founding Fathers" in the world of science was Benjamin Franklin. As noted earlier, not only was Franklin America's first man of science, but he also ranks among the greatest of all scientists. He was a man of varied interests, but in his own words, the interests that were dearest to his heart were those of a scientific nature. He always hoped to be free to devote himself "to those subjects (science) that are more agreeable to me than political operations. Then I will be happier than all the grandees of earth."

Franklin possessed the spirit of a true scientific investigator. He was first a philosopher, but he went beyond the theoretical, the speculative stage. His range of scientific interests took him into many fields of research. He made contributions to agriculture, engineering, medicine, ventilation, meteorology, aeronautics, hygiene, nutrition, shipbuilding, canal construction, and most important, to electricity.

Franklin always emphasized the useful, the utilitarian. His constant concern was to discover the cause, the why of things, to determine the meaning of each discovery, and to put it to practical use. Here in America, he said, "it is what a man does that counts, not who he is." He pointed out that the artisans, the husbandmen, and the mechanics held positions of honor in America because their employments were useful. He told his friends abroad that the American people had a saying that, "God Almighty is Himself a mechanic, the greatest in the Universe."

In his *Discoveries* (1736), written when he was only thirty years old, Franklin noted that until shortly before his time, man had had

war without powder, shot, cannon, or mortars; nay, the mob made their bonfires without squibs or crackers. They went to sea without compass, and sailed without needle. They viewed the stars without telescopes, and measured latitudes without observation. Learning had no printing press, writing no paper, and paper no ink . . . They were clothed without manufacture, and their richest robes were skins of the most formidable monsters. They carried on trade without books, and correspondence without posts . . . , They had surgery without anatomy, and physicians without materia medica.

The rich had no commerce, the poor no employment; war and the sword was the great field of honor, the stage of preferment; and you have scarce a man eminent in the world but for a furious outrageous falling upon his fellow creatures.

The world is now increasing in experimental knowledge; but let no man flatter the age with pretending we have arrived at perfection of discoveries.

Study of Heat. When, at the age of forty, Franklin had achieved sufficient fortune to enable him to retire from the printing trade and embark upon a number of scientific ventures, the discovery of scientific principles that could be put to man's use became an obsession with him. He made a careful study of the physics of heat, air-conditioning, ventilation, and the circulation of hot and cold air. He observed that air was rarified by heat and condensed by cold. He discovered that air rises when heated. He noted that fire gave off heat, light, and fumes, and that the fumes or smoke moved only when carried by the rarified air. He noted that heat was transmitted through an iron plate as well as through air, and that heat and light are given off in all directions, but the greatest heat is over the fire where the rarified air rises. The result of these studies was the famed Franklin stove. (See Chapter 3.)

Clothing. Franklin's studies of the science of heating also led to practical suggestions in regard to clothing. Studying the effect of color on the transmission of heat, he noted that dark colors absorb more heat than light colors. He came to this conclusion after a simple experiment in placing pieces of cloth of dark and light colors on the snow in direct sunlight. The dark cloth sank deeper in the snow than did the light cloth. This finding was also turned to practical use. He recommended that one should wear light colors, in the tropics, and during the summer months, in the temperate zones.

Hygiene, Nutrition. In the field of hygiene and nutrition, subjects which had not yet been recognized as a science, Franklin observed the value of sunshine and fresh vegetables to health. He wrote treatises on the advantages of bodily cleanliness, of fresh air, and of sun baths. He conducted an experiment in reviving prostrated insects by focusing the rays of sunshine on them. He learned from personal experience the nutritive value of vegetables.

Practical Science. In every field of endeavor Franklin was alert to the possibilities of useful improvements that would promote the comforts and joys of living. In this he set the pattern for America's scientific progress. As postmaster general, he designed a device which, when attached to the wheel of a coach, measured the distance of the mail routes. He drew up plans for paving streets, using the center drain rather than side gutters to carry off water. Even as an old man, instead of being annoyed with two pairs of glasses—one for close reading and one for long distance—he inserted a half lens of each kind into one frame, thus devising the first bifocals. He experimented in crop rotation, studied the effects of winter frosts on various kinds of seeds, and urged the introduction of the science of agriculture in the curriculums of schools.

Pioneer in Electricity. As in the field of statesmanship, Franklin extended his services in science throughout Europe and America. He was a pioneer in the field of electricity, and the many successful experiments he performed brought him recognition as the world's leader in electrical science. Many of his discoveries received greater acclaim in Europe than they did in America. Every schoolboy knows the story of Franklin and the kite, but these discoveries were only a small part of his contribution to the field of electricity. His greatest achievement was the formulation of a workable theory as to the nature of electricity. It is on his research in electricity, which forms his most permanent and lasting contribution to scientific thought, that Franklin's scientific reputation rests.

Franklin demonstrated to the peoples of both continents the nature of electrostatic induction and the importance of grounding and insulation in electrical experiments. He explained the action of the newly discovered condenser in the Leyden jar. The words *armature, plus* and *minus, positive* and *negative* (as applied to electricity), and electrical battery were first used by Franklin. He proved that the electricity which could be created by so-called electrical machines, and the electricity of lightning were identical.

Lightning Rod. In this field also, usefulness came to be an objective. Franklin invented the lightning rod that, like the Franklin stove, was widely adopted in Europe and throughout the American colonies. The lightning rod had been little improved upon since Franklin's first invention. It was of immediate economic value in that it saved property worth millions of dollars from destruction by fire.

Publications on Electricity. Franklin's observations on electricity were set forth in a series of letters which he wrote to Peter Collinson, a friend in England. They were read before the Royal Society, and later published. They proved so popular that they went through five editions and were translated into French and German. Joseph Priestley declared nothing had been written on the subject of electricity that had been so widely read, and Sir Humphrey Davy, noted British scientist, declared that Franklin's work made the whole world aware that the American colonies had "produced a scientist second to none." Franklin's reputation as a scientist spread throughout all Europe, and when he was sent abroad a few years later to represent the American colonies, he was acclaimed both as a scientist and a diplomat. Science had paved the way for statesmanship.

Diversified Accomplishments. Franklin was not only a "founding father in science" in the realm of inventions and discoveries, but in other fields as well. As a printer and publisher, he pioneered in disseminating scientific information, not only to scientists but also to the public. He fostered schools, published his famous almanac, essays, treatises, newspapers, and he organized scientific societies. As early as 1743 he conceived the idea that Philadelphia should have a school devoted to teaching practical, vocational subjects. He worked over his plans diligently, and in 1749 the first board of trustees for his academy was organized. Franklin was elected president. This academy later became the University of Pennsylvania, and Franklin served on its board for the remainder of his life.

Franklin's *Almanac* was filled year after year with notes on the weather, astronomical phenomena, phases of the moon, comments on scientific truths, the useful arts, and numerous axioms of advice.

It served as a poor man's college. Franklin established the first library in the United States, laid plans for the first hospital, organized the first fire company, established the first insurance company, sponsored the first voyage to the Arctic.

Influence on American Life. Franklin set a pattern for the American way of life. The homely philosophies of "Poor Richard" continued to guide the thinking and the practices of Americans for many generations. Practical, useful, experiments became the foundation upon which America's economy was built, and Franklin became the pattern for many generations of scientists.

George Washington, 1732–1799

One of the reasons for George Washington being hailed as "first in war, first in peace, and first in the hearts of his countrymen" was because of the service he rendered in the field of science. His chief interest lay in farming. His inquiries were directed toward improving farming methods and in ascertaining exact knowledge in farming procedure. Living in an age when there were no schools of agriculture, no farm journals, no seed catalogues, Washington could resort only to experiment and investigation and to the exchange of findings with others of similar interests. He carried on extensive correspondence with two pioneers in English argriculture, Arthur Young and Sir John Sinclair. They exchanged ideas regarding the best methods for seeding grains and curing clover, discussed the value of the Albany pea, the field pea of Europe, the depredations of the weevil to the wheat crop, and other topics of agriculture.

Agricultural Experimental Stations. Washington's interest in his farm was dearest to his heart, and he became one of the great landholders of his day. While he owned thousand of acres of land in Virginia and in western Pennsylvania, his chief interest was centered on his estate at Mt. Vernon, where his five farms became veritable experimental stations. He carefully tested seeds, tried out new varieties, introduced new methods, kept careful notes on the planting time of crops, the best seasons for harvesting, and the mechanical improvements used on his farms. During a period of almost forty years he compiled agricultural records of great value.

Tobacco was the one money crop for Virginia farmers, and early in his career Washington grew large amounts of it, but after noting the way in which it depleted the soil, he turned to other crops. He experimented with lucerne, alfalfa, clover, wheat, and buckwheat, and imported numerous varieties from England. After receiving an order of English field peas, which he noticed had been eaten by bugs, he sent instructions on the best methods of shipping seeds. He accepted only the best seed and specified, "They must not be carried in the hold of the vessel which is too damp."

Wheat Growing. Washington's greatest interest was in his wheat crops, although eastern Virginia was not a good wheat region. He imported seed wheat from England, Siberia, Poland, and even from the Cape of Good Hope. He sowed wheat in dry soil, "with favorable results." In 1769 he exported over 6,000 bushels, sending large quantities to the West Indies in sacks stamped, "G. Washington." His wheat sold at a premium. Much of his wheat was ground in his own gristmill. Washington's interest in growing and developing wheat was so great that he proudly added stalks of wheat to his coat of arms on his book plates.

Soil Conservation. Washington practiced soil conservation long before that term was in use. He fertilized the soil by having tons of mud hauled from the Potomac River bottoms and spread over his fields. In pasturing his stock he moved them from field to field, with the view of manuring the soil. Shortly after the close of the Revolution he needed a new farm overseer and demanded that "above all he must be Midas-like, one who can convert everything he touches into manure as the first transmutation toward gold."

Improved Farm Implements. Washington was always eager to try out new farm implements and introduced many improvements of his own. He noted in his "Diary" (1760), "I fitted a two-eyed plow instead of a Duckbill plow, and with difficulty made my chariot a wheel-horse plow." In March of that year he spent several days with his blacksmith perfecting a new plow of his own design. He tried it out several times before telling his neighbors about it, and reported "she worked very well."

During the years of the American Revolution, Washington was unable to spend any time at his Mt. Vernon estate. Following the treaty of peace in 1783, he returned to the tranquil acres where he hoped to spend the remaining years of his life. Although Washington had given seven years of his life to winning American independence, he was the first to realize the need for keeping in touch with England on matters of agriculture and manufacturing. In 1796 he resumed correspondence with Arthur Young and asked him to send over seeds, tools, and "two of the simplest and best constructed plows for land which is neither very heavy or sandy." He specified that the plows were "to be drawn by two horses—to have spare shares and coulters, and a mould on which to form new irons when the old ones are worn out, or will require repairing." This was the forerunner of the system of interchangeable parts soon to appear in the technology of American toolmaking.

Meanwhile, Washington was busy designing new farm implements of his own. He designed an automatic seeder by boring holes in a barrel which was mounted lengthwise between two wheels. He regulated the flow of grain by plugging alternate holes. He made an improvement on the harrow by devising a spiked roll which could be fitted with either iron or wooden teeth.

Breeding Livestock. Washington is credited with being the first American to introduce mules into this country. When he asked the American consul in Spain to procure for him a jack of the best breed, he was informed that there was a law forbidding the export of livestock from that country. In 1785, however, the King of Spain sent him two jacks and two jennets. One of the jacks died in crossing, but the other one arrived safely, and Washington allowed him to be taken on a tour through the southern states, thereby introducing mule breeding in the south. In 1786 General Lafayette sent Washington a jack and two jennets of the Malta breed.

Washington took special pride in improving his livestock. He was interested in producing sheep that would yield better wool. The wool from his sheep averaged from 5 to 5½ pounds per fleece. He also succeeded in improving the breed of horses, cattle, hogs, and dogs. Over a period of years, he raised the average weight of his cattle, "from 500 pounds to 800 pounds the four quarters, by proper feeding and shelter." He imported red swine from England which, because of their "short legs and plump backs" were content to stay in the sty, in contrast to the razorbacks that roamed the woods. His imported hogs produced a more tender and tasty meat.

Even after taking office as first President of the United States (1789) the happenings on his Virginia plantation continued to be of great concern. When he received word that his flock of sheep had produced the highest yield of wool yet recorded, he sent some samples to his friend Arthur Young to have them examined by English woolen manufacturers. It was found to be equal in quality "to the best Kentish wool."

During the second year of his presidency, while still in New York, Washington learned of an improved threshing machine designed by Baron Poeffnitz of Murray Hill, located near the present Grand Central Station. Washington rode out to see it operate. The machine, a Winslow thresher, separated the heads from the straw and by running the heads through the mill shattered out the grain and threshed a half bushel of clean wheat in 15 minutes. Washington observed that two boys could turn the wheel, feed the mill, and remove the threshed grain. He noted that this new method afforded an easier and a much cleaner way of removing the grain than the usual method of threshing. It was vastly to be preferred to treading, "which is hurtful to the horses and filthy to the wheat. . . ."

Encouragement to Agricultural Sciences. Although Washington sensed the divided political opinion that members of Congress might have on such questions as the tariff, trade, Indian affairs, and other issues, he hoped they would all unite in supporting agriculture, commerce, manufacturing, and science. The mutual interest in science and agriculture that existed between Washington and Jefferson continued throughout the years of their political differences. In 1795 the two were exchanging letters recounting the latest experiments each had made on his farm. Washington reported that he was having "no better success with lucerne and alfalfa, than you." He advised Jefferson how to sow his buckwheat in order to get the best results, and he was anxious to learn from Jefferson about his experiments relative to winter vetch and peas.

Washington wanted to see agricultural societies organized throughout the states, and in his last annual message to Congress (December 7, 1796) he urged the creation of a board of agriculture. Its duties would be to collect and diffuse information, to offer premiums and small pecuniary aid, and encourage a spirit of discovery and improvement. The results would be twofold. It would stimulate further experiments, and it would assemble in one common center the results of individual skill, and would diffuse this information over the whole nation. Congress failed to act on his recommendation, however, and the office of a commissioner of agriculture was not established until 1862. Not until 1889, almost a century after Washington's recommendation, was the office raised to Cabinet rank.

Retirement Activities. After having given the greater part of two score years to the service of his country, Washington retired to the quiet acres of his Mt. Vernon farm, March 4, 1797, "where seated under my vine and fig tree," he longed for abundant leisure. But within a few weeks he addressed a letter to his English friend, William Strickland, declaring that he had never been so busy as he was in repairing the ravages of eight years absence. He took up intensive farming saying that, "a half, a third, or even a fourth of what we mangle [land] well wrought and properly dressed, would produce more than the whole under our system of management; yet such is our system of habit that we cannot depart from it."

Washington took special delight during his last years in experimenting with methods for increasing his yield of oats. By careful selection of seeds and by fertilizing the ground, he reported that he was producing "upward of fifty pounds the Winchester bushel." He discovered that clover seed sowed on top of a light snow in late February or March would sink in with the snow and take root. Mixing orchard grass seed with clover seed, he discovered that the crop "bloomed precisely at the right time, rises quick after cutting, stands thick, yields well, and both horses and cattle are fond of it, green or in hay."[1] In building fences Washington discovered that the old worm fence was too expensive and wasteful, so he substituted hedge fences of varying heights. He worked out a careful system of crop rotation which he planned to follow many years, extending into the nineteenth century. But it was not given to him to see his plans fulfilled. Washington died December 14, 1799.

Washington's Service to Science. For young America, Washington was truly a "founding father of science" as well as a founding father in other fields. In his devotion to the land he set a standard that characterized the American way of life as much as his admonitions to guard against the "baneful effects of party," and "to steer clear of permanent alliances with any portion of the foreign world." Late in life he wrote, "I know of no pursuit in which more real and important service can be rendered to any country than by improving its agriculture, its breed of useful animals, and other branches of a husbandman's care."

Thomas Jefferson, 1743–1826

Thomas Jefferson went farther than either Franklin or Washington in his devotion to science as a service to the nation. Jefferson had abiding faith in the common people, and his hope for democracy's success was founded upon the creed that the promotion of the general welfare depended upon advances in science and technology. Jefferson had a wider knowledge of the value of science in serving the welfare of the young nation than any other American of his time. In this belief, he anticipated the philosophy set forth a century and a half later in America's famous Point IV: namely, that democracy's hope rests upon a nation's ability to be self-sustaining.

To Jefferson, like Franklin and Washington, the study of science was personally appealing. He believed that nature had intended him "for the tranquil pursuits of science" and wrote, "science is my passion; politics is my duty." While Jefferson made no such basic contribution to science as Franklin did in his experiments in electricity, yet Jefferson's classical education led him into wider fields of investigation and, like Franklin, he believed that the discoveries of science should be applied to man's use.

Jefferson became interested in the science of agriculture at an early age. His agricultural creed was expressed in these words: "Those who labor in the earth are the chosen people of God. Cultivators of the earth are our most valuable citizens. They are the most vigorous, the most independent, the most virtuous; they are tied to their country, and wedded to its liberty and interests by the most lasting ties."

Agricultural Experiments. Jefferson's farms around Monticello became experimental stations. New crops, new methods of rotation, new and improved machinery were tried out year after year. He drew plans for a threshing machine, a fulling machine, and a spinning jenny. He designed a seeding device that would sow four rows at a time. He invented a hemp break which he attached to his threshing machine, powered by a single horse. It broke as much hemp in a day as ten men could break. In his methods of cultivation, he worked out several improvements. For hillsides, he recommended horizontal plowing, a system modern agronomists term "contour plowing." He established a definite rotation of crops extending over a seven-year period. From Du Pont de Nemours he secured some fine Merino sheep with which he improved his breed.

Scientific Publications on Virginia. Jefferson had first gained recognition as a scientific observer when his famous *Notes on the State of Virginia* (1781–1782) were published. The *Notes* contained a mass of scientific data which he had collected over a period of several years. He described in detail Virginia's physical features, its rivers, mountains, climate, its forests, plants, and mineral deposits. But he went further; he classified the plants according to their ornamental, medicinal, or edible values. He noted those indigenous to America and those brought in by the colonists.

In his observations upon the climate of the mid-Atlantic region, his statistics compiled over a period of five years (1772 to 1777) showed the amount of rainfall, the direction of prevailing winds at different seasons, wind velocity, varying temperatures, barometric pressure in different seasons, the amount of frost, and seasons of heaviest dews. During the five-year period he made a total of more than three thousand observations on wind directions, and then reduced his tables to averages. His work in this field placed Jefferson, the meteorologist, in a rank above that of the amateur, and placed the study of climatic geography on the level of exact science.

He made a careful study of the Appalachian Mountains, and their several peaks. He estimated the highest peak to be slightly more than 4,000 feet, which he observed was just about one-third high enough to preserve ice throughout the year. He noted the location of lead, copper, iron, coal, marble, and limestone.

Jefferson's Monetary System. Jefferson's scientific interests extended far beyond his studies in natural history. In 1784, at the request of Congress, Jefferson headed a committee to consider the question of coinage in relation to a monetary plan advanced by Robert Morris. Jefferson proposed a much simpler money unit. He drew up a plan making the dollar the unit for payment and its subdivisions a decimal ratio. He pointed out the ease with which multiplication and divisions could be made by using the ratio of ten and the decimal. He further proposed the minting of four coins; a gold piece equal in value to ten dollars, the dollar unit in silver, the tenth of a dollar in silver, the one-hundredth part of a dollar in copper. Jefferson's system of decimal coinage was adopted in 1785 and remains to this day a tribute to the mathematical soundness of his computations.

Scientific Investigations in Europe. On Jefferson's diplomatic mission to France (1784–1789) he was accepted not only as a diplomat, but also as a scientist. These were years of great stirring throughout Europe. Jefferson was an eyewitness to the birth of modern chemistry initiated by Lavoisier, to the early experiments in steam power in France, and in England to the improvements being made in agriculture, in stock breeding, and in canal transportation. Jefferson witnessed all these developments with the view of introducing them into America. He secured seeds and specimen plants for his friends back home. He made detailed sketches of numerous mechanical devices so that he could explain them to his scientific friends upon his return.

While abroad, Jefferson carried on extensive correspondence, not only with his political friends, but also with college presidents and men of science. Ezra Stiles, president of Yale, Joseph Willard, president of Harvard, Reverend James Madison of William and Mary, David Rittenhouse, Charles Thompson, and others, all received letters on scientific and educational matters. Jefferson urged President Willard to assign his college students to the study of natural history, mineralogy, and botany, and to develop these vast untouched fields. "We have spent the prime of our lives in procuring them the precious blessings of liberty," declared Jefferson, "let them spend theirs in showing that it [liberty] is the great parent of science and virtue; and that a nation will be great in both always in proportion as it is free."

On a tour of southern France and northern Italy Jefferson studied the crops, the implements of agriculture, the types of soil, the varieties of wine, the breeds of cattle, the wages of laborers, the prices of food, the climatic zones of olive and orange trees, the use of canals, and the construction of mills for cleaning rice. He sent to his Charleston friends some seeds of malta grass, acorns from the cork oak, and several species of olives. He attempted to send a supply of rice grains from the crops grown in northern Italy, but discovering there was a law against its exportation, he dropped a few pounds in his pockets which he succeeded in getting to his Charleston friends who introduced its culture there.

An improved rice mill in northern Italy, a corn husker and an improved rice beater in France, a Dutch wheelbarrow, a device whereby one horse worked three pumps at one time, and a screw bridge over a canal, were among the mechanical inventions Jefferson studied with a view to introducing them to America.

Inventions and Adaptations. Observing the clumsy plow used by peasant farmers in France, he sketched an improved mold board that would operate on a mathematical principle with a regular inversion to a fixed depth and with the least application of force. Jefferson's device could be carved out of wood by any farmer and fixed to his plow. He called it the "mold board of least resistance," and he regarded it as one of his most important agricultural inventions. He planned to cast this mold board in iron, but that was left for Jethro Wood and others.

For this invention the French Academy of Agriculture, in 1805, elected Jefferson a foreign associate and awarded him a gold medal. The Academy noted in its citation that, "America received the plow from Europe; it is admirable that America should give it back to Europe, its gift perfected." Models of the plow that Jefferson devised were placed in the French National Museum of Natural History and in the National Conservatory of Arts and Crafts in Paris.

In 1788 Jefferson learned of a pedometer, an instrument of delicate construction carried in the watch pocket or vest and attached by a tape to the knee band of the breeches, designed to compute distance accurately. He was so impressed with the usefulness of such an instrument that he sent one to James Madison. Because of its delicate mechanism he warned Madison never to turn the hands backward, but to note the number at which the hands stood when he began to walk and to compute the distance from that point.

Pioneer of Patent Protection. Returning to the United States, Jefferson served as Secretary of State in Washington's Cabinet, later as Vice-President, and finally as President of the United States. In all these positions he found many opportunities to continue his scientific studies. When the first patent law was passed, it was administered by a committee of which Secretary of State Jefferson was the head. In this capacity he made what is believed to be the earliest investigation of a chemical nature reported by an official of the federal government. He set forth his findings in a carefully prepared scientific paper, which he submitted at the request of the House of Representatives, on a memorial of one Jacob Isaak relating to a plan for converting salt water into fresh water. Jefferson proved that it was by a distillation process that had been known for years and not by the solution Isaak had proposed.[2]

Conflict of Scientist and Statesman. Dividing his energies between his passion for science and his duty to the state, Jefferson arrived in Philadelphia in 1797 to take the oath of office as vice-president, carrying with him a bundle of bones which he had collected from the regions "beyond the Blue Mountains of Virginia." They were the remains of the *Megalonyx,* or "Great Claw," which were later defined as a gigantic extinct edentate quadruped related to the sloths. Jefferson's *Megalonyx* was later placed in the Philosophical Society's museum. That same year, 1797, also marked Jefferson's election as president of the American Philosophical Society.

During the tense weeks of February, 1801, when Congress was trying to break the tie vote between Jefferson and Aaron Burr, and elect a president, Jefferson was carrying on correspondence with Dr. Caspar Wistar, professor of chemistry in the College of Philadelphia. He also wrote to William Dunbar about some recently discovered fossils, and corresponded with Dr. Williamson on climatic differences in various regions of the United States.

As President of the United States, Jefferson was forced to curtail his voluminous correspondence on scientific interests, although he managed to find a few spare moments. He wrote several letters to John Taylor describing an improved grain drill, and to M. Sylvestre of Paris on the merits of French and American sugar culture. While President of the United States he organized the nation's first scientific expedition and planned the famous Lewis and Clark westward march in minutest detail. He dispatched William Dunbar to the Red River region, and for his excellent work there, proposed him for membership in the Philosophical Society. He sent Zebulon M. Pike to explore the headwaters of the Mississippi and the Colorado. After serving two terms as president, Jefferson told his friends that he longed for the tranquillity of Monticello, where he spent the remaining years of his life.

Jefferson's Faith in Science. Jefferson's work in science was, above all else, directed toward utilitarian value. His contributions were practical. He promoted education in science in the University of Virginia, in the founding of which he greatly rejoiced. "Science," he declared, "is more important in a Republic than in any government . . . indeed, science is important to the preservation of our Republican Government and it is also essential to its Protection against foreign powers."

In a letter to Benjamin Waterhouse, March 3, 1818, Jefferson eloquently proclaimed his confidence in his country by declaring: "When I contemplate the immense advances in science and discoveries in the arts which have been made within the period of my life, I look forward with confidence to equal advances by the present generation, and have no doubt they will consequently be as much wiser, as we have been wiser than our fathers were, and they, than the witch burners." Jefferson died July 4, 1826.

Benjamin Rush, 1745–1813

Dr. Benjamin Rush, another of the "founding fathers in science," like Franklin and Jefferson, pioneered in many fields of learning. He was a distinguished physician, a great teacher, and a social reformer as well. Although Dr. Rush had many scientific interests, they were all directed toward one goal—the betterment of mankind.

Father of Experimental Medicine. In his chosen profession of medicine, his search for the cause and cure of diseases won for him the title, "Father of Experimental Medicine." He was appointed professor of The Theory and Practice of Medicine in the University of Pennsylvania, the first chair of its kind in America; head of the staff of the Pennsylvania Hospital, first in America; director of the Clinic of the Philadelphia Dispensary, first in America. He was the author of the earliest medical books in America. In Europe he was hailed as the "Sydenham of America." During the Revolutionary War, he served as surgeon-general for the armies of the Middle Division, and as surgeon in the Navy.

Soon after the close of the Revolutionary War Dr. Rush wrote that, while the "war" was over, "it is far from being the case with the 'Revolution'; only the first act of the great drama is closed." He sensed the need for bringing about another kind of revolution, a revolution in medical science and in the entire educational system of young America.[3]

Medical Pioneer. Rush was a century ahead of his time in diagnosing diseases. Noting that the North American Indian was not a victim of tuberculosis, he advised his tubercular patients to simulate the Indian by breathing fresh air, eating only vegetables, and taking cold baths. He pioneered in the science of sanitation, and fought for the purification of drinking water. His practice of inoculating soldiers against smallpox by using the newly discovered *kine pox* greatly reduced the mortality in the ranks of the Continental Army. He pioneered in investigating the effects of using alcohol and tobacco. He pronounced drunkenness a disease. He set forth his views in "An inquiry into the Effects of the Ardent Spirits upon the Human Body and Mind," and handed down advice as important for his country's welfare as any found in Washington's Farewell Address.[4]

Diversified Scientific Interests. Rush's interest in science was not confined to medicine alone. In 1769 he was elected to membership in the American Philosophical Society to which he contributed many scientific papers. One of these was a formula for making maple sugar and a plan for the establishment of a maple sugar industry which he believed could be developed to exceed in volume the cane sugar industry of the West Indies. He was the earliest advocate of the study of veterinary science in America, and in 1807 he recommended that such a course be introduced into the curriculum of the schools in the country. In the field of chemistry, he holds several "firsts": the first professor of chemistry in the United States, appointed in 1769, and the compiler of the first syllabus in a lecture series on chemistry, described as "the first textual contribution in chemistry from the hand of an American." [5]

In the field of education, Dr. Rush, like Franklin and Jefferson, laid great stress on the study of the mechanical arts. He discounted the study of Latin and the classics and recommended a thorough study of the English language, saying that, "above all others it will probably be spoken by more people in the course of two or three centuries than any one language at any one time since the creation of the world." [6]

Along with Franklin, Washington, and Jefferson, Dr. Rush helped to shape the American way of life by emphasizing the role of science and technology in developing the resources of this young nation. Following his death, April 19, 1813, Jefferson wrote, "another of our friends of 'seventy-six' is gone . . . , another of the co-signers of the independence of our country. And a better man than Rush could not have left us, more benevolent, more learned or a finer genius. . . ." [7]

David Rittenhouse, 1732–1796

David Rittenhouse, born 1732 (the same year as Washington), was descended from a line of forebears known for their accomplishments in the mechanical and technical arts, especially in printing and papermaking. David had the ill fortune of having a delicate constitution, but the good fortune of acquiring his uncle's chest of tools and books. He started early to tinker with tools and took up clock making, and his excellent workmanship soon won him recognition.

Rittenhouse's chief interest was in the field of mathematics and astronomy. Early in life he mastered Newton's *Principia*. He rendered practical service by surveying the boundary between Pennsylvania and Maryland. This was so accurate that it was later accepted by Mason and Dixon, surveyors who established the historic Mason-Dixon line. Rittenhouse also surveyed boundaries for five additional states. Interested in the worldwide discussion regarding the transit of Venus in 1769, he constructed his own telescope, believed to be the first made in America, and made careful observations. With this data he made exact calculations, and reached findings described as among the best that were made.

Astronomer, Orrery. While still in his thirties Rittenhouse set out to construct an orrery (planetarium) which was to be the most accurate reproduction of the solar system possible. It was the only instrument of its kind in the colonies and was pronounced the finest in existence. When he completed it (1770), he sold it to the College of New Jersey in Princeton. The same year he constructed another one for the College of Philadelphia. The facsimile of this orrery was later incorporated in the official seal of Pennsylvania.

Military Technology. During the Revolutionary War, he served as engineer for the Committee of Safety. When the armies needed lead he went about Philadelphia collecting lead clock weights, and persuaded the owners to use iron weights instead. As noted earlier, Rittenhouse had supervised the casting of cannon and the manufacture of saltpeter. He was an adviser to gunmakers, and designed a chain to protect harbors from the landing of enemy ships.[8]

Inventions. Rittenhouse's work in astronomy led to the observation of the solar and lunar eclipse, the transit of Mercury and Uranus, and to the discovery of a comet in 1793. From all these observations he developed a method of determining true time. He computed the solar parallax, and made astronomical calculations for the publishers of almanacs. He kept careful meteorological records over a period of several years. In making his barometer he used

Stiegel (American) glass, which he declared was as fine as any imported glass. He fashioned a reading glass and a pair of spectacles which he gave to President Washington. He invented a collimator tube for his telescope that enabled him to fix a meridian mark without going great distances in the field. He invented a metallic thermometer and experimented with the action of heat on steel springs. He devised an astronomical clock and experimented with the variations of heat and cold on the oscillations of the pendulums.

Recognition as a Scientist. Rittenhouse's discoveries in his many experiments were published in the *Transactions of the American Philosophical Society.* He became president of this society after Franklin's death. He was elected to membership in the Royal Society of London, and the American Academy of Arts and Sciences. In 1792, President Washington appointed him first director of the United States Mint.

Following his death in 1796, Jefferson wrote: "As an artist he has exhibited as great proof of mechanical genius as the world has ever produced. He has not made a world, but he has approached nearer its Maker than any man from the creation to this day."

Scope of Early Scientific Progress

The founding fathers in science pioneered in virtually every field of learning—agriculture, archeology, biology, chemistry, medicine, engineering, electricity, mathematics, and astronomy. Their discoveries and inventions fostered a spirit of science in the new nation. They believed with William Smith, provost of the University of Pennsylvania, who wrote in 1790: "The man who will point out a new and profitable article of agriculture and commerce will deserve more from his fellow citizens and from heaven than all the Latin and Greek scholars that ever existed in any age or country." [9]

Notes for Chapter 9

1. Everett E. Edwards, "Washington, Jefferson, and Lincoln and Agriculture," *Agricultural History Series,* No. 7 (1943), p. 35.
2. *American State Papers,* XXXVII (Nov. 24, 1791), 24, 44.
3. See Nathan Goodman, *Benjamin Rush, Physician and Citizen* (1934).
4. See Yandell Henderson, *A New Deal in Liquor* (1934).
5. Edgar Fahs Smith, *Chemistry in America* [*Old Philadelphia*] (1914), p. 77.
6. Harry G. Good, *Benjamin Rush and His Services to American Education* (1918), p. 176.
7. Quoted in Goodman, *op. cit.,* p. 349.
8. *Dictionary of American Biography,* XV, 630-632.
9. Quoted from Whitfield J. Bell, "The Scientific Environment of Philadelphia; 1775–1790," *Proceedings,* American Philosophical Society, XCII (1948), 10-11.

Works relating to Franklin's scientific interests are extensive. The following are especially helpful: John Bigelow, editor, *The Complete Works of Benjamin Franklin,* 10 vols. (1893); Albert Henry Smyth, Jr., *The Amazing Benjamin Franklin* (1929); James Parton, *Life and Times of Benjamin Franklin* (1897); Paul Leicester Ford, *Franklin Bibliography: A List of Books Written by or Relating to Benjamin Franklin* (1894); *Benjamin Franklin's Experiments,* edited by I. Bernard Cohen (1941); Nathan C. Goodman, editor, *The Ingenious Dr. Franklin* (1931); Carl Van Doren, *Benjamin Franklin* (1938); James G. Crowther's *Famous American Men of Science* (1937).

Following is a partial list of works relating to Washington's scientific interests: W. E. Brooks, editor, *Agricultural Papers of George Washington* (1919); Moncure Daniel Conway, editor, *George Washington and Mount Vernon: A Collection of Washington's Unpublished Agricultural and Personal Papers* (1889); Douglas Southall Freeman, *George Washington: A Biography,* six completed volumes 1948-1954; Paul L. Haworth, *George Washington, Country Gentleman* (1925); Joseph Meredith Toner, *George Washington as an Inventor and Promoter of the Useful Arts* (1892).

For Jefferson's many scientific interests see: Andrew A. Lipscomb, *Writings of Thomas Jefferson,* 20 vols. (1905); Paul Leicester Ford, *Writings of Thomas Jefferson,* 10 vols., 1892–1899; and most valuable of all, *The Papers of Thomas Jefferson,* currently appearing volume by volume, under the editorship of Julian P. Boyd. See also Thomas Jefferson's *Farm Book with Commentary and Relevant Extracts from Other Writings,* edited by Edwin Morris Betts, (1953).

PART II

LAYING THE FOUNDATIONS: from the American Revolution to the Civil War

During the seventy-year period from 1791 to 1861 (outbreak of the Civil War) young America astonished the world by her rapid growth. Territorially she expanded to the Pacific Coast. In manufacturing, her mills and factories turned out products that challenged the output of the Old World's seasoned industries. In technology, her inventive genius was developing improved tools, techniques, and new machines. In agriculture, her reapers and plows were acclaimed the world over.

In transportation, she built up the greatest system of river navigation on earth; her clipper ships were the envy of seamen throughout the world. On land, her network of railroads exceeded the combined mileage of all other countries. In communication technology, she gave the world the electromagnetic telegraph system and extended it over more miles of territory than all the lines of Great Britain and the Continent combined.

In the field of science, national associations were organized to exchange ideas, to pool the discoveries of scientists, and to place them at the government's disposal.

Charles Goodyear had vulcanized rubber, and laid the foundation for an important new industry. Eli Whitney and Samuel Colt put into practice the system of interchangeable parts and thus paved the way for America's unique contribution to world technology—mass production.

It took a Civil War to prove that the side which possessed a superiority in scientific, technological, and industrial power would ultimately triumph.

Chapter 10

AN AGE OF AUSPICIOUS BEGINNINGS—1789–1815

The early days of the American Republic were testing days, testing whether this young nation could endure. The Thirteen Colonies had won their political independence, had adopted a constitution, and had established a new republic. However, there were many pitfalls into which the nation could stumble. These were both political and economic. They emanated from conditions both at home and abroad.

New Theories, Political and Economic

The French Revolution and Napoleonic Wars threatened not only western Europe but also the future of young America. In France new political theories and new technological experiments were being tested. But while the forces of reaction on one side, and the forces of the extremists on the other side, were delaying the acceptance of new ideas in the Old World, this was not true in America. Here was a virgin country that offered auspicious beginnings for a new way of life—"liberty, fraternity, and equality." Our government had been founded upon principles which up to that time had reposed only in the minds of philosophers. Its future would be determined not only by political theorists, but also by practical men who believed that by the promotion of science, theories could be transformed into realities.

The scientific and technological developments of those first years of the young republic mark this as a period of auspicious beginnings for generations of Americans. The very forces that threatened our political success strengthened the scientific and technological. The riots and unrest in the eastern colonies caused thousands of settlers to start moving west, carrying their skills and techniques with them. The stoppage of foreign trade forced ingenious Americans to invent new tools and new devices of their own. The curtailment of maritime enterprise diverted what capital there was into channels of industry.

Important Developments. In the midst of these economic threats there were a number of scientific and technological developments that helped to strengthen the young nation. Oliver Evans of Philadelphia, and Camden, New Jersey, was constructing his Red Clay mill which was to revolutionize the gristmill industry and to benefit the entire country. While the "Founding Fathers" were meeting in Philadelphia in the summer of 1787 to write a new constitution and form a more perfect union, John Fitch took a number of the delegates on their first steamboat ride up the Delaware, and thus clinched the argument that Congress should be given power to regulate commerce between the states. When President Washington and his political advisers gathered in New York (temporary capitol) to inaugurate the new government, Samuel Slater in nearby Rhode Island was setting up the first factory system in America.

Numerous forces combined to bring about this era of auspicious beginnings: an abundance of natural resources, a people trained for more than a century to depend upon their own ingenuity and inventiveness, and a new government, eager to encourage the useful arts and industries. Tench Coxe aptly expressed the spirit of the times in 1790 when he wrote, "We should harness the force of nature to the Chariot of human progress."

President Washington gave official sanction to the movement in his first annual message to Congress, January 8, 1790, when he said:

> I cannot forbear intimating to you the expediency of giving effectual encouragement to the introduction of new and useful inventions from abroad, and to the exertions of skill and genius in producing them at home. . . . Nor am I less persuaded that there is nothing which can better deserve your patronage than the *promotion of science*. . . .[1]

First Patent Laws

Washington's message was cordially received by both Senate and House, and both promised to give it "early and respectful attention." That the First Congress was eager to "encourage new and useful inventions" is seen in the passage of the first federal patent law in 1790.

Patent Petition. The idea of the government giving aid to new inventions was not new. Long before the federal Constitution was adopted, a number of states had granted patents, and when the federal Constitution was drafted in 1787 this right was delegated to Congress. Scarcely had the new government been established when petitions for patents began to pour in requesting exclusive rights for special inventions. As early as May 4, 1789, Alexander Lewis of Pennsylvania applied for a patent on his method for "impelling boats of 25 tons." John Fitch wanted a patent for making nails. John McPherson applied for a new type of lightning rod. Still others were petitioning for patents on other new devices.

Patent Act of 1790. To meet these demands the first Patent Act was passed, and signed by President Washington, April 10, 1790. The Act was short, simple, and easy to administer. "Any useful art, manufacture, engine, machine or device or any improvement thereon not before known or used" was patentable. The Act was to be administered by a special committee of which the secretary of state was to be a member. Thomas Jefferson, serving in that position, became administrator of the first Patent Act. He had great faith in the inventive ability of his fellow Americans and became a strong defender of the rights of the patentee. He declared that, "an inventor ought to be allowed the right to the benefit of his inventions for a certain time . . . , nobody wishes more than I do that ingenuity should receive liberal encouragement." Later in life Jefferson wrote that the Patent Act (1790) had given a "spring to new inventions beyond my conception. . . ."

As the applications for patents on numerous inventions and discoveries began to pour in, Jefferson and his committee carefully examined every petition for a patent. The first one granted was to Samuel Hopkins of Vermont for "a new method of making pot and pearl ashes." The second was for a special process of making candles. The third, granted to Oliver Evans, was for new improved flour milling machinery. One inventive genius, Samuel Milliken, was granted four patents in one day.

Patent Revision. The rush for patents became so heavy that the committee found it impossible to handle all the petitions. A second bill, passed in 1793, provided that patents could be granted upon the oath of the applicant that his device or invention was original and did not infringe upon the existing patents. While the aim of the act was to simplify proceeding, it resulted in crowding the courts with patent suits. The 1793 Act, with some amendments, remained in force until 1836.

Encouragement to Inventors

The interest of the new government in encouraging "the introduction of new and useful inventions" from home and abroad during this period of auspicious beginnings was again proven when

Alexander Hamilton, Secretary of the Treasury, upon the request of the House of Representatives was asked to compile a report on the nation's manufactures.

Hamilton's Report on Manufactures, 1791. Hamilton had spent almost two years in compiling the report that has come to be regarded as one of the most important documents in American history. It went beyond a mere statistical compilation of the nation's manufactures and products. The introduction of new machines in manufacturing, he declared, would increase production, which in turn would increase employment. Increased employment would ensure a steady demand for the surplus products of the soil. Our nation would never be prosperous said Hamilton, until there was great diversity of labor and a high degree of skills developed. He believed that the labor shortage in this country could be solved, for, "there is in the genius of our people a peculiar aptitude for mechanical improvement."

Demand for Artisans and Mechanics. Hamilton wanted to encourage immigrants to come to America to help supply the labor for our factories and shops. Even though they might be lured away to farms later, industry would benefit by reason of their skill while employed within its ranks. On this point Hamilton and Jefferson reached their one (and perhaps their only) agreement. Jefferson was never sure that immigration would solve the problem of the scarcity of labor, but he did allow one exception—he welcomed *useful* artificers. "Spare no expense in obtaining them," he wrote. "They will after a while go to the plow and the hoe; but in the meantime they will teach us something we do not know."

Supply and Demand. Hamilton noted that in the matter of household manufactures there was not only a sufficient supply for the families of America, but also enough goods for export. A number of industries had developed to the point where they were carrying on an extensive trade. Among these were the manufacture of nails, rods, implements of husbandry, stoves, pots, kettles, anchors, weights, and scale beams. Boots, shoes, harness, saddlery, trunks, portmanteaus, leather breeches, and gloves were being produced in large quantities. In the production of articles made from wood, a growing trade in cabinet making and in shipbuilding had developed. Advances had occurred in the manufacture of hemp and flax, of tiles, bricks, and potter's ware; of writing and printing paper, refined sugars, hats, copper, brass, tinware, and gunpowder.

Hamilton's *Report* helped to inspire a feeling of confidence. A new era was about to dawn. Steam power was just being introduced. New machinery was coming into use. The new government was showing a benevolent attitude, and a lusty population was eager for work. Joseph Priestley, chemist, who arrived from England, wrote in 1794, that this republican government, by "encouraging all kinds of talents is far more favorable toward the sciences and the arts than any monarchial government has ever been." "A free people," he declared, "will in due time produce anything useful to mankind."

Agricultural Developments

Agriculture was given impetus as the new era dawned. Inspired by the reports of the agricultural societies in England, a few leaders proceeded to organize similar societies here in America.

Societies and Publications. Among the first groups was the Philadelphia Society for the Promotion of Agriculture, 1785; another was organized in Massachusetts in 1792. Articles on farming began to appear in pamphlets, periodicals, and almanacs: *Poor Will's Almanac,* 1787; Squibb's, *The Gardener's Calendar,* 1787; McMahan's *Gardener's Calendar,* 1806, (this latter publication continued on down to 1857); and Silas Dean's book, *The New England Farmer,* 1790.

Agricultural Fairs. An important agency for disseminating advances in agricultural technology was the local fair. Organized by the farmers themselves, these fairs rendered practical service. Elkanah Watson of Pittsfield, Massachusetts, often called the father of agricultural fairs, sponsored one of the large exhibitions in 1807, and in 1810 he organized the Berkshire Agricultural Society. At these fairs farmers exhibited their prize livestock, poultry, and the best yields of their fields. The farmers' wives exhibited homemade butter, home-grown fruits and vegetables, and samples of needlework such as quilts, coverlets, and home-spun woolen goods. These early fairs served a twofold purpose: they developed a spirit of competition on the part of farmers and they were the first agricultural agency to offer mass education to the farmers.

New Agricultural Implements

American farmers began early to discard the heavy, clumsy English tools and implements and, with the aid of local blacksmiths, designed their own. They demanded implements, whether wooden or metal, that were simple to construct and easy to operate.

Variations of the Plow. Charles Newbold of Philadelphia invented a cast-iron plow in which the moldboard, share and landside were cast in one piece. Thomas Jefferson and some of his gentlemen farmer friends became interested in Newbold's plow, but most of the farmers were afraid to try it for fear that iron might poison the soil. Another cast-iron plow made by Robert Smith and his brother, Joseph, of Bucks County, Pennsylvania, aided by Charles Newbold, incorporated Jefferson's principles of moldboard construction and was patented in 1800. The Smith plow was used in Pennsylvania and New Jersey for the next half century.[2]

Jethro Wood patented a plow in 1819 of which every part (the point, moldboard, and landside) was replaceable. Here was the forerunner of the system of interchangeable parts in agricultural implements. Wood's improvement on the moldboard made possible the pulverizing of the upturned sod and reduced friction. His plow was easier to guide, and with it one man could plow an acre a day. But the prejudice against the use of metal moldboards was so strong that Wood was forced to give away his plows. He died in poverty.

An improved hay rake, introduced around 1800, was constructed on the principle of a huge comb, 10 or 12 feet wide, and fitted with fifteen to twenty teeth. Handles similar to plow handles were attached to the frame and, while the horse dragged it across the field, the farmer held the row of teeth firmly in place and gathered the hay into small windrows. This was the forerunner of the wheeled hay rake that came later.

Livestock Breeding

Improved livestock breeding was stimulated by the patronage of the gentleman farmer class. There was need for it, since livestock were poorly bred, poorly fed, and received little care. They roamed at large and often during the winter months had to shift for themselves. Razor-back hogs, scrubby cattle, and scrawny sheep with poor fleece resulted. The British, meanwhile, had greatly improved the grade of their livestock by selective breeding. Robert Bakewell of England, pioneer leader in animal husbandry, sent over a few well-bred cattle which were purchased by some gentlemen farmers in Philadelphia and New Jersey. Shortly after the Revolution, a number of the Shorthorn cattle were imported to stock some farms in Maryland and Virginia. Purebred Holsteins, "Dutch Cattle," were introduced into Pennsylvania in 1794 by the Holland Land Company.

Cattle. An important step was taken in 1809 when the "Pennsylvania Society for Improving the Breed of Cattle" was organized in Philadelphia. Premiums of one hundred dollars were offered for prize bulls of foreign breed, another hundred dollars for milk cows of a superior quality. Weight came to be a standard of excellence in breed. In 1811 a hundred dollar prize was awarded to a Mr. Carpenter whose ox weighed 1668 pounds. Other records were published of animals weighing from 2200 to 2500 pounds.[3]

Sheep. Notable advances were made in sheep husbandry during this period. The Napoleonic Wars in Europe and our own embargo and nonintercourse acts led to numerous attempts to increase both the quality and quantity of wool production. American sheep growers had for years been attempting, unsuccessfully, to import some of the prize wool-growing Merino sheep from Spain. They hailed the small beginnings made when DuPont de Nemours finally succeeded in smuggling out of Spain a full blooded Merino ram, "Don Pedro," and introduced this prize breed into America.

Following the invasion of Spain by Napoleon's armies, 1809–1810, Spain permitted wholesale exportation of sheep to the United States, and a veritable "Merino Craze" spread throughout the eastern states. Some twenty thousand Merino sheep were shipped from Lisbon within the next two years, and according to Fletcher, they "created a furor." Merino wool jumped to seventy-five cents a pound, then to three dollars a pound, whereas common wool was selling for less than forty cents. Merino rams were reported selling for one thousand to fifteen hundred dollars each.[4] But the Merino craze soon passed. No sheep of any breed was worth such fantastic prices. A number of sheep breeders lost small fortunes.

Meantime, wealthy farmers were upbreeding their own native stock. The offer of special premiums for superior wool caused farmers to improve their breed. At a George Washington Parks Custis annual sheep-shearing contest in Arlington, Virginia (1811), first prize for the longest staple wool and for the largest amount of fleece from one animal went to John Taylor whose sheep yielded 8¼ pounds of fleece of finest fiber. A second prize went to another sheep raiser who exhibited staple wool 12 inches in length.

Industrial Growth

The most significant technological development during this period of auspicious beginnings was the introduction of the factory system by Samuel Slater in Providence, Rhode Island, 1790. The factory system had been building up in England during the Industrial Revolution, where the most significant developments had occurred in textile technology. The invention of the flying shuttle and Hargreaves spinning jenny, the application of rollers for spinning, Arkwright's water frame, Crompton's spinning "mule," and Cartwright's power loom had revolutionized the textile industry in England. Parliament, however, had prohibited the exportation of any of these inventions to America, and even forbade the migration of experienced mechanics who had worked in English factories.

Beginning of the Factory System. Samuel Slater, who had served as an apprentice in an English cotton-spinning plant, upon hearing of the bounties and rewards offered by American legislative bodies to experienced mechanics, migrated to America. He arrived in September, 1789, with no drawings or machinery, but with his head crammed with plans and designs of textile mills. He had been advised by the ship's captain to seek the acquaintance of Moses Brown of Rhode Island, owner of a cotton-spinning establishment. Brown engaged Slater to reproduce in every detail Arkwright's cotton machinery. Within a year Slater had erected a mill of three 18-inch carding machines, draw heads with two rolls, roving cases and winders, and a thistle spinning frame with seventy-two spindles. Four days before Christmas, 1790, power (first hand power, then water power) was applied, and the machine began to operate. Hamilton in his *Report* referred to Slater as, "The first to introduce into the United States the celebrated cotton mill."

Production of American-made cotton goods was limited, however, by the small amount of available raw cotton. But this need was soon met by a new technological development destined to change the entire course of American history. Eli Whitney's invention of the cotton gin stands as one of the major inventions of all times.

Whitney's Cotton Gin, 1793. Before the cotton gin was invented, cotton fabrics were luxury articles. Whitney's invention, the model of which was completed within two weeks after he first conceived the idea, brought cotton within the reach of everyone. The invention was simple. It consisted of a revolving cylinder with horizontal rows of teeth, brushes, and a revolving fan to produce a current of air to throw the cleaned cotton away from the rows and bushes, all essential to the modern gin. Whitney's gin solved the difficulty of removing the seeds from the cotton boll by hand, the most tedious and time consuming operation in the processing of raw cotton.

Cotton soon became the poor man's cloth. Of the four first known staples that provided man's clothing (cotton, silk, wool, and flax), cotton, which came to be the cheapest and most indispensable, was the last to come into general use. It was Whitney's cotton gin that made it available to all. It transformed the entire economy of the South. It led to the development of large plantations, to a rapid increase in slavery, to a one-crop system that continued until cotton became "King," and to a Civil War. Lord Macaulay's oft-quoted remark, "What Peter the Great did to make Russia dominant, Eli

Whitney's invention of the cotton gin has more than equaled in the relation to the power and progress of the United States," is not an exaggeration.

Textile Expansion

Increased production of cotton throughout the South resulted in the erection of more cotton factories in the North, and to improvements in textile technology. Improvements in carding and spinning machinery followed. The manufacture of machinery for textile mills became a thriving business.

Machinery for Cotton Production. In 1804 a factory in Pittsburgh was building carding machines consisting of twenty to thirty-two cards, drawing frames that would draw 20 to 25 pounds of cotton daily, roving frames with a daily capacity of 20 to 25 pounds, water-spinning frames of eighty-four spindles, spinning mules with sixty to two hundred spindles, twisting frames to make yarn stockings, and looms with flying shuttles.[5]

Advances in Woolen Technology. Meanwhile, the increasing demands for woolen clothing led to numerous experiments in factory production. The two Schofields, John and Arthur, experienced woolen manufacturers in Yorkshire, England, arrived in the United States in 1793 and established a factory at Byfield, Massachusetts. The factory was equipped with carding machines driven by water power, described as the first practical carding machines used in making woolen cloth in the United States.

The Schofield wool-carding machines were introduced into various parts of the country, and woolen factories followed in their trail. The census of 1810 reported 1,776 wool-carding machines in operation, mostly in New England, although one was reported as far west at Lexington, Kentucky. Gallatin's *Report* of 1810 stated that "water driven carding machines are in use everywhere."

Other improvements in textile woolen technology included napping and shearing machines invented during the 1790's. A machine invented by Walter Burt, 1797, raised the nap on cloth, while improvements in a cloth-shearing machine, operating with an oscillating motion, greatly speeded up the production of woolen cloth. One workman could operate two to four shearing machines driven by water power and could do as much work in one day as a hand shearer could do in three weeks. With these technological improvements woolen factories sprang up in the New England states, in New York, New Jersey, Pennsylvania, Delaware, and Maryland. The *Census* of 1810, though not entirely accurate, reported twenty-four woolen mills in the country. By 1816, power-driven machines were in general use. A congressional committee reported in 1816 that twelve million dollars had been invested in buildings and machinery engaged in the manufacture of woolens, and that fifty thousand persons were employed in this rapidly growing industry. From that date, factory-produced woolens began to exceed homespun goods.

Tanning Technology

Notable advances in tanning technology occurred during this age of auspicious beginnings. New techniques and the discovery of special tanning liquors for improving the quality of the leather speeded up the process. Farmers, instead of preparing hides and skins at home, took them to experienced tanners and exchanged them for ready-made shoes and leather goods. Shoemaking became a shop industry instead of being carried on by itinerant cobblers.

Shoe Shops. As settlements grew, the master cobbler set up a little shop where he made shoes on order. These shops, called "tenfooters" because of their size, included four to eight benches for workmen, some of whom were apprentices "bound out" to learn the trade. By 1800 the countryside was dotted with these little shops, which mark the first definite effort to put shoe production on a factory basis. When orders were slack, shoes were made to be sold on the streets on market days.

Leather Processing. Other technological advances in the manufacture of shoes occurred shortly after 1800. Samuel Parker in 1809 invented a machine for splitting leather. Before this time the leather was shaved very much as a carpenter shaves a board, a wasteful process compared with splitting the leather. The new splitting machine doubled the amount of leather that could be used. About the same time David Randolph devised a method for fastening the uppers to the soles by using brads which were clinched against metal lasts. In 1810, M. I. Brunel invented a machine that enabled the operator to use short nails to fasten the uppers to the soles, and a few years later a shoe-pegging machine was invented. Prior to these inventions, shoes were either handsewn or pegged with hand-driven wooden pegs. In 1815, Thomas Blanchard, famous toolmaker, introduced the use of the lathe in the leather industry.

Grain Processing

The forerunner of the modern assembly line appeared during this age of auspicious beginnings. Before the close of the eighteenth century, Oliver Evans (1755–1819) had designed and built an automatic gristmill. Grinding grain, and processing flour and meal, had for thousands of years been a slow, tedious task. But Evans devised a mechanical system that carried the grain through all operations—from the farmer's wagon to the manufactured flour—without the use of manual labor. Before Evans' invention, power had been used only to operate the millstone.[6]

A few of the principles in Evans' milling process had been tried out before, but he combined them into one continuous operation in milling grain. His grain elevator consisted of an endless band, to which were attached small wooden (later metal) buckets, capable of elevating 300 bushels of grain or meal an hour. This did the work of two men. A conveyor carried the grain along in troughs, or pushed it along a slightly inclined trough to chutes. The device was an endless band with wooden rakes and pulleys revolving as in an elevator. The descender, which ran by gravity, consisted of a broad strip of leather, canvas, or flannel running over two pulleys. The "hopper boy," so named because it took the place of a boy, was a large revolving rake which spread the meal evenly on the upper floor of the mill (the meal loft) and guided it to the chute over the bolting hopper, which passed it into the bolting chest.[7] These inventions, and the manner in which they were applied by Oliver Evans, won for him the title "Father of Mechanized Flour Milling."

Evans made a number of improvements in manufacturing millstones. Before this time, stones built from imported French burr (buhr) were considered superior to any made in America. They consisted of small stones dressed and cemented with plaster of Paris and bound with iron hoops around the edges. They were expensive and were not always perfect. In 1796, Evans secured a patent for making millstones. By properly balancing his stones, and by carefully carving out furrows and grooves, he was able to grind and crush grain to any desired size. America no longer depended upon imported millstones.

Introduction of Steam Power

Meantime, the nation's expanding industries were demanding more power. A new source of power, other than animal and water power, had to be provided: the development of steam power was the answer. When Thomas Newcomen developed his atmospheric engine in the early eighteenth century, the ground was laid for all subsequent steam engines. The separate condenser developed by James Watt in 1765, with improvements, inaugurated the rotative type, low-pressure steam engine. But the demands of industry set other inventors to work, among whom was Oliver Evans, a pioneer in developing the reciprocating engine.

Evans' Steam Engine. Evans' high-pressure engine was a new departure in that it dispensed with the vacuum and worked solely by the pressure of steam above the atmosphere, into which the steam was discharged "after it had done its office." Evans' engine, constructed in 1804, was of the double-acting, vertical type, with a speed of 30 rpm. With a 15-foot copper-shell boiler constructed to withstand a pressure of 50 pounds per square inch and fitted at each end with cast-iron headers, it compared favorably with the Bolton and Watt type. It vastly extended the range of industry that could be served by steam power. With this engine, Evans was able to break and grind 300 bushels of plaster of Paris in 24 hours. He used it to saw stone in Market Street, Philadelphia, "where the driving of twelve saws in heavy frames, sawing at the rate of 100 feet of marble in twelve hours, made a great show."

Evans was the first to attempt steam navigation on the lower Mississippi River. In 1802 he shipped an engine and boiler by water to New Orleans, intending to use it for propelling a small boat. But a disaster caused by a heavy flood floated the craft and engine a half-mile inland, and when the waters subsided the engine was stranded. It was then rigged up to run a sawmill and was soon sawing 3,000 feet of lumber daily. A demand for steam-driven sawmills followed. Steam-driven saws in turn led to the introduction of the circular saw, following the close of the War of 1812.

Steam-Operated Machinery. In 1803 Evans was operating his machine shop in Philadelphia by steam power. The following year he put his new high-pressure steam engine on the market, declaring that it "would commence a new era in the history of the *STEAM ENGINE.*" In 1808 Evans ventured westward to Pittsburgh where he built a steam flour mill that excelled the water mills in the western country. Two pairs of stones driven by steam power ground 10 bushels of rye or 9 bushels of wheat in one hour. The erection of this steam flour mill was hailed by the local press as ushering in a new era for the wheat and rye farmers of the west.[8]

By 1811 Evans was building steam engines of 70 horsepower. Three years later steam power was used to operate printing presses in New York City.

Beginning of Mass Production

This period witnessed the introduction of another major development which became the very essence of American technology, namely, the application of the principle of interchangeable parts. Although Jethro Wood is credited with the origin of this idea, Eli Whitney was the first to employ the system in the United States when he began to manufacture firearms for the federal government in 1798.

Whitney's "Jigs." Whitney made a metal mold, which he called a "jig," for each part of the gun. The worker was taught to file pieces of metal until each part was identical in shape. Each part in each gun was a duplicate of the similar part in all the other guns. Whitney and his men assembled the separate parts into finished guns, and when completed, the guns were pronounced satisfactory by the war department experts.[9]

Increased Production. This system eliminated the need for highly skilled gunsmiths. A simple worker in metals could be taught to file the desired part according to the pattern. Precision gauges were worked out to determine exact measurements, and the manufacture of guns developed rapidly. The 1810 census showed one hundred and forty establishments producing more than forty thousand firearms. By 1820 the system had spread to the other armories in the country.[10]

Whitney later applied this system of interchangeable parts to clockmaking. Eli Terry, an early clock manufacturer, set up a plant for the manufacture of wooden clocks at Windsor, and later at Plymouth, Connecticut, where he began to turn them out by the thousands. Mass production reduced the cost and made clocks available in greater numbers. They were bought up as rapily as they could be produced. The names of three workmen in his plant later became synonymous with clockmaking in America: Seth Thomas, Joseph Ives, and Chauncey Jerome, whose clock works survive until this day.

The Steamboat Era

The steamboat, symbol of American river transportation, made its appearance on inland waters during this age of auspicious beginnings. America needed the steamboat for transportation more than any other nation. Her vast inland waterways had afforded the American settler his cheapest method of transportation during the keelboat and flatboat age. But such crude, clumsy craft could not keep pace with the young nation's expanding industrial development. A steamboat that could be propelled upstream as well as downstream was the answer.

First Steam-Propelled Craft. Like many other inventions, the steamboat was the product of many minds. As early as 1784 James Rumsey of Virginia built a small boat driven by jet action, powered by a steam pump which drew water in at the bow and expelled it at the stern. After making a number of trial runs, Rumsey went to England where he sought aid for his new invention. He died there, but his friends succeeded in introducing his boat on the Thames River where it attained a speed of 4 miles an hour.

John Fitch has been called the first successful builder of a steamboat in America. His idea was to propel a row boat with oars operated by steam power. In 1786 he built a small craft large enough to hold two men, equipped with six side paddles and propelling machinery. The engine had a double-acting brass cylinder, 3 inches in diameter, which propelled the boat with ease against the current. In 1787 Fitch built a boat 40 feet in length, which carried sixty passengers, and in 1788 he built "The Thornton," driven by three paddles placed at the stern, which sped along at 8 miles an hour. By 1790 he had established regular steamboat passenger service on the Delaware River.

In 1797 Fitch was in Bardstown, Kentucky, where he built a paddle-wheel model with a steeple engine, but he was never able to obtain sufficient financial backing either in the United States or in France, which he visited in 1793. He returned to the United States and died in Bardstown, Kentucky, 1798, a disheartened, disillusioned genius, "at war with the world, but loving it, and verging on monomania for steamboats." [11]

John Stevens was another pioneer steamboat builder. He and his brilliant son Robert L. Stevens contributed greatly to American transportation technology both on water and land. In 1802 the senior Stevens built the "Little Juliana," a steamboat equipped with a high-pressure multitubular boiler, which was driven by twin-screw propellers. In 1806 he started building a 100 foot steamboat, the "Phoenix." When completed it was taken by sea from New York to Philadelphia and, "established for itself the record of being the first sea-going steamboat in the world." [12]

Fulton's Steamboat. Next on the list of pioneer steamboat inventors comes Robert Fulton. Born in Lancaster County, Pennsylvania, in 1765, he early showed traits of genius as an artist, engineer, gunsmith, and inventor. When only fourteen years of age he devised a mechanism for propelling a boat by paddle wheels. He went abroad in 1786, where he remained for twenty years supporting himself by his paintings. All this time he was observing engineering projects and devising mechanical equipment of his own. He took out British patents on devices relating to canal and water transportation. In 1797 he crossed over to France, where again he obtained a number of patents. He developed a submarine diving boat called the "Nautilus," and tried in vain to get Napoleon to add it to his navy.

In 1806 Fulton returned to New York where, with the aid of Charles Brown, a shipbuilder, he built the "Clermont." Equipped with a Boulton and Watt engine (which he purchased while in England), two side paddles, and "with various additions of bell cranks and flywheels," he started his boat on a successful voyage up the Hudson River to Albany, August 17, 1807. This is the date usually given as marking the beginning of steamboat navigation on American rivers, although as Gilfillan points out, some thirty steamboats had already been built, three of which had been put into practical use before Fulton's "Clermont."

In 1811, four years after the launching of the "Clermont" on the Hudson, Fulton, Robert Livingston (who had been instrumental in the Louisiana Purchase), and Nicholas Roosevelt built and launched the "New Orleans" on the Ohio River at Pittsburgh. This was the first steamboat on western waters. The occasion was acclaimed as a great event, "a boat moving without appearance of sail, oar, pole, or any manual labor—moving within the secrets of her own mechanism and propelled by power undiscoverable." Commanded by Captain Nicholas Roosevelt, it made the trip from Pittsburgh to New Orleans and marked the beginning of inland river transportation in America.

The appearance of this steamboat on inland rivers led the pioneer boat builder to envision a great future for this new technological achievement. Robert Fulton believed that the introduction of steam navigation throughout the entire world was only a few years away. In a moment of enthusiasm he wrote to Jefferson informing him that within four or five years he would have a line of steamboats operating on the St. Lawrence, the Hudson, the Delaware, the Ohio, and the Mississippi Rivers.[13] The same year that Fulton and his partners launched the "New Orleans," Fulton was attempting to secure from the Russian Czar the exclusive privilege of operating steamboats between St. Petersburg and Kronstadt. The following year he entered into an agreement with a friend in England to introduce steamboats on the Ganges River in faraway India.

Land Travel and Communication

Improvements in transportation technology included the construction of new roads and bridges, and the opening of a new postal system. Throughout the Colonial period, transportation had been chiefly by water. Roads leading to isolated inland settlements were scarcely passable.

Roads and Highways. The first improved road built after the establishment of the new federal government was the Philadelphia-Lancaster Turnpike, 1792.[14] A few other roads were built before 1800, and efforts were made to lay out and construct roads according to engineering specifications. Improved roadbeds were from 16 to 20 feet wide, with an 18-inch crown in the center, sloping to a depth of 12 inches along the edges. Screened, crushed stone was used for the top surface. A short road of this general construction was built leading out of Alexandria, Virginia. Thomas Jefferson described the road as "a notable improvement." In Massachusetts the Boston–Salem Turnpike, the Essex Road, and a highway to Newburyport were all built with crushed stone before Telford and McAdams had introduced their system of macademized road building in England.

Bridges. Thomas Paine is given credit for designing the first iron bridge in the United States. In 1787 he drew up plans for a 400-foot bridge and built a model which was exhibited in London, where it attracted considerable attention. In the same year Lemuel Cox, a shipwright, performed an engineering feat by building a bridge across the Charles River in Boston, described as the longest bridge in the world at that time. Timothy Palmer experimented with truss-type bridges, and in 1794 constructed a large wooden bridge across the Piscataqua at Portsmouth, New Hampshire. In 1801 he built a wooden covered bridge across the Schuylkill, forming a necessary link in the Philadelphia–Lancaster Turnpike.

An engineer from Fayette County, Pennsylvania, James Finley, built America's first iron suspension bridge. In 1800 he erected a bridge across Jacob's Creek in western Pennsylvania, which marked a technological triumph. His design was different from any other in that the roadway was laid nearly horizontal, hung from iron chains by vertical rods. The chains themselves passed over towers anchored in rock. Here was the forerunner of our modern suspension bridge. Finley built many bridges, one as far north as New Hampshire, and one across the Potomac.[15]

The Postal System. President Washington in both his second and third annual messages urged Congress to provide for the establishment of more post offices and post roads, and the postmaster general was instructed to provide for carrying the mails by horse or stagecoach. A Postal Act was passed which created in embryo the United States post office system of today. State and local authorities were urged to improve their roads and keep them in repair. In 1790 there were only seventy-five post offices in the country and less than 2,000 miles of post roads. Pittsburgh was the only mail route to the west. By 1800 there were 14,000 miles of post roads, extending south to Knoxville, Tennessee, and west to St. Louis, Missouri. By 1815 there were 12,000 miles of post roads in the western territory alone, and much of the travel was by stagecoach.

Results of Western Expansion

From around 1800 to 1812 the Napoleonic Wars imposed severe hardships upon American commerce. Many industries along the eastern seaboard closed down, and a number of artisans and mechanics began to move west. Many of them settled in Pittsburgh, a midway town between the Atlantic Ocean and the Mississippi River. Large numbers of diversified industries sprang up there. Toolmakers, blacksmiths, gunsmiths, cabinet makers, saddle and harness makers, and a number of small iron manufacturers and glassmakers developed a flourishing trade at the forks of the Ohio. The emigrants who passed through that gateway town had to have supplies, and they waited until they reached Pittsburgh to outfit themselves for the journey down the Ohio and out into the Northwest Territory.

Frontier Industries. In 1807 the traveller, Cummings, noted a cotton factory in Pittsburgh equipped with a mule of one hundred twenty threads, a spinning jenny of forty threads, four looms and a wool-carding machine, all under one roof. There were also two glassworks, one making white glass products and the other green products. Four nail factories were noted, one of which cut and hammered out 100 tons of nails yearly. Four printing offices were turning out newspapers, almanacs, and doing miscellaneous jobs of printing. One important manufacturing plant was making machines for cotton and woolen factories. The town's most advanced technological process was found in a new copper-plate printing press.

John Melish reported in 1811 that Pittsburgh had a steam mill built of hewn stones

> . . . which can drive three pairs of stones capable of grinding upward of five hundred bushels of grain in twenty-four hours . . . four glass houses . . . several air furnaces . . . two cotton manufactories . . . a wire drawing

factory wrought by a steam engine, an iron grinding mill . . . and many others. . . . I have no hesitation in hazarding it will become one of the greatest manufacturing towns of the world.

At nearby Harmony, Pennsylvania, was a communal settlement in which a number of technological advances were occurring. In 1811 Melish noted that these frontier settlers had more than one thousand Merino sheep, which were producing wool that was manufactured into the finest fabrics, "as fine as any ever made in England."

Yankee Ingenuity Moves West. The settlements along the Atlantic Coast no longer had a monopoly on Yankee ingenuity. A number of settlements located in the western country were also attracting skilled artisans, inventors, and mechanics. In western Virginia around Charleston there was a colony of highly skilled workers in pottery and queensware. Wheeling, in 1811, boasted of two potteries, a nail factory and rope walk, a boatyard, four iron foundries, four woolen and cotton mills, four steam engine plants, two paper mills and eight glass plants. In 1815, Steubenville, Ohio, had one of the seven largest woolen mills in the country, and two years later James Flint noted that Steubenville had a gristmill, a woolen mill, and a paper factory, all operated by steam.

In Lexington, Kentucky, there were three nail factories which produced 60 tons of nails a year by a "new process." There were also two copper and tinware factories. The saddle and harness shops in Lexington in 1818 were employing thirty skilled workers, and their yearly products amounted to almost twenty thousand dollars. Four cabinet-maker shops were making furniture entirely from native wood, and the womenfolk preferred it to the expensive mahogany of the eastern shops. Other manufacturing concerns in Lexington during these years included rope walks, printing shops, powder mills, brickyards, and distilleries. Tanneries were producing leather products valued in excess of thirty thousand dollars a year. An umbrella factory was also busy, as were the brushmakers, hatters, dyers, and chairmakers.

In 1818, in the penitentiary at Frankfort, Kentucky, a new warden, a Captain Taylor, was selected on the basis of his mechanical ability. The penitentiary had been running at a financial loss until Captain Taylor put the prisoners to work making nails, barrels, chairs, turning wood, cutting stones, and performing other mechanical jobs. The warden, aided by his most skilled workmen, designed and built most of the machinery used. The penitentiary was soon reporting a surplus to the state treasury.

The Great Migration. Following the close of the War of 1812 there occurred a great mass population movement, generally described as "The Great Migration." Emigrants poured into the old Northwest and the Southwest in sufficient numbers to cause five states to be admitted into the union: Indiana, 1816; Mississippi, 1817; Illinois, 1818; Alabama, 1819; and Missouri, 1821. Included in this mass movement were thousands of skilled workers, artisans, mechanics, inventors, and scientists who assumed a prominent role in the political, industrial, and cultural life of the frontier.

The quarter-century following the establishment of the young republic was an era of auspicious beginnings in science and technology. A basic pattern had been developed for America's later economic and industrial growth.

NOTES FOR CHAPTER 10

1. Richardson, *Messages and Papers of the Presidents,* I, 88. (Author's italics.)
2. See Stevenson Whitcomb Fletcher, *Pennsylvania Agriculture and Country Life, 1640–1840* (1950), pp. 93 *ff.*
3. *Ibid.,* p. 178.
4. *Ibid.,* pp. 190-195.
5. Pittsburgh *Gazette,* April 27, 1804.
6. For an excellent account of this early American genius, see Grenville and Dorothy Bathe, *Oliver Evans, A Chronicle of Early American Engineering* (1935).
7. *Ibid.,* p. 80. President Washington in 1791 purchased milling parts from Evans for his mill at Mt. Vernon, and so did Jefferson in 1808.
8. Pittsburgh *Gazette,* Dec. 6, 1809.
9. For an excellent brief study of Whitney's life, see Jeannette Mirsky and Allen Nevins, *The World of Eli Whitney* (1952).
10. Fitch's "Report on Manufacture of Interchangeable Mechanics," *U. S. Census,* 1880.
11. S. G. Gilfillan declares that Fitch also built the world's first model locomotive engine, 1795–1798, the forerunner of the marine "steeple" (crosshead or square) engine; see Gilfillan, *Inventing the Ship* (1935).
12. *Dictionary of American Biography,* XVII, 615; Gilfillan, *op. cit.,* p. 89.
13. For the most recent and most thorough account of the technological advances in early river transportation, see Louis C. Hunter, *Steamboats on the Western Rivers, An Economic and Technological History* (1949).
14. Roadbuilding in England and America was first undertaken by private companies which collected tolls. Poles, studded on one side with spikes (pikes) were placed across the road to halt traffic. When the toll was paid, the pole was turned; hence the origin of the term "turnpike."
15. For reference to early bridges see Mansfield Merriman, *Roofs and Bridges* (1912); Richard Shelton Kirby, *The Early Years of Modern Civil Engineering* (1932); Henry Grattan Tyrrell, *Bridge Engineering, A Brief History* (1911); Arthur Jakkula, *A History of Suspension Bridges* (1941); David B. Steinman and Sara Ruth Watson, *Bridges and Their Builders* (1941).

For some other general histories on early steamboats see: James Thomas Flexner, *Steamboats Come True* (1944); Thomas Alexander Boyd, *Poor John Fitch* (1935); Clara Judson, *Boat Builder: The Story of Robert Fulton* (1940); Ella May Turner, *James Ramsey, Pioneer In Steam Navigation* (1930).

Chapter **11**

BUILDING AN EMPIRE OF SCIENCE

> Here, then let us found An Empire—An Empire of Science, . . . , let us erect an edifice which in all coming time will prove the glory of the American name; thus shall we serve our country, the country of our fathers and the country of our children.
>
> *Anniversary Address* by Samuel L. Stoddard to the Columbian Institute, 1827.

Hopes for building "an empire of science" were expressed by a number of scientists and statesmen throughout the country. Their aim was to introduce courses in science in colleges and universities, to establish schools and institutes of applied technology, and to organize scientific societies, lyceums, and lecture courses.

Science, Hope of the Nation

The need for building such an empire was plainly evident. The young republic had thus far weathered a number of financial and political storms. While remaining aloof from the warring nations of Europe, it had fought "a second war of independence." It had survived its first financial depression when "famine stalked the land."

The Era of Good Feeling. Although Monroe's "Era of Good Feeling" had brought with it a wave of nationalism and a cessation of party differences, sectional differences that were to culminate in the great Civil War were already disrupting the nation. One group of statesmen believed that the westward expansion movement, internal improvements, and a program of compromise and appeasement would bring about a strong, united nation. On the other hand, there was a group of scientists who believed that the destiny of young America lay in "the promotion of the useful arts." They believed that the nation which promoted the useful arts was, "of all nations, most certain to attain prosperity." For this group, science pointed the way. The application of science would make this young nation great, and in the glowing language of one of their spokesmen, science would achieve for America, "more lasting advantages than did the conquests of Alexander, Caesar, or Genghis Khan in their lust for power." [1]

Role of Science and the Useful Arts. In 1808 James Cutbush of Philadelphia wrote a series of essays proclaiming the advantages of applied science to industry. He believed that the nation's prosperity would be greatly enhanced by the general diffusion of scientific knowledge. "Science," he declared, "is useful to man not only because it dignifies the mind and fills it with exalted sentiments, but on account of its importance in all the affairs of life, whether we employ it for the good of others or for our own benefit." A proper understanding of science would benefit the mechanic as well as the philosopher.

Federal Aid for Inventors. In 1812 Adam Seybert, scientist, physician, and member of Congress from Pennsylvania, advocated the use of Federal funds to encourage inventors and to promote the useful arts. He also proposed the creation of a Home Department as a Cabinet position. The secretary of this department would be expected to display models and patents which would become the foundation for establishing a school of useful arts. While ancient republics had patronized the fine arts and had paid honor to the genius of Pericles and Phidias, "it was our duty," declared Mr. Seybert, "to honor our Whitneys, Fultons, and Evans', the real benefactors of the human race, the formidable pillars of national independence." Then, becoming more grandiloquent, he expressed the belief that, "American inventions will make this infant republic the Hercules of all nations."

Government's Policy. Ideas somewhat similar had often been expressed in the messages of our early presidents. Washington and Jefferson particularly had advocated the promotion of science as a means for making America independent of all other nations. President Madison in his first inaugural address, March 4, 1809, declared that it was his desire "to favor the advancement of science and the diffusion of information as the best aliment to true liberty." [2] In his second annual message, Madison recommended the establishment of a national seminary of learning, "a temple of science," as a "means of diffusing and improving useful knowledge." President Monroe, near the close of his second administration, reviewed the achievements of his two terms, and in comparing the conditions of the Union in 1820 with those at the close of the Revolution, declared, "the history of the world furnishes no example of a progress in improvement in all the important circumstances which constitute the happiness of a nation which has any resemblance to ours."

Trend Toward Scientific Training

This interest in science extended beyond the halls of Congress. Others interested in promoting science joined in the movement to build an "Empire of Science." They turned to the colleges and the universities. They organized learned societies, published books, and contributed articles to newspapers and periodicals. They appeared on lecture platforms and in lyceums. A few colleges introduced courses in the practical arts. The term "technology" was revived from the classics and given new meaning by applying it to the arts, industry, manufacture, and agriculture.

Lectures on Technology. The Count Rumford chair in Harvard College,[3] established in 1814 to teach "The Application of Science to the Useful Arts," was filled by Jacob Bigelow, distinguished scientist, physician, and professor of medicine. Dr. Bigelow's lectures on "The Elements of Technology" became popular, both with his students and with Boston audiences. He discussed such subjects as mechanics, chemistry, mineralogy, engineering, architecture, and domestic economy.[4]

Dr. Bigelow would have his students understand the workings of newly invented machines that had increased many times the output of the cotton, woolen, and iron products. He pointed out the differences between the technologies of ancient people and those of his time. The ancients, in order to survive, had developed quick eyes and expert hands and had seldom wearied in applying skill. Modern man, however, was supplementing all these with "preparatory science, economical computation, and mechanical power." And, whereas primitive man had used the skill of his eye and hand in fashioning a war club, modern technology (he used that word throughout his lectures) turned out the same thing on a lathe, invented by someone else, in a hundredth part of the time: "the work of a hundred artificers is now performed by the operation of a single machine." Young America, with its knowledge of recent technological developments, could never return to conditions that existed even fifty years earlier, declared Dr. Bigelow.

Referring to the steam engine, yet in its infancy, Bigelow related this incident. When George III had visited the works of Boulton and Watt and was told that they were manufacturing an article of which kings were fond—namely power—the king was struck with "the force and disadvantageousness" of the comparison. "But now," declared Bigelow, "that new source of power which has become beautiful and intricate has crossed the ocean, and is weaving the fabric of our clothing and introducing new energies on every hand." Bigelow believed that as long as the useful arts continued to yield such rich rewards there would never be a lack of competent hands and minds to carry forward their advancement. The time was near, he thought, when a knowledge of our natural elements and of the mechanical arts would be an essential prerequisite to a good education.

Chemistry and Medicine

America's First Chemists. While Bigelow was lecturing and writing about the recent developments in the useful arts and in technology, there appeared on the scene a group of scientists who became the vanguard of the builders of an "empire of science." They were physicians, medical practitioners, who became the nation's first chemists. Among these was Dr. James Woodhouse who was graduated in medicine under the distinguished Dr. Benjamin Rush. His dissertation, "On the Chemical and Medicinal Properties of the Persimmon Tree and the Analysis of Astringent Vegetables," won for him immediate acclaim. Three years later he occupied the chair of chemistry in the University of Pennsylvania, where he set up a small laboratory and performed many valuable experiments. He attacked the doctrine of phlogiston held by Priestley, isolated the metal potassium, determined the conduct of metals toward nitric acid, determined the superiority of anthracite over bituminous coal, pioneered in plant chemistry, and in the development of chemical analysis. He wrote the first laboratory manual, *The Young Chemist's Pocket Companion,* in 1797. He attracted such able students as Robert Hare, and Benjamin Silliman, two chemists who helped to make chemistry the handmaid of early American industry.[5]

Practical Chemistry. Thomas Cooper was another physician who advocated the practical application of chemistry. English born and educated in Oxford, he followed Joseph Priestley to the United States in 1794. Here the two men continued their interest in science, with an occasional venture into politics. In 1811, Cooper was appointed to fill the chair of chemistry in Carlisle College (now Dickinson). In his *Introductory Lectures on Chemistry,* he emphasized the dependence of all life on the science of chemistry; declaring that, "it enters every workshop, every factory, every home . . . , and we must look to chemistry for future improvements in agriculture, in manufactures, and in all that tends to render human existence more desirable." A knowledge of chemistry was even necessary in building a home, if one was to understand the composition of cement and stucco. In furnishing the home, Cooper declared it was necessary for one to know some chemistry before he knew the value of bleaching, dye stuff, varnishes, paints, and pottery.

Cooper's services to chemical education and to industry lay in the stimulus he gave to the new science, and to its introduction into the curriculum of other colleges. In 1816 he was appointed to the chair of applied chemistry and mineralogy in the University of Pennsylvania. Jefferson was a great admirer of Cooper, and he was desirous that he become the "cornerstone" of the new University of Virginia. Instead, however, he accepted the chair of chemistry in South Carolina College (now the University of South Carolina), and later became president of that institution. Although active in politics, he continued his scientific publications, and ranks as one of the pioneers in helping to build an empire of science.[6]

Chemistry in Mineralogy. Another distinguished scientist of the early years of our republic was Dr. Samuel L. Mitchell, professor at Columbia College, editor, oceanographer, mineralogist, and a member of Congress, both in the House and in the Senate. He established the first medical journal in the United States, was a member of some fifty learned societies, and carried on a vast correspondence with scientists in America and abroad. He was one of the first to apply chemical analysis to minerals. He assisted the United States Navy Department in detecting defective saltpeter in gunpowder. He published voluminous reports of experiments with different chemical fertilizers, contributed helpful advice in soap making, and advocated the use of chlorides as disinfectants. He analyzed the salt content in the waters of Saratoga Springs and published his finding. He gave great impetus to the early education of Americans in the field of science.[7]

Chemistry in Crystallography. Among other early physician scientists whose education and travel placed them among the greatest in Europe and America were Benjamin S. Barton, Adam Seybert, and John MacLean. Barton and Seybert were graduates of the University of Pennsylvania, and continued their studies at Edinburgh, Paris, and Göttingen. Barton became a professor of natural history and botany and, later, of materia medica in the University of Pennsylvania. Seybert, after studying mineralogy in Paris, returned to Philadelphia and became known as the "Father of Crystallography" in America. He set up a laboratory for the manufacture of chemicals, and was said to have produced the first mercurials made in the

United States. MacLean, who was Scotland born and educated in Glasgow, after coming to the United States was given the chair of chemistry in Princeton. He became the first professor of chemistry in an American college other than a medical institution.

Griscom, Teacher of Chemistry. The physician-scientists, followed by the teachers of science, paved the way in developing a new interest in both science and the useful arts. John Griscom of Rutgers College gave up medicine to devote his life exclusively to the teaching of chemistry. For over a quarter of a century he was the acknowledged head of all chemistry teachers in the country. He went abroad, associated with the leading chemists of England, and made a firsthand study of numerous industries, with special reference to their fuel power supply. Returning to the United States he studied the formation of anthracite of eastern Pennsylvania, and published reports emphasizing the importance of this fuel for use in the mills and factories of America. He attributed England's superiority in manufacturing to her use of coal for fuel.

Chemistry in Heat Study. Robert Hare, professor of chemistry in William and Mary College, and later in the University of Pennsylvania, was one of the first to make science a vocation, instead of an avocation. He became known best for his discovery of the oxyhydrogen blowpipe, the source of the highest degree of heat then known. This discovery enabled him to fuse the most refractory substances, and led to the founding of the platinum industry and to the development of limelight and allied illumination. Interested in electricity, he built a calorimeter, a powerful galvanic battery, and the deflagrator for generating a high electric current. He was noted for his originality in scores of experiments, and contributed some two hundred papers to the magazines of his day.[8]

Popularization of Science. Benjamin Silliman, described as "the most prominent and influential scientific man in America during the first half of the nineteenth century," was appointed to a newly created professorship of chemistry and natural history in Yale College at the age of twenty-three. Three years later he went abroad and studied with the distinguished scholars and scientists in England, Scotland, Wales, and Holland. Returning to the United States he assisted Robert Hare in his newly established laboratory and worked with MacLean at Princeton.

Silliman's greatest service was in popularizing and in diffusing knowledge of science. He established *The American Journal of Science* (commonly known as *Silliman's Journal*) in 1818, the forerunner of American scientific periodicals. Regarded as the most popular lecturer on scientific subjects, he addressed eager audiences throughout all the eastern states and journeyed as far south as New Orleans, and as far west as St. Louis. He reached thousands of listeners. The marvels of science as presented by Silliman awed the multitudes. He was elected first president of the Association of American Geologists (1840), and in 1863 became a charter member of the National Academy of Science. A man of deep religious conviction, he regarded the study of science as "a duty to learn the wonderful manifestations of God in the natural world." He attempted to place the study of science on an equality with the classics, and "he made a whole nation conscious of its value to mankind." [9]

Others who aided the movement to popularize science included William McClure, a Philadelphia philanthropist, a patron of science, and a pioneer geologist. McClure contributed to the practical arts by preparing a geological map of the United States. His surveys led to the discovery of new minerals and new natural resources. Another was Doctor Charles Thomas Jackson of Harvard, who pioneered in the science of electricity and magnetism, and mineralogy and geology. He made a geological and mineralogical survey of Maine in 1829. He made a chemical analysis of sorghum, and called attention to the many economic values that could be realized from the by-products of cotton seed.[10]

Professional Training

The interest in building the new empire of science became so widespread that a movement was undertaken to establish institutions where the students' time would be divided between study in the laboratory and practical work in the technical field of instruction. Washington had realized the need for this kind of training during the years of the Revolutionary War. Congress, in 1802, met this demand in part by providing for the establishment of a military academy at West Point. Jefferson had often proposed the establishment of schools where "technical philosophy" could be taught. Such schools would be attended by "mariners, carpenters, shipwrights (he enumerated some twenty professions), who would study the sciences of geometry, mechanics, statics, hydrostatics, hydraulics, and a number of other professions."

Training to Use Science. The most important program undertaken during this period in providing courses in the practical arts was sponsored by Stephen Van Rensselaer, a wealthy up-state New York patroon on the Hudson. In 1824 he announced the establishing of Rensselaer Polytechnic Institute, a school that still bears his name. The school was founded "for the purpose of instructing persons who may choose to apply themselves in the application of science to the common purposes of life." Emphasis was to be given to the natural sciences, agriculture, domestic economy, the arts, and the manufactures.

The First Institute for Practical Study. The plan of organization and the real task of establishing the school was entrusted to Amos Eaton, brilliant teacher from Yale and a noted scientist. Eaton was graduated in law from Williams College in 1799 and had a successful practice, rising to the position of Attorney General of New York State. He later came under the influence of Dr. Samuel Mitchell and Benjamin Silliman, and found the sciences more attractive than law. After a few years of study at Yale, he was appointed to give lectures on chemistry and geology. His fame spread beyond the classroom, and he traveled over the country addressing popular audiences in much the spirit of a religious evangelist, translating everything in science into familiar language. He simplified his lectures by illustrating "the most abstruse parts by a dish, a kettle, a warming pan, a bread tray, a tea pot, a soap bowl, or a cheese press." [11]

Eaton's lectures on mineralogy and geology attracted the attention of Governor Clinton of New York and led to the establishment of the New York State Agricultural Society, 1819. Stephen Van Rensselaer became president of the Society and employed Eaton to make a geological survey of Albany and Rensselaer Counties, one of the first geological and agricultural surveys ever made in the United States. This was followed by a survey of the Erie Canal route from Albany to Buffalo. These surveys placed Eaton at the head of the nation's geologists, and won for him the title "Father of American Geology."

A Model for Learning. When Eaton opened the Rensselaer Institute in January, 1825, educators from a distance came to Troy to study his plan. "Learning by doing" was the motto for all students. He introduced the laboratory experimental technique. Courses were offered in botany, geology, astronomy, chemistry, and civil engineering. The latter course included work in land surveying, methods for constructing water works, and principles of milling. He also introduced courses in agriculture, which, according to McAllister, led to the establishment of "the first school of agriculture in the United States, antedating by nearly thirty-five years its nearest rival, Michigan Agricultural College." [12] Eaton lectured almost daily for thirty-five years, delivering some six thousand "loud and long lectures," according to his own statement.

Scientific Societies and Institutes

World's First Chemical Society. Meantime, there were a number of scientifically minded individuals who had been carrying on chemical experiments with the idea that they might make discoveries that would be of practical use. As early as 1792 a group of chemists, led by Dr. James Woodhouse, organized the Chemical Society of Philadelphia, claimed to be the first chemical society in the world. The members early called attention to the many potential uses of applied chemistry.

Chemistry had emerged in the late eighteenth century as a distinct science. A second chemical society was founded in 1811, also in Philadelphia, with Thomas Jefferson as patron, and James Cutbush as president. In the same year, Dr. F. Aigster, noted chemist and an honorary member of the Philadelphia society, traveled to the frontier town of Pittsburgh and delivered a course of lectures on chemistry. Two years later the Pittsburgh Chemical and Physiological Society was organized.

Growth of Scientific Societies. During the first quarter of the nineteenth century a number of other scientific societies were organized with the avowed purpose of disseminating information on science and the useful arts. These societies were both learned and popular. Some were devoted to the general interests of science, while others were established for promoting specialized scientific interests. However, they all had the same objective—that of stimulating the practical and the useful—and they were truly democratic, with membership open to anyone interested. In this respect they differed from the scientific societies of Europe, where membership was restricted to the intellectual elite. In America the societies sprang up in all parts of the country, from New England to South Carolina, and as far west as Cincinnati, St. Louis, and New Orleans. In Pittsburgh, a Philosophical Society was founded in 1820.

Aims of the Columbian Institute. The most prominent of these early nineteenth century scientific societies, and the first to receive national recognition, was the Columbian Institute, organized in 1816. It was granted a charter two years later. Due to its distinguished membership, and the recognition given to it by Congress, it served to some extent as an advisory body to several agencies of the federal government. Its roster of members included statesmen and scientists in the nation's capital: Henry Clay, John C. Calhoun, Daniel Webster, Charles Wilkes, Richard Rush, Joel R. Poinsett, and others of equal distinction. Former Presidents John Adams, Thomas Jefferson, and James Madison became members, as did two future Presidents, John Quincy Adams and Andrew Jackson. There were fifteen members of the Senate, twenty-eight members of the House, and six cabinet members enrolled.

The idea of establishing the Columbian Institute originated with Thomas Law, a scholarly English emigrant, who married Anne Custis, George Washington's ward. Dr. Edward Cutbush was elected first president of the Institute. The aim of the Institute was to develop the practical arts and the sciences, aid agricultural experiments, ascertain new species of plants that would yield oils suitable for manufacturers, procure ores from which pigments might be manufactured, stimulate discoveries in the mechanical arts and sciences, and place the findings and discoveries of its members at the service of the federal government.[13]

Institute's Support and Achievements. The Institute hoped to receive from the federal government pecuniary aid sufficient to enable it to extend its benefits to all the citizens of the United States, and thereby to serve the country and to "elevate it to the highest pinnacle of glory." Congress permitted the organization to hold its public meetings in the House of Representatives, and its regular meetings in a large room under the library in the Capitol building; Congress also donated a tract of land in the mall to be used as a botanical garden.[14]

Samuel L. Southard, in an address before the members of the Institute in 1827, reminded his fellow scientists that the purpose of the Institute would always be that of aiding the march of science and promoting the things that were of practical use. "The cause of science," he continued, "is the cause of human happiness; the claims of science are the claims of liberty and the social order." Men of science, he declared, had an obligation to promote happiness and to advance intelligence, "not of our own country alone, but of man everywhere throughout the globe."

The distinguished Edward Everett read a paper, excerpts of which were quoted in newspapers and periodicals for several months. He declared that science rather than politics would make this nation great; that lack of scientific knowledge was responsible for most of our evils, and that an understanding of science would become a prerequisite for those "who traverse the ocean, construct canals, build steam engines, work our mines, or set up manufacturing establishments." He concluded by saying that "the mechanician, not the magician, is now the master of life." [15]

Although the Columbian Institute never succeeded in achieving its original aim, it became a recognized learned society where papers were read, discussed, and then filed away in its archives. In 1838 its members united in forming a new organization called the National Institution, which in turn became the chief factor in forming the Smithsonian Institution in 1846 (see Chapter 17).[16]

Mechanics Institute. Other scientific institutes of a more practical nature were established during this period. Among these was the Mechanics Institute, organized in Boston (1795), with the aim of assisting the members to master the mechanical and technical arts. Similar institutes were founded in other cities to train artisans and craftsmen. Lecture courses, libraries, and museums were established by some of the institutes, and schools for the children of the artisans were provided. Many of the mechanics' institutes held demonstrations, exhibited the latest machines, and taught apprentices how to operate the machines. The institutes were wholly voluntary organizations on the part of the artisans, but they received the patronage of leading men of science such as Eaton, Silliman, and Edward Everett.[17]

Franklin Institute, 1824. The most important of all these organizations was the Franklin Institute, founded in Philadelphia in 1824, as a memorial to Benjamin Franklin. Prior to the 1820's the only way one could obtain a technical education was through the apprentice system. There was always a shortage of skilled workers. Because of these conditions a situation developed in the early 1820's that led to the founding of a new kind of a society, one that set a pattern for later institutions of a similar nature.

When Samuel Vaughn Merrick, a young man lacking technical training, received from a rich uncle a factory for manufacturing fire engines, he was unable to operate it. He sought technical aid from some of his acquaintances, including Professor William Keating of the University of Pennsylvania, and William Kneass, a copper engraver, and sought the financial help of Nicholas Biddle. Together, they organized an institute where technical and vocational training would be given. When the doors were opened, four hundred fifty men and women (special courses were provided for women) applied for admission. No previous formal educational training was required. Courses were offered in agriculture, architecture, chemistry, geology, mineralogy, mechanical arts, and natural science.

It was difficult to find qualified lecturers at first, but within a few months lecturers were recruited from the more brilliant students, who in turn taught other students. A committee on inventions was appointed to aid young inventors in developing their special ideas. The committee members pointed out to inventors what previously had been accomplished, and encouraged all those whose inventions were deemed worthy of support.

Journal of Franklin Institute. Thomas P. Jones, professor of mechanics at the University of Pennsylvania, a member of this committee and later head of the United States Patent Office, began the publication of a journal, *The American Mechanic Magazine.* In 1826 this journal merged with the publication of the Franklin Institute, and became known as the *Journal of the Franklin Institute.* It became the official organ of the United States Patent Office, and it published the American, English, and French patents, supplying the only continuous record of American patents from 1828 to 1843.

The *Journal* aided greatly in developing an interest in science. Under the editorship of Thomas P. Jones, it served as a clearing house for all kinds of mechanical, scientific, and technical information. It carried accounts of new devices, new processes, and new techniques that would increase the people's comforts and raise living standards. It helped to make the young nation machine-conscious. It helped to usher out the age-old secrets in trades and skills, and thereby opened up all available technical knowledge. Also, it extolled the mechanic and the workman. "In America," Jones wrote, "science proffers freely her aid to the artisan and the mechanic, and in return, asks for his services." This was the keynote of the many advances made in the rapid industrial development of the years that followed.

A New Age of Science

The builders of the empire of science had discovered that theirs was not an easy venture, but it was worth the effort. The growth of scientific societies and the publication of scientific journals devoted to American discoveries and inventions proved that the United States no longer had to rely solely on Europe for commodities or for ideas. Science, engineering, and technology would give America its own way of life.

There were many dangers, national and international, threatening the young republic during the second quarter of the nineteenth century. At home, sectional strife stirred up by the institution of slavery, was becoming more ominous. The young nation was destined to resort to blood and battle before that evil was abolished. Abroad, thousands of liberty-loving citizens, fearful of their future in their own country, began flocking to America.

Foreign observers pictured America as a paradise for the artisan, the mechanic, the inventor. Such a picture described by a German-born inventor, John Adolphus Etzler, won a wide audience. He envisioned America as a technological wonderland. Addressing himself to "all intelligent men" (1833), he gave a detailed description of "the paradise within the reach of all men, without labor, by powers of nature and machinery." [18] That paradise was to be found only within the borders of young America.

Notes for Chapter 11

1. Adam Seybert, *Report of the Patent Office Annals of Congress,* 12th Congress, 1st Sess. (1812), XXIII, 2179.

2. Richardson, I, 453. (Messages of the Presidents.)

3. Benjamin Thompson (Count Rumford), 1753-1814, was born in Massachusetts. During the American Revolution he sided with the Loyalists and fled to London. He is best known for his theory of heat. He advanced the theory that heat is not a material substance, but "a mode of action." He refuted the current theory that the manifestations of heat were due to a fluid named "caloric." Rumford reached his conclusions after his celebrated experiment in boring a cannon during which the heat generated by friction brought cold water to the boiling point.

4. The lectures were later compiled and published under the title: Jacob Bigelow, *Elements of Technology, A Course of Lectures on the Application of Science to the Useful Arts* (1829).

5. *Dictionary of American Biography,* XX, 491-492; Edgar Fahs Smith, *James Woodhouse, A Pioneer In Chemistry* (1819).

6. See Edgar F. Smith, *Chemistry in America* (1914). Also, Eva B. Armstrong, "Thomas Cooper as an Itinerant Chemist," *Journal of Chemical Education,* XIV (April, 1937), No. 4, 153.

7. Courtney B. Hall, "A Chemist of A Century Ago," *Journal of Chemical Education,* V, 253.

8. *Dictionary of American Biography,* VIII, 263-264; and Edgar F. Smith, *The Life of Robert Hare* (1917).

9. See John F. Fulton and Elizabeth H. Thompson, *Benjamin Silliman* (1947), the best biography of this early American scientist.

10. Charles A. Browne, "History of Chemical Education," in *Journal Of Chemical Education,* IX (1932), 696 *ff.*

11. The latest, most definitive biography of Amos Eaton is by Ethel M. McAllister, *Amos Eaton* (1941); see also Ray Palmer Baker, *A Chapter in American Education* (1924).

12. McAllister, *op. cit.,* p. 516.

13. Edward Cutbush, *Memoirs from Library of Congress,* Miscellaneous Pamphlets (V. 488), No. 2, 2 *ff.*

14. Richard Rathburn, *United States National Museum Bulletin,* No. 101 (Washington, 1917), states that 177 papers and addresses were given at various meetings of the Institute.

15. Edward Everett, *Orations and Speeches,* I (1836), 246 *ff.*

16. G. Browne Goode, "Genesis of the National Museum," *Smithsonian Institute Report,* XLVI (1890–1891), Part II, 273.

17. Other groups of scientists were organizing societies of their own: The Charleston Botanical Society, in 1805; the Linnean Society in Philadelphia in 1814; the Washington Botanical Society in 1817; the New Haven Geological Society in 1819; the New York Horticultural Society, 1818; the Pennsylvania Horticultural Society, 1827; and a similar one in Massachusetts in 1829.

18. The author is indebted to Hugo A. Maier for this item. It is included in a paper entitled: "Technology and The Idea of Social Responsibility, 1800–1850," read at the meeting of the American Historical Association, New York, December 28, 29, 30 (1951).

Chapter 12

PASSING OF THE HOMESPUN AGE

The "homespun" age was on its way out during the second decade of the nineteenth century. The cold facts were revealed in the tables of the Census Bureau and the records of the Patent Office, but more dramatically its passing was marked by the sound of the chugging steam engine, the whirring of spindles, the clatter of the looms, and by the black coal smoke rising from factories where wood was formerly used as fuel. Its passing also coincided with the growth of towns and cities, the introduction of comforts and luxuries in living, and the arrival of an increasing flood of immigrants. The homespun age disappeared gradually. It prevailed along the frontier as settlers moved west long after its accompanying hardships were forgotten in the industrial centers of the East. The homespun age was not a period of time, but a way of life. It was the age in which production passed from the home and the handcraft system to the factory and the machine.

America, Land of Mechanics

The Industrial Revolution in America differed greatly from what it was in England. In the United States the machine was welcomed by the workers. The shortage of labor both on the farms and in the cities made it necessary to introduce machines to do the work which the abundant hand labor produced in other countries. Moreover, as machines were installed in factories, mechanics, toolmakers, and millwrights were prepared to operate them. For well over a century Yankee ingenuity had been developing the simpler labor-saving machines, so that by the 1830's the average American mechanic knew the principles of power transmission by shaft, chain, or belt drive. He knew that power could be directed by spiral gears, worm gears, universal joints, ratchet wheels, friction wheels, and eccentric crown wheels.[1] He also knew the use of cams, cranks, pistons, connecting rods, universal levers, toggle joints, couplings, and clutches. He had worked out these principles when machines were clumsy affairs made of wood and operated by hand power, and at slow speed. He had learned how to repair his machine, to improve and adapt it to his own use.

Skilled and Efficient Workers. As machines became more intricate, driven by steam power, they presented no obstacle. Gibb declares that the Industrial Revolution in its infancy "produced surprisingly few basic technical skills not already familiar to American mechanics." [2] When Francis Lowell introduced the power loom in 1814, he found great numbers of skilled workmen, endowed with "the training of many generations of farmer-mechanics in the workshops of New England." [3]

Transition from Hand to Machine Work

The factory system in the United States meant the substitution of the machine for hand labor, the introduction of mechanical power for human power, and the assembling of all these forces under one roof. In America, as in England, it was in the field of textile technology that the factory system made its first major advance in replacing homespun products.

Textile Industry. The early history of the New England cotton industry is largely the history of the early factory system in the United States. Here labor-saving machinery was first introduced. Also, here the economist saw the corporate form of organization applied to industrial ends. Here the sociologist observed women take their first step toward economic independence, and workmen become factory hands rather than craftsmen.

Work for the Unskilled. It was the introduction of the power loom in 1813–1814 that brought about the new era in the American cotton industry. It brought about the greatest technological change, since it required fewer attendants and could be operated by relatively unskilled laborers. The beginning of this revolution in American textile technology occurred at Waltham, Massachusetts, when the joint efforts of Francis Cabot Lowell, Patrick Tracy Jackson, and Paul Moody, with the financial backing of Nathan Appleton, organized the Boston Manufacturing Company and brought together an effective business organization, superior technical skills, and competent executive leadership.

The Waltham Mills. These mills marked the beginning of American big business. The plant was organized for mass production. All processes were integrated. When the huge power loom was installed and the doors were opened in the autumn of 1814, Nathan Appleton wrote, "I well recollect the state of admiration and satisfaction with which we sat by the hour, watching the beautiful movement of this new wonderful machine, destined as it evidently was, to change the character of all textile industry." [4]

Complete Factory System. Lowell's plant, described as the first mill in the world to combine all factory operations under one roof, geared the spinning, carding, drawing, roving and weaving processess to a large water wheel. Belts of leather transmitted power from shaft to shaft. Raw cotton was introduced at one end of the mill, and yard goods emerged from the other.

Effects of the Factory System

Out of the power loom grew Lowell, Massachusetts, which became the Manchester of America. With the Waltham people behind it, the Merrimack Manufacturing Company was formed with a capital of six hundred thousand dollars which laid the foundation of the city of Lowell. Within a few years, other companies were formed. Locks and canals were constructed on the Merrimack river to furnish water power to operate the mills. In 1824 the town was incorporated with 15,000 inhabitants. By 1827, there were twenty-five factories in operation. Land prices jumped to one dollar a square foot. By 1835, the plants were producing almost 40,000,000 yards of cloth annually, and the total monthly payroll amounted to almost one hundred thousand dollars.

Women Employees. In the field of labor relations, the factory departed from the family system prevalent in England and on the Continent. Girls from middle-class families were employed and lived in respectable boarding houses. Special attention was given to providing social, religious, and recreational opportunities. Charles Dickens paid glowing tribute to the employer-employee relations in the Lowell mills, declaring that he observed "no face that bore an unhealthy or an unhappy look." This, he noted, was in striking contrast to labor conditions in his own country. Another English observer noted that of the nearly five thousand females employed, all appeared clean, neat, and fashionably dressed. "How different," he wrote, "from the manufacturing class in Great Britain."

Introduction of Steam Power. Improvements in textile technology occurred in other plants throughout New England. The Slater textile mills in Rhode Island introduced the English power loom and dresser (machine used to apply sizing to the warp threads in the loom), installed a "live" spindle of the Scotch type (one that rose and fell with the spool in contrast to the fixed spindle of the Lowell mills). More important was the change made by the Slater mills from water power to steam power in 1828. Within a few years a number of textile mills were powered by steam, but it was not universally used in the cotton mills until the railroads were sufficiently developed to transport coal cheaply.

Danforth's Cup Spinner. While the machinery in these factories was all constructed on principles borrowed from European designs, it was not long before America was making contributions of her own to textile machinery. Arkwright's water-twist frame (1769) had been gradually superseded by the throstle, which was a continuous spinning machine. It drew and twisted the roving at the same time. While this had its advantages, the drawbacks were that thread of finer quality could not be drawn, because if the machine were speeded up the spindle would vibrate and the thread would loop.

Charles Danforth of Massachusetts introduced in 1828 an improvement known as the "cup spinner." Danforth made the spindle stationary, placed an inverted cup over the top to receive the spool, and then put a freely fitting sleeve around this to rotate the cup and spool.[5] With this invention the bobbins could be run seven thousand turns per minute, and thus produced from 40 to 50 per cent more yarn.

Thorpe's Ring Spinner. Another invention, that of John Thorpe of Providence, Rhode Island, laid the foundations for cotton spinning machinery for the next hundred years. It was entirely American in its origin. It was used extensively in this country for many years before its adoption in England, but it later became the basic system for continuous cotton spinning machinery in that country also. Its chief feature was the substitution for the flyer (the contrivance that carries the rove from the drawing rolls to the spindle) of a smooth ring in the center of which was the spindle. A thin wire, the "traveller," sprung from the ring, held the roving at the winding point as it passed from the drawing rollers to the spool on the spindle. This enabled the spinning to proceed continuously and at great speed, eleven thousand revolutions of the spindle per minute. No other basic device in textile machinery has been the subject of so many patents. From it many other improvements have developed.

Other Improvements. Other improvements included leather belts instead of shafts used to drive the spindles and iron instead of wood construction for the drums of the carding machine. An improved roving machine was introduced. Plant construction was improved, and greater power facilities were provided, including a dam built across the Connecticut River which stood 30 feet in height, and supplied adequate power for a number of mills.

New Weaving Processes. The first power looms were heavy, clumsy affairs that had to be stopped frequently for repairs on the warp. In many cases the weavers were paid by the week instead of by piece work (as was the custom) because the breakdowns in the machines limited their output. Improvements facilitated smoother operation of the machines. The objective was to increase the speed so that the output could be greater per operator, and thus compete with European loom workers who were paid lower wages. A self-acting temple was devised to keep the cloth distended to its full width as it was woven. This machine reduced the number of operators required. One operator was able to attend four looms at a time, while in English factories two was the limit. Moreover, a weaving speed of one hundred and twenty picks a minute was averaged, as compared with ninety in English shops. The product of American mills was, however, little more than coarse white sheeting, rather than the finer fabrics.[6] Quantity production was the first objective, with quality secondary. Prior to 1860 many mills were still operated by handlooms, since the great change in weaving did not come until after the Civil War.

These improvements in spinning and weaving stimulated the introduction of further new processes. Cyrus W. Baldwin, of Massachusetts, invented a unique process for weaving seamless bags, a novel and an important article for the sacking of grain. The loom was threaded with a double warp thread, and was so constructed that the "filling" or weft traversed both sides and united the warp at the edge. It proceeded until the number of "picks" required to make up the length of the bag were thrown in; then it closed the bottom, and proceeded on to the next. It was an entirely automatic process. All the weaver had to do was to tend the machine and cut off each bag.[7] Erasmus Bigelow's loom for weaving ingrain carpet, introduced in 1844, marked the beginning of the great Bigelow carpet industry that later captured the world's markets.

Bleaching and Dyeing. With the machines rapidly increasing in output, it became necessary to step up the slow, tedious bleaching and dyeing operations. Here the American chemists entered the textile field and soon reduced the time required from several days to a few hours. The cost of finishing and bleaching was reduced to one cent per yard. Dye houses and printing houses became a part of all large factories. The dyeing operation, however, was dependent upon natural dyestuffs, and the science of chemistry did not eliminate the guesswork until after the Civil War.

America Challenges Britain. A British industrialist, writing in 1835, noted that the English people were more indebted to Arkwright and Watt for Britain's world supremacy than to Wellington and Nelson; but he warned Britain that her supremacy was in danger because of the advances made by the textile mills of the United States. America, he said, is a land where "spinning machinery is equal to that of England, where ingenuity and enterprise mark the national character, and where the finest cotton is grown. . . ." He said that the rapid introduction of machinery was the most remarkable feature in America's progress, and that labor-saving machines were replacing the hand labor operations that still persisted in England.

A contemporary American writer noted that the mechanical and industrial technicians of the country had made it possible to "supply our own demand for the most important fabrics necessary for our comfort, and even carry the war of commerce and manufacture into the country of the enemy." [8]

Woolen Technology

Just as Samuel Slater has been called the "Father of American Cotton Machinery," so the Schofields (Arthur and John) may be called the "Fathers of Machinery for Manufacturing Woolen Goods." Woolen production did not reach the factory stage as early as cotton. It continued as a semihousehold industry. Schofield's carding machine, while widely adopted, led to the establishment of community carding shops rather than factories, and the wool prepared in the local community shops was returned to the household for spinning. This relieved the home of an arduous task. Likewise, local fulling shops were established, and itinerant weavers tended to postpone further the passing of the homespun age as far as woolen manufacture was concerned. It was not until the 1830's that the cleaning of the wool, carding, spinning, weaving, and finishing were combined and carried on as a single industry. Two states, Pennsylvania and Massachusetts, took the lead in the woolen textile industry. In isolated instances woolen cloth was successfully produced, and of a quality as fine as any imported. Such was the case with the Harmony Society in western Pennsylvania.[9]

Burr Picking Machine. In 1834 a burr-picking machine, invented by Michael H. Simpson, proved of great value in producing better cloth. Before this, wool had to be cleaned by hand. This was a slow process. Simpson's device contained a rapidly revolving blade which picked the burrs from the wool. It proved an instant success.

Card-Making Machine. Elisur Smith, of Walpole, Massachusetts, invented a card-making machine described as the marvel of the age. It inserted wire in cards, making possible rapid card manufacture. The iron fingers of the machine seized the wire, bent it, punched holes in the cloth or leather, and then inserted the wire. If the slightest derangement occurred, it stopped and waited until the difficulty was remedied. Daniel Webster, after watching the machine operate, said, "It seemed to be more nearly endowed with human intelligence than any machine ever invented," and John Randolph was so impressed by its efficiency as to declare, "it possesses all but the immortal soul."

The Goulding Condenser. The greatest contribution to carding wool was made by John Goulding in 1826 when he introduced the Goulding condenser that still bears his name. The condenser consisted of a series of carding machines with intermediate mechanism that made it possible for the filaments of wool as they left each machine to be delivered automatically to the next. The filaments were finally transferred to a rover where they were drawn and twisted and made suitable to be spun directly into yarn on a jenny or a mule. This invention eliminated the labor of children in twisting the strands by rubbing them between their hands (slubbing) at the delivery end of the carding machine. It made it possible to increase the number of spindles on the jack or mule from one hundred and twenty to two hundred or more, because the danger of breaking the strands was lessened. The Goulding condenser enabled the American woolen manufacturer for the first time to compete successfully with foreign woolen goods.

Crompton's Loom. Another important American contribution to textile technology was Crompton's loom. In 1836 William Crompton of Massachusetts designed a loom to weave goods in a pattern, or a "fancy loom," as it was called. Although he first patented his invention in England, it was not until some major improvements were made by Francis Knowles (1856), and George Crompton (1857) that it was adopted here. The feature of Crompton's invention was an endless pattern chain which allowed any number of "harnesses," usually twenty-four, to be used so that any desired figure could be woven. This marked one of the first attempts to weave fancy woolens on power machinery.

Basic Improvements by 1840. A unique fringing machine involving an ingenious contrivance for twisting the threads for fringes on shawls was introduced in the 1830's. The task of fringing and napping, formerly carried on by hand, was now transferred to machines. By the 1840's the woolen industry had installed almost all the basic technical features in use today. The Census of 1850 showed that twenty-four of the thirty-one states had set up factories engaged in some phase of woolen manufacture. The total number of male employees exceeded 22,000 and the female employees numbered more than 16,000.

Mechanization of Leather Production

The leather industry passed out of the homespun age in a slow, gradual, almost imperceptible manner. One of the most ancient arts known to man—the curing, drying, and tanning of leather—had been carried on almost without change until well into the nineteenth century. First, it was a home industry. Next, it was transferred to small, specialized shops, and still later to small factories where a few machine processes were introduced. Machine-made boots and shoes came even later.

Gideon Lee, "Father of Leather Trade." An early step in establishing the tanning industry as big business was taken in 1809 with the incorporation of the Hampshire Manufacturing Company in Boston. Eight years later Gideon Lee of Green County, New York, often called the "Father of the American Leather Trade," incorporated all the processes of tanning, from leaching the hides to finishing and currying the leather, into one business and all under one roof. In 1822 Seth Boyden of Newark, New Jersey, introduced patent leather in America.

New Leather Splitting Machine. Improving on Samuel Parker's leather splitting machine, Elias Putnam of Danvers, Massachusetts, contrived a machine (1838) that both split and pressed the leather at the same time. In this machine the leather was propelled through two rollers and pushed against a knife by the action of the rollers.

Colored Leather. Herman Hibbert of New York made two important contributions in coloring leather, that of using ferro-prussiate for black leather, and sulfuric acid and nitrate of copper in combination with vegetable oils, for other colors. These processes removed the uncertainty from the dyeing operation. Hibbert also introduced a new finishing operation in which the skins were oiled and passed through revolving heated cylinders which kneaded them and made them more pliable. This method hastened the drying process. All these improvements tended toward greater concentration of the tanning business under one roof and led to the passing of another home industry.

Boot and Shoe Industry. The development of the factory system in shoe manufacture followed a slightly different pattern from other industries, in that shoemaking was taken out of the home when still in the handcraft stage. Small factories were established before machines were introduced and centralization of the work in these small factories made possible the introduction of the machine.

Development progressed from the itinerant cobbler to the establishment of small shops, called "ten-foot" shops, where four to eight benches were established, apprentices trained, shoes made to order, and "sale" shoes made to be sold on the market. These developed into "central shops." With the central shop system, craftsmanship became unnecessary and specialization was adopted. A workman did not need to know how to make a pair of shoes, but only how to perform one phase of the work. He sewed on uppers, made buttonholes, or cut sole leather. The central shop provided the materials and was responsible for the sale of the shoes.[10]

The business depression following the panic of 1837 struck a severe blow to the central shop proprietors, and they were forced to look for improvements. Labor, material, patterns, and tools were all taken into account. The technical advances in shoemaking, which had been developing all through the period, were combined into one system. The central shop was equipped to perform more of the work. New machines were introduced and each invention meant that one more operation was consolidated in the central workshop so that it gradually evolved into the modern shoe factory.

New Shoemaking Processes. A number of important technological improvements brought about these rapid changes. One of these was Samuel Preston's invention (1833) of a machine for driving pegs into soles rather than inserting each peg by hand. Another important development occurred in 1832 when a full set of diagram patterns for cutting uppers was introduced. Following this innovation patternmaking became an industry in itself. The patterns were used for shaping soles and patternmaking machines were devised for cutting. It was not until the 1850's that "crooked shoes," that is, shoes cut for the right and left foot became general.

The cutting of the sole leather was made easier and more uniform by the invention of the stripper, a heavy blade operated by foot power, which cut lengths of leather to be fashioned into soles. In 1845 a leather rolling machine did away with the cobbler's old-fashioned method of pounding leather on a lapstone with a flat-faced hammer. With the rolling machine, the leather could be rolled in one minute as effectively as it could be processed by a half-hour of pounding.

Boot and Shoe Specialization. The development of specialization in the boot and shoe industry gave rise to a number of allied industries. Besides patternmaking and lastmaking, already mentioned, a shoe-thread mill was introduced in 1835. Shoe boxes and cartons, not generally used before 1830, had by 1840 become the basis of a new business organized to package the shoes. Most of these developments originated at Lynn, Massachusetts.

In the second quarter of the nineteenth century shoe manufacturing increased rapidly. In 1830 the entire boot and shoe jobbing business exceeded only a little over a million dollars, but by 1860 the output of the shoe factories totaled almost sixty million dollars. The application of the sewing machine and the Goodyear welt shoes were to bring about the next major technological developments.

Iron Industry Becomes Big Business

The transfer of the iron industry from the blacksmith shops, the local forges, and foundries to mill or factory-stage production marked another step in the passing of the homespun age. The enormous demands created by the many different uses of iron served as an incentive for increased production. Prior to the 1830's these increased demands were met by increasing the number of establishments rather than by any new technological advances. But in the next couple of decades, changes were so radical that they constituted a revolution.

Puddling and Rolling Process. The first major technological improvement in American iron manufacture, the rolling and puddling system for refining iron, was developed in the new region west of the Appalachian Mountains. The rolling and puddling method, introduced by Cort in England in 1784, had been known to American ironworkers for several years, but it was not introduced here until after the close of the War of 1812, when Isaac Meason erected a puddling furnace and a rolling mill for refining iron at Plumstock in western Pennsylvania.

Pittsburgh, "Iron City of America." With the westward expansion of the industry, Pittsburgh assumed an increasing importance, "particularly in articles of iron." There the first foundry was erected in 1805, the first rolling mill for sheet metal in 1818, and the first puddling and rolling mills for refining iron in 1819. Peter Schoenberger built the Juniata Iron Works in Pittsburgh in 1824, where he made bar iron, sheet iron, boiler iron, and cut nails. By 1829 there were eight rolling mills and nine foundries manufacturing products from pig iron brought in from surrounding communities.

The first blast furnace in Pittsburgh was not built until thirty years later.

Cast-Iron Products. Casting, which had hitherto taken place only at the furnaces, became a separate industry and foundries were established for special castings. These were used for frames for manufacturing machinery, for printing presses, for pipes, hydrants, and lamp posts, as well as for pillars and store fronts for what were then termed "fireproof" buildings. Cast iron was used for ornamental purposes, railings, fountains, and lawn ornaments. It supplied hardware for builders, bath tubs, kitchen utensils, sugar kettles, shafts, and casings for pumps. American ironworkers developed unusual skill in making light, strong, ornamental castings. The foundry supplanted the forge in providing machine parts. Foundries for making iron for stoves were established. Iron castings were also used for heavy parts, such as cannon, mill irons, rollers for slitting, and sugar mills. A few cast-iron rails to meet the demand created by the coming of the railroad were manufactured.

Rolled Iron Products. Specialization included the introduction of wire nails for wire fences and wire netting. Other mills produced rolled boiler plate and rolled sheet iron for engines. The bloomery and the forge gave way to the improved foundry and puddling and rolling mills.

Coal Fuel for Furnaces. Of still greater technological significance in the iron industry was the passing of the charcoal era and the introduction of anthracite for fuel in the blast furnaces, and later of bituminous coal. David Thomas, an immigrant from Wales, is called "The Father of the Anthracite Industry." While working at the Lehigh Crane Company furnace in eastern Pennsylvania, during the summer of 1840 he succeeded in operating a blast furnace by using anthracite as fuel.

Thomas' success was based on the application of the hot blast to blast-furnace firing, a system which James Neilson of Scotland had discovered some years earlier. Dr. Frederick Geissenhainer, a Lutheran clergyman of New York City, had also been granted a patent in 1833 for the manufacture of iron and steel with the use of anthracite by employing a strong hot blast. In 1836 he succeeded in making a few tons of iron in a pilot plant in Schuylkill County, Pennsylvania. Geissenhainer died before his project was completed and its execution was left to George Crane who, with the help of David Thomas, carried it to a successful culmination.

The depletion of the forests along the Atlantic seaboard, in the early iron-producing communities, gave impetus to the use of this new fuel. A number of furnaces were built for the use of anthracite during the late 1840's, and by the mid-1850's the amount of iron products made from furnaces using anthracite exceeded that made by furnaces using charcoal as a fuel.

Introduction of Bituminous Coal. The reign of anthracite was brief, however, since further experiments in fuel technology proved that a new product, coke made from bituminous coal, was more satisfactory as a fuel than anthracite. In 1835 the Franklin Institute offered a gold medal for the manufacture of not less than 2 tons of iron to be produced by using coke or bituminous coal as fuel. With this incentive, a number of experiments were undertaken. Henry Oliphant, a successful ironmaster at Fairchance, Pennsylvania, reported to the Franklin Institute that he had produced more than 100 tons of iron from coke made from bituminous coal.[11] However, the use of coke did not gain immediate favor, since an efficient method of coking coal was not discovered until after the Civil War.

The Iron Age Emerges. The iron age had already arrived, however. The isolated forges and furnaces of the Colonial era had passed. The village blacksmith was no longer able to supply the community with tools and implements. The production of pig iron, which in 1810 had amounted to slightly more than 50,000 gross tons, had by 1850 risen to almost 600,000 tons, produced in three hundred seventy establishments. While this growth was small in comparison with the unparalleled increases that occurred during the following decades, it is nevertheless significant in explaining the passing of the homespun age in the iron industry.

New Industries—New Patents

A number of other industries began to be transferred from the home to factories during this period. The manufacture of sugar from sugar cane, which had always been a family industry, was taken over by sugar cane mills and factories in the 1830's. The manufacturing process involved three stages: crushing the cane by passing it through rollers, evaporation of the juices in a series of open pans, and the crystallization and refining of the sugar.

Sugar Refining. The modern era of sugar refining was inaugurated in the 1820's when the steam engine was used to operate the crushing mills. Improvements made in the rollers and the vacuum-pan system (invented in England) prevented the over-evaporation and caramelization which resulted in the old open-pan methods. An

American innovation, the use of bone charcoal instead of blood and clay in the refining process, was made by a Mr. Figuier of Vermont in 1813. These improvements were incorporated by the Archibald and Belafield Company in New York in 1833, where the first modern sugar refinery was built. Archibald has been called the "Father of Modern Sugar Technology." New York became the center of the industry and boasted of having eleven establishments in 1830. By 1850 one refinery reported a melting capacity of 35 tons a day. These developments marked the transference of another ancient industry from the home to the factory.

New Patent Act, 1836. The flood of patent applications during this period indicates better than anything else the passing of the homespun age. The Patent Office was not able to pass on all applications and the courts were bogged down with cases that were pending on appeal. Senator John Ruggles of Maine, an inventor in his own right, succeeded in having a new Patent Act passed, July 4, 1836, which marked a turning point in patent history. The Act provided for compulsory examination of all patent applications. The title, Commissioner of Patents, was provided for and the appointment was to be made by the President, subject to Senate approval. The commissioner was authorized to appoint a staff of special examiners and clerks. A library was established to house works of science and scientific periodicals (both foreign and American), and a special patent office building was provided for.

A *Register of Patents* was inaugurated. Of special interest was the granting of Patent No. 1 for an invention of "a locomotive steam engine for rail and other roads, designed to give a multiple tractive power to locomotives, and prevent the evil of the sliding of the wheel." Appropriately enough, the patent was issued to Senator Ruggles, author of the new Act. The patent system of 1836 has continued in principle to this day. "It came at a time," declared Senator Orville H. Platt, "when the young nation had begun to develop its vast natural resources . . . , at a time when the American brain evolved and the American hand fashioned labor saving machines that made this nation throb with new energy and new life." [12]

Interchangeable Parts. The growth of the system of interchangeable parts was another development in technology that hastened the passing of the homespun age. First introduced by Eli Whitney in 1798 in the manufacture of guns, this new process was applied in a number of fields. Samuel Colt's invention of the revolver, patented in 1836, was a notable application of this principle.

The Colt Revolver. Samuel Colt was only sixteen years old when he whittled out a model from a wooden block of what was to become the revolver. A Connecticut Yankee from Hartford, he was apprenticed to sea, and got his inspiration from watching the pilot at the ship's wheel. He conceived the idea of a revolving cylinder, consisting of a number of chambers that could hold charges. His device rotated a many-chambered breech in a single barrel firearm by the action of cocking a hammer. He automatically locked the cylinder by the same action, that of cocking the hammer. He placed partitions between the nipples, and devised a way to bring the revolving cylinder in line with the hammer (and eventually in line with a single barrel) through which the shot (cartridge) was discharged.

Unable to get financial backing for manufacturing and marketing his weapon in the United States, Colt sailed for England and France and took out patents for his revolver in both countries. Upon returning to the United States he took out a patent in 1836. Despite the endorsement of President Andrew Jackson and the recommendation of some army officers in the Seminole War, Colt was unable to get his revolver adopted by the War Department during Van Buren's administration. As a result, his first factory closed after being in operation for six years.

Then came the Mexican War. Orders began to pour in for this newest invention, a repeating firearm. It was just what the Texans needed to fight the Mexicans and Indians mounted on their ponies. A weapon that fired six rounds as fast as the finger could pull the trigger wrought terror to the enemy. The "six-shooter," as it came to be called, played a role in the saga of mid-nineteenth century conquest similar to that played by the Pennsylvania (Kentucky) rifle in the eighteenth century.[13]

Rising Standards of Living

The passing of the homespun age brought about many changes in the American way of life. It created new occupations, made possible more leisure, stimulated the growth of urban communities, and brought new improvements to the cities. It also brought about changes in the homes and in the living habits of people. Families were no longer entirely self-sufficient, nor did everyone look to the soil for his livelihood. Men and boys went into the factories as mill hands or operators. Machinists, engineers, and firemen gained recognition as specialized groups. In the new occupations there were those who tended the toll gates on the turnpikes, and those who drove mules along the tow paths of canals.

Changing City Requirements. The influx to the cities created demands for more houses and a desire for improved facilities. Modern conveniences, such as gas lighting, central heating, and a running water system, began to be introduced into the homes during this period. Boston and New York had gas lighting in the late 1820's. In 1835 a gas mantle made of lime was introduced which greatly improved the quality of street lights. Stoves supplanted the open fireplace. Then came a central heating system. This led to the warm-air furnace, which in its early stage consisted of large stoves encased in brick. Cold air was admitted through a casing at the top or bottom, and the heated air delivered from the top. Coal, when available, was preferred to wood as a fuel.

Running water was brought into a few homes during the 1830's. The water pipes were usually made of wood, although cast-iron pipes were occasionally used. A few could afford an even more costly pipe made of sheet copper, tinned before it was cut and soldered. Lead was used where it was necessary to make a bend in the pipe. Bathrooms were installed as early as 1820 in a hotel in Philadelphia. The Tremont House in Boston in 1828, and the Astor House in New York in 1832 took the lead in introducing plumbing and toilet facilities.

Home Furnishing, Food Preservation, and Clothing. In the matter of furnishings, carpets began to replace sanded floors and the occasional rag rugs. Dishes, cutlery, and porcelain were no longer confined exclusively to the wealthy. Cabinet furniture and other "luxury" articles, fashioned by a factory power lathe, began to compete with the products of the home cabinetmaker.

There was also a change in food habits during this period. The canning process, developed by the Frenchman Nicholas Appert, had spread to America, and in 1819 Ezra Daggett and Thomas Kent began canning oysters in hermetically sealed jars. The following year the firm of Underwood and Mitchell was selling pickles, jams, and jellies preserved by the same process. Sylvester Graham, in his new treatise *Science of Human Life,* was expounding the use of the special type of flour that today still bears his name.

In the matter of clothing, the homespun age was also passing. Master tailors began to advertise their excess suits for sale as ready-to-wear clothing. In 1831, one George Updyke opened a small tailoring establishment in New York, and a branch shop in New Orleans where ready-made clothing could be procured. When tailors began to use sewing machines, following Elias Howe's invention of 1846, another step was taken in the passing of the homespun age. And hatmaking, which involved a few complicated machines, was being transferred from the back room of homes to shops equipped with small steam drums, dryers, and hand presses.

Triumph of the Machine Age

It was not only the number of labor-saving inventions, but also the accelerated pace at which they were introduced that characterized the passing of the homespun age. An English observer wrote in 1836 that the Americans were producing machines for every purpose, "a machine to peel apples, another to beat eggs, a third to clean knives, and a fourth to wring clothes." And the scholarly Virgil Maxey appropriately wrote in a Phi Beta Kappa paper in 1839, "These ingenious Americans with their new machines are rapidly gaining control over the power of the elements, and are making fire, air, earth, and water their ministering servants." He declared that an understanding of science was the key to American prosperity. "They (scientists) stand as it were on the confines of nature, and by their magic touch give life and energy to the brute forces of the world."

The philosopher Ralph Waldo Emerson, impressed by the technical skill of young Americans, was moved to "pity for our fathers for dying before steam and galvanism . . . , and before we borrowed the might of the elements. These forces now serve us day to day," he added, "and cost us nothing." Salmon P. Chase in an article, "The Effects of Machinery," written for the *North American Review* (January, 1832), commented on the revolutionary changes in American life because of the machine: "With the introduction of pumps, gears, and other machinery," he declared, "we force the very atmosphere we live in to raise our water from the wells and from rivers, and to aid in a countless variety of operations." And he noted that man was also harnessing the running streams, and using their power to spin, weave, and grind.

While certain foreign techniques were introduced during the passing of the homespun age, they were consciously altered to meet home conditions. One by one the products were made cheaper in the fast growing factories than they could be made in the home. Mechanical power, as Alexander Hamilton had prophesied, was overcoming the handicap of the scarcity of labor, and America was becoming geared to a system of manufactures.

Notes for Chapter 12

1. Jacob Bigelow, *Elements of Technology* (1829), p. 28 *ff*.
2. George Sweet Gibb, *Saco-Lowell Shops: Textile Machinery in New England (1813–1949)* (1950), p. 10.
3. *Ibid.*, p. 10.
4. *Ibid.*, p. 14.
5. The same idea is employed in the device that keeps the modern film in a motion picture projector from looping.
6. Up until this time the only products were blue jeans and coarse cotton muslin.
7. George Wallis, *Special Report of New York Industrial Exposition* (London, 1854), p. 11 *ff*.
8. George S. White, *Memoir of Samuel Slater* (1836), p. 186.
9. The Harmonists, followers of George Rapp at Economy (Ambridge), near Pittsburgh, Pennsylvania, developed a number of industries including woolen, cotton, and silk manufacturing.
10. Blanche Hazard gives an excellent analysis of the industry in *The Organization of the Boot and Shoe Industry in Massachusetts Before 1875* (1921).
11. Frederick Overman, *The Manufacture of Iron* (Philadelphia, 1850). This production was not continued. It is not known that the Institute's offer was ever claimed. The most notable success was achieved in three furnaces in the Frostburg basin area of western Maryland.
12. *Congressional Record, 48th Congress,* 1st Sess. (March 31, 1884), p. 2408 *ff*.
13. See William B. Edwards, *The Biography of Col. Samuel Colt* (1954); Charles T. Haven, *History of the Colt Revolver* (1940); and Walter Prescott Webb, *The Great Plains,* chap. V (1931).

For general reading, see Jared Van Wagenan, *The Golden Age of Homespun* (1953).

Chapter 13

TRANSPORTATION TECHNOLOGY TO 1850

Improvements in transportation were essential in effecting the passing of the homespun age. Where homes were isolated, domestic technology continued in the handcraft stage, each household being self-sufficent. But specialization in industry necessitated an improved transportation system, both for exchange of products and for procurement of the essentials of food and clothing.

Expansion of Highway System

The building of roads, the first improvement in opening up transportation routes, was undertaken in the early days of the republic (see Chapter 5). The macadam roads built by turnpike companies proved the most durable of the improved roads. Although this type of stone construction came from England, and was based on the principles developed by Telford and Macadam, the American roadbeds were deeper than the English design, and road surfaces were more sharply convexed.

The Cumberland Road. The most famous of the macadam highways was the Cumberland Road, which the English traveler David Hume declared, "was made forever, not like the flint roads in England, nor soft and dirty like the gravel roads, but smooth and hard."

The Cumberland Road was authorized by Act of Congress in 1806, and extended from Cumberland, Maryland, to Wheeling, Virginia (now West Virginia). The period of construction extended over thirty-two years. Specifications called for a roadway 20 feet wide to be constructed of crushed stone 18 inches deep in the middle, and 12 inches on the sides. The roadbed was to be constructed with good solid shoulders, and a ditch or water course was to be provided on each side. The size of the stone used was specifically stated: for the bottom layer, stones must be no larger than would pass through a 7-inch ring; stones on the top layer had to pass through a 3-inch ring.

Problems of Construction. Almost insurmountable barriers were overcome in the construction of this highway. Virgin forests were felled, and the ranges of the Appalachian Mountains were scaled. Innumerable streams were bridged, and swamps were filled. Severe weather conditions hampered progress. Only primitive tools were available, such as carts, picks, shovels, and crude blasting tools. A round 2-pound hammer was used for breaking the stones. The surveyor's only instruments were a Gunter's chain and a circumferentor. Nevertheless, contractors like Mordecai Cochran with his crew of several hundred Irish workers pushed forward and built a road that commanded the admiration of the nation. It was one of the few early roads in the United States laid out and built prior to 1890 according to scientific and engineering principles.

Turnpike Development. Hard-surfaced roads were built with varying degrees of perfection in a number of states. Pennsylvania, which had pioneered this type of construction with the Philadelphia–Lancaster pike in 1792, led in the number of miles of turnpike construction. Other states had similar projects. Maryland ranked second in extent and value of her highways. In Connecticut as many as fifty turnpike companies were incorporated in the 1830's. Not all the road construction was of stone, however.

A variation of the stone road was the stone-rail road, in which stone slabs were laid for wheel tracks and the center filled with cobblestone. The outstanding example of this type was the Albany–Schenectady Road, erected in 1834. Such roads were especially conducive to speed and provided smooth riding. The ease in securing traction on these slabs led to an increase in loads, until 4 and 5 tons became the ordinary load for two horses, and 10 tons for four. Some sections of roads built with these stone rails were in use as late as 1900.[1]

Plank Roads. The decade from 1845 to 1855 witnessed a mania for the construction of plank roads. The planked-road movement began in New York State, spread as far south as Alabama, and extended westward to the Mississippi. These roads were built as toll roads by private companies. The plank roads offered easy riding and were easier on horses and carriages than the hard roads. Ten or twelve miles could be covered on a plank road in less time than five or six miles on a common road, and heavier loads could be hauled.

In building a plank road the bed was constructed similar to any other road. Then two rows of wooden stringers, 3 inches thick and 6 inches wide, were laid 3 feet apart, with the intervening space filled with earth. Across these stringers was laid a plank floor, 3 inches thick, and from 8 to 12 feet wide. When finished, the plank road afforded the smoothest road ever constructed prior to the concrete roads of the nineteenth century. The plank roads could be used twelve months in the year. But the excessive cost of maintenance doomed them to failure. While the original cost of construction was about two thousand dollars a mile, the lifetime of a road was but five or six years without repair. The era of the plank road passed soon after the close of the Civil War, although a few were in use as late as 1900.

Roadmaking Machinery. Little improvement was made in road-making machinery until 1858, when Eli Whitney Blake's stone-crusher was introduced. Blake's machine "operated on the principle of a strong mechanical jaw, which crushed the stone by power" operated by a belt from a steam engine. No successful grading machinery was introduced until many years later. Road building had not yet become a profession, since most engineers considered road construction as nonprofessional work. Their only contact with road construction was in the erection of bridges. Meantime, poor as the roads were, a few light horse-drawn implements were introduced. In 1831 a sulky scraper was patented. This was little more than a drag to scrape the dirt from the ditch to the center of the road. A roller drawn by horses was devised as early as 1843 on a principle in use in Germany, but it was not until it was made applicable to steam power that it became effective. It should be observed that the lack of road construction in the United States was not unique. Other countries had made few improvements, and in no country did the roads compare with the old Roman roads built two thousand years earlier.

Bridge Building. In bridge construction, however, considerable progress was made, and America soon became famous for some of the finest bridges in the world. The construction of permanent bridges in this country attracted the interest of engineers everywhere. Many of the fine stone bridges on the Cumberland Road remain standing today.

Some noted American bridge builders were Theodore Burr, who erected the Trenton bridge in 1805, and Ithiel Towne, who in 1820 invented a special type of truss and erected some covered bridges for highways. Despite these pioneer attempts however, bridge construction could hardly be described as a scientific project at this early date. It was not until the middle of the century that the first correct analysis of stresses appeared in a work by Squire Whipple. Thereafter, mathematics was substituted for rule-of-thumb methods in bridge construction.

Modes of Vehicular Travel. While American roads were often citicized, European travelers were generous in paying tribute to the superiority of American vehicles. These were lighter and afforded easier riding than European models. Coaches, wagons, buggies, carts appeared on the highways, and in the wintertime a variety of sleighs was used. For public transportation the stagecoach was adopted. These vehicles had three transverse seats to provide a seating capacity for nine passengers. In the boot at the rear was space for luggage. Leather spring suspension, while it produced an oscillating motion, was decidedly easier riding than the heavy jolting of the body resting on the wheelbed. By 1835 the flat roof had supplanted the convex, and thus additional baggage and passenger space was afforded.

Improved coaches, together with improved roads, facilitated speed in transportation. In 1820 it took seventy-eight days to carry a presidential message from Washington to Little Rock, Arkansas, but ten years later the feat was accomplished in fourteen days. In 1826 John Quincy Adams reported that the journey from Boston to New York consumed four days. In 1832 the trip could be made in forty-one hours.

The most famous of the stagecoaches was the Concord coach introduced by Lewis Downing in Concord, New Hampshire, in the late 1820's. It became the forerunner of the coaches that were destined to roll across the western prairie region for the next half-century. These coaches were noted for their lightness, durability, and elegance of finish. They were fashioned entirely by hand, made from well-seasoned, carefully selected wood, each part of a special variety. The coaches weighed from 1,800 to 3,000 pounds.

Skilled craftsmen at Troy, New York, were the makers of another line of stagecoaches. Here Charles Veazie, Erasmus Eaton, and Uri Gilbert manufactured coaches that plied the highways of almost every state in the Union. Travel by stagecoach appealed to the masses. In 1830 more than seventy coaches were running out of Boston. Simon Cameron, in 1833, ventured a bold prophecy when he said, "A child born within this decade will, before he dies, be able to eat breakfast in Harrisburg and take supper in Philadelphia the same day, almost one hundred miles away."

Accessories on the stagecoaches included mud boxes and sand boxes, leather curtains, brakes, fluid lamps, side candle lights, boot and bag footboard, and corner plates. The coaches were gaily painted and bore names vying in distinction with Pullman cars of a later period. Landaus, barouches, and carriage coaches were made for private ownership, while for the average farmer there was that distinctly characteristic vehicle, the buggy. This four-wheeled conveyance, drawn by one or two horses, had a collapsible top that could be raised and lowered at will. It was America's distinct technological contribution to light, speedy vehicular transportation. Rubber tires, which were invented by Robert Thompson of London in 1846, were seen on the streets of New York the next year, mounted on the wheels of a black, low-bodied buggy.

Cabs made their appearance in New York in the 1840's and created something of a sensation. Crowds gathered in the streets to view this new vehicle, and a New Yorker was heard to remark that their appearance was evidence of the city's determination "to keep pace with the onward march of improvements." The New York *Daily Mirror* chided its readers for not having introduced this particular type of conveyance ahead of the Europeans. "America is not in the habit of waiting for such improvements," the article stated, "especially in matters pertaining to mechanical devices."

Passing of the Toll Roads. The coming of canals and railroads had a disastrous effect on toll roads. The income dwindled until there were scarcely sufficient funds from which to pay the toll-keeper's salary. The national roads were transferred to the states within whose borders they lay, and the building and maintenance of roads became a local project. In local hands the fear of increased taxation led to the adoption of the old English system of requiring the personal services of the rural population on the highways to keep them in repair. This led to the custom of maintaining the highways by "working out" the road tax. In many communities roads became little more than rough trails for communication between neighboring farms.

Canal Age

The canal age extended from 1817, the date of the beginning of the Erie Canal, to the late 1840's. Although there was nothing new about the technology of canal construction (it having been developed abroad), the second quarter of the nineteenth century was marked by such a frenzy of canal development in the United States that it has been characterized as the Canal Age. Dunbar described it as "a sudden, exotic, sporadic, and forced phenomenon." It was short-lived, not because America had outlived the usefulness of the canal, but because its development was premature and because the country lacked an appreciation of its future needs.

First Canals. Canal construction was advocated as early as 1772, when Franklin issued his *Canal Papers.* George Washington later proposed a canal to connect the Potomac and the Ohio Rivers. A ditch called the "Dismal Swamp Canal" was constructed in Virginia in 1787. This was soon followed by the Middlesex Canal in Massachusetts in 1795–1809. Gallatin's *Report* in 1808 proposed a complete system of inland waterways connecting the North and the South, and extending by way of Muscle Shoals to the Mobile River. He suggested a second system between the East and the West, and in addition numerous interconnecting canals. These, Gallatin believed, should be built at federal expense.

Canal construction involved the building of artificial waterways. These were usually 36 to 40 feet wide at the top, 24 to 28 feet wide at the bottom, and 4 feet deep. They were provided with a system of locks and dams to maintain a constant water level. Canal boats and barges were privately owned, and the canal construction company charged a toll for permission to use the canal. The canals were originally designed for the transportation of freight, but they came to be used extensively for passenger travel, even for the transit of houseboats bearing whole families to new western homes.

The most important canals, and those that had the greatest effect on the development of the country, were the Erie Canal, connecting the Hudson with Lake Erie; the Pennsylvania Portage Canal from Philadelphia to Pittsburgh; the Chesapeake and Ohio Canal which extended from Georgetown to Cumberland, Maryland; the Ohio Canal from Portsmouth to Cleveland; the Miami Canal from Miami, Ohio, to Toledo; and the Wabash and Erie Canal which extended the system out into Indiana.

Erie Canal, 1817–1825. The Erie Canal was America's first great man-made waterway, and its construction shaped the future of the nation's history. During the first quarter of the nineteenth century, New Orleans was contending with New York for the position of America's leading port. Downriver navigation favored New Orleans and, but for the daring undertaking of canal engineers inspired by Governor Dewitt Clinton, New Orleans might have eclipsed New York City as America's leading port. Steamboats had captured the inland trade and were directing it down the river to New Orleans. The East and the West were being divided instead of united, and sectionalism was threatening the country's future.

When the Erie Canal was completed, the trend was changed. The East and the West were united. It cost one hundred dollars to transport a ton of freight from New York to Buffalo in 1810; after the Canal was completed in 1825, it cost only five dollars. Travel was cut from twenty days to eight days. In one month in the year after the Canal was completed, more than eight hundred barges left Albany for the West. New York City soon surpassed New Orleans in the South and Philadelphia in the East in both population and wealth, a position which she has never surrendered.

Pennsylvania Portage Canal. Unique in construction was the Pennsylvania Portage Canal, which extended about 400 miles from Philadelphia to Pittsburgh. It was completed in 1835. This portage canal project involved a system which included a railroad, cable transportation, and canal passage. The state built and owned all the links of the traffic system and installed and operated the motive power. The first unit of the system consisted of rail transportation from Philadelphia to Columbia. This involved horse-drawn, and later steam-propelled, railroad cars fitted with iron-flanged wheels and operated over iron-topped wooden rails. From Columbia to Hollidaysburg transportation was by a canal. From that point to Johnstown, the Allegheny mountains were surmounted to a height of 2,300 feet. This was effected by a series of inclined planes. The portage over the mountain involved one of the world's outstanding engineering feats of that day.

To conquer the mountain barrier a series of levels, with five inclined planes on one side of the mountain, and four on the other, was employed. Canal boats were drawn up to the top of the mountain, and lowered down the inclined planes by means of rope cables made of hemp. A stationary engine installed on each plane level supplied the power. The ropes, 7 inches in circumference, ranged in length from 3,500 feet to over 6,000 feet. An English observer described the Pennsylvania Portage Canal System as the greatest engineering feat since the building of the pyramids of Egypt.

Failure of the Canal System. The canal age was short-lived. It left ruined investors and states with heavy debts, some of which were later repudiated. The railroad presented a challenge to the slow-moving canal boats and soon drove them out of existence. The tow paths became overgrown, and the waterways sank into mere ditches. Before many decades passed the only reminder in many towns of the once-prosperous canal days was the name of a street called Canal Street. The canal was at best a temporary and an inadequate answer to the need for inland transportation. Its picturesqueness is preserved in the nation's literature, and the slogan

"From tow path to the White House" helped to elect James R. Garfield to the presidency of the United States in 1880.

The Railroad Age

The coming of the railroad marked the greatest advance in transportation technology of the pre-Civil War period. The "puffing billy" with the cloud of smoke and shower of sparks signaled the doom not only of the slow-moving arklike canal boats, but of lumbering stagecoaches as well. By 1850 the advances in railroad engineering had achieved the two great American objectives in transportation technology, speed and power. Railroads could be extended wherever they were wanted: they could "go where a mule could go." They were dependable, not seasonal as were canals and highways; nor were they obstructed by snow and ice, or stranded by drought or mired down in the mud of the rainy season. Emerson had written in 1846 that "the Americans take to this little contrivance, the railroad, as if it were the cradle in which they were born."

Railroad Engineering Technology. It is significant that, during these early years of railroad building, virtually all fundamental technological principles in railroad construction were developed. Engineers second to none were produced despite the dearth of opportunities for education in engineering, and the vision of a nation held together by bands of steel was conceived. From such engineers as George Washington Whistler, Wm. Gibbs McNeill, John A. Roebeling, Benjamin A. Latrobe, and Moncure Robinson, plans were designed to tunnel through mountains, bridge valleys, construct grades over rugged terrain, and alter streams in their courses.

Engineering Problems. The first task in constructing a railroad was surveying the route. The profession of civil engineering was just beginning at the dawn of the railroad era. During the early stages, engineers were so hampered by the expense of the project and so hard pressed to complete their tasks that they were forced to sacrifice scientific surveys for expediency. Quick, low first-cost construction was more important than permanence in construction. Sharp curves, steep grades, and winding courses were characteristic of the early railroads. "Lay the tracks first and build the road later," became the slogan. Despite these obstacles, most of the main lines in the East were laid out in the paths they follow today. Along many roads the old tracks can still be seen, and the same crossings are made, and the mountains are scaled at the same places that the early engineers planned.

At the opening of the railroad era no one believed that a locomotive with smooth wheels on a smooth track could pull cars up more than a trifling grade. The same skepticism existed regarding curves. Ascents had always been made by rack and pinion or by cable methods. These beliefs were a challenge to the early engineers. They discovered that curves considered impossible to negotiate were really advantageous, because they could be used to climb grades that had been thought too steep. The prime purpose of Peter Cooper's "Tom Thumb" was to demonstrate that a railway locomotive could negotiate curves; and railway engineers soon learned how to use curves.

Moncure Robinson built more railroads in the first two decades of the era than any other man, and much of his work endures today in location of routes. Robinson foresaw the day when railroads would put an end to canal construction. He knew the engine could out-pull the horse, and that railway wagons could carry more tonnage than a canal boat. Stephen Girard, the canal builder, engaged Robinson to build railroads and to make them double-tracked.

Early construction of the roadbed was performed with pick and shovel and mule-drawn scrapers. Some early machines were introduced, however. A steam pile-driving machine was used on the Utica and Syracuse Railroad in New York in 1840. In Massachusetts a steam excavating machine that lifted the earth and dumped it on the railroad cars was introduced about the same time. William S. Otis developed a steam shovel which was used on the Baltimore and Ohio Railroad in the 1830's and which introduced the fundamental principles of the modern steam shovel.[2] In general, however, the early railroads were constructed by hand labor.

Improvement in Construction Methods. One of the first advancements came in the improvement of the rail. Robert L. Stevens, the son of John Stevens, and a pioneer in railroad building, was sent to England to purchase locomotives. On shipboard he whittled out the design for a new rail, the "T" rail, that soon became standard equipment in America. Stevens' rail differed from the British in that it had a broad base, designed to be attached to the ties, and was not set in an "iron chair," as was the British custom. Stevens persuaded an English manufacturer to produce a number of "T" rails, which were shipped to the United States. They were first used on the Camden and South Amboy Railroad in 1832.

The spike, a long square-headed pin used to attach the rail to the tie or sleeper, also designed by Stevens, proved surprisingly successful, since the wooden ties supplied the resiliency desired in track construction. Another device, which became permanent, was the fish plate, or joint bar, used to connect the abutting rails. The idea of filling between the rails with stone ballast was introduced by Moncure Robinson. Other railroad builders followed these principles, which became standard features of track construction on American railroads.

In the meantime, other technical devices were being tried. Because of the prevalence of wood, strap rails made of wood with a bar of iron on the surface first came into general use. These were known as "snake rails" because the iron would often break away from the wood and come through the floor of the cars, much to the discomfort of the passengers. Iron rails were made in other patterns, such as the "U" rail, the "ridge" rail, and the "H" rail. Rails were made of cast iron as well as rolled iron. At first all iron rails were imported, but in 1844 iron rails were produced in the United States at the Savage Iron Works in Maryland. By 1850 there were fifteen mills in the United States producing iron rails, with a total production of more than 45,000 tons.

Conquering Natural Hazards. The pioneer surveyors and engineers made the railroads go where they wanted them to go, despite hills, rivers, or swamps. Bridges were flung across the streams, trestles stretched out across the low places, and tunnels were bored through the mountains. Many of the bridges and trestles were flimsy wooden affairs and sometimes collapsed from their own weight. Trestles were more or less temporary, because the railroad builders preferred an earthen foundation and planned for a fill to replace the trestle.

The basic features in tunnel construction had been established long before the railroad era. Tunnels had been employed in the construction of canal routes, and the construction was continued in railroad use. The most ambitious tunnel attempted was the one in the Berkshires begun in 1849, the famous Hoosac Tunnel. It was 4.7 miles long and required twenty-five years to build.

More advances were made in bridge construction than in tunnels. While most of the bridges were of wood, erected without thought of permanence, others were built to endure. Such was the masonry arch at the Carrolton Viaduct in Maryland, erected in 1829 by Benjamin H. Latrobe; it is in use today carrying loads a hundredfold greater than was originally expected. Robinson designed the first iron railroad bridge, erected just outside Philadelphia. (A segment of this old bridge is today in the Smithsonian Institution.) Plate girder construction, an assembly of riveted plates and angle irons that served as support for the rails in spanning viaducts, was devised.

The truss form of bridge construction emerged in the 1840's, when

William Howe introduced the timber truss. In 1844 Caleb and Thomas Pratt constructed the truss framed with timbers. Whipple, three years later, used the timber and iron truss, and by mid-century Wendell Bowman had built an all-iron truss. There were streams so wide and so deep, however, that no combination of trusses and girders could span. Such was the Niagara River, and for this John Roebling built the world's first railroad suspension bridge in 1855.

The Railroad Locomotive

The first locomotives used in this country were Stephenson engines imported from England. In 1829 the Delaware and Hudson Canal Railroad imported a British locomotive, the "Stourbridge Lion." Horatio Allen, the engineer, thought it was too heavy for the track, especially the 9 miles of trestle on the road, but risking his own life, he drove it over the road without a mishap. After the initial trial, the "track smasher"—as it was dubbed—was stabled, and the "Stourbridge Lion" became only a memory.

Adoption of Steam Engines. When the Baltimore and Ohio received authority from the Maryland Legislature in 1827 to build a railroad, the company began experimenting with various types of propulsion such as horse-drawn vehicles, sailing cars, and steam-propelled cars. Peter Cooper made a trial run with his miniature engine, the "Tom Thumb," which drew a car with twenty-four passengers, on a 13 mile track from Baltimore to Ellicott City at the rate of 6 miles an hour. The event was highlighted by a race with a horse-drawn vehicle over a parallel track. Unfortunately for Mr. Cooper, something went wrong with the engine and the horse won.

It was left to a railroad in South Carolina to be the first in the country to adopt formally the use of steam, and build a road with the special purpose of operating a steam locomotive. Horatio Allen of "Stourbridge Lion" fame was placed in charge of construction, and the first American-built locomotive, the "Best Friend of Charleston," began operation on this line in January, 1831. This road also has the distinction of having the first locomotive explosion, for the engine blew up the following June. The engine was made without a tender; it carried its own wood and water. The wheels were of wood, with spokes like those in a wagon wheel, but were rimmed with wrought-iron tires. The engine attained a speed of 21 miles an hour, pulling four cars loaded with passengers.

In August, 1831, another American-built locomotive made history for the Mohawk and Hudson Railroad on a successful run from Albany to Schenectady. This engine, called the "Dewitt Clinton," was 10 feet long, had large wheels, a lofty smoke stack, and a central dome. A flatcar attached to the locomotive served as a tender and carried a couple of barrels of water and a pile of wood. The engine attained a speed of 30 miles an hour on part of the route. These early trials of American locomotives and railroads opened the way for later improvements that brought about the great American railroad system.

Development of American Locomotives. Improvements came rapidly in American-built locomotives. John B. Jarvis, of the Mohawk and Hudson Railroad, who had so showered his passengers with sparks on the trial run of the "Dewitt Clinton" that fire companies had to be called out, decided that the English locomotive was not suitable for American tracks. After a number of new technical changes were made, an American type of locomotive was built.

Jarvis rebuilt the English-made "John Bull." He placed a separate four-wheeled truck under the front part of the boiler, attaching it to the frame by a king bolt in the center. He moved one pair of driving wheels to the rear of the firebox. With the front truck, which operated as a swivel, he could manipulate the curves on the American tracks. The drivers distributed the weight and the engine was balanced so that the track was not so badly battered. By increasing the size of the drivers, greater speed could be attained. On this "boggie" truck, as it came to be called, the position of the wheels on the frame became the distinguishing feature of what developed into the most popular engine in the country.

Isaac Dripps, working for the Camden and Amboy Railroad about the same time, introduced another idea. He steadied his engine with a plowlike contrivance which ran on two wheels ahead of the engine, and developed the device called the "cow catcher."

Coleman Sellers and his two sons, Charles and George, working with John Brandt of the Philadelphia and Columbia Railroad, invented a system of counterbalances which offset the disturbing forces that made a locomotive hard on tracks. Their design was universally adopted and it also became an essential part of the American type of locomotive. Finally, when Joseph Harrison invented an equalizing beam by which the weight was distributed equally to each driver, regardless of how much the engine rocked or jumped on the fragile track, the major mechanical features of the American locomotives were established.

William Harris of Philadelphia contributed another major technological feature when he built the grade-climbing locomotive "George Washington" in 1835. This little engine, equipped with a single pair of drivers, hugged the rails and climbed a 7 per cent grade, hauling two passenger cars bearing fifty-three people. The feat became more spectacular when the engine stopped on the grade and started again on its own power. Although the English journals declared the accomplishment impossible, British railroad designers had more confidence and placed orders for American-built locomotives. The first engine of this order was shipped abroad in 1839 and proved satisfactory on the famed Lackey incline.

The "Gowen and Mars," a passenger locomotive of the Reading Railroad, designed by Moncure Robinson, pulled a train forty times its own weight, consisting of 101 four-wheeled freight cars, a distance of 58 miles from Reading to Philadelphia, at an average speed of 10 miles an hour. When the Czar of Russia heard of this feat, he made an offer to Mr. Robinson that caused the firm of Harrison and Eastwick to send engineers to Russia to set up a branch of their locomotive works.

Engineering Refinements. Thomas Rogers of Patterson, New Jersey, who, like Coleman Sellers, had been tackling the problem of the locomotive's battering the tracks, made his contribution in attaching weights to the wheels to counterbalance the momentum of the reciprocating parts. To facilitate this he introduced hollow cast-iron wheels instead of wooden ones with iron rims, as had been used theretofore. Another development by Rogers led to the institution of the link device that facilitated stopping, starting, and reversing the locomotive.

Two other pioneers who contributed immeasurably to the development of the American locomotive were watchmakers by trade. They found that the accuracy and precision they employed in designing, shaping, and processing the delicate springs, balance wheels, and jeweled movements in watches qualified them above all other artisans to turn to the mightier, but none the less exacting job of boring cylinders, making pistons, and gearing the huge drive wheels of America's locomotives.

One of these watchmakers who took up the designing of locomotives was Phineas Davis of York, founder of the firm of Davis and Gartner. In 1831 Davis was the successful competitor for the four thousand dollar prize offered for a successful locomotive by the Baltimore and Ohio Railroad, after Peter Cooper had demonstrated the possibilities of steam in the "Tom Thumb" race. The "York," built by Davis, had a vertical boiler. The engine was mounted on a flatcar having four wheels, 30 inches in diameter. It weighed 3½ tons and could pull a load of 15 tons at a speed of 15 miles an hour on a level track. One marked feature of this locomotive was that anthracite coal was successfully used as fuel. The "Atlantic," which Davis built the next year, was similar in appearance, although it had two road wheels 36 inches in diameter, while the drivers were only 28 inches. This locomotive had the distinction of remaining in service until 1892.

Matthias W. Baldwin, another watchmaker, while not the first to build an American locomotive was, more than any other person instrumental in developing it, improving it, and making it one of the great forces in railroad transportation. Although Baldwin carried into the building of locomotives the same precision he gained at the jeweler's bench, his experience in other lines also proved to be valuable training. He had made tools for book publishers, copper cylinders for printing calico, and a small horsepower, vertical type of stationary steam engine.

Baldwin started to work on his first full-sized locomotive in 1832. Following the plan of the English locomotive, and studying the knocked down "John Bull" for specifications, he built the famous "Old Ironsides." It had a horizontal boiler, four wheels, and was modeled after the British type. Baldwin found it necessary to devise tools with which to operate. The cylinders were bored with a chisel fixed in a block of wood and turned by hand. Even though the locomotive performed successfully, it was termed the "fair weather" locomotive, since the printed schedule announced that it would run from Philadelphia to Norristown, and be operated by steam power "if the weather was fine," but would be drawn by horses in case of bad weather.

After building "Old Ironsides" Baldwin decided to use more of his own ideas in his later locomotives. He adopted the four-wheel swivel truck and provided a single pair of drivers behind the firebox. He employed other mechanics, and by 1839 he had built one hundred and twenty locomotives. His plant was geared to turn out one complete engine a week. Before his death in 1866, Matthias Baldwin had built fifteen hundred locomotives for American and foreign railroads.

Baldwin's technological contributions to the locomotive industry were such as would be expected of a watchmaker. He ground the joints for his steampipe connections, making possible 120 pounds pressure instead of half that amount used in British locomotives. He fitted the coupling rods, which transmitted the power to the wheels, with flexible connections so that there would be no binding as the truck rounded the curves. He improved the design of the driving axles and improved the water pump. He was among the first to construct a locomotive with a cab. He devised the sand box for slippery tracks. The great objective in Baldwin's construction was to combine power to draw a maximum load, with sufficient flexibility to operate on sharp curves.

Although Peter Cooper's "Tom Thumb" and Davis' "York" burned anthracite, most American locomotives continued to be built as woodburners. In most sections of the country wood was cheaper than coal, but as the country developed and the forests disappeared the roads on the seacoast sought a substitute fuel. Ross Winans, who had introduced many new features in his construction of engines for the Baltimore and Ohio Railroad, pioneered in the use of Cumberland coal. The Reading Railroad, located in the heart of the anthracite region, built a coal-burning engine in 1847. Balloon stacks continued to belch forth their shower of sparks on most roads, however, until after the Civil War.

Locomotive Design. With the improvements in the technical features of the locomotive came improvement in design also. By 1850 the American locomotive no longer looked like a cook-stove-on-wheels, or a boiler mounted on a flatcar. The upright boiler of the "Tom Thumb" and the "York," stuffed with rifle gunbarrels for tubes, gave way to the horizontal placement. Tenders were introduced instead of the makeshift flatcar with a pile of wood and a couple of barrels of water carried behind the engine. The closed cab was added for protection of the engineer and fireman, and the cow catcher had acquired a standard design.

The bell rope, the whistle, the headlight became standard equipment. Wheels and tires were improved. Speed was increased by increasing the size of the drivewheels which often measured 7 feet in diameter. Greater power was obtained by enlarging the boiler, so that locomotives came to weigh 10, then 20, and by 1850, 25 tons. C. H. Gregory in 1841 introduced the semaphore to regulate traffic. A system of lights was used at night on the ends of the arms of the semaphore, although the signal code was not yet universal. The telegraph was first used to determine the locations of trains in 1850.

The superiority of American locomotives was recognized at an early period, and foreign countries began placing orders with locomotive builders in the United States. The Norris Works in England placed an order in 1837 for eight American-built locomotives, with an order for eight more the following year. Baldwin's first shipment of locomotives was made in 1838 to Cuba. Orders came from Austria and Germany. Karl Gerstner, noted traveler, advised European countries to buy all their locomotives, engines, and tenders from America, and added that it would be of advantage for every railroad company in Europe to procure at least one train from America.[3] He declared that Russia and Germany would learn more from the United States railroads than they would by installing the British system.

Improved Train Design. Improvements were also made in railroad coaches. At first, passenger cars were little more than modified stagecoaches. These were coupled together with chains that allowed 2 or 3 feet of slack. Taking up this slack subjected the passengers to a violent jerk when the train started or stopped. The development of the long eight-wheeled cars, each provided with fifty to sixty seats, was gradual. Since Americans frequently made long trips, comfort in travel was demanded. On some roads special cars, carpeted and equipped with wide comfortable sofas, a dressing table and a washstand, were provided for ladies. Though the diner had not yet been designed, some cars had bars where refreshments were served. The cars were lighted with candles and heated with stoves.

The original freight cars consisted of plain flatcars on which the freight was piled and tied down securely with ropes. In some cases the sides were built up and a canvas supplied to protect valuable merchandise. By the 1840's a few crude boxcars had been introduced.

Economic Effect of Railroads. The economic and industrial development of the United States has been largely dependent upon the development of the railroads. When the railroads first appeared, the population numbered less than thirteen million. By 1850, with a population of more than twenty-three million, there were more than 9,000 miles of railroad binding together the twenty-five states east of the Mississippi. New York State with its thriving Erie Canal trade had the greatest expanse of railroad, more than 1,300 miles. Pennsylvania with its busy Schuylkill Canal, the Portage Canal, and a number of waterways, had 1,240 miles of railroad. Massachusetts was third, with slightly more than 1,000 miles. Tennessee was the only eastern state in which no railroad had been constructed. Florida had only 21 miles of road.

From the beginning, the railroad proved not only a means of communication, an outlet for products, and a means of securing materials; it was also a trail blazer in opening up unsettled territory. It made vast regions accessible to the farmer, the lumberman, the miner, and the manufacturer. It carried the torch of civilization into the unknown territory. It was not merely a means of transportation and communication between civilized points; it promoted civilization.

The railroad had arrived by the mid-nineteenth century, and while a number of major developments were to be added during the next ten years, they were not revolutionary. By means of highways, canals, and railroads, the vast sprawling sections of the country were being linked together.

Notes for Chapter 13

1. Seymour Dunbar, *A History of Travel in America* (1915). This publication remains the most complete of all histories of early railroads.
2. For a good description of these early developments, see W. Barnes, *Excavating Machinery* (1928).
3. "Gerstner Letters," *Journal of the Franklin Institute,* IV (April 1839), 289.

While there are scores of books dealing with the history of American railroads, there is no satisfactory history of the major technological developments that have brought about the vast network of railway transportation.

Chapter 14

WATER TRANSPORTATION TECHNOLOGY TO 1850

There were numerous advances in water transportation technology during the first half of the nineteenth century. These included the introduction of the steamboat on inland rivers, the improvement of transportation on the lakes, the development of a new technology in trans-Atlantic commerce, an enlargement of the coastal trade (involving communication with California during the days of the Gold Rush), and a program for building a United States Navy. A high point in river and ocean transportation was reached before the Civil War.

Inland Water Travel

While Colonial America had established water communications with the rest of the world and had built up a prosperous coastwide trade, there had been no development on her inland waterways. Settlers moving inland followed the many rivers and streams until the Appalachian Mountains presented a barrier. The more venturesome made their way over the divide and were already established around the headwaters of the rivers that led to the West when the new nation was established. The first Federal Census (1790) showed that nearly all the settlements west of the Appalachian Mountains were located on rivers—Pittsburgh, Wheeling, Marietta, Cincinnati, Louisville, and others with lesser population. To them the broad stretches of the Ohio and the Mississippi and their tributaries were of paramount interest.

Rivercraft. In 1782 Captain Yoder is credited with taking a flatboat from Pittsburgh to New Orleans, the first commercial venture on the western rivers. In 1793 one Jacob Myers advertised in the Kentucky *Gazette* that he would carry passengers, letters, and cargo from Pittsburgh to Limestone, Kentucky. The same year a line of passenger boats (flatboats) was opened from Cincinnati to Pittsburgh. Regulations and timetables were posted, assuring a trip every week.[1] This line received encouragement from the government when, in 1794, it contracted to carry mails from Wheeling to Cincinnati. Even seagoing vessels constructed at Pittsburgh and Marietta, Ohio, found their way via New Orleans and the Gulf of Mexico to European ports. In 1801 commerce down the river netted almost 40,000 tons, and seaboard cities became interested in the western waterways.[2]

The early rivercraft included flatboats, arks, keelboats, and barges. The flatboats and arks were used for passage downstream only, for on arriving at their destination they were sold for lumber. The keelboats and barges made the tedious upstream transit propelled by poles, sweeps, and ropes. The keelboats were in themselves a technological achievement. Long and narrow, with sharp keels, their well-modeled hulls were constructed to pull a light draft. They continued in use long after the advent of the steamboat, especially on tributary streams. As late as the 1840's there were still more than a half-hundred keelboats plying the Mississippi.[3]

Steamboat on Western Waters. While the early experiments with the steamboat took place on the Atlantic seaboard, those identified with the beginnings of steam navigation realized that its greatest possibilities would be on the western waters. Rumsey, Fitch, Evans, and Fulton all had in mind the use of the steamboat on inland rivers. As noted earlier, Oliver Evans in 1802 built a boat equipped with a steam engine for use on the lower Mississippi.

It was left to Robert Fulton and Robert Livingston to make the first successful steamboat venture on western waters with the launching of the "New Orleans," built in Pittsburgh in 1811. The boat was a 371-ton vessel, 116 feet long, and cost thirty-eight thousand dollars. Commanded by Captain Nicholas Roosevelt, it arrived in New Orleans two and a half months after leaving Pittsburgh—not a very speedy trip, but it marked the beginning of the nation's most notable technological achievement in inland water transportation.

The "New Orleans" of 1811 bore little resemblance to the floating palaces immortalized by Mark Twain in his *Life On The Mississippi.* Like all new inventions the steamboat during its experimental stage had to have a number of improvements added. These were made in the design and structure of the craft, in the type of engine, in the location of the paddle wheel, and in its size and potential speed. Early boatbuilders discovered that what was most desired was a vessel that would travel in shallow water, at faster speed, and be capable of carrying a heavy cargo.

In Brownsville, Pennsylvania, located on the Monongahela river, near Pittsburgh, Daniel French and Henry Shreve began building boats that possessed several improvements over the "New Orleans." In 1812 they launched a small craft, "The Comet," of 25 tons, the second boat on the Ohio, which steamed down as far as Louisville. Next, they built "The Dispatch"; then "The Enterprise," a 75-ton vessel. The latter made its way back up the river from New Orleans to Louisville and then on to Pittsburgh. When, in 1817 "The Washington," another Shreve-built vessel, made an upstream voyage, all doubt about the success of steamboats being able to move against the current was removed.[4]

Passage from Pittsburgh to New Orleans in 1800 by flatboat and keelboat cost $160.00; and freight rates were $6.75 per hundred pounds. When the steamboat appeared passenger and freight rates were cut in half. By 1817 seventeen boats had been built; by 1830 the number jumped to two hundred and twenty-eight. Moreover, the steamboat was the principal instrument by which steam power was introduced throughout the United States. By the mid-1830's, steamboat engines supplied three-fifths of the steam power in use in the entire country. They had an average power capacity three times that of engines used in industry and on railroads.[5]

West Surpasses East. The building of steamboats in Pittsburgh and other frontier towns marked the beginning of a distinctly American technological development. Also, it brought about a revolutionary trend in water transportation. Throughout the Colonial and Revolutionary periods, New England had been the maritime capital of the nation, but following the appearance of the steamboat the tonnage on western rivers soon doubled that of all the Atlantic ports. In 1839 shipments on inland rivers amounted to more than 100,000 tons compared with 54,000 tons registered in the eastern ports. New Orleans ranked highest with a registered tonnage of 61,213 as against New York's 30,348; [6] and Pittsburgh, still a frontier settlement, rose to the nation's third ranking city in shipping tonnage.

To Henry M. Shreve goes credit for introducing the accepted design in steamboats, a vessel of shallow draft and low, flat hull with towering decks. Shreve built his boats to sail *on* the water, not *in* it. He rejected the shiplike hulls of the eastern-built boats and mounted his engine on a keelboat model. "The Washington," a 400-ton vessel, the largest boat afloat in 1816, possessed all these features. The boat had two decks, was equipped with luxurious cabins, and was driven by an engine of 100 horsepower.

Western Boat Design. Equally important as the shape and design of the hull was the type of engine built for the western river boats. Daniel French set the pattern in designing the simplest engine then in use. His engine weighed less than the flywheel on Fulton's engine, and employed neither beam nor flywheel, but acted directly on the stern paddle by means of a connecting rod and gears. French installed a high-pressure engine that became standard equipment on boats navigating western rivers. Many types of boilers were tried out, some upright, some horizontal, some plain cylinders, and still others with internal flues. The horizontal, cylindrical boiler, with two internal flues and with the firebox placed under one end, became the accepted type. The early achievement of the Pittsburgh rolling mills in producing iron high in tenacity and forging qualities was an important contribution to the development of these early steam boilers.

Experiments were made also in side wheelers and in stern wheelers. The Fulton type of boat employed side wheelers, while the Shreve-built boats used stern wheels. By 1830 the stern wheelers had all but disappeared, for stern wheels were applicable only to smaller vessels. For the larger boats, side wheels measuring 20 feet in diameter and 12 feet in width became standard equipment. With all these features, the objectives of cargo space and speed were achieved.

Importance of American Boatbuilding. The influence of the American steamboats began to be felt in many countries throughout the world—in Central America, Mexico, Canada, Russia, Manchuria, India, Egypt, and in Belgian Congo. One Pittsburgh boatbuilding firm received orders from each of these countries. The French government sent its distinguished engineers to the United States to study the construction and operation of American-built steamboats. British naval architects came to inspect American steamboats, one of whom wrote, "they stand in every respect in science, in beauty, in magnitude, in speed, unparalleled by the river steamers of our own or any other country." [7] While the steamboat was not America's first contribution to world technology, yet the extent of its economic and social influence was in many ways the most notable achievement of our infant republic.

Although most of the technological advances in steamboat navigation were first introduced in boats plying the western rivers, there were some major advances on the boats that traversed the eastern rivers. These boats were made larger and more luxurious, and were equipped with improved boilers. Tubular boilers were introduced on the steamboat "Novelty" in 1830. Anthracite was used as fuel by

1840, and by 1844 iron was beginning to be used in the construction of the hull.

The railroads that spanned the country and bordered the rivers eventually robbed the river trade of much of its cargo, but the steamboats in a less dramatic way continued to play an important part in inland commerce. By the time of the Civil War the steamboat had made its contribution to the building of the nation. It had made possible the settlement of new territories and the opening up of vast natural resources. Moreover, the technological know-how acquired in steamboat development was carried over to other modes of transportation and industrial development.

Ocean Travel

A number of technological developments in ocean-going vessels were also made during this period. The year 1818 saw the beginning of the world's pioneer packet lines (sailing vessels) established between the United States and Europe, the posting of scheduled sailing dates. The vessels of the packet lines achieved speed and had stability to ensure regular runs. They had carrying capacity for both passengers and cargo, and provided comfortable accommodations for passengers.

Packet Lines. In the 1830's there were a score or more packet ships plying between New York and Liverpool, a dozen others to London, and another dozen to Le Havre. Regular lines were running from Boston to Philadelphia. "A new era in commerce had arrived," declared a British publication. The new ships were described as "probably the finest and fastest sailing vessels in the world . . . , beautifully modeled and of the best workmanship, fitted with every convenience for passengers so that they have no occasion to provide anything but personal apparel." [8]

To keep pace with the growing demands for ocean travel and commerce, these packet ships developed a design differing from the old kettlebottoms of earlier sailing days and from the small, speedy Baltimore clippers. To increase the cargo space, the drag aft was diminished and the ship made to sail on a more nearly even keel. Following the lines adopted by the steamboats plying the lower Mississippi, ships with a shallow draft and an almost flat floor were designed. This feature was an innovation that had hitherto been considered a deterrent to speed and sailing qualities. The "Roscious," a striking example of this marine architecture, was the most remarkable ship of her day. She registered 1,000 tons, had an immense spread of canvas, and her main mast towered 160 feet above the deck. She was a blunt, chunky craft built for speed and to carry a good-sized cargo.[9]

The Clipper Ships. The accomplishments of the packet ships, together with the success of the light brigs and schooners, were the factors that led to the development of that romantic type of vessel, the clipper ship. The clipper ship, immortalized in song and story, and synonymous with speed, daring, and American enterprise, held the center of the stage in the drama of American technological transportation during the "fabulous forties," and the "fermenting fifties." The glory of the clippers has receded into relative unimportance, but in the heyday of their adventure their exploits crowded crime, politics, and wars from the headlines. They brought fame to United States seamen and wealth to the coffers of the merchant princes.

Clipper Design. Technically, the term *clipper ship* might be applied to a variety of sailing vessels. The several clippers were different, some complete opposites in particular features. But to the deep-water sailor, the term *clipper* had a significant meaning. It was the "thoroughbred racer of the seas, clean, long, smooth as a smelt, with arching head, thin hollow bow, convex sides, light, round and graceful stern." [10] Aloft, it was large, with towering masts. It had long proportioned spars, "spreading clouds of canvas to the gale." The clipper did not come as a revolution in shipbuilding, but rather as an evolution. Old principles of shipbuilding were put to new uses. Cold, calculated reasoning and applied science rather than nostrums and rule-of-thumb methods were used in construction. The clipper was brought into being by the trade requirements of the period.

Baltimore Clipper. The Baltimore clipper, described as the forerunner of the real clipper, came into prominence during the War of 1812 as a "pilot boat." It was called the "Virginia model." These boats were produced all along the coast, however, and were characterized by their easy lines, long bow, and deep drag aft. They were sparred to the limit of safety, for speed outweighed all other considerations.

The transition from the old line vessels to the clipper type came gradually. The "Anna McKim," 1833, a 493-ton craft, has been termed the last of the old line models, and the "Rainbow," launched in January, 1845, the first of the real clippers. John Griffith, naval architect, designed the model ship "Rainbow" in 1841 that called for a lengthening of the bow into graceful upward curves, making it concave rather than convex in contour, and thus enabling it to slice like a knife through the waters. This feature spelled doom to the "cod's head and mackerel tail" design that had existed to that day. The "Rainbow" was a 750-ton vessel. Other famous clippers were the "Sea Witch," 908 tons; the "Flying Cloud," 1,783 tons; the "Challenge," 2,006 tons; the "Sovereign of the Seas," 2,421 tons; and the greatest of all, the 4,000-ton clipper, the "Great Republic," built in 1853.

Era of Famous Clippers. The most famous builder in the history of clipper ships was Donald McKay, of Newburyport, Massachusetts. His ships were conceded to be the fastest, the sturdiest, the largest and the most beautiful of all clippers. In the four years after gold was discovered in California, his yards turned out one hundred sixty ships, including the notable "Flying Cloud," the "Sovereign of the Seas," and the "Great Republic."

Though each was individual in design, certain trends dominated all extreme clippers. The hull was longer and sleeker. The bow was concave rather than convex; the ship rode *on* the water, not *in* it. The stern was rounded and full, increasing the cargo space. The most spectacular characteristic was the expanse of sail. These ships were equipped with canvas out of all proportion to their size and were thus enabled to catch the slightest breeze. Some sails towered more than 200 feet above the deck of the craft, almost equaling in height the entire length of the ship. The "Great Republic," 325 feet long, largest afloat, spread 15,000 yards of sail.

The clippers were famous for their speed. The "Lightning," which was engaged in the Australian trade, traveled 436 miles in 24 hours, or an average of 18 miles an hour. On one special occasion she made a voyage to Liverpool in thirteen and one-half days. "The Flying Cloud" sailed from New York to San Francisco in eighty-nine days, covering 433 miles in one day. The "Sovereign of the Seas" logged 495 statute miles in one day, and piled up an average of 378 miles a day for four consecutive days. The "Oriental" won the record from New York to Hong Kong, making the passage in eighty-one days. On ocean runs covering thousands of nautical miles, the clippers carried all possible sail at all times. Such feats were not without reward, for they enabled the American clippers to deliver their cargoes in foreign ports more quickly than their competitors, and therefore they could charge higher rates.

Effect of Clipper Ships. The clipper ship with its technological improvements had profound effects upon maritime history. New trade requirements were developed. The opening of the treaty ports in the Far East offered new opportunities. Trade in tea and silk, together with other oriental luxuries, presented a challenge to the

American shipper, and the discovery of gold in California provided a new avenue of trade. The success of the clippers was due not solely to skilful designing and craftsmanship in building, but also to the skill and courage shown by their commanders and crewmen in getting the utmost out of them. The clippers possessed two outstanding advantages, speed and spaciousness; they carried larger cargoes in record time.

Oceanography. The success of the clipper ships was greatly enhanced by the contributions of science and technology in the form of maps and charts of the sea, showing the ocean currents and prevailing winds. Matthew Fontaine Maury, Superintendent of the United States Naval Observatory in the 1840's, used the logs on file to compile wind and current charts, and sailing directions to aid mariners on the high seas. Maury had mastered Nathaniel Bowditch's classic study, *The New American Practical Navigator,* first published in 1801 (it went through some sixty editions), which had set a pattern for later American surveyors, astronomers, and mathematicians. Maury's charts, which he called *Sailing Directions,* were first issued as a little pamphlet, but they were soon expanded into a ten-volume publication and continued through eight editions. These led to the publication in 1854 of a *Physical Geography of the Seas,* the first study in oceanography ever issued.[11]

Maury's Mariner's Guides. Commodore Maury enlisted the cooperation of ship captains plying the seas and collected observations from more than a thousand mariners. His project was first limited to the North Atlantic, but later it comprised every navigable sea. His data were acquired on a strictly scientific basis and prepared from careful observations.

Maury's charts and maps were hailed by mariners everywhere, for they were time- and money-savers. By following his directions the passage to the Equator was cut by ten days; to California, it was reduced from one hundred and eighty-three days to one-hundred and thirty-five days, then to one hundred and seven. A few of the speedier ships made it in ninety days. The charts cut the trip to Australia from one hundred and twenty-four to ninety-seven days. Moreover, ships followed the same path and could be located if trouble arose. Hunt's *Merchant Magazine* declared that no such boon had ever been given to commerce, and through it to civilization, since the invention of the mariner's compass and the great adjunct of the chronometer.[12]

The Secretary of Navy noted in his annual report of 1851 that England, although its hydrographic office was venerable with age, had created a new department to carry out the objective of Maury's charts. A British magazine stated that of all the benefactors to the commercial world in respect to the transit of merchandise from one port to another, the name of Maury stood first. The estimated annual savings to the commerce of the world amounted to several million pounds sterling. Admiral Fitzroy of England noted that wagon loads of Maury's charts had been distributed in England and many editions of his *Physical Geography of the Sea* were sold there.[13]

Maury was signally honored both at home and abroad. He was elected to membership in more than twoscore learned societies, was awarded honorary degrees by many universities, and received numerous gold medals—many from the crowned heads of Europe. The honor he loved most was the appellation "The Pathfinder of the Seas." [14]

The Passing of the Clipper. The clipper ship had its inception in the silk trade in the East and reached its culmination in the years of the California gold rush. It passed out rather ignominiously to the half-clipper, and next to the slow freighter. At the same time it was meeting a rival in the oncoming steam vessel which presented none of the fascination or romance of the clipper. Sailing vessels continued to ply the waters even into the twentieth century, but the age of the clipper was gone and so was America's supremacy on the seas. The clipper ship was America's chief contribution to marine transportation technology during the nineteenth century, a demonstration that art and science could be put to practical use.

Steamship Building. Although the United States pioneered in the development of the ocean steamship and was the leader in many phases of steamship development, she never attained the ascendancy in the use of steam on the high seas that she had achieved in its use on inland waters. Nor did the United States acquire the same fame in steam propulsion that she won in the development of her packet and clipper ships.

America Pioneers in Steam Propulsion. The development of the uses of steam on the ocean included three technological revolutions: first, from sail to steam; second, from wood to iron; third, from iron to steel construction. The United States was the first nation to advocate the building of ocean-going steam vessels, including a steam-propelled navy. The possibilities of steam navigation on the ocean were demonstrated in 1809 when John Stevens captained a coastwise sea voyage under steam power. A few years later steamboats were used in the coastal trade from Philadelphia to New York and down to Charleston. In 1819 the "Robert Fulton" was launched on ocean travel, calling at various ports along the Atlantic. These first ocean steamers were merely sailing vessel hulls, fitted with the side wheels used on inland river boats. They hugged the shores, Phoenician-like, to avoid stormy seas, since the paddle wheels did not operate successfully on rough water.

The adaptation of the steam engine to maritime use came gradually. At first steam did not supplant sails, it merely supplemented them. Steam vessels were equipped with sails, and for years sails continued to be used in fair weather, thus conserving fuel. Such practice, however, necessitated the employment of two crews—one to operate the sails, the other to operate the engine and boiler. Such equipment was provided for "The Savannah," the first steam vessel to cross the Atlantic Ocean, in 1819. It was captained by a Yankee, and financed by a Georgia businessman. Although steam was applied only occasionally on the voyage, the vessel entered the Irish harbor under steam, much to the consternation of the harbor master who thought she was on fire.

During the next fifty years, steam development in all its phases was more advanced in the United States than in all Europe. By 1850 six steam vessels were carrying mail on the Atlantic, and five on the Pacific. In 1849 the Collins Line launched its first steam vessel, "The Atlantic." During the early 1850's fifteen companies owning sixty-three steamers, with a combined tonnage of 110,000 tons, were operating steamship lines out of New York. The Pacific Steamship Line had thirteen steamers; the Vanderbilt Line, which operated by way of Nicaragua, had ten; the New York–San Francisco, four; and the Empire City, three.[15]

Steam for Ships of War. Plans for converting steam-propelled vessels into ships of war received considerable attention in the early 1840's. The tension over the Oregon boundary issue posed a threat of war with England, and the issue was used to gain Congressional favor. President Tyler in a message to the House of Representatives, February 23, 1844, urged Congress to hurry the building of steam vessels. He reminded the Congress that other nations were giving attention to the subject of steam-powered vessels and, "it would be criminal in the United States, the first to apply to practical purposes the great power which has been brought into use, to permit others to avail themselves of our improvements while we stood listlessly and supinely by." He reminded Congress that the world "is indebted to America for that great discovery of the means of applying steam power which has in the last quarter century so materially changed the condition of the world." [16]

Government Subsidies for Shipbuilding. The debate for federal aid continued for several years. Senator Stockton of New Jersey brought up the question at every opportunity. "If the government would only support our shipbuilders," declared the Senator, "American artisans and mechanics would build steam powered ships that would win future battles fought on the ocean, in the factory, and on the railroad; steam would dictate the question of superiority in all the pursuits of human life." [17] Senator Bayard of Delaware was another ardent champion of federal aid for building steam vessels. He pointed out that such vessels would make excellent transport ships in case of war and would be practical as cruisers. "Their speed would enable them to overtake or escape any vessel afloat." Senator Bayard's warning proved a prophecy when he told his colleagues that, "if you abandon this idea for want of government aid, you abandon to Great Britain the entire transportation of passengers except emigrants, and you abandon a valuable freight." [18] While the debate revealed considerable support for the subsidy, the measure failed to pass and the proposal was finally dropped. American goods began to be carried under the British flag.

Technological Problems in Steam Navigation. Ocean steam navigation presented many technological problems. Since the paddle-wheel method of propulsion proved unsatisfactory on rough seas, the screw propeller was devised. Lubricants presented a problem, since animal oils, the principal lubricant available, became rancid. The daily task of blowing out the boilers was a greater hazard on the ocean than on inland waters, because the vessel idled during the process. Fuel was also a problem, since the supply of coal needed for a transoceanic voyage occupied valuable cargo space. Marine engines were poorly developed. The one-cylinder engine was used at first, then a direct-acting, two-cylinder engine was applied; but it was not until the mid-1850's that the compound engine was devised. A final problem was the selection of construction material: whether to use iron or wood. The relative advantages of each was long in debate.

In the matter of propulsion, many experiments had to be made to determine the best position of the paddle wheel. Some advocated placing it at the stern of the vessel; others, on the side. All paddle wheels were inefficient in rough seas, for they would often be found spinning in midair.

Ericsson's Screw Propeller. The principle of screw propulsion had been known since the days of Archimedes, but its practicability had never been demonstrated. In this country, the ingenious Franklin had observed the advantages of using a small steam engine to pump water at the bow and force it out at the stern, thus providing a means of propulsion. In 1782 a boat was driven by James Rumsey at the rate of 4 miles an hour by a method called "jet propulsion." Colonel John Stevens had devised several model propellers of this type. But it was left to John Ericsson, a native of Sweden who came to America in 1839, to make screw propulsion practical. Ericsson gave credit for the basis of his plan to Benjamin M. Smith of Rochester, New York, who had secured a patent in 1829 for propelling vessels by means of sculling wheels, placed at the stern of the vessel. Ericsson combined Smith's plan with his own plan for a screw steamer, and the successful screw propulsion resulted.[19]

Ericsson superintended the building of "The Princeton" (1844) for the United States Navy. She was a full-rigged ship of 954 tons, with steam power of the screw-propeller type. The vessel was built of iron, designed to burn anthracite, and was pronounced a marvel in shipbuilding technology. But on one of her early runs a fatal disaster occurred. While President Tyler's Secretary of Navy Upshur, Secretary of War Gilmer, and others were aboard the "Princeton" on a pleasure trip, a gun exploded, killing the two secretaries and four others. The shock to the nation delayed temporarily the building of steam naval vessels.

Coal-fired, Screw-propelled Steamships. Within a few years, the program was resumed. The screw propeller had the advantages of costing less, both in construction and in operation, of allowing increased storage space for freight, and ensuring a more equable performance of the engine. Screw propellers also made possible the construction of larger ships, for two engines could be coupled, thus giving the screw propeller more power by enabling it to run at higher revolutions. By 1850 one hundred fifty vessels had been equipped with screw propellers. On many occasions, however, these early screw-propelled vessels failed to come up to expectations on rough seas.

Fuel for trans-Atlantic steam voyages also presented a problem. English scientists argued that ocean steamship transportation was impossible because the vessels could not carry sufficient fuel for the voyage; and when wood was to be used, they were correct. In the late 1840's, however, anthracite coal came into use. The fuel problem continued until improvements were made in the engine. Realizing the obstacles the fuel problem presented, Congress authorized marshalling the forces of science, and secured the services of Professor Walter Johnson of Philadelphia to make an analysis of the several coals and their properties of combustion. Gradually all steam vessels were burning coal for fuel.

Iron Ships for Peace and War. The revolution in maritime transportation involved not only a change from sail to steam, but also from wood to iron as construction material for ships. Tall masts and wooden hulls could not withstand the vibration caused by the throbbing steam engine. Moreover, the development of explosive shells, an invention by the Frenchman Henri J. Paixhans, necessitated the sheathing of vessels used in armed conflict. This fact was dramatically emphasized to the whole world in the destruction of the Turkish fleet at Sinope, 1853.

The use of iron began in a small way. As early as 1825 a small iron steamer, the "Codorus," was launched on the Susquehanna. By 1836 a few iron steamers were operating between New York and New Orleans. The use of iron was introduced in a big way with the construction of an iron revenue cutter, "The Saranac" in 1841, the "McLane" in 1842, and the "Michigan" built in 1842. This latter steamer was in service on the Great Lakes where, under the name of the "Wolverine," it continued until 1900. The "Princeton" was built of iron, as was the "Allegheny" in 1843, and the "Water Witch," 1844. In 1842 a line of small iron steamers was plying between Philadelphia and Hartford.

The advantages in the use of iron, as pointed out by the Navy, were: greater strength, fire resistance, lower maintenance costs, shell resistance, freedom from vermin and from decomposition of vegetable matter, and a great saving in paints. Iron construction also made it possible to construct watertight bulkheads. But despite all these technological advantages, the wooden vessels continued to be used until late in the nineteenth century. Nevertheless, the introduction of steam into naval operations revolutionized naval warfare as much as gunpowder had changed land fighting centuries earlier. Shell guns, the screw propeller, rifled ordnance, and armor, all were introduced in the mid-nineteenth century as additional revolutions in naval construction.

The United States government was the first to recognize the value of steam for war purposes. The "Princeton" with her machinery below the waterline and safe from the enemy's fire, marked a major step in naval technology. The nations of Europe followed America's lead. The armored ship led to a race between the development of guns that would pierce the armor and construction of armor that would resist the projectiles, a race that continues to this day.

American Shipbuilding Declines. At the close of the Civil War, however, silence fell on the busy dockyards of the Navy, a silence that was unbroken for twenty years. Shipbuilders of the Old World took up the task where America left off and developed the fundamental ideas that had originated in the United States. Nevertheless, the technological developments in water transportation during the first half of the nineteenth century had played a prominent part in America's expansion.

Notes for Chapter 14

1. *The Sentinel of the Northwest Territory,* I, No. 1, 1793–1794.
2. "Report On Internal Commerce Of The United States," *House Exec. Doc.* No. 6, 5th Congress, 1st Sess., XX, Part II, 181.
3. Leland D. Baldwin, *The Keelboat Age on the Western Waters,* is the standard work on keelboats (1941).
4. See Louis C. Hunter, *Steamboats on the Western Waters* (1949). A monumental work, covering in detail the many aspects of early river transportation technology.
5. *House Exec. Doc.* 21, 20th Congress, 3rd Sess., p. 3.
6. *Journal of the Franklin Institute,* XXX (1840), 86-87.
7. James Reese and Sons, *The Reese Company Papers,* on file in Western Pennsylvania Historical Society.
8. Quoted in *McCullough's Commercial Directory for 1836* (London).
9. Carl C. Cutler, *Greyhounds of the Sea* (1930), p. 98.
10. *Ibid.,* p. 114.
11. Plate I in the *Physical Geography of the Sea* represents more than one million separate observations, or the findings of a fleet of ten sails over a period of more than one hundred years.
12. Hunt's *Merchant Magazine and Commercial Review,* XXX (May, 1854), 53 *ff.*
13. *Eclectic Magazine,* XVIII (July, 1873).
14. For sympathetic accounts of Maury, see John W. Wayland, *"The Pathfinder of the Sea"* (1888); Charles L. Lewis, *Matthew Fontaine Maury* (1930); Diana Fontaine Maury Corbin, *Matthew Fontaine Maury* (1888); and *Dictionary of American Biography,* XII, 428-31.
15. T. C. Purdy, "Steam Navigation in the United States," *U. S. Census* (1880).
16. Richardson's *Messages and Papers of the Presidents,* V, 2132.
17. *32nd Congress,* 1st Sess. (May 1, 1852), p. 1323.
18. *32nd Congress,* 1st Sess. (May 10, 1852), p. 1304.
19. See James Phinney Baxter III, Introduction of *The Iron Clad Warship,* (1933), Appendix A, Affidavit of John Ericsson (March 8, 1845); see also Captain Edgar Smith, "First Twenty Years of Screw Propulsion," *Transactions: The Newcomen Society,* XIX (1938–1939).

See also such works as Frank Charles Bowen, *America Sails the Seas; The History and Romance of America on the High Seas from the Fifteenth to the Nineteenth Cen-*

tury (1938); Edward Keeble Chatterton, *Ships and Ways of Other Days* (1913); by the same author, *Sailing Ships and Their Story* (1923); Henry B. Culver, *The Book of Old Ships and Something of Their Evolution and Romance* (1935); Martin D. Stevers and Captain Jones Pendlebury, *Sea Lanes: Man's Conquest of the Ocean* (1935). See also William Conant Church, *Life of John Ericsson,* 2 vols. (1890); the John Ericsson Memorial Commission's *Proceedings at Unveiling of Statue,* U. S. Printing Office (1929).

Chapter 15

THE COMMUNICATIONS REVOLUTION TO 1860

The demand for speed in the realm of physical communication that had resulted in improved roads, steamboats, railways, and steamships had, by the 1830's, entered the field of communication of ideas and information. It brought about a need for speedier printing presses, faster and cheaper typesetting methods, a more abundant paper supply, and a revolutionary method for transmitting news—the telegraph. All these demands brought about many technological developments in printing and in the mechanics of publications. A change in the function and position of the newspaper in American life also occurred.

Communication Methods

Early Newspapers. From the time of the first printed publication in America, the *Freeman's Oath* in 1639, printing had developed as the country expanded. Newspapers sprang up wherever there were printing presses. But, because printing presses were slow, heavy, and cumbersome in operation, and because weeks or months passed before the news arrived, the papers carried little current news. Local happenings were rarely reported. One looks in vain in the New York papers for the account of the trial voyage of Fulton's "Clermont" in 1807.

Accounts of crime were seldom published, although the papers often carried advertisements of rewards for escaped criminals. Foreign news was obtained from private correspondence and from foreign papers which had been as slow in publishing the events as were those in this country, and which were still tardier here because of the time lost in transit.

The Poor Man's College. In the early years of the nineteenth century the newspapers were small, blotchily printed four-page sheets, devoted to political polemics, with paddings from private correspondence and clippings from foreign papers brought by the mails. In 1800 there were approximately one hundred fifty newspapers being published; by 1830, eight hundred and sixty-three; and by 1850, twenty-eight hundred. The early papers carried advertisements that ran by the month, usually unchanged, shipping news, and a plenitude of editorial comment.[1] By mid-century, however, mechanical and technological improvements had so facilitated printing that the press had become "man's most trusted servant," and the newspaper was hailed as "the poor man's college." So vast had its influence become that America was referred to as having a "print-made civilization."

Early Printing Presses. Before these innovations could be effected, however, the slow clumsy printing presses had to be replaced by more easily operated and speedier machines. The mechanics of printing as devised by Gutenberg in the fifteenth century remained hardly changed for more than three hundred years. Benjamin Franklin used a press that differed but little from the original Gutenberg design. The printer stood over cases and fumbled crude type into sentences, and then tugged on the lever of a heavy wooden press for each impression. Such methods required arduous labor, and there was no opportunity for speed. To the Yankee mind these were two serious limitations. To put out a paper with a circulation of one thousand (although few had that many subscribers), ten hours of press operation were required. As a result, news was never timely and circulation was limited.

Development of the Printing Industry

Christopher Sauer (Sower), a German publisher, is usually given credit for constructing the first American-made press, 1760. It was not until the eve of the Revolution that good presses were made in America, but by 1800 the manufacture of printing presses had become an established business. The first presses were designed after the English makes. Improvements were soon introduced, however, such as those made by Adam Ramage of Philadelphia, one of the earliest manufacturers of printing presses, so that by 1800 foreign importation had ceased.

George Clymer's Press. Another marked American improvement was introduced by George Clymer in 1813 in his Columbian press. This was an iron machine with the platen raised and lowered by a spindle, operated by a lever. The ornate, heraldic embellishments, surmounted with a cast-iron eagle that gave it its name, were not merely decorations—Yankee utilitarianism demanded more than that. The press was exceptionally sturdy and the eagle served as a counterweight, which facilitated the operation of the press. Clymer was honored by the King of the Netherlands and the King of Prussia for his improvements.

The invention of the toggle joint (elbow joint) by Otis Tufts, which made the operation less arduous, and Peter Smith's improvements on this feature, represented other contributions to the development of the American hand press. And, when Samuel Rust in 1827 produced the Washington press, it was believed that the ultimate in hand-press design had been achieved. With only a few improvements it has continued in favor to this day for fine press work.

The flatbed press, as it was first used, was operated by lowering a heavy platen that was suspended from two uprights over a bed of type. This was done by adjusting a screw to a moving bar. Then the awkwardness of this task was overcome by substituting a lever, which raised and lowered the platen. There still remained the task of moving the bed of type backward and forward to a position under the platen. A device was installed for rolling the bed in and out on the horizontal frame. Substituting iron for wood gave greater power and made the press less cumbersome. Spring adjustments were introduced, and coupling bars and balance weights were added. These contributed to the simplicity and accuracy of the press. Later, a clamshell motion was contrived by raising the bed of type to meet the platen as it was lowered. Another device raised the type bed to meet the stationary platen. These improvements contributed to the simplicity and accuracy of the presses. Their success not only preempted the domestic market, but also created a demand for American-made presses in Europe.

Improving Domestic Presses. These early presses were all hand-operated, but the search continued for a method of saving labor and utilizing power in press operation. A number of inventors pioneered in both fields. In 1822 Daniel Treadwell, by attaching a treadle, made it possible to use his foot in operating the press instead of pulling with his hands. This device was later improved so as to permit the application of horsepower, then water power, and finally steam power to the treadle. By 1827 Treadwell had built a huge power press of enormous weight, consisting of cams, gears, and cranks, all operated by a wilderness of belts. It was capable of putting out five hundred impressions an hour (when it did not break down) and in some establishments its use continued until 1845.

The Adams Press. Improvements on Treadwell's press were made by Phineas Daw, Stephen F. Ruggles, and by two brothers, Isaac and Seth Adams. The latter two built the famous Adams' press, which proved to be a favorite with printers for more than fifty years and for some purposes has never been improved upon for a power platen press. The Adams press was capable of seven hundred fifty impressions an hour, and it had the added feature of the "fly," a contrivance used in delivering the printed sheet from the machine. This press became popular with book printers. The introduction of the power press was by no means general, however, nor was its adoption rapid. In 1837 Harper and Brothers, New York, had thirty-seven hand presses and only one power press, which was operated by a mule. The press was located on the third floor of the printing shop, and every morning and evening the mule was taken up and down by a hoist outside the building. As late as 1835 newspapers were generally printed by handpower, with crank men supplying the force.

Introduction of the Cylinder Press. The flatbed press, regardless of the number of improvements, could not meet the demands of the modern newspapers. British printers were the first to recognize these limitations, and from them came the initial steps that led to a revolution in printing. What is often called the greatest improvement connected with printing since the discovery of the art itself was the substitution of a cylinder for the flat platen to make the impressions. The idea was developed by a Saxon mechanic, Frederick Koenig, who apparently got the idea from William Nicholson. The use of the cylinder was applied by the London *Times* in printing its issue of November 29, 1814. The type, still set in a flatbed, passed back and forth by reciprocating motion under a revolving cylinder, which made the impressions on paper placed on the type.

Some fifteen years passed before the American printers became interested in the cylinder press. But with the rapid development of cities and improved means of transportation there came a demand for both greater speed and an increased output of newspapers. It was often necessary to hold back the forms until nearly daylight for the latest news, and in the hurry which ensued, the presses often broke down. There was a constant demand for technological improvements that would permit speedier printing. This demand was met by Robert Hoe and his son Richard.

Hoe's Rotary Cylinder Press, 1844. In 1828 Robert Hoe, a carpenter by trade and a skilled press repairman, built the first cylinder press in this country, an adaptation of the English Napier press, which was an improvement on Koenig's invention. Hoe did not stop here, however. Together with his son Richard he made a number of improvements. Instead of the single small cylinder, making two or three revolutions over the flatbed of type, the Hoes used two cylinders. They made a number of other improvements, and by 1844 brought out the rotary combined cylinder press that used any number of cylinders.

While this marked a significant advance in printing technology, it still proved too slow for those papers with the largest circulation. In 1830 in New York City, two thousand was considered a large circulation for a newspaper; but by 1838 the New York *Sun* boasted a circulation of thirty-eight thousand copies daily, the largest in the world. This new spirit of enterprise taxed the crankmen who laboriously turned the cylinders, and the rivalry among the newspapers in being first with the news created a demand for a speedier press.

The Hoe Revolving Press, 1847. Again the two Hoes came forward with a new principle in printing technology—the revolving press, patented in 1847. Applegath in England had been working on the same idea, and each brought out a new machine differing slightly in principle. Hoe's idea proved the more practical and was later adopted by the British publishers, even by the London *Times.* The principle of the new machine was the elimination of the flatbed of type, and the substitution of a revolving-type cylinder, as well as revolving impression cylinders. The problem had been to find a way to secure the type on such a cylinder, since the cylinder stood upright and stereotyping had not yet become practical for newspaper work. (Stereotypes were made at this time, but the material used was plaster of Paris, which required a day's time to set.) Hoe's machine, known as the "Lightning" press, consisted of a huge type of cylinder 15 feet in diameter on which the type was arranged in column-wide flat surfaces. It was held in place on the curved surface by prismatic construction of the type and by wedge-shaped column rules. The Philadelphia *Ledger* installed this press in 1847. By 1860, newspapers in all the larger cities had installed a Hoe Lightning press.

This machine, with four impression cylinders, was capable of printing eight thousand papers an hour on one side. By increasing the number of impression cylinders, the number of papers could be increased two thousand per cylinder. A ten-cylinder machine turned out twenty thousand impressions an hour. But even this huge machine had its limitations. The addition of cylinders involved adding more pressmen and more space. A ten-cylinder machine was as large as a two-story house and weighed over 60,000 pounds. The paper was fed in separate sheets and was printed on only one side at a time. The need grew for a perfected self-feeding machine, but this demand was not met until after the Civil War.

Papermaking

The scanty output of print paper was always a handicap. Editors were continually pleading for old rags, and the method of manufacturing rags into paper was a time-consuming process.

Gilpin Papermaking Machine, 1817. In 1817 Thomas Gilpin of Delaware produced what was probably the first machine-made paper in America. Gilpin's machine included a cylinder and rolls. The cylinder, an arrangement of wires and felt, dipped into the pulp vats and passed the partially formed paper along to the rolls, which did the work of the screw press in flattening it out. This machine produced a paper of any desired length in a continuous band, instead of single sheets as was the case with handmade paper. Although Gilpin tried to keep his device a secret, other papermaking machines appeared. In 1822, John and David Ames in Springfield, Massachusetts, produced a machine on the same principle and this was pronounced much superior to the Gilpin device. Other improvements were made so that by 1830 the entire process from wet pulp to dry finished paper, cut to size, could be performed on one mechanical train of rollers, wire nets, and dryers.

Growth of Papermaking Industry. The introduction of these machines made a vast improvement in the process of paper manufacture. The new machines had a speed capable of 45 feet per minute, and when operated by two men and four girls could produce 24 miles of paper per day. In time, rolls heated by internal fires, later by steam, were placed in line with the cylinder. This eliminated loft drying. The machines also provided a quick-drying process. Under the old hand method the wet paper was dried on poles, a process that required at least five days in favorable weather—ten times longer than the time required to make it. By the machine method the product was dry within minutes and when cut in sheets was ready for the printer. The manufacture of paper increased, and by 1830 the United States was the greatest paper-producing country in the world, as well as the greatest paper-consumer.[2] Paper establishments were numerous but small. The Census of 1850 reported four hundred and forty-three paper mills in the country, with an average of fifteen employees each. Virtually all paper was machine made.

While machinery had facilitated the manufacture of paper, there still remained a dearth of material. A search was made for a substitute for rags. Straw, grasses, and many other plants were experimented with. William Morgan in 1828 succeeded in manufacturing print paper from straw, and a year later an edition of *Niles Register* was printed from such paper. But a satisfactory substitute was not found until a much later date.

Typecasting and Typesetting

Hand Typecasting. Typecasting developed slowly. At first all type used in this country was imported. The process of making type involved two main operations: making the mold and casting the type. To make the mold, a type face was designed and a pattern punch was cut. From this mold, or matrix, the type was cast in a foundry.

Typecasting Machines. Until 1836 all type was cast by hand, but many type foundries had been established by 1850. They produced all the type that was used in this country, and enough for export to the British provinces and to South America. The first improvement in typemaking was the introduction of machines for casting type in the smaller font. One such machine operated by one workman could produce ninety brevier type a minute. All larger type continued to be cast by hand, however, until a later period. The most successful of the typecasting machines was made by David Bruce, Jr., of New York, who devised a machine to pour the molten metal in the matrices. His invention was in use until the 1890's.

Typesetting Machines. Typesetting, one of the most exacting and arduous of all printing operations, enlisted the genius of many inventors. As early as 1840 Frederick Rosenberg devised a typesetting machine to imitate the hand operation, but no significant progress was made until a much later date. Expert hand typesetters could set from twelve hundred to fifteen hundred ems of type an hour. Stereotyping processes continued, but were not applicable to newspapers.

Newspaper Rivalry and Growth

The many technological developments in printing were reflected in the development of the nation's newspapers. Increased demands of the newspapers in turn stimulated greater developments in printing.

Modernizing the Newspaper. By 1830 improvements were made in the content of the newspapers. Greatly improved facilities for newsgathering were developed, and papers printed the news as soon as it happened. Bitter rivalry developed among the several papers in the larger cities. In their efforts to increase circulation, editors printed sensational news stories which were read by rich and poor alike. The make-up changed. Display heads were used, and illustrations were introduced. Newspapers and printers began to pride themselves on the speed with which they could serve the public. Improved transportation facilities stimulated this rivalry and demands for daily newspapers increased.

Speeding the News. The difficulties encountered in gathering and dispatching news were among the greatest handicaps faced by newspapers. In the early days the editors had no special facilities for getting the news. They had to wait until some traveler brought it, and on occasions even advertised for news. One New York editor was especially anxious to inform his readers about the peace terms at the close of the War of 1812, but was unable to do so because the only London paper in town was "being kept from the public eye." "We will not," he announced, "conjecture what the contents are." To speed up the newsgathering service, various devices were instituted. Ships, horses, and fleet-footed reporters were employed. Riders of the "pony express" were organized.

Famous "Scoops." Late in the year of 1830, three New York newspapers joined forces to employ a series of expresses for furnishing their readers with the earliest possible report of President Jackson's annual message of that year. As soon as the President's message was completed, the expresses dispatched their horse relays out of Washington for Baltimore. From there a steamboat carried the dispatches to Philadelphia, where the copy was picked up and speeded on its way to New York.

Rival journalists resorted to many ingenious devices, such as employing carrier pigeons, chartering special locomotives and steamboats when these became available, and employing relays of spirited horses. Henry J. Raymond of the New York *Tribune* outwitted his rivals in reporting one of Daniel Webster's great speeches. While on board a boat returning from Boston, Webster wrote a digest of the speech and passed the sheets to printers who had brought the type and cases on board for that special purpose. When the boat docked in New York the story was all set up, ready for the press.

Other ingenious methods were employed to speed up the reception of news from abroad. Small, swift sailing vessels were engaged by different newspapers to cruise the harbors in order to intercept incoming vessels and race back to shore with the latest news. Carrier pigeons were used, especially on foggy days. The New York *Sun* paid tribute to pigeons as auxiliary newscarriers, declaring they could fly at the rate of 60 miles an hour; they could leave Washington at daylight in the morning and reach New York by noon, thereby beating the regular mail by 8 to 10 hours. While these early methods seem quite amateurish when compared with present day network of news-gathering agencies, they were in advance of the methods used in other parts of the world.

Increase in Circulation. The appearance of the newspapers began to change radically as a result of these technological developments. These changes effected a social revolution in the function of the newspaper in American life. The city daily grew from the four-page, two-column sheet to a large seven-column journal, called derisively a "blanket" paper. The New York *Sun,* in referring to the first large-size issue put out by the *Courier* and *Examiner,* described it as "large enough for a blanket and two pairs of pillow cases." But soon the *Sun* followed suit and enlarged the size of its paper. The *Sun* took another step forward when it began to print woodcuts of all the leading ships. When the newly built "British Queen" arrived in New York, the readers of the *Sun* opened their paper to behold a large half-page picture of the vessel, described as the "finest ship that had ever been built in Great Britain."

With all these technological changes in printing, and the improved facilities for gathering and disseminating the news, there followed a rapid increase in newspaper circulation. The United States Census for 1840 took cognizance of the newspaper industry, and reported over fourteen hundred newspapers and periodicals in circulation. The New York *Sun* announced a circulation of almost forty thousand issues daily. An "extra" was issued on June 1, 1839, in honor of the arrival of the "Great Western," and fifty-seven thousand copies were sold. Newsboys delivered the papers and sold them on the street, an innovation as compared to the old delivery by mail and distribution at the counting house or club to subscribers only. A number of great names appeared in the newspaper world—Horace Greeley, James Gordon Bennett, Charles A. Dana, Benjamin Day, and William Cullen Bryant.

The greatest boost to newspaper circulation of this period was the appearance of the "penny press." The *Sun* in 1833 started issuing a daily that sold for one cent, which meant that the newspaper was no longer confined to the seclusion of the bar room, the political salon, or the counting house. It now became the people's paper. The Philadelphia *Ledger* and the Baltimore *Sun* soon followed the New York *Sun,* and within a few years the penny press was being read in several cities. By 1840 the population of the United States showed an increase over the 1830 Census of 32 per cent, while the increase in newspaper circulation was estimated to be 187 per cent.

American journalists looked upon their newspapers as the best in the world, and they had good reason for doing so. They wondered how other nations of the world had been able to disseminate news of happenings and to describe the culture of their people before the coming of the newspaper. Edward Everett, in addressing the literary societies of Amherst College in 1835, declared that there was probably not a newspaper of any character in the United States "which does not in the course of a year convey more useful information to its readers than is found in the twenty-one folios of Albertus Magnus, light of the thirteenth century."

New Postal System

Developments in the postal system proved another technological triumph in the communications revolution of this period. The early attempts of the expanding frontier to secure post offices and post roads met with considerable derision on the part of many Easterners. Daniel Webster, in a speech before the United States Senate early in the 1830's, in discussing a bill proposing a postal route between Independence, Missouri, and the headwaters of the Colorado River, declared the proposal utterly impractical and visionary. "What do we want with this vast, worthless area?" he asked. He pictured it as a region of savages, wild beasts, deserts, shifting sands, and whirlwinds of dust; of cactus and prairie dogs. He concluded by asserting that he would never vote one cent from the public treasury "to place the Pacific Coast one inch nearer Boston than it now is."

Demand for Post Offices. While Webster, unfortunately, voiced the views of many eastern statesmen, yet there were at that moment technological forces at work which would within a few years cause men of vision to reshape their whole political thinking on the subject.

The pioneer postman who rode the trails of the first settlers, and the later networks of routes that spread across the plains, popularized a slogan in the 1830's which had been paraphrased from Herodotus: "Neither snow, nor rain, nor heat, nor dark of night stays these couriers from the swift completion of their appointed rounds." While adherence to this motto was more theoretical than actual in the 1830's, yet definite strides were taken toward speeding up the mails.

Pioneer Postman. Mail was carried on horseback, in carts, in stagecoaches, on steamboats and railroads, and by every means of transportation that existed at the time. The government made contracts with individuals or companies to transport the mail by common roads, on horseback, in sulkys, on railroads, or by steamboats. In 1838 there were approximately 100,000 miles of post roads; and when there was a surplus in the post office at the end of the year, it was used to open up new roads.

Railroads Carry the Mail. The postmaster general in 1837 began to award government contracts to railroads for carrying the mail. In 1838 the Post Office Department appointed a "travelling agent" to handle the mail on the trains running between Philadelphia and New York. The post office authorities stipulated the requirements for a railroad car. It had to have two compartments, one to store bags of mail; the other to contain boxes with labels bearing the names of all the small post offices on, or near, the railroad.

In 1838 Congress declared a number of railroads to be post roads, and in 1840 railway mail clerks were employed on the larger roads to sort the mail on board the trains. While most of the mail continued to be carried by stagecoach, yet the railroads and steamboats were rapidly taking over a great portion. A report submitted by the Ways and Means Committee of Congress in March, 1845, showed that the steamboats and railroads provided a "most expeditious mode of transportation. . . ." This Committee insisted that it was the duty, as well as the right, of the government to avail itself, "even at heavy additional expense, of the powerful agency of steam for the purpose of accelerating the mails." The first night trains operated by American railroads were run because of the insistence of the postal service that the mail should not be delayed by stopovers.

Telegraphy Speeds the News

The greatest development of this period in communication technology was the invention of the telegraph. The idea of using electricity for telegraphic purposes was as old as the science itself. Franklin had been able to direct a current of electricity on numerous occasions. Bishop Watson had sent signals more than a mile, by using a Leyden jar. Lesage, as early as 1774, had erected a telegraph in Geneva, with twenty-four wires, one for each letter of the alphabet then in use. But a number of scientific discoveries had yet to be made, and new apparatus and equipment had to be devised before a satisfactory system of telegraphic communication could be introduced.

Scientific Background. Briefly, these were the major discoveries. Static electricity, with its instantaneous discharge from the Leyden jar, was supplemented by the flowing current from a battery. Alessandro Volta had developed a simple, reliable method for producing the continuous flow of a current in "the Voltaic pile." André Marie Ampere had distinguished between static and galvanic (Luigi Galvani) electricity—electricity created by the action of chemicals. Hans Christian Oersted discovered the action of a galvanic current on a magnetic needle; and William Sturgeon of England had developed a crude, horseshoe-shaped electromagnet.

In this country, Joseph Henry, working independently, had discovered many of these same basic principles—but he went further. Henry converted electrical energy into mechanical energy. With the battery and the electromagnet to work with, Henry discovered a way to increase the intensity rather than the quantity of electricity. By increasing the amount of coiled wire around the magnet, he observed that the power of the single battery was increased. He then designed a battery that supplied sufficient power to lift a weight of 2,000 pounds.

Henry was able to send currents from a multiple cell battery through a mile of wire to an electric magnet that rang a bell. He observed that when a wire was wrapped around a piece of iron, and an electric current was passed through the wire, the iron instantly became a magnet. But, if the iron was annealed and soft, it lost its magnetism the moment the circuit was interrupted. It was on this principle of "the attractive power of the electro-magnet that the science of the electric telegraph was born." There yet remained the factor of uniting science and technology, and of designing the equipment that would utilize these principles in dispatching messages.

Early System of Signaling. While science was making these discoveries the tempo of American life was creating a demand for an improved telegraph. The forms of telegraphy then in use—the system of rockets, bells, and signals employed by the navy, and the semaphore wigwag system devised by the Frenchman, Claude Chappe, and used by the French government—were not applicable at night or during storms and fogs. So great was the need that the demands were taken up in Congress, and in 1837 a committee was authorized to investigate and recommend the adoption of the most improved method. Men of science, men in the armed forces, and those who had been carrying on experiments in telegraphy were consulted.

S. F. B. Morse, 1791–1872. It was then that Samuel Finley B. Morse, son of the geographer Jedediah Morse, and grandson of

Princeton's early president, Samuel Finley, joined science with technology and developed the Morse system of telegraphy. His interest in science was aroused by his college professors, Benjamin Silliman and Jeremiah Day, and by the lectures of James F. Dana on electromagnetism.

In 1832, during the leisure afforded by the long sea voyage on his return from Europe, Morse devised a plan for transmitting letters and words by means of the electromagnetic needle. He sought the scientific guidance of Joseph Henry, Professor L. D. Gale of Columbia, and the technical aid of ironmaster Alfred Vail. After twelve years of patient struggle, Morse perfected his invention and introduced it to the public.

Introduction of Morse's Telegraph. Morse's equipment was noted for its simplicity. The whole apparatus occupied no more than 6 cubic feet. It consisted of acid and zinc batteries (with some eighty small plates, 1x3½ inches), the Cruickshank battery (which, by 1837, was in common use), a simple mechanism for making and breaking the circuit, which was controlled by an operator, and a register on which the transmitted electric signals were recorded. He used two complete sets of wires, one for each way, placed on elevated poles. The committee reported that Morse's apparatus represented "in construction and execution, the height of mechanical skill and ingenuity." [3]

In 1837 Morse gave a successful demonstration before a group of scientists at New York University, and a later demonstration in the Hall of the Franklin Institute in Philadelphia. The next year he gave an exhibition in Congress before President Van Buren, several members of Congress, and other leading citizens. Although the Committee of Commerce made a favorable report following this demonstration, victory was not yet won. Congress busied itself investigating other telegraph systems. Morse, meantime, sought help abroad, but without satisfactory results.

A later petition to Congress bearing the endorsement of Professor Joseph Henry of Princeton, and of the National Institute, met with the much desired success. Professor Henry affirmed, "that science is now fully ripe for this application." He was certain that the invention would prove successful. "The time has arrived," he declared, "when the developments of science are able to furnish the means of certain success, and devise a plan for carrying it into practical operation." Henry was sure that Morse had established a just claim to scientific reputation, "and deserved public patronage."

The American Institute also acclaimed Morse's success and declared that the electromagnetic telegraph was admirably adapted for the purpose for which it was intended. It was capable of forming words, numbers, sentences, "nearly as fast as they can be written in ordinary characters; and of transmitting them to great distances . . . , it is a most important application of high science brought into successful operation by the exercise of much mechanical skill and ingenuity." [4]

Telegraphy Put to Work. Despite these high recommendations it was with much misgiving and with sarcastic raillery of "mesmerism, Millerism," and hoaxes in general, that Congress passed an Act (March 3, 1843) appropriating thirty thousand dollars to provide for a demonstration of a system of telegraphy. Lines were set up between Washington and Baltimore. After fourteen hectic months, 40 miles of copper wire spanned the intervening distance between the two cities. With Morse presiding at the key in the Chambers of the Supreme Court in Washington, the famous first message, "What hath God wrought," was sent over the electromagnetic telegraph, May 24, 1844.

The fact that the Democratic Convention was meeting in Baltimore the day after this first message was transmitted had much to do with making the public aware of the new invention. This was the occasion of the nomination of James K. Polk for the presidency, and the telegraph performed miracles on the floor of the convention. The proceedings were transmitted to the assembled congressmen in Washington, and their messages wafted back, "like the ancients received the responses of an oracle," the *National Intelligencer* observed. Other newspapers also heralded the triumph. The *New York Herald* declared, "It [the telegraph] has originated in the mind a new species of consciousness, and is unquestionably the greatest invention of the age." Morse had been proclaiming for twelve years that the telegraph would startle the world, but the world was only now convinced that it was not a hoax.[5]

Use of Telegraph Expands. New York and other cities immediately demanded telegraph services. By September, 1846, hundreds of miles of telegraph lines had been established, reaching from Boston to Washington, then to Harrisburg. By the beginning of 1847 the telegraph had reached Pittsburgh, and then stretched south to Petersburg, Virginia, and to Cincinnati and Louisville in the West. By 1848 Florida was the only state east of the Mississippi that was untouched by the telegraph. Polk's message to Congress in December, 1848, was carried by wire to St. Louis, where it appeared in print 24 hours after its delivery.

Within a few years railroads began to recognize the telegraph as a natural ally in both communication and transportation technology.

The New York and Erie Railroad installed a line in 1851, and the Pennsylvania Railroad began construction of a telegraph line in 1855. This became a part of the Western Union main line, which entered the field in 1856. Joseph Henry, who had become Secretary of the Smithsonian Institution, secured a Congressional appropriation for the use of the telegraph in making meteorological observations. The success of the telegraph gave credence to other ideas for improved transportation. The promoters of the Atlantic cable did not have to wait twelve years, as did Morse, to gain recognition for their project.

Communication Builds America

Just as the exchange of commodities was effected by the development of land and water transportation, with its attendant industrial and economic improvements in the country, so communication in ideas was brought about by a number of new technological improvements. These occurred in improving the printing press, in newsgathering, and in the invention of the telegraph. The technological developments on land and water transportation were changing the American way of life to one of specialization and the accompanying emphasis on centralization of industry.

The nation's newspapers, with their improved methods for gathering and disseminating news, not only recorded the events that were making history, but also became makers of history. They helped build a nation; they promoted a local consciousness. The wide circulation of newspapers became a constant stimulus to literacy. Even the least ambitious was spurred to learn to read the newspaper at hand. The French had long ago recognized the power of the press when they termed it "the fourth estate." Technological developments made the American press universal and helped to spread the influence of this "estate" to everyone. As a French philosopher once said, "that with twenty-six leaden soldiers he could conquer the world," America was early in taking steps to arm the nation.

Notes for Chapter 15

1. Allan Nevins, *The Evening Post, A Century of Journalism* (1922), has an excellent review of early newspapers.
2. S. N. D. North, "Manufacturing," *Census of 1880*.
3. "Letter from Joseph Henry to Professor Morse," Feb. 24, (1842), U. S. *Globe House Report,* No. 17, 27th Congress, 3rd Sess., Dec. 30, 1842, Appendix A, p. 5.

4. *Ibid.,* Appendix B, "Extract from the Minutes of the American Institute," N. Y., (Sept. 12, 1842).

5. See Carleton Mabee, *The American Leonardo, A Life of Samuel Morse* (1943), for an excellent account of Morse's many experiments.

Valuable information relating to the early telegraph is found in Robert Luther Thompson, *Wiring A Continent* (1947); Thomas Coulson, *Joseph Henry, His Life And Work* (1950); Walter Kellog Towers, *Masters of Space* (1917); and Edward Lind Morse, *Samuel F. B. Morse, Letters and Journals* (1914).

Chapter 16

THE REVOLUTION IN AGRICULTURAL TECHNOLOGY TO 1860

Jefferson's belief that the United States would remain an agricultural nation, and that "the charms of agriculture" would attract everyone who could engage in it, was well founded. When the industrial revolution arrived, agriculture was not abandoned; rather it developed step by step with industry. Jefferson was also correct in the belief that the development of the cotton and woolen industries would not turn the people from farming to factories, for these manufactures were based on the production of the farm, and their very existence depended on farm development.

Problems of Agricultural Expansion

As other industries developed, the demand on the farmers increased. The passing of the homespun age did not deplete the farm population; it merely brought about a corresponding expansion of agriculture. This expansion involved the change from a self-sufficient agriculture to commercial farming. Agriculture, instead of opposing industry, entered into partnership with it.

Beginning of Mechanized Agriculture. This revolution in agricultural technology got under way in the 1830's. A generation earlier the cotton gin had made possible extensive cotton culture. Now it was the invention of the mechanical implements for the farm—the reaper, the mower, the seed drill, the steel plow, the threshing machine—that made farming posssible on an extensive scale, particularly in the prairie regions of the Midwest where old style farming practices had proved impracticable. As farming became a large-scale operation, it became at the same time a commercial enterprise and the process resulted in an agricultural revolution.

Factors Influencing Agricultural Changes. Many factors brought about this revolution. The character of the farmers themselves was a determining factor. The American farmer, unlike the European, was neither the traditional feudal landlord nor the peasant type. He was a mechanic. When convinced that a new tool or a machine would save him labor, he bought it—and he knew how to keep it in repair. Then, too, the opening up of the West, with its abundance of cheap land; the coming of canals, railroads, and steamboats; the development of commercial centers in the West, accessible to the new farmlands; the growth in cities; and the rise in price of farm products, especially wheat, all contributed in bringing about many changes in agriculture.

The most important factor, however, was the introduction of labor-saving machinery. The beginnings of the agricultural revolution, like the industrial revolution, were due to this improved machinery and to advances in agricultural technology.

Development of the Plow. Advances in agriculture had from the very beginning been curtailed by the shortage of labor. America, again unlike England, had no influx of cheap Irish labor for the harvest seasons. As the demand for greater production increased, the American farmer looked for mechanical aid to lighten his labors. This came first in the improvement in hand tools. They were made lighter and more efficient. The heavy, clumsy wooden or iron-plated shovels, hoes, plows, and forks were replaced by lighter and better designed tools with blades and tines made of cast steel.

The development of the plow, which had greatly concerned Thomas Jefferson when he devised his improved mold board, progressed along three lines. The change in construction material passed from wood to cast iron and then to steel. The means of construction passed from the local blacksmith to the factory system, accompanied by standardization of parts. Motive power developed from oxen to horses. Still later, attempts were made to introduce steam power.

Daniel Webster's "Bull plow." Names outstanding in the early improvement of the plow were those of Joel Morse, John Lane, Daniel Webster, and most important of all, John Deere. Morse made improvements on the cast-iron plow by developing a mold board so that it would be less subject to breakage, and by polishing it so that it would scour better. Lane devised a steel mold board made from a saw blade and braced with iron, but not realizing the importance of his discovery, he failed to patent it. Daniel Webster, desirous of a plow that would go into the ground deep enough to cut the roots of stumps, designed what he called a "bull plow." It was a huge affair, requiring four yoke of oxen to pull it. It cut a furrow 12 inches deep and had a beam made of tough seasoned oak 12 feet long. Mold board, share, and coulter were made of iron. Webster had a local blacksmith fashion it for him and took great pride in its operation. It proved too ponderous, however, to be widely used.

John Deere, 1804–1886. The farmers of the Midwest were never satisfied with their plows. The soil of the western prairie, lacking in stone and grit, failed to scour the mold boards. John Deere attempted to overcome this defect by fashioning a steel mold board from an old broken mill saw and attaching it to an ordinary plow. The experiments proved successful, and Deere placed an order with a Pittsburgh steel firm for a supply of high-grade steel. This resulted in manufacturing (in 1846) the "first slab of cast plow steel" ever rolled in the United States. These experiments started Deere on a successful career as a manufacturer of farm equipment. In 1858 the John Deere factory made thirteen thousand steel plows for the prairie trade.

Throughout this period, factory production of plows increased and gradually displaced the blacksmith's plow. As early as 1836 there were two plow factories in Pittsburgh that boasted an annual output of thirty-four thousand plows. In 1845, in Massachusetts, twenty-three plow factories made more than sixty thousand plows in one year. Ten years later, twenty-two factories in Massachusetts turned out more than one hundred and fifty thousand plows.

With the introduction of stronger and lighter plows, horses were used to supplant the slow-moving oxen. But even the faster-stepping teams of horses did not satisfy the farmers' demand for more speedy plowing. Farm organizations, state and county agricultural societies, and railway companies began offering cash awards to anyone who would invent a successful steam-plowing engine.[1]

Steam Plowing. In response to these offers, Obed Hussey of Baltimore exhibited a steam-plowing outfit at the Indiana State Fair in 1856, and J. W. Fawkes of Lancaster, Pennsylvania, built a 3-horsepower giant which he exhibited at the Illinois State Fair in 1858. It pulled eight plows and traveled at a speed of 4 miles an hour. Its successful performance caused the Chicago press to exclaim, "the goal is won. Steam has conquered the face of nature and the steam plow has become a fact." The *Wisconsin Farmer* declared that the age of steam plowing had arrived, and "Fawkes is immortal." [2] Abraham Lincoln was also impressed and declared in 1859, "our thanks and something more than our thanks are due every man engaged in efforts to produce a successful steam plow." The hopes for the early use of steam-power plowing, however, were not realized. A steam engine was too cumbersome to be hauled over the fields, and too expensive for the individual farmer to own. A steam-drawn plow, unlike the steam threshing machine, did not lend itself to group ownership. Although steam power was producing a revolution in transportation technology, it had little effect in bringing about a revolution in agricultural technology.

Inventions of Planters and Harvesters

Corn Planters. Improvements made in light machinery for agricultural purposes were reflected in devices for planting corn and sowing grain. The old method of dropping grains of corn, three or four "to a hill," and using a hoe to cover the grain was a slow time-consuming task. To speed up the planting operation, a number of drills called "corn droppers," both single-row and double-row models, were invented during the decade preceding the Civil War. These corn drills consisted of a dropper, a furrow opener, and a steering device. They included a hopper with a row of holes from which the grain fell into a boot, which carried it into the ground. With these drills, one man with a team of horses could plant from 6 to 8 acres of corn a day.

Wheat Drill. Moses and Samuel Pennach of Pennsylvania were granted a patent in 1841 for a drill equipped with hoes that was used "for the more rapid sowing of wheat, oats, peas, turnips, etc., etc." It consisted of a seed box mounted on a frame supported by two large wheels and equipped with seven hoes set 9 inches apart. The hoes could be thrown into the ground and raised by a hand lever. One man with a two-horse team could sow 8 to 10 acres a day. The general use of grain drills, however, did not come until the last quarter of the nineteenth century.[3]

Early Reapers. The most significant development in agricultural technology was the invention of the reaper. From time immemorial, grain had been cut with a hand sickle. The cradle, a late eighteenth century invention, marked a slight improvement over the sickle, but it, too, was a hand implement. For centuries men had dreamed of a device that could be used for the rapid cutting of grain, and one that would operate so gently that it would not shake the ripened grain from the stalks. Centuries ago Pliny described a crude contrivance operated by horsepower, that was used to harvest grain as early as 70 A.D. Interest in such a machine was stimulated by Arthur Young's *Annals of Agriculture* (1784), in which bits of information about this ancient machine were given. Every few years thereafter some ingenious mechanic announced a new grain-cutting machine. Some of these contained essential features that later proved successful.

The McCormick Reaper, 1831. It remained for Cyrus Hall McCormick to combine in one machine all the principles needed to produce a successful reaper. McCormick's reaper, patented in 1831, included seven vital principles. The motive power was made to draw, not to push the machine. The machine was operated by side draft. The cutting instrument was a horizontal knife set at right angles to the machine. A divider, made of wood and iron and constructed in a triangular shape, separated the swath from the standing crop and turned the stalks toward the cutting blade. A cutter bar arranged above the knife and projecting guards of fingers below it held the straws in vertical position while they were being cut. A reel was constructed to bend the heads of grain toward the knife and push the cut stalks back upon a platform. This platform, or horizontal surface of boards, built behind the cutter bar and blade, served as a gathering place for the cut sheafs until they were raked to the ground, ready to be bound by hand. A main wheel, called the "bull wheel," carried the weight of the machine, and as it turned it generated the power that was transmitted by cogs and bands to the reel and the cutter.[4]

Sales of the McCormick reaper were slow during the first few years; that is, while he remained in the hilly country of Virginia. But in 1844 when he moved to Chicago and opened a small factory in a region that was soon to become the center of a great agricultural empire, the sale of his reapers began to increase. Farmers realized that they could now expand their wheat acreage, for they could harvest crops many times larger than they could with the old hand cradles. Within five years McCormick had sold close to five thousand reapers.

Many factors contributed to this development. The Crimean War increased European demands for American wheat. The gold rush, and later the Civil War, made the shortage of farm labor more acute. The building of railroads made possible the expansion of the wheat-growing territory. The triumph of the McCormick reaper over European makes at the London Exposition in 1851 (see Chapter 18), where it was awarded the grand prize, greatly enhanced its popularity.

The Hussey Reaper. Other reaping machines began to appear on the market. The Hussey reaper was contemporary with the McCormick reaper, and for a time a contender with it for popularity. Hussey's machine was a crude two-wheeled implement, consisting of a mower with a platform attached to the rear of the frame. The sickle consisted of a series of triangular plates, riveted to a flat bar of iron. It was pulled by two horses, with the driver of the machine riding one of the horses. It had a heavy draft, its cutting mechanism was imperfect, and it was not provided with a reel. It had to be driven faster than normal speed so that the cut grain would fall properly on the platform.

A number of other inventors came up with model reapers and tried desperately to persuade farmers to buy them. Salesmen representing the different kinds of reapers toured the wheat-growing states, each proclaiming the advantages of his particular make. One observer noted that in 1854 agents selling reapers "were as thick in Iowa as Yankee tin peddlers."

By 1855 the reaper was no longer a curiosity. The effect upon the nation's agrarian economy was as far reaching as was that of Whitney's cotton gin. By 1860, some eighty thousand reapers were in operation on farms west of the Allegheny mountains. The same year the nation's wheat crop totaled close to two hundred million bushels. Most of this grain was harvested by machinery.

Mowing Machines. There was also a need for a machine to harvest the hay crop as well as the wheat crop, and efforts were made to design one that could be used to cut both hay and wheat. But the rigid cutting bar of the reaper was not suitable for mowing close to the ground except on very level land. In 1856 Cyremus Wheeler brought out a separate mower, a two-wheeled machine with a flexible cutter bar. With this device it was possible to raise the bar when obstructions were encountered, and to turn corners without damage either to the machine or the grass. By 1860 the mowers had been so improved as to be thoroughly practical. A Missouri farmer declared that at the prevailing cost of labor, "a mower will pay for itself in forty days of cutting."

Threshing Machines. Other machines which aided in bringing about a revolution in agricultural technology were headers, clover hullers, hay forks, hay bailers, and most important, the threshing machine. Early interest in threshing machines had been confined largely to gentlemen farmers. A few English threshers invented by a Scotchman, Andrew Meikle, had been imported, but until 1820 these were used only by a few wealthy, amateur farmers in the eastern states. The lack of good roads prevented bulky machinery from being hauled to the Western states, where they were most needed.

The Pitt Thresher. The first practical machine for large-scale threshing was the Pitt thresher, which came on the market in the late 1830's. Pitt's machine was an innovation, in that it was the first to combine all three threshing operations: threshing, separating, and winnowing the grain. This invention of Hiram and John Pitt of Maine was first produced in 1836. The bundles of wheat were fed, heads first, into a rotating cylinder fitted with projections that acted as beaters. The cylinder was set over a concave case which caught the grains as they were beaten from the heads of the stalks. The straw was carried off by a traveling apron. A fan attachment was added later, which winnowed the chaff from the grain. With this latest feature the machine was converted into a complete thresher-cleaner, delivering the grain to the waiting sacks or baskets. The Pitt machine was powered by horses operating on a treadmill, and the entire outfit could be transported from one farm to another by a two-horse team. It could thresh from 200 to 250 bushels a day.

A number of other manufacturers entered the field and placed thresher-cleaners on the market. Hiram Pitt, like McCormick, saw a great future in the wheat-producing areas of the Midwest and opened a factory in Illinois, while his brother John built a plant in Buffalo, New York. By the early 1850's most of the threshing was performed by itinerant threshermen who owned and operated the larger improved machines. Wheat threshing became something of a specialized industry, one of the first in the history of American agriculture.

Grain Storage Problems. A new problem in agricultural technology appeared with the establishment of the threshing machine practice. Heretofore threshing had been a winter operation, and the task was performed when it best suited the individual farmer. The market was able to absorb all the grain as it was threshed. However, the threshing machines and fanning mills brought about some radical changes in planning the season's work. As soon as the harvest was over, large amounts of grain were threshed and rushed to the mills. There were no facilities for storing such huge quantities. Buffalo, New York, was one of the first cities to experience this difficulty. Grain arrived at that port by lake boats from the Midwest for transshipment via the Erie Canal. If transportation were not immediately available, then the recent mechanical advantages afforded by the improved harvesting and threshing facilities would be of no value.

The Grain Elevator. But once again a new development occurred in agricultural technology which affected the entire grain and wheat industry. In 1842 a small beginning was made in what came to be known as the "grain elevator," when Joseph Dart patented a granary designed to be "heat and moisture proof," and which could be filled by bucket elevators from the hold of a ship docked in the harbor. The grain was emptied by gravity through a hopper-bottom in desired quantities for railroad or canal transportation. Within a few years Chicago wheat buyers were constructing a system of grain elevators that soon outgrew the original ideas of Joseph Dart. When the railroads reached Chicago, that city surpassed Buffalo as a grain center. It should be remembered that many of the ideas incorporated in these early grain elevators were merely the application of a standardized technique that Oliver Evans advanced before the close of the eighteenth century.[5]

The Dairy Industry Specializes

Just as the threshing machine had helped to bring about a specialized industry for the farmers, farmers themselves began to concentrate on special phases of agriculture. One such development was the dairy industry. The dairy business, including the factory production of butter and cheese, the large scale production of ice cream, and the canned milk industry, began before the Civil War.

Need for Increased Supply and Distribution. Factors promoting this development were reciprocal. The growth of industrial centers with large populations created a demand for a greater supply of milk than the nearby farmer who kept a cow could provide. The development of farm machinery made it possible for the farmer to have time to pay more attention to his cattle and to increase his herd. The coming of the railroads facilitated the rapid transportation of milk from greater distances. The improvement of the breeds of cattle resulted in an increased milk supply. The shifting of the grain-growing centers to the Midwest enabled the East to develop small, rich pastures, ideal for dairy cattle. With the introduction of better feeding methods for cattle and better stabling, milk production was made possible the year round. Dairying became more profitable than raising cattle for beef. All these developments had occurred before 1850, and with them came the dairy industry.

Early Distribution Methods. For many years the city dwellers had received their milk supply from farms located on the outskirts, or from town dwellers whose cows roamed the streets. This latter custom in New York City developed into what was known as swill dairies, where the cattle were fed on the discarded mash from the breweries. Families had become so accustomed to this quality of milk that when the first large shipment of whole milk from the country was received in New York in 1842, customers complained of the yellow scum (cream) that had formed on the top. The first supplies of milk where shipped in churns. By 1852 shipments were made in milk cans, the design of which has remained unchanged to this day. A system of cooling pits established at the shipping points, consisting of deep holes in the ground filled with ice, kept the milk cool until it was placed on the trains. In some of these pits the water became so solidly frozen in winter that the ice continued in the bottom throughout most of the summer. During the 1850's a few iced cars were introduced.

Butter. Factory production of butter was slow to be adopted. Butter was produced at the creamery in much the same manner as on the farm, only under less favorable conditions. Barrel churns were used, with dogs or sheep providing the motive power by working a treadmill; in some places water power was used. The butter was packed in tubs. The farmers in Orange County, New York, and in the Berkshires, took the lead in butter production, and the eastern sections of the United States long held the supremacy in this field. As late as the 1880's New York butter was being shipped to Chicago. The United States Navy for years specified Orange County butter for its crew members, since it kept better in the tropics.

Cheesemaking. Cheesemaking, which also began as a farm industry, with a reputation opposite to that of creamery butter, gained both in quality and in popularity when it was transferred from the home to the factory. The Cheddar cheese industry, which began around 1850, had by 1860 reached such proportions that great quantities were shipped abroad. Jesse Williams of Herkimer County, New York, had learned the art of making Cheddar cheese in England, and when he returned to the United States he introduced a cheesemaking machine of his own and trained others in the craft. Cheese from the Williams factory quickly established a favorable reputation. New York State became famous for its Cheddar cheese, long known as American cheese or New York State cheese.

Ice Cream. Ice cream making as a commercial industry was another by-product of the specialization that was occurring in the dairy business. Unlike the butter and cheese industry, it did not originate as a household undertaking. The manufacture of ice cream was made possible when the hand freezer was invented. The freezer consisted of a can fitted with a beater that rotated by means of a crank, similar to the modern hand freezer. In 1851 Jacob Russell of Baltimore began the commercial production of ice cream. His establishment was equipped with pewter pots for freezing, molds for packing, a tin box for storage, a few pails for ice, mortar and pestle for breaking the ice, and shovels for ice and salt. Due to the lack of refrigeration, however, the growth of the ice cream industry was slow and it was not until near the turn of the century that it consumed any considerable part of the farmer's milk supply.

Gail Borden, 1801–1874. The canned milk industry (evaporated milk) also helped bring about the revolution in agricultural technology. Gail Borden, ex-school teacher, surveyor, editor, and one of the founders of the Texas Republic, had been working on the idea of concentrating different foods so that they could be packaged and shipped long distances and kept indefinitely. He produced a concentrated meat biscuit which won a prize at the London Exposition in 1851. His biscuits soon became a common ration for explorers and sailors. In 1853 Borden applied for a patent for preserving milk by adding sugar, and heating it in a vacuum. His petition was denied at first on the grounds that it offered nothing new, but three years later, when he demonstrated that the vacuum phase of his process was both novel and essential for the successful production of canned milk, he received a patent. Plants were established in Connecticut and New York before 1861. Borden was ready to supply the Civil War soldiers with canned milk and thus achieved one of the greatest contributions to dairying.[6]

Because of these many by-products—the production of fluid milk, butter, cheese, ice cream, and condensed milk—dairying came to be a major factor in American agriculture. By 1860 America's dairy industry was greater than that of any other country in the world, and exports of cheese continued to increase.

Experimental Agriculture

Accompanying the revolution in agricultural technology, and contributing to its success, was the growing interest of the American farmer in scientific agricultural experiments. With the rise of the factory system and the higher wages paid for labor, the farmer often found his own sons leaving for the cities to accept higher paying jobs. Facing an even greater shortage of help than they experienced in previous years, the more progressive farmers welcomed not only the introduction of labor-saving machinery, but also the aid of science. They began to join agricultural societies, read newspapers and farm journals, and attend agricultural trials and fairs, where the latest and most improved machines and methods were demonstrated.

The farmer's acceptance of the principles of scientific agriculture was a slow but definite part of the revolution in agricultural technology, as was also the discovery of these scientific principles.

Chemistry Applied to Agriculture. It was a German chemist, Justus von Liebig, who first introduced the science of chemistry to the American farmer. His classic study, *Chemistry and Its Application to Agriculture,* was published in this country in 1841, and its influence was widespread. Liebig demonstrated that growing plants, by taking chemicals from the soil, deplete it; and that the addition of chemicals to the soil through fertilizers promotes the growth of plants. Before this it was thought that plants received much of their nourishment from the air, and little was known about the chemical principles of fertilization. Liebig's work marked the beginning of the study of soil science and paved the way for the great commercial fertilizer industries of a later day. American students of chemistry made the University of Giessen, Germany, where Liebig taught, the mecca of their pursuits, and a number of them completed their education there.

Miraculous results were expected from these new ideas of scientific farming, but though the progress was marked and basic, it was gradual. Henry L. Ellsworth, Commissioner of Patents, predicted in 1842 that the value of the nation's agricultural products would exceed all anticipation "now that science has come to be applied to husbandry, and agricultural chemistry should become our most popular and useful study." Even Emerson was moved to comment on the possibilities offered by the application of chemistry to agriculture. "The chemist," he wrote, "with a teaspoonful of guano could turn a sand bank into a corn hill." Such predictions did not come true for many years.

Government Participation. The annual reports of the Commissioner of Patents on agriculture during the 1850's contained important papers on soils, fertilizers, crops, and the industrial utilization of farm products. They were prepared by men out in the field, and then compiled by the Washington office. All these developments paved the way for the creation of a Commissioner of Agriculture, established by an Act of Congress in 1862.

Schools of Agriculture. The growing interest in agricultural chemistry led to a demand for the establishment of schools of agriculture. In 1855 the Farmers High School, later to become Pennsylvania State University, was founded. Michigan State Agricultural College was established two years later. The Sheffield Scientific school at Yale, under the direction of William H. Brewer, a one-man college of agriculture, early introduced lectures on the importance of fertilizers in agriculture. This new venture caused the *Scientific American* to comment, "To see Yale College stepping out from among the mists of antiquity and the groves of dead languages and taking up the shovel and the hoe is certainly one of the signs of our times." [7] Cornell early established a department devoted to the chemistry of dairying.

Land-Grant Act of 1862. But the greatest impetus to the development of technical training in the field of agriculture came with the Land-Grant Act of Congress in 1862, establishing the land-grant colleges. Provision was made for giving federal aid to colleges to encourage the study of agriculture as a science. This Act, passed during the dark days of the Civil War, climaxed the long series of developments in agricultural technology of the preceding quarter-century.

Effect of Publications and Societies. Magazines and newspapers aided greatly in bringing about the agricultural revolution. They publicized the discoveries of science, the invention of new machines, and the benefits of improved livestock breeding. One of the most prominent farm journals was the *American Farmer,* published in Baltimore by John S. Skinner and Gideon Smith. Another was the *Plough Boy* published in Albany, New York, by Solomon Southwick. Another was the *New England Farmer,* published in Boston. Other prominent journals included the *Southern Agriculturist* of Charleston, South Carolina, the *Genesee Farmer,* and the *Cultivator* in Rochester, New York. Some seventy such magazines had been started before 1850.

Farm Fairs. State and county fairs, which had their beginning with Elkanah Watson, also contributed greatly to bringing about the agricultural revolution. The golden age of the agricultural fairs dates from the 1850's. At the Indiana State Fair in 1853, organized for the purpose of "promoting science in agriculture," Horace Greeley, editor of the *New York Tribune,* gave the keynote address. Speaking on the subject, "What the Sister Arts Teach as to Farming," he called upon the farmer to become educated in "the science of his calling." He lamented the fact that there were no schools yet established to give farmers special education, but expressed the hope that such institutions would soon open in every state of the land.

Competition and Developments in Farming. Contests were held at these fairs to demonstrate the latest developments in farm machinery, and farmers had an opportunity to see for themselves how efficient the different machines were. Notable among the fairs were those held at Geneva and Syracuse, New York. A score or more of reapers were entered in the various contests and farmers watched each machine as it took its turn. Following the contest they compared notes, listened to the salesmen as they explained the advantages of their particular machine, and then some of the more progressive ones placed an order for next year's delivery.

In 1857, according to the *Scientific American,* a prosperous northern or western farmer who owned 160 acres of land was supposed to possess the following list of farm implements: a combination reaper and mower, horse rake, seed planter and sower, a hand-operated thresher and grain cleaner, portable gristmill, corn sheller, a two-horse plow and three other plows, three harrows, a roller, and a cultivator. Probably not one farmer in a hundred possessed all these implements.

Economic Aspects of Agricultural Advances

In the 1860's farming continued to be the nation's basic industry. Eighty per cent of the population lived on farms, and the farms represented one-half of the nation's wealth. The shift from hand husbandry to machine farming had been established. Farming had become a part of the nation's industrial system, tied up with banking, railroading, and manufacturing.

Farming, and National Wealth. Farming as a way of life was giving way to farming as a business. "The listlessness that characterized the farmers of twenty years ago," wrote one observer in 1869, "has disappeared and given place to a feeling of interest and a display of enterprize and energy."

The agricultural revolution contributed greatly to the wealth of the country. The value of farm machinery increased from less than seven million dollars in 1850 to more than twenty million in 1860. The manufacturing of farm machinery had been added to the nation's roster of big industries. The output of grain was greatly increased. Five million bushels of barley in 1850 had increased to fifteen million in 1860. Corn had increased from less than four hundred million bushels to more than eight hundred million in the same period; rye from fourteen million bushels to twenty-one million: wheat from 44 million bushels in 1840, to almost 200 million in 1860.

Machines Replace Manpower. Aside from the material increase, the agricultural revolution was to be a determining factor in the Civil War. Just as it was a technological development in agriculture that created the social and economic conditions that brought on the conflict (Eli Whitney's cotton gin) so agricultural technology was one of the major factors that determined the outcome of the war. During the years of conflict when manpower was scarce, the agricultural machinery enabled the North to produce ever increasing amounts of corn and wheat, with a diminished labor supply. Soldiers of the Union Army were not constrained to desert and go home to harvest their crops as their forefathers had done during the Revolutionary War. Men who were no longer needed to swing scythes and flail grain helped swell the armies of the North. Secretary of War Stanton declared, in a moment of enthusiasm following the close of the Civil War, that the McCormick reaper had won the war by freeing farm labor for the Northern Army. Cyrus McCormick, a native of Virginia, was acclaimed as having done more for the cause of free men than the eloquence of William Lloyd Garrison or Wendell Phillips.

Superiority of American Technology. These early years of the agricultural revolution were years of great promise for the farmers. Still the respected sons of the soil rather than the "hay seed" or the "hick" of a later generation, their position was an enviable one. Horace Greeley noted (1871) that the half-century which bounded his recollections had witnessed greater improvements in farming than in any industry. "We use far better implements than did our grandfathers with corresponding increase in efficiency," he wrote. "Belgium, Holland and Northern France can teach us neatness and thoroughness in cultivation, the British Isles may boast of surer crops of wheat, oats, potatoes, and grass," he added, "but in selection of implements and average efficiency of labor our best farmers are ahead of them all." He prophesied that the next forty years would witness agricultural advances far greater than all those of the previous decades and declared that the recent advances had been the greatest since ancient times.

Notes for Chapter 16

1. R. M. Wik, "Steam Power on the American Farm," paper read at a meeting of the Mississippi Valley Historical Association, 1949. Quote is from manuscript copy kindly loaned to the author.
2. Quotes by permission from Wik's manuscript.
3. For an excellent study on this subject, see Russel H. Anderson, "Grain Drills Through Thirty-nine Centuries," *Agricultural History,* X (1936), 157 *ff.*
4. William Thomas Hutchinson, *Cyrus Hall McCormick* (1930), is still the standard work on the McCormick reaper.
5. See Guy A. Lee, "Historic Significance of Chicago Grain Elevator System," *Agricultural History,* XI (January, 1937).
6. See Joe B. Frantz, *Gail Borden, Dairyman to A Nation* (1951), one of the latest and best histories of the dairy industry.
7. *Scientific American* (N.S.), II (Feb. 11, 1860), No. 7, 102.

Chapter 17

A NATION ALERT TO SCIENCE

Scholars and statesmen, alike, were becoming aware of the rapid advances that were occurring in science and technology during the second quarter of the nineteenth century. Recalling the scholarly essay of Virgil Maxey, on the necessity for understanding science as the key to American prosperity,[1] Job Durfee, two years later in another Phi Beta Kappa essay, proclaimed that, "human progress is dependent on the scientific and inventive mind of our people." Even Senator John C. Calhoun, the philosopher-statesman, impressed by the number of patents that were being issued during this period, remarked late in the year 1841, "We have, Senators, reached a remarkable point in the progress of our civilization and in the mechanical and chemical arts, which will require a great change in the policy of our civilized nation."

Writing on a little less scholarly level, and with unrestrained enthusiasm, a city daily newspaper declared in an editorial, September 18, 1840, "Democracy is the cradle of science and only an unfettered mind can expand to its utmost, for discoveries and inventions are the natural result of an atmosphere of free men."

Science—Key to America's Progress

This idea that science would flourish more rapidly if embedded in the "cradle of democracy" had also found an ardent spokesman in the French philosopher, Alexis de Tocqueville. While studying American society of this period, he was impressed by America's advances in science and in industry. He observed that there was a close relation between these recent advances and the social mobility of America's democratic system. He noted that Americans were never satisfied with their position, that they always felt free to leave any job and seek a better one. "To minds thus predisposed," he continued, "every new method which leads by a shorter road to wealth, every machine which spares labor, every instrument which diminishes the cost of production, seems to be the grandest effort of the human intellect."

Such expressions reveal the growing interest in science during the period preceding the Civil War. Two presidents of this period, Fillmore and Pierce, were impressed by the recent nationwide interest in science and technology. President Fillmore, in his third annual message, December 6, 1852, noting the myriad inventions and recent scientific discoveries, declared, "We live in an age of progress. . . . Our territory is checkered over with railroads, and furrowed with canals. The inventive talent of our country is excited to the highest pitch. And the numerous applications for patents for valuable improvements distinguish this age and its people from all others. . . . The whole country is full of enterprise." [2] And the following year, President Pierce in his first inaugural address requested Congress to give favorable consideration, "to those men of genius who by their inventions and discoveries in science and arts have contributed largely to the improvements of our age. . . ." [3]

That science and technology, as well as her democratic form of government, were to be America's contribution to world civilization was well established before the outbreak of the Civil War. In their attitude toward technological improvements, scientists, differing from the craftsmen of the early guild age, did not attempt to keep their discoveries secret. On the contrary, they began early to exchange ideas, join forces, and combine their discoveries. To this end they organized scientific societies, held national conventions, founded scientific publications, and established schools of science.

Scientists Unite

First Chemical Convention, 1831. An early example of this union of scientists was the chemical convention held in New York City in 1831. It was attended by some five hundred delegates, consisting of both chemists and manufacturers and representing such major industries as iron, steel, glass, sugar, salt, tanning, and chemical industries. Industrialists and scientists, alike, were disturbed over the influx of British goods being dumped on the American markets. The keynote of the conference was the economic and industrial welfare of America. The announced objective was the establishment of this nation's economic and industrial freedom. Specifically, these early men of science declared they could make the nation independent in chemical products if Congress would only pass an adequate protective tariff.

They presented a report on the chemical industries of America, pointing out their importance to the defense of the nation, and memorialized Congress to give liberal aid to industry. The only way to combat militarism was by means of an efficient industrialism. Napoleon's defeat at Waterloo was due, said the scientists, not to Wellington and Blücher's artillery fire, but to the mightier machinery of the power loom and the spinning jenny—machines that had made England strong. They emphasized the importance of chemistry in modern life and cited the waste products that could be made useful by the chemical industries. Among these were the by-products from the slaughter houses, the horn shavings from the combmaker shops, and the trimmings and scraps from the leather manufacturers.

America's First Scientific Congress, 1844. Of greater importance than the chemical convention was America's first Scientific Congress, held in Washington in 1844, attended by scientists from a number of colleges and universities. After President John Tyler had officially welcomed the delegates, Senator Robert J. Walker praised America's men of science for "enlarging our knowledge and providing man with increasing comforts." He said that "the scientists could make this country the greatest and freest nation of the world."

Sectional meetings were held on such diverse subjects as astronomy, geology, physics, chemistry, meteorology, paleontology, and physiology—a long roster of sciences for that day. Matthew Fontaine Maury reviewed the recent studies that had been made of the Gulf Stream. Alexander Dallas Bache read a paper in which he compared the development of science in the United States with that of Europe. Secretary of War John C. Spencer expressed the hope that America's contribution in original inventions "and her new discoveries and applications of science would greatly enrich the nation's treasury of useful facts." This first scientific congress set a pattern for the later meetings of the nation's scientists.

Goodyear's Vulcanizing Process

Meantime, a number of talented young men who had never attended a meeting of scientists were carrying on experiments (without knowing chemical formulas) that were destined to affect the entire industrial life of the country. One of these was Charles Goodyear, who discovered a method for vulcanizing rubber in 1839 and patented the process in 1844. The many possible uses of rubber had been known for years. Priestley had applied the term "rubber" to the product of the caoutchouc tree because it could be used as an eraser. But its uses were limited (it was too hard in winter, and too soft in summer) until Charles Goodyear discovered a simple chemical process of heating rubber with sulfur. The result was a soft, pliable, elastic substance that became at once an indispensable item in American industry.

Rubber began to be widely used for footwear, for waterproofing garments, for fire hose, for rubber bands, door springs, and for inflated beds. In 1851 Nelson Goodyear, brother of Charles, discovered a process for making hard rubber. He applied it in the manufacture of imitation whalebone, in great demand at the time, and in making jewelry. The growth of the rubber industry was rather slow for several years, however. The next great impetus to the industry was brought about by the bicycle age during the last quarter of the nineteenth century.

Photography. Another benefit to mankind that resulted from the alertness to science was the invention of photography. Photography has been compared to printing, in that it multiplies at will the visual representation of facts and events. No one person can be heralded as the sole inventor of photography, since a number of persons, both in Europe and America, were experimenting simultaneously on image reproduction. A pioneer American photographer, John William Draper, made a number of early improvements, chief of which was the development of a method for shortening the time of exposure used by the celebrated Frenchman, Daguerre. Draper was born in England, but came to the United States in 1833 and was educated at the University of Pennsylvania.

The Daguerre photographic process, using the iodine-treated silver plate method, required an interminably long exposure. Draper used fumes of bromine and discovered that a mixture of iodine and bromine was especially sensitive to light. This greatly reduced the time exposure. A photograph that he made of his sister Dorothy is reported to be "the first complete portrait of a person ever recorded by the sun." Draper was a scientist of considerable attainments, a distinguished chemist, physician, and historian. He made original studies in the chemical action of light, and is best known for his work on radiant energy.

Alter's Spectrum Analysis. Another pioneer scientist of this period was Dr. David Alter, distinguished physician and physicist of Freeport, Pennsylvania. Inspired by reading the achievements of Benjamin Franklin, Dr. Alter developed a homemade electrical telegraph as early as 1836, produced coal oil (kerosene) from coal, and pioneered in the field of spectrum analysis.

The Smithsonian Institution

Another indication of the nation's alertness to science was the establishment of the Smithsonian Institution in 1846. By the will of James Smithson, a British chemist and mineralogist, America became the recipient of more than a half-million dollars to be used to "found at Washington under the name of the Smithsonian Institution, an establishment for the increase and diffusion of knowledge among men."

The money was turned over to the United States government in 1838 and deposited in the United States mint. As the nature of the establishment was not specified, a debate extending over a period of eight years ensued in Congress. Every type of institution had its advocates. Some wanted a library, some an observatory, others wanted a laboratory, a lecture lyceum, an art museum. It was not until 1846 that Congress passed an Act establishing the Smithsonian Institution. A board of regents was provided with an executive officer in charge, and specifications were prepared for a museum and a library.

Functions of the Institution. While the Smithsonian Institution became known as the "American Treasure House of Science," it has never been merely a repository for collections. The function of the Smithsonian has been to explore every field in the realm of knowledge, and to be the custodian of objects of art, science, and invention. This function was well defined by Joseph Henry, the first secretary, and a great scientist, in laying the foundations for the work of the Institution. The program included research in pure science with the aim of using the results in practical, everyday things; a museum and an exhibit hall for the education of the public and the preservation of collections; and the issuance of publications for "the diffusion of knowledge among men."

Professor Henry established a series of publications that has continued to this day. "The worth and importance of the Institution are not to be estimated by what it accumulates within the walls of its building," declared Henry, "but by what it sends forth to the world." The greatest mission of the Institution was to facilitate the use of all the implements of research and to inform the nation of the value of all new discoveries. Henry inaugurated the custom of issuing annual reports which presented in popular language the accomplishments of the Institution for the current year. A more specialized series of publications was established in 1848 to publish the discoveries of staff members in the many fields of science. In 1860, *Smithsonian Miscellaneous Collections* appeared, which included shorter papers on all phases of scientific investigation.

Scientific Repository. The acquisition of scientific information that could be put to practical use was the chief aim of the Institution. Fossils were collected, not as curiosities, but as a guide in estimating the age and structure of rocks from which the possibility of valued mineral deposits could be determined. Of like value was the determination of glacial deposits from which the famed Agassiz had evolved his theories in glaciology. Specimens brought from explorations in the far west were classified and catalogued. These rendered valuable information to petroleum and mining geologists.

Indian Archeology. One of the earliest requests sent out by the Institution was for collections from Indian mounds. Professor Henry was desirous of preserving these for future study, and he wanted to prevent their destruction as the western frontier was being settled. The Smithsonian scientists also engaged in the science of taxonomy, the classification of plants and animals according to natural relationships. This obviated the confusion resulting from the variety of names given in different countries to the same object. The Institution fostered the establishment of the Weather Bureau and the Bureau of Fisheries, and it also became the medium for a great system of international exchange of scientific information and materials.[4]

U.S. Weather Bureau

While the Act of Congress officially organizing the Weather Bureau was not passed until 1870, the groundwork for weather forecasting was established before the Civil War. Meteorology, as applied to weather forecasting, had intrigued astronomers from ancient times. But it remained for Joseph Henry and his staff of the Smithsonian Institution to pave the way by organizing a weather forecasting system and by using telegraphy to put it into operation. It was the scientific approach that prevailed in the mid-nineteenth century that instituted the method, and it was the invention of the telegraph that provided the technological means of communicating the information about the weather.

Study of Meteorology. Among those who laid the ground work for meteorology by studying the characteristics of storms and the accompanying atmospheric disturbances were William C. Redfield, who advanced the rotary theory of storms, and James Henry Coffin of Massachusetts, who made an extended study of the movements of the winds. Redfield's study of winds, storms, and whirlwinds, entitled, *"Observations on the Hurricanes and Storms of the West Indies and the Coast of the United States"* (1833), was described as a meteorological classic. James H. Coffin, professor of mathematics in Lafayette College, was selected by the Smithsonian Institution in 1846 to make a scientific study of wind directions and wind velocity. The Institution published two of the studies, *"Winds of the Northern Hemisphere"* and *"Winds of the Globe"*; the latter was described as, "the most exhaustive collection and the fullest analysis of wind data that had ever been made."

Another pioneer meteorologist was James Pollard Espy, known as the "Storm King." Espy, a native of Bedford, Pennsylvania, after teaching school for a time, began giving lectures on meteorology. He was appointed as meteorologist by both the War Department and the Navy. In the latter capacity he issued daily weather observations, and in 1843 Espy put out the first *annual* weather report.[5]

Specialization of Scientific Groups

The trend that had been established in late eighteenth century, beginning with the American Philosophical Society, to form organizations devoted to science, assumed nationwide proportions during the second quarter of the nineteenth century. The societies became more specialized and more democratic in character. Unlike the earlier societies, which represented a small leisure class, they included men from all walks of life. They were scientists, teachers of science, and those who were interested in applied science.

West Point Society. One of the early organizations was the Associate Society of West Point, formed in 1829, which became the "American Association for the Promotion of Science, Literature, and the Arts." Organized with the idea of becoming another Royal Society, or French Academy, within a year it had a membership of eighty-eight professors and cadets from the National Military Academy. Their aim was to study the *practical sciences* by the method of laboratory experiments. Emphasis was given to the study of the useful arts and inventions, and its members advocated a nationwide program, similar to the one later adopted by the Smithsonian Institution, for disseminating the useful discoveries.[6]

American Association for Advancement of Science. The first of the national scientific societies to become permanent was the Association of American Geologists and Naturalists, formed at Philadelphia in 1842. Six years later it expanded its activities and changed its name to the American Association for the Advancement of Science. The newly named organization was "designed to embrace all laborers in Physical Science and Natural History."

The first meeting of the American Association for the Advancement of Science (AAAS) was held at Philadelphia in September, 1848. The leading scientists in the nation were included in its membership. William C. Redfield, meteorologist, became its first president. William B. Rogers, professor of chemistry and natural philosophy at William and Mary College, and one of the four Rogers brothers of whom Senator Robert J. Walker said, "they are scarcely to be excelled for native powers and acquirements in the history of science," was chairman of the organizing committee. (This same Professor Rogers later established Massachusetts Institute of Technology.)

Distinguished Scientist Members. The activities of the AAAS were from the first departmentalized. The first fields to be explored included the physical sciences and natural history. At the initial meeting the speakers included Benjamin Peirce, acclaimed as the nation's greatest mathematician, and Louis Agassiz, the Swiss naturalist and geologist, who came to the United States in 1846 to deliver a course of lectures at the Lowell Institute at Boston and decided to remain here. Among the early presidents were such distinguished scientists as Joseph Henry, who gave to the world the theory of the electric telegraph, Alexander Dallas Bache, Superintendent of the United States Coastal Survey, and later first president of the National Academy of Science, and John Torrey who, with his collaborator Asa Gray, developed a new system of nomenclature in botany.

Early Accomplishments. The time was ripe for such an organization of scientists. New and revolutionary developments were occurring in the field of biological sciences. The cell theory of the structure and functioning of living organisms was being evolved, and the principles of organic evolution were taking form in the minds of Darwin, Wallace, and Spencer.[7] In this scientific awakening, United States scientists played a part. It was to an American scientist, Asa Gray, that Charles Darwin of England wrote his famous letter (September 5, 1857) in which he first outlined his theory of evolution of the species by means of natural selection. And the same Asa Gray was one of the scientists to receive an advance copy of Darwin's *Origin of Species* (1859) in testimony of the respect that was being accorded to American men of science.

One of the original aims of the AAAS was "to hold periodical and migratory meetings, and to promote intercourse between those who are cultivating science in different parts of the United States." Within six years, by 1854, the membership numbered more than a thousand scientists. The publication of the *Proceedings* was from the first an important function of the Association. With the outbreak of the Civil War in 1861, all national meetings were canceled and were not resumed until August 1866, when a meeting was held in Buffalo, New York.

Engineering Societies. It was also during this same period that a number of engineering societies were organized. In 1848 the Boston Society of Engineers was founded. In 1852 it became the American Society of Civil Engineers. This was the first of "The Founder Societies," although it was not incorporated until 1877.

Early Schools of Science

Concurrent with the formation of engineering societies was the establishment of a number of schools of science. One factor in this movement was the accumulation of wealth by some of the nation's industrialists, who inaugurated the practice of contributing to institutions of higher learning. Prominent among these were two brothers, Amos and Abbott Lawrence of Boston. By a gift of fifty thousand dollars in 1857, the Lawrence Scientific School was established, which later became affiliated with Harvard. Similarly, Joseph E. Sheffield, attracted by the scientific work of the two Sillimans, father and son, and of John P. Norton, made a generous contribution to Yale in 1847. This marked the beginning of the Sheffield Scientific School. The Chandler School of Sciences and Arts was established in 1851 by Abiel Chandler of New Hampshire, as a part of Dartmouth College. The Polytechnic Institute of Brooklyn was opened in 1855.

The year that witnessed the outbreak of the Civil War also witnessed the founding of the Massachusetts Institute of Technology with William B. Rogers as its first president. The Act of incorporation was passed by the Massachusetts Legislature in 1861.

University Curricula Changes. Agitation was widespread in the years from 1840 to 1860 for introducing the study of the sciences and engineering in the curricula of established colleges and universities. Chemistry, geology, and biology were the subjects in most demand. At the Western University of Pennsylvania (University of Pittsburgh) the interest in the science courses, 1843–1849, was the cause of a heated intramural battle. Norwich University gave the Bachelor of Science degree in 1845. Harvard granted it in 1851. The University of Michigan granted it in 1855, and the University of Wisconsin in 1859. The University of Michigan and the University of Pittsburgh both gave the Civil Engineering degree in 1855, and Union College in 1857.

Cooper Institute. Unique in its appeal to the artisan and mechanic class was the Cooper Institute, endowed by Peter Cooper of New York City in 1853 and opened for classes in 1859. For the benefit of those who were engaged in mechanical pursuits, courses "in the most useful and practical sciences" were offered day and night free of charge. The sons of mechanics and laborers were invited to enroll. The Institute was designed to open the avenues of scientific learning and to encourage the acquisition of useful knowledge. Instruction was given in the physical sciences, chemistry, architecture, free-hand drawing, mechanical drawing, and vocal music.

Foreign Study. It was during this period of national awakening in science that Americans began going abroad for study, especially to German universities. They sought such noted teachers of science as Justus von Liebig and Frederick Wöhler. Oliver Wolcott Gibbs studied under Liebig, whose preachment was education through research. Gibbs adopted Liebig's laboratory methods and introduced them in his own teaching while occupying the Rumford professorship in Harvard College. John Addison Porter of Yale also studied in Germany under Liebig, and became the first dean of the Sheffield Scientific School. Charles Frederick Chandler, who became one of America's most distinguished industrial chemists and sanitary engineers, studied under Wöhler and received his Doctor's degree at Gottingen. A number of other scientists also went abroad to learn at first hand the laboratory method of studying science.

Early Chemical Laboratories. This emphasis on research and on the importance of the laboratory technique led to new methods of teaching the sciences, particularly chemistry. Emerson had complained that when he went to college he listened to lectures, took notes, but performed no experiments of any kind. It was in private laboratories that American students first found opportunities to make chemical analyses. Charles T. Jackson, brother-in-law of Emerson, established a private laboratory in Boston, and there the great philosopher, though reared in the classics, became imbued with the wonders of the scientific world. In marveling over the accomplishments of chemistry, Emerson later penned one of his best known lines (but which is usually quoted only in part) when he wrote poetically, "the wisdom of man has brought him to hitch his wagon to a star and see his chore done by the gods." He declared that man had mounted science but warned that some day science might have the existence of mankind in its power and that the human race might possibly commit suicide by blowing up the world. And prophetically, he wrote, that the next war, a war of principles, "would be fought in the air!"

Scientific Expeditions

Other factors that aided in developing an alertness to science during the pre-Civil War decades were the exploring expeditions sent out by the United States government, and the geological and coastal surveys authorized both by the federal government and by several states. Awareness of the vast expanse of the nation and curiosity as to its natural resources were felt during the second quarter of the nineteenth century. Several reasons account for this growing interest: the early military expeditions conducted by the United States government, the explorations made by private parties, the overland expeditions during the Mexican War, the government boundary surveys that followed the war, the gold rush to California, and the surveys for transcontinental railways.

Government Sponsorship. The federal government sponsored most of these expeditions, which included geologists, botanists, ethnologists, and paleontologists. Their findings provided a vast workshop for scientists in identifying and classifying the rich discoveries of the western country. Because the material was new, great reservoirs of information were made available. Though these expeditions were largely conducted by military officers with the gun and sword, the accompanying scientists with their notebooks and canvas kits and with their appreciation of natural phenomena, were able to preserve a vast and precious heritage that might otherwise have been despoiled.

Schoolcraft and Fremont Expeditions. Henry Rowe Schoolcraft, who had accompanied the Lewis Cass expedition to the Upper Mississippi and the Lake Superior copper region (1803–1806), remained in the Lake Superior country as Indian agent and married the daughter of an Indian chief. He spent many years collecting geological and ethnological data, which were later published in six large volumes. The expeditions by John C. Fremont, topographical engineer of the United States Army, 1841–1845, dispelled the general ignorance of the main features of the Northwest and opened up possible lines of communication to the Pacific.

Wilkes Expedition. A wealth of scientific material was secured on the expedition of Captain Charles Wilkes when he encircled the globe, 1838–1842. He explored the coasts of southern South America, New South Wales, the Antarctic, the Fiji Islands, Hawaii, the northwest coast of the United States, the Philippines, and Polynesia, returning to New York by way of Cape of Good Hope. Accompanying the expedition were a galaxy of naturalists, botanists, taxidermists, and philologists whose findings were invaluable to science. Wilkes himself issued a nineteen volume report of the expedition.

Elisha Kent Kane. Of special value to science were the three volumes prepared by James D. Dana on the zoophytes, geology, and the crustacea. Accounts of Dr. Elisha Kent Kane's Arctic explorations (1856) recording the hardships and discoveries of the Grinnell expedition, which was sent out in search of Sir John Franklin, became the nation's best seller, and for a decade lay with the Bible "on almost every parlor table in America." [8]

Scientific Publications

The growing interest in these scientific and technological developments was accompanied by the founding of numerous magazines and journals of science. "These publications (scientific) had reached a vast affluence," observed E. B. Hunt in a survey covering the period 1845 to 1858. He listed more than six thousand volumes of scientific memoirs and periodicals, and noted that there was not a single library in the country "equal to the demands of the great army of scientific investigators." [9]

Magazines and Textbooks. The most popular of these journals was the *Scientific American,* founded in 1845. It soon claimed the largest circulation of any magazine of its kind in the world. Its announced purpose was to be an advocate of industry, a journal of science, an aid to mechanical improvements, and a Patent Office register. The editors welcomed the assistance of those "who can elevate and instruct" its readers in these fields, and declared that an intimate knowledge of science and art, "would conduce more to the honor and elevation of man than any other cause." [10] The *Scientific American,* unlike the *American Journal of Science,* did not limit its pages solely to scientific material, but frequently carried a poem or a literary production that had popular appeal.

Equally important in making the nation alert to science were the textbooks that began to appear during this period. Robert Hare, for many years professor of chemistry at the University of Pennsylvania, partly filled the need for textbooks with his *Compendium* (1840). There were no texts in organic chemistry, but Hare made up for this lack by enlarging the section of animal and vegetable chemistry in his book. James and Robert Rogers published a textbook on chemistry in 1846 which became a standard work and led to the introduction of courses in chemistry in a number of colleges. James Dwight Dana, successor to Silliman at Yale, published a number of manuals and texts on mineralogy and geology which had wide circulation. Dana was a member of many learned societies and was the joint editor of the *American Journal of Science and Arts.*

Torrey and Gray. Recognized as modern in its viewpoint and a standard for subsequent works was the *Flora of North America* (1843), written by John Torrey and Asa Gray. John Torrey was the nation's recognized authority on the classification and arrangement of botanical specimens. He owed his inspiration as a boy to Amos Eaton, and his training while a student of medicine to the celebrated Dr. Samuel Mitchell of Columbia College. He taught chemistry and geology at West Point, and botany at the College of Physicians and Surgeons, New York. Because of Torrey, New York became the focal point for assembling the botanical data of the nation. Plants were sent to him from all parts of the world for identification and classification. No man in the history of botany had studied, organized, and classified as many genera and species of American plants.

Asa Gray, a pupil of Torrey's, collaborated in the production of *Flora of North America.* Gray, like Torrey, was educated in medicine and followed in the field of botany. As early as 1836 he produced a book on the *Elements of Botany,* and in 1839 he wrote a textbook on botany for use in colleges. His most important individual work was his *Memoir, The Botany of Japan and Its Relation to That of North America.* Gray established the department of botany at Harvard and held the Fisher professorship of natural history. His *Flora of North America* proved more than a botanical classification, since in this work he outlined the evolutionary development of the plant kingdom. Botanists were no longer satisfied with merely collecting specimens, assembling herbariums, and arranging and assorting them into species, genera, and families, as in the old Linnaeum system. Botany became a science based on philosophical truth and theory that helped man explain his own origin.

Effect of Science on Literature. Those publications that aided in creating an alertness to science on the eve of the Civil War opened their columns to authors not usually classified as scientists. Science began to have an influence on the literature of the period. The habit of thinking scientifically had developed, and literary geniuses began to look at things realistically. Steam, electricity, the physical forces, the study of the composition of the earth, all had an effect on contemporary literature, which in turn increased the alertness of the nation to science. Emerson's writing continued to reflect his growing awareness of science, especially in the field of chemistry. "Chemistry has served the apothecary, the physician, the cook, the manufacturer," he observed, "and now is itself a science."

Walt Whitman taught his fellowmen to comprehend the democratic spirit of science, and the "larger religion" to which the modern world was being led. He was keenly interested in books and lectures of a scientific nature, and was one of our first poets to refer to astronomy, chemistry, geology, phrenology, and to the controversial topic of science versus religion. He was interested in machinery, too, as noted in "Ten Minutes in the Engine Room of a Brooklyn Ferry Boat." He appreciated the value of fairs and expositions to display the achievements of American science and technology.

Edgar Allan Poe, aware of the march of science, was a little disturbed. Science, by revealing the secrets of nature, was robbing the poets of their dreams.

In his *Sonnet—To Science,* he wrote:

> Science! true daughter of old Time thou art!
> Who alterest all things with thy peering eyes.
> Why preyest thou upon the poet's heart?
>
> * * * *
>
> Hast thou not dragged Diana from her car?
> And driven the Hamadryad from the wood
> To seek shelter in some happier star?
> Hast thou not torn the Naiad from her flood,
> The Elfin from the green grass, and from me
> The summer dream beneath the tamarind tree?

Even the classicist William Cullen Bryant borrowed from science in the minuteness and exactness of his descriptions of nature.

America Honors Its Early Scientists

By the mid-nineteenth century the achievements of American men of science were attracting nationwide attention. A contemporary steel engraving by Christian Schussele, entitled "Men of Progress," appeared in 1851. Among those portrayed were: William Morton of anesthesia renown; Samuel Colt, inventor of the revolver; Cyrus McCormick, who brought forth the reaper; Joseph Saxton, known for his mine machinery; Charles Goodyear, to whom the world was indebted for rubber; Peter Cooper, hailed by the public for his Cooper Union; Jordon L. Mott, founder of the Mott iron works on the Harlem river; Professor Joseph Henry, famous for his pioneer experiments in electricity and for his work at the Smithsonian Institution; Dr. Eliphalet Nott, pioneer in the practical studies in heat; Captain John Ericsson of screw propeller fame; Frederick E. Sickles, whose mechanism for steering vessels revolutionized the utility of steam power; Samuel F. B. Morse, renowned for the telegraph; Henry Burden, who was applying steam to agricultural machinery; Richard Hoe of the Lightning press; Erasmus Bigelow of the carpet loom; Isaac Jennings; Thomas Blanchard of power lathe fame; Elias Howe with his sewing machine; and James Bogardus, the father of iron construction in buildings.

Here, in a friendly grouping, were the portraits of the men who were regarded as the leading inventors and scientists of their age. Reproductions of this galaxy of scientists hung on the walls of thousands of homes throughout the country—in farm houses and cottages, in school houses and public buildings—and gave evidence of the nation's interest in its men of science.

An exuberant contemporary, in writing of America in the 1850's, said:

Yes, America is the land of experiment. One of her sons drew the lightning from the clouds; another created wings out of steam for all the people of the earth; a third has discovered the means of mitigating life's bitter enemy, bodily suffering. [11] All of this has been done in the early morning of the country's life. . . . The day lies before it as the future. What will not this people accomplish during their day?

Notes for Chapter 17

1. For an earlier reference to Virgil Maxey's address, see Chapter 12, p. 172.
2. Richardson's *Messages and Papers of the Presidents,* VI, 2717.
3. *Ibid.,* p. 2750.
4. See Webster P. True, *First Hundred Years of the Smithsonian Institution* (1946).
5. By invitation Espy addressed the French Academy and the British Association for the Advancement of Science. He was also awarded the Magellan prize, offered by the American Philosophical Society.
6. Sidney Forman, "West Point and the AAAS," *Science* (July 19, 1946).
7. See AAAS *Summarized Proceedings, June 1929 to January 1934, Including a Chapter on the Background and Origin of the Association . . .* (Washington, 1934), p. 29. See also, *A Brief History of the Association (AAAS) from Its Founding in 1848 to 1948* (1948).
8. *Dictionary of American Biography,* X, 257.
9. Compiled in H. C. Bolton, *Catalogue of Scientific and Technical Periodicals,* a Smithsonian Institution Publication (1895).
10. *Scientific American,* II (1846), 413.
11. The reference here to "mitigating bodily suffering" related to the recent discoveries and applications of anesthesia by Doctors Crawford Long, Horace Wells, William Morton, and James Jackson.

Chapter 18

AMERICAN TECHNOLOGY GOES ABROAD

By the mid-nineteenth century the developments in American technology were attracting attention from abroad. Products marked for their utility, comfort, and for their low cost appealed to the American masses, and also attracted attention of the peoples in Europe to whom such commodities had never been available. Labor-saving and time-saving tools and machines were welcome innovations to manufacturers and producers abroad. The raw products that had been our main exports in the early years—such as skins and hides, naval stores and timber, iron ore, and baled cotton—had shown this country to be rich in natural resources, but such exports did not reflect the nation's advances in technological skills. Gradually, however, as a result of Yankee ingenuity and inventive skill, a reputation for American products was established. Europe's technological superiority began to diminish.

American Enterprise in Europe

For approximately two hundred years European technology had been transported to America. Men brought their skills with them: their patterns and models were smuggled in. But by the middle of the nineteenth century, the stream began to flow the other way; American technology started going to Europe. Some of this technology was a return of Europe's own talent, improved and altered. Some had originated here.

Building European Railroads. In 1829 the "Stourbridge Lion" was brought from England and gave America her first view of a locomotive. Nine years later American locomotive builders were shipping their engines abroad. The same year a shipment of the Baldwin Locomotive Works was sent to Cuba. In 1840 the Norris Company delivered a locomotive to the Birmingham and Gloucester Railroad in the very heart of the locomotive building district in Great Britain. An American-made engine was able to perform a feat impossible with British-built engines, that is, to haul trains up the famous Lackey incline.

In the early 1840's, Russia had sought the aid of American engineers to build a railroad from Moscow to St. Petersburg. Czar Nicholas I and his advisers had heard glowing reports about the extensive railroad developments in the United States. The achievements of young American engineers in surveying the wilderness, bridging streams, tunneling mountains, and laying tracks to carry huge locomotives appealed to the Czar, and he sought their aid in opening up the vast expanses of western Russia. Moncure Robinson and Colonel George W. Whistler, two distinguished engineers, were engaged to go to Russia and undertake the task. The sum of forty million dollars was voted by the Russian government to finance the project. Mr. Robinson was unable to go, but Colonel Whistler sailed for Russia and took with him a group of able engineers experienced in railroad construction.

They surveyed the 500-mile track between Moscow and St. Petersburg, laid out a right-of-way 400 feet wide, upgraded a road bed from 6 to 8 feet above the ordinary level, and laid the tracks for a 5-foot gauge railroad. A special crew of bridge engineers was assigned to construct wooden bridges across all streams over 20 feet wide. A Philadelphia firm, Harrison and Estwick, was awarded the contract for supplying equipment for the road. It was a huge order for that period, consisting of one hundred and sixty locomotives, one hundred and seventy-two passenger cars, two "Imperial Salon" cars for the Czar, and twenty-five hundred freight cars.

The company built the rolling stock in Russia, with American engineers supervising the construction and providing the equipment used in the shops. Some thirty thousand Russian laborers, working under American supervision, supplied the manpower. American engineers taught the Russians how to operate the locomotives and run the trains after the road was completed. Colonel Whistler received a special award, the Cross of St. Anne, by order of the Czar in 1847. This railroad formed the nucleus of Russia's ambitious railroad dream, and around it the major part of her heavy industries was later developed.

Railroad construction in other countries profited by American skill and experience. Friedrich List, the celebrated German economist and the father of the German railway system, acquired his technique in America under Moncure Robinson. At his instigation, Prussia sent a delegation to this country to observe our methods and to acquire our technical knowledge. American engineers also provided stimulus for the development of the French railroads; and Austria-Hungary negotiated with American engineers for the construction of roads in that country. The Irish Railroad Commission also made frequent appeals to American engineers for their latest plans.

London Exhibition 1851

The most striking proof of America's technological advances by the mid-nineteenth century was demonstrated by the number of prizes and awards given to American inventors and their products at the London Exhibition in 1851. This "tournament of industry, parent of all succeeding international exhibitions," because of the nature of the building in which it was housed, was called the "Crystal Palace Exhibition." An American newspaper boasted it was first to suggest the idea of holding an exhibition of progress, which was developed by Prince Albert and a group of promoters. The *New York Herald,* under date of June 17, 1851, declared it to be an undoubted fact, "that the origin of this idea with Prince Albert and its promoters was suggested by and began with this newspaper."

The *Herald* explained that a long historical article pertaining to fairs in general, from Biblical times down through the medieval period, which had appeared in its columns in connection with the Industrial Exhibition at Castle Garden in New York in 1848, had been copied extensively in London papers. From this article, it was claimed, Prince Albert and the promoters of the London Exhibition got their idea. But whoever the originator, no project had ever been undertaken comparable in boldness or elaborateness of detail. Great Britain undertook a daring feat while the rest of the world looked on.

Purpose of the Exhibition. The purpose of the exposition was emphasized when Prince Albert noted in his opening address that the world was living in a period of transition, "which tends rapidly to the accomplishment of that great plan, and to which indeed all history points—the realization of the unity of mankind." He pointed out that vast distances which had always separated mankind were vanishing before the achievements of modern invention, and that "thought is communicated with the rapidity and even by the power of lightning." He believed mankind was rapidly approaching a more complete fulfillment of that great and sacred mission to which it was destined: "the discovery of laws by which Almighty governs His creation."

Prince Albert declared that science had discovered the laws of power, motion, and transformation, and that industry was now applying them to the raw matter which the earth yielded in abundance. "This Exhibition," he concluded, "is to us a true and a living picture of the point of development at which the whole of mankind has arrived in this great task, and a new starting point from which all nations will be able to direct their future exertions. . . ." Queen Victoria voiced a similar hope. She asked for the cultivation of all those arts that were aimed to foster and maintain peace.[1]

The Glass Building. A revolutionary idea was introduced in planning the exhibition hall by designing a huge glass building to house the thousands of exhibits. Horace Greeley described it as, "The most capacious, convenient, economical, healthful, and admirable structure ever devised for any kindred purpose." Invitations had been sent to many governments, inviting them to take part in the exhibition. Lists of articles acceptable for exhibition were enumerated under the general heads: raw materials, machinery, manufactures, and fine arts. London bustled with activity, and Prince Albert declared that, "This city has become the commercial metropolis of the world! The foreign invasion has begun."

American Participation. American reaction was varied. The federal government, while it duly appointed a national committee to act officially for the United States, failed to appropriate any funds to aid American exhibitors. The people were skeptical; they were sensitive to European prejudices and attitudes of superiority; they feared it was a speculation promoted by John Bull for his own ends. American newspapers, however, were filled with advertisements for passage to England, and American industrial leaders recognized that an exhibition of American industry at London would give Europe an appreciation of the true value of the American republic if the project were supported.

The American committee was headed by President Millard Fillmore; it included such scientists as Joseph Henry, Secretary of the Smithsonian Institution, Jefferson Davis, a member of the Smithsonian Board of Regents, Matthew Fontaine Maury, Superintendent of the National Observatory, Thomas Ewbank, Commissioner of Patents, and others of equal rank. State commissioners were appointed in twenty-eight of the thirty-two states. The United States Navy placed the Frigate "St. Lawrence" at the disposal of those who wished to send exhibits. The departure date was ill-timed, however, and the proffer of the various shipping interests to carry free all articles designed for the exposition was more acceptable.

United States Awards. Despite the fact that it was a costly undertaking to transport goods across 3,000 miles of ocean and set up displays, American participation was gratifying. More than five thousand Americans sailed from New York, Boston, and Philadelphia with the purpose of visiting the Crystal Palace. While the number of exhibits in the American sector was not flattering, the relative importance of America's entries was most gratifying. Less than 3 per cent of the twenty thousand exhibits placed in the exposition were from the United States, but America took more than her share in awards. She won more prizes than many of the Continental nations and relatively more than Britain herself. Of the total one hundred and seventy Council Awards given for unique and ingenious design, the United States captured five—the Borden meat biscuit exhibit; the Dick engine tools and presses; the Bond device for observing astronomical phenomena; the Goodyear India rubber; and the McCormick reaper. Of the 2,918 prize medals given, one hundred and two went to the United States and fifty-six other exhibits received honorable mention.

Agricultural Exhibits. The American agricultural exhibits attracted most attention, with the McCormick reaper leading them all. When the reaper first arrived it was the butt of all jokes. It was described as a cross between a chariot, a flying machine, and a treadmill; as a huge, unwieldy, unsightly, Yankee contrivance, unworthy of serious study. Some labeled it "the ugly duckling." But the attitude changed when a practical demonstration of farm implements was held in a farmer's wheat field a few miles out of London, before a jury of three men, and witnessed by some two hundred farmers. The wheat was still green and heavy. With a single span of horses pulling the American reaper, "it went through the grain, green as it was, cutting all before it." The crowd that had come to mock broke forth in cheers. The jurors asked that the machine cut another swathe, so that it might be timed and checked again. Again the machine was put in motion; this time it cut a swathe 74 yards long in 70 seconds "to the satisfaction of every one present." [2]

An American observer noted that the McCormick reaper not only cut down the grain in a way never seen before, but it also "mowed down British prejudice and opened the way for the bringing of our countrymen and their contributions before the public in a proper light." [3]

Salute to McCormick's Reaper. When the reaper was returned to the exhibition hall it attracted more visitors than the famed Koh-i-noor diamond. Condescendingly, American inventors were then deemed by an English spokesman as "our brethren across the sea, descendants of *this* (British) country." The London *Times* announced, "The reaping machine from the United States is the most valuable contribution from abroad. . . . It is worth the whole cost of the Exposition." [4] The McCormick reaper was hailed as bringing about as great a revolution in agricultural technology as the English spinning jenny and the power loom had produced in textile technology. Both the Grand Medal and the Council Medal were awarded to McCormick.

American Plows. American plows stood up equally well under the tests. The Prouty and Mears draft plow won the Grand Prize over contestants from five other countries. The American plows attracted attention because they were inexpensive, and because of their strength, elegance, lightness, and the adaptability of the running coulter. Other American agricultural tools such as axes, shovels, hoes, pitch forks, and scythes were judged to be superior to the European-made implements.

American Firearms. Another ranking American exhibit was that of firearms. Because of the emphasis of the Exposition on peace, no prize medals were given for firearms, but the United States won three honorable mentions with the Colt revolver, Palmer's target rifle, and the military rifle manufactured by Robbins and Lawrence. The Colt revolver attracted more attention than any other article in the Great Exhibition. It was described as one of the most important inventions in the entire history of firearms.

The London *Times,* September 2, 1851, stated that the Colt revolver threatened to revolutionize military tactics as completely as the original discovery of gunpowder. It was described as "the most popular and famous invention of American industry . . . , a weapon which will kill eight times as quick as any weapon formerly in use." [5] The Committee on Military Affairs of the Exhibition reported that one hundred cavalry armed with this repeating pistol "would be as effective as three hundred armed in the ordinary way." The military and scientific journals naturally praised Colt's ingenuity, and the British government for the first time in its history removed the prohibition against the importation of firearms in favor of Colt's revolver.

The American rifle also won distinction for the rapidity with which it could be loaded and fired, and for the self-cleaning operation of each cartridge. John Bright, member of Parliament, praised his countrymen for inviting American gunmakers to come to England, and said that his country was happy to receive such modern inventions as the Colt revolver and the Marston rifle. The superiority of these two weapons led the *London Evening Chronicle* to write: "The scepter is fast passing from England; westward the Star of Empire takes its way."

Sewing Machines. The sewing machine that received a prize medal was the exhibit of Blodgett and Lerow of Boston. Their machine involved Howe's shuttle principle. The machine was small, occupying a space only one foot square, and was simple in construction. Throughout the exhibition it was operated by a small girl, and it sewed at the rate of six hundred stitches per minute, a feat which caused a British newspaper to comment on its "astounding velocity." The publicity given to the sewing machine enabled other companies like the Singer to establish agencies abroad. By 1860 more sewing machines were being sold in Europe than in the United States.

Goodyear's Achievements. Charles Goodyear won the Council medal for his India rubber products, "for originality and novelty of invention," and Goodyear was hailed as a benefactor to mankind because of his many novel contributions. Among the rubber goods exhibited were rubber boots, shoes, life boats, pontoons, rubber handles for special instruments, waterproof fabrics, hose trappings, trunks, fire buckets, knapsacks, and cartridge belts. One English observer, Dionysius Lardner, noted that, "the gigantic (rubber) boots, really seven-league boots, are fitting emblems of Jonathan who, when he walks a step, necessarily makes the strides of a giant."

American Food Products. American food exhibits were in a class by themselves. They consisted of such articles as Indian corn, ground, hulled, and in the ear; rye, oats, barley, many grades of wheat, rice, beef, pork, lard, and hams. Cincinnati hams, both plain and sugar cured, received a prize medal by a unanimous vote of the Committee on Foods.

The highest premium, a Council medal, was awarded to Gail Borden, Jr., of Texas, for a specially prepared meat biscuit. This preparation astonished the food specialists when they subjected it to numerous chemical tests. When they discovered that it retained for an indefinite time all the nutritious and succulent character which it originally contained, they pronounced it the best ever prepared. They recognized its usefulness for seamen and crew members, especially since it preserved all its qualities in all kinds of climate. It was not affected by heat or cold. The British saw in it a great new American enterprise that would one day become one of the nation's leading industries.

American Mechanism Exhibits. Among the exhibits that aroused considerable competition were the American locks and safes. The Paratouptic Bank locks of the firm of Day and Newell, received the prize medal. The intricate construction of the lock baffled everyone, including the ingenious Mr. Garbutt, hailed as "the celebrated English mechanic, proficient at lock picking." Queen Victoria, Prince Albert, and the Duke of Wellington, after examining the lock, were all quoted as saying, "the great American lock has no rival for mechanical skill and for the security it affords the whole area of the Crystal Palace." This American-made lock was the only lock in the Great Exhibition submitted to the test of burglars and picklocks that was not successfully picked.

In a special lock-picking contest, British merchants and bankers were mortified to see the only lock that they had described as impregnable (the famous Chubb-Bramah) picked before their eyes, and by an American picklock. Upon Mr. Hobbs, the champion, was conferred the questionable title, "the world's renowned lockpick." Horace Greeley observed, "This small, young, shrewd, quiet-seeming Yankee would have made fewer enemies in England had he proved less potent a master of his calling." The editor of *Punch* referred to the event as, "The Pick of The Exhibition," and announced that, "We have not been able to sleep since Mr. Hobbs picked Mr. Chubb's lock." *Punch* then questioned, "Is the Koh-i-noor (diamond) safe?"

The Dick antifriction press, exhibited by David Dick of Meadville, Pennsylvania, especially designed for the manufacture of heavy tools and machine parts, was acclaimed "the noblest and most perfect arrangement of power ever discovered." Its punching power was so great that a boy, by operating a small crank, could punch a half-inch thick cold plate of iron with ease. The huge press was continually surrounded by crowds witnessing it in operation in crushing wood or punching metal. It was awarded the Council prize. English firms were eager to buy Dick's patent rights for producing his machine in England.

The Bond astronomical clock was another winner of the Council prize. This was an exhibit of William C. Bond and his son, George, makers of precision astronomical instruments. George Phillips Bond was the director of the Observatory at Harvard. While this exhibit was little noticed by the crowd at the exhibition, its merit was recognized by British astronomers. It eliminated the variance in accuracy of the eye and ear by recording simultaneously both the clock beats and the exact time of observation. This astronomical clock was regarded by one observer as "the most wonderful achievement of science since the days of Newton."

Photographic Exhibits. America's photographic display attracted wide attention, with three exhibitors of daguerreotypes receiving prize medals. Mathew R. Brady with his "Likenesses of Illustrious Americans" received the prize medal. He pioneered in introducing portraits on ivory, which combined the accuracy of daguerreotypes with the finish of miniature painting. The London *Times* observed that "So numerous and life-like are the daguerreotypes that we could gaze upon the United States Worthies as effectually as if we had got them within the reach of Rosse's telescope." J. Whipple, with a remarkable daguerreotype of the moon, and W. Lawrence were prize medal winners also.

Vehicular Exhibits. Among the articles of transportation that attracted special attention were a coach, several wagons, a buggy, a four-wheeled gig, a trotting sulky, a slide-top buggy, and a phaeton. The wheels of the phaeton were made of carefully selected materials, and the joints perfectly fitted; an American-type axletree, light springs, and thin rimmed tires combined to produce a vehicle that brought many orders from London buyers. In all these, lightness in weight as contrasted with the lumbering bulkiness of the English carriages was the characteristic feature.

An entirely new vehicle, the sporting wagon, another American production, was so light that it appeared to be made of paper. One observer declared that it could be moved with one finger. Nothing like it had ever been seen before in London or in Continental Europe. With this "break-neck affair," the Americans were pictured as traveling like an arrow, never raising the dust, ever going ahead, and "they will go farther still." [6]

Textile and Iron Awards. In those products which England had long held supremacy, America was able to win recognition and even carry off a few prizes. The cotton and woolen goods, sheetings, woolens, and carpetings represented the skill and enterprise of the Yankee manufacturer, and compared favorably with the best European productions. Articles made of native wool, the flannels, blankets, and tickings, all captured prize medals. Oilcloth printed by a special American process was remarkable for its beauty and durability, and won a prize medal. A specimen of cast iron (it was really steel), a product that had been recently made in America, after being put to a severe test proved superior to any that had been seen before and was awarded the prize medal. Rolled and boiler-plate iron received prizes as well.

Machine Tools. In machine tools America rated high. Whitney Hayden's cotton-drawing frame was exhibited in operation and was constantly surrounded by an admiring crowd. Its labor-saving features increased production by 10 to 15 per cent. Lowell's self-acting lathe and power loom "was a wonderment" in its automatic features. Woodbury's wood-planing machine was superior in power, in safety, and in quantity and quality of work accomplished.

American Arts Recognized. Though America's exhibits were generally utilitarian, those commodities reflecting the refinements of life, which were becoming increasingly possible as the wealth of the country increased, were exhibited also. The pianofortes exhibited were acknowledged to be second to none in the world. Their fame was already familiar to professional and amateur musicians in England and on the Continent. Three American firms exhibiting pianofortes received the prize medal, and three others were awarded honorable mention.

The fragile and fairylike creations in the flint glass exhibit were conspicuous among the finest in the English and French art. The Brooklyn Flint Glass Company received a prize medal. The lamps exhibited revealed improvements unknown in Europe. The development was attributed to the fact that lamps were used more extensively in America than in Europe.

Scientific Instruments. America was not lacking in her contributions to the scientific exhibits. Professor Alexander Bache won the prize medal for his invention of a balance used for the adjustment of standard weights and measures. William A. Burt entered his solar compass that was widely used by surveyors in this country but had been unknown beyond the United States until the Exhibition gave it worldwide recognition. Captain John Ericsson, long known to the nautical and mechanical world as an inventor of valuable instruments and lauded by learned bodies in England and America, had many entries. His pyrometer, which was scaled to measure heat to the melting point of iron, received the prize medal. Other exhibits of his included a marine barometer with alarm attachment, an instrument for taking soundings, an instrument for measuring distances at sea, a nautical compass, and a model gas engine.

Novelties and Gadgets. Yankee ingenuity was evidenced in a number of novel devices. There was the trick umbrella that could be folded to a pocket size. There was a gold pen that prevented corrosion and permitted the ink to flow more readily than a steel one;

there were lightning rods of extra strength, easily erected; mechanical reclining chairs, the forerunner of the patent furniture vogue. All these were among the exhibits that attracted the attention of the six million visitors who passed through the halls of the Crystal Palace during the one hundred forty-one days that the Exposition was open.

American Yacht Wins Exposition Race. The most coveted prize won by America during the great Exhibition was the triumph of the schooner "America" in the international yacht race at Cowes. Attended by Queen Victoria, Prince Albert, lords and ladies, captains and admirals, and thousands of eager spectators, it was regarded as the greatest sporting event of the season. At the beginning of the race, the "America" got off to a slow start, but after the first few miles of the 60-mile course she began to move ahead of the other sixteen vessels and finished 3 miles (one report said five) in advance of her nearest rival. This was the most stunning defeat for all the Britishers. "The little yacht 'America' had done more to humble the pride of England than anything that had happened since the War of 1812." [7] She had won over the picked models of naval architecture that were owned and personally directed by the proudest of her nobility and veteran yachtsmen, the Marlboroughs and the Angleseys. A reluctant adversary made the prediction that thereafter every vessel in the Cowes squadron "would be trimmed to the very image of the 'America.' "

Following the humiliating defeat, *Punch* ran an announcement which read, "Timber for Sale: A great quantity of Planks, Sticks, Masts and Spars to be had cheap. Inquire at The Royal Yacht Club." And Daniel Webster at a dinner in Boston, celebrating "America's" victory, paid tribute to the skill and ingenuity of a people "who may soon command the ocean, both oceans, all oceans."

Results of the Exhibition. The close of the Exposition saw young America gain in immediate commercial advantage in the introduction of American goods into European markets. There was the long-time advantage of the recognition of American technology and the realization of the potential forces of America in technological developments. American patents were sold to European manufacturers. Agencies for the distribution of American goods were set up in European centers. European skilled artisans and unemployed European capital were to a greater degree attracted to America. Europe adopted American innovations, and exports from this country were stimulated.

The far-reaching effects were even greater. Horace Greeley pointed out that the diffusion of scientific and industrial information among the rest of the nations was alone worth the cost of the Exhibition. "Among the agencies which aided the march of invention, and which impelled the Car of Industrial Progress," declared Greeley, "I doubt not that our children, looking back on that progress from heights whereof we can but vaguely dream, will honorably distinguish the World's Exhibition of 1851."

Technological Invasion Continued

Although the London Exhibition closed late in 1851, American technology continued throughout the decade to penetrate other foreign countries. The number and variety of American-made products shipped abroad during the 1850's increased rapidly. In 1852 an American clock company exported over one hundred thousand clocks to England, where they were sold at an average price of three dollars each. America was making clocks cheaper than any other country in the world. The British in turn resold these clocks to the people in her colonies and dependencies.

Paris Exhibition, 1855. In 1855, Paris held a Grand Exhibition and again American technology was displayed, although on a much smaller scale. One of the highlights of the Paris Exhibition, like that of the London Exhibition, was a contest of the agricultural machines exhibited by the various nations. The contest attracted widespread attention, attended by notables from England, Germany, Holland, Belgium, and the United States—including former President Fillmore. Hundreds of machines, plows, drills, reapers, mowers, and threshers were entered in different contests. The greatest interest here, as in London, centered about the reapers and threshers. In the contest of reapers, McCormick's machine swept the field. He was awarded the Grand Medal of Honor. The Hussey reaper won second prize. The French Emperor Napoleon III purchased the McCormick reaper and had it placed in the Conservatoire des Arts et des Métiers.

Threshers versus Flails. Another contest was one in which four threshing machines—one from Belgium, one from France, one from England, and one from the United States—competed against six sturdy men with flails in a threshing event. The six men with flails threshed 60 liters of wheat in 30 minutes; the Belgian machine threshed 150 liters; the French machine threshed 250 liters; an English machine threshed 410; while the American-made Pitt machine threshed 740 liters and received a Grand prize.

Other American Triumphs. Goodyear received the coveted cross of the Legion of Honor for his India rubber, and other American exhibits received special premiums and awards. Recognition was given to Tousley and Reed's oscillating engine, which M. Périer, railroad king of France, purchased and installed on one of his four railroads; to Samuel Colt's revolvers; the Singer sewing machine; Wheeler and Wilson sewing machine; Lieutenant Matthew F. Maury's wind and current charts; Thomas Blanchard's wood bending machine; and to Bache and Saxton's weights and measures system. These were a few of the American exhibits that received special awards.[8]

America Speeds European Progress. Each succeeding year witnessed new triumphs for American technology in foreign fields. In the summer of 1857, in an agricultural exhibition in Budapest, Hungary, an improved McCormick reaper won first prize over all other machines. In the following year, American railway carriages were adopted exclusively in Austria, and the first steamboat launched on the Danube was equipped with American-made engines. The Morse telegraph monopolized the telegraph wires of Austria. "There seems to be hardly an American invention of the higher class," wrote a correspondent, "which has not found its way into Austria." He observed that Americans, because of their inventions and mechanical skill, were gaining peaceful victories on the continent of Europe, "more important to the world than Austerlitz or Waterloo. Reaping machines are greater civilizers than swords, and Yankee unpickable locks give greater security to property than jails or gallows." [9]

In 1859 the Pasha of Egypt ordered an American-built locomotive to be installed on the Alexandria–Suez Railroad, and after a 200-mile journey completed in 12 hours on a train pulled by this locomotive, he exclaimed, "God is great, but those Yankees are very near perfection." [10] This same year saw American engineers building railroads in Brazil, Chile, and Russia. American locomotives soon followed. Large shipments of American steam engines and sawmills were sent to Russia in 1859, upon orders from the czar, to be set up on the Amur River in northeast Siberia. Even the flagship of the Russian Navy, "The Grand Admiral," was built in America.

A contest between British and American-made locomotives held in Chile in 1859 resulted in a victory for the American steam giant. The American locomotive took a train of thirty-five cars, weighing 587 tons, a distance of 11⅔ miles in 41 minutes—a feat which the English locomotives could not perform at any speed. In a grade contest the American engine pulled fifteen cars of 288 tons to a summit in $26\frac{1}{10}$ minutes, as compared with the English record of $37\frac{5}{6}$ minutes.[11]

Enfield Rifle Adopted. America's potential superiority in the manufacture of guns and firearms during the years preceding the Civil War was also recognized throughout Europe. France had held the distinction in this field throughout the seventeenth century, and Britain had captured it by the close of the Napoleonic Wars. But during the Crimean War, 1854–1856, English soldiers had suffered great losses due to ineffective firearms. A commission was appointed to investigate the American method of manufacturing rifles, with the result that a large factory was established at Enfield near London to manufacture rifles. The English company secured the services of the superintendent of the arsenal at Harpers Ferry, Virginia, who went to England in 1855, took with him a number of experienced gunmakers, and began to manufacture guns modeled after the American make. They became known as the Enfield rifle.

This rifle became the general weapon of the British infantry until it was converted into the breechloader in 1867. In its manufacture the British adopted interchangeable parts, in what they termed, "the American system." On the shiny new machinery introduced at the Enfield plant were the names of American manufacturers, Ames of Massachusetts, Robbins and Lawrence of Vermont. Within twenty years American machine toolmakers established a trade with the governments of Russia, Prussia, Spain, Turkey, Sweden, Denmark, and Egypt, and in addition furnished thirty-five thousand small arms to Denmark.[12]

"Americanizing England." By mid-nineteenth century American technology was being acknowledged abroad. An article in *Living Age* (1859), entitled "Americanizing England," ridiculed John Bright, a member of Parliament who was engaged in a fight to extend the franchise to a greater number of British voters, for leading a movement aimed at "Americanizing" England. Undeterred, Mr. Bright willingly admitted that his countrymen could profit greatly by adopting American political and business methods, particularly American inventions. He said that it was American machines and American workmen that were giving England her Enfield rifles: that the London and the Manchester newspapers were being printed on machines either made in America or, if manufactured in England, constructed on an American patent design. "There [in America] we find a population enjoying physical comforts and abundance such as are not known to the great body of people in this country,

and which have never been known to any country in any age of the world." [13]

Charles Reade, English author, writing in 1860, stated, "American genius is at this moment ahead of all nations in mechanical invention." He noted that since 1850 America had led Great Britain in the number of patents, and that Europe was teeming with the products of America's mechanical skill. He informed his readers that he was sitting in London, dressed in a shirt that was sewed on a Singer sewing machine; and that two of three English newspapers on the table were printed on Hoe (American) printing presses. When he walked along the Liverpool docks and inquired from what home port "the biggest and smartest ships" came, the answer was always, "Yankee, sir—Yankee." [14]

Notes for Chapter 18

1. *The Industry of All Nations* (London, 1851), *Introduction.*
2. Quoted from Charles T. Rodgers, *American Superiority at the World's Fair; Report of Commissioner* (Philadelphia, 1852), p. 62.
3. *Ibid.,* p. 82.
4. *The Times* (London, June 9, 1851).
5. Quoted in the *New York Herald* (October 20, 1851).
6. Dionysius Lardner, *The Great Exhibition and London in 1851,* (London, 1851), pp. 591-592.
7. Rodgers, *op. cit.,* p. 77.
8. *Scientific American,* XI (Oct. 27, 1855), No. 7, 51-52.
9. *Scientific American,* XV (March 5, 1859), No. 26, 211.
10. *Scientific American,* XIV (Feb. 12, 1859), 184.
11. *Scientific American,* (N.S.) I (July 1, 1859), 217, 240.
12. "Report on Manufacturers," *U. S. Tenth Census,* p. 620.
13. *Living Age,* LX (1859), 192.
14. *Scientific American,* (N.S.), III (Oct. 27, 1860), No. 18, 280.

Chapter 19

ON THE EVE OF THE CIVIL WAR

On the eve of the Civil War, two conflicting social philosophies became so vigorous in opposition to one another that they evolved political issues. The South, amply supplied with slave labor, rejected the advantages of mechanized agriculture and manufacturing. The North, its mills and factories far outstripping the productive capacity of the South and seeking expanded markets, decried the evils of slavery. The result was a socio-economic issue: manpower versus machine power. The conflict was soon to prove which was the stronger.

Socio-Economic Aspects. A review of the country's industrial growth during the two decades preceding the Civil War revealed a multitude of new technological developments, particularly in the northern and eastern states, that were adding to the nation's power. These recent developments pointed the way for utilizing the wealth and natural resources of the land, for offsetting the shortage of labor in those areas, and for new methods in transportation technology that could cover vast distances and rugged terrain. A national pride in recent technological achievements had developed, and the young nation had gained recognition abroad.

America was no longer a distant isolated nation known only for her fisheries, and her exports of natural products. A contributor to the London *Economist,* writing in 1854, drew a comparison between the English and American mechanical achievements of that period. He described the American as being more ingenious, more skillful, more industrious, more steady, sober, methodical and energetic than his own countrymen. "Here," he said, "was a whole nation of engineers, mechanics, clever workmen, Franklins, Stevensons, Watts, and similarly gifted persons." He declared that, "the prospect is something wonderful for other nations to contemplate." [1]

Sir Charles Lyell, distinguished British scientist, while on a visit to the United States in 1853 had expressed the same view to President Pierce, when he declared that Britishers were struck by the many labor-saving inventions in this country, "to which far more than its soil or any other cause, we [Britishers] ascribe the great wealth that has accumulated here."

Parade of American Technology

Many of the nation's technological achievements were revealed at the New York Exposition in 1853. Those Americans who had attended the London Exposition two years earlier had been so impressed by the excellent showing of the United States' exhibits that they decided this country should hold a world's fair of its own.

N.Y. Crystal Palace Exposition, 1853. The New York Legislature granted a charter in 1853 to "An Association for the Exhibition of Industry of All Nations," and a "Crystal Palace" Exposition was planned. It was privately financed. A huge edifice, erected entirely of iron, was built to provide 3 acres of space for exhibits. President Pierce, in officially opening the Exposition, declared that "the American people are living in an age where science has become tributary to the arts, manufacturing and agriculture." The minister who preached the opening sermon on "Glorifying God as Man Triumphs Over Material Things" declared that man had been given dominion over the form and forces of all material things, and that his discoveries and inventions were finally bringing him into closer harmony with the Almighty's plan of governing the universe.

The exhibits at the Crystal Palace included just about everything, with agricultural tools and machines taking the lead. Iron and steel products, glass, pottery, printing presses, photographic exhibits, firearms, household utensils, mineral displays, and other exhibits crowded every inch of space and led to the construction of an additional building. Over two thousand exhibits arrived from England, France, Austria, Italy, Norway, Sweden, and other countries. Gas lights permitted the huge halls to be open at night. It was the hope of the managers that the Crystal Palace would become a permanent museum, "a school that would spread abroad knowledge of America's great scientific, industrial development." This hope, however, was not realized.

Patent Activity. America's rapid stride in the industrial, scientific, and technological developments of the pre-Civil War decade was dramatically reflected in the Patent Office records. In the one year (1850) 2,183 patents were issued. Most of the patents were awarded to inventors of labor-saving machinery and machines of a utilitarian character.

A steam excavator, called a "Yankee Geologist" eased the work for the man with the pick and shovel. This machine, built by the locomotive firm of Eastwick and Harrison at Philadelphia, was used in railroad construction. It could move a cubic yard of dirt in one bite, or 1200 cubic yards in one day, the equivalent of the work of a hundred men. A number of patents were issued for new washing machines. By 1860 twenty-nine factories were producing washing machines in ten states, all striving to lessen the chores of washday. A mania for inventing swept the country. Even Abraham Lincoln was granted a patent in 1848 "for a device for buoying vessels over shoals and sand bars."

Walt Whitman wrote in 1856 that the list of one week's issue from the national Patent Office "illustrates America and American character about as much as anything I know." The new Patent Office building that was authorized in 1836 was already becoming overcrowded by 1850. Applications accompanied by models were pouring in and the commissioner of patents reported (1850) that over seventeen thousand models had accumulated during the past fourteen years. Congress voted ninety thousand dollars in 1850 to complete two wings of the building.

Birth of the Steel and Iron Age. The advances in iron and steel technology were a fair measure of the nation's scientific and industrial progress on the eve of the Civil War. This period witnessed the introduction of the pneumatic (often called the air-blown) process for making iron (or mild steel as it came to be called) by the Kelly and Bessemer process. At Eddyville, Kentucky, in 1846, William Kelly, seeking a better way to utilize an inferior grade of ore, observed that the iron could be heated by blasts of air. After experimenting with the idea he succeeded in making a metal that was used in steamboat boilers.

Working independently, Henry Bessemer in England discovered the same principle in 1854. He introduced a special smelting vessel that came to be known as a "converter." Two years later, Robert Mushet, also an Englishman, made the pneumatic process practicable by adding spiegeleisen (pig iron containing manganese and carbon) after the "blow" in the Bessemer process. From these discoveries the Bessemer process of making steel was evolved.

This decade also saw the establishment of the Cambria Iron Works at Johnstown, Pennsylvania, which became the proving grounds for Bessemer steel, the Jones and Laughlin firm in Pittsburgh, and the Bethlehem Iron Company at Bethlehem, Pennsylvania. Pittsburgh began to take the lead in iron production when, in 1859, a blast furnace, the first since the short-lived furnace of 1792, was built there. Also, as a forecast of what was to come, the first ore from the Lake Superior region was smelted just before the outbreak of hostilities. John Fritz, later to receive the Bessemer medal for his accomplishments in the iron industry, invented his pneumatic hoists for blast furnaces and his famous three-high rolling mill during the pre-Civil War decade.

The improvements and refinements in iron technology brought the use of steel within the reach of every one, even the ladies. A new process for tempering a cheaper grade of steel was invented which provided steel hoops of light weight, great toughness, and elasticity, which were used for hoop skirts. This process proved a great factor in the crinoline age, enabling the women from families of limited income to be dressed in the height of fashion. The output of tempered wire for crinoline skirts amounted to several tons annually.

A few of the ironworks began to specialize during the pre-Civil War period. At the Cambria Iron Works in Johnstown, Pennsylvania, Captain Robert Hunt in 1860 became a pioneer chemist—he claimed to be the first to be employed in an ironworks laboratory. Most of the manufacturers were still firing their furnaces with charcoal, but coal was gradually being introduced as fuel where the forests were being depleted. In 1860 coke was used as fuel in the Clinton furnace in Pittsburgh, a practice that led to making Pittsburgh the steel city of America. With the development of railroads, the distribution of coal and coke became nationwide.

Manufacture of Clothing. One of the most important inventions during the pre-Civil War years was the sewing machine. This invention marked as great a revolution in clothing manufacture as did the power loom in weaving. It ushered in a new era in the history of textile technology, both in the United States and throughout the world. Due to an army of energetic salesmen, there was no land on the face of the globe in which the American-made sewing machine was not acclaimed, from the Far East (Japan) that had just been opened to the world, to Africa's tribal communities, and even in the lonely tents of the Arabs.

In the evolution of the sewing machine there had been many attempts to mechanize the dexterity of the seamstress. Contributory ideas came from many sources; but in the final language of the courts, to Elias Howe was due the honor of "giving to the world a practical sewing machine." In 1846 Howe succeeded in putting out a workable machine that combined the shuttle for forming a stitch and the intermittent feed for carrying the material forward. Howe's patent covered the mode of making the stitch, termed the "back-stitch," and the use of the eye-pointed needle. The original machine, with the needle working horizontally and with the cloth suspended upon pins projecting from the baster plate, bore little resemblance to the sewing machine of a few years later, but it involved the basic principles in a workable combination.

A number of improvements were made by A. B. Wilson, who with Nathaniel Wheeler entered into manufacture under the firm name of Wheeler and Wilson. In 1851 Isaac Singer patented a sewing machine in which he introduced a vertical pressure foot that was held down by a spring and the treadle method of operation which left both hands of the operator free to guide the cloth. Singer's dramatic sales ability and showmanship in publicizing his invention helped to give his machine worldwide recognition.

Approximately $2,500,000 of capital was invested in the manufacture of sewing machines by 1860, and more than three thousand salesmen were engaged in selling machines throughout the country. Between 1840 and 1860 the cotton goods industry increased nearly twice as fast as the population, and the woolen industry increased more than 50 per cent in ten years. The sewing machine began to invade the American home and enabled the housewife to provide extra garments for the entire family.

The invention of the sewing machine brought about a new development in the manufacture of ready-made clothing for men and women. In the four cities of Boston, New York, Philadelphia, and Cincinnati, the manufacture of men's and women's clothing almost doubled in three years. The value of ready-made clothing jumped from approximately forty million dollars in 1850 to more than seventy million dollars in 1860.

During the pre-Civil War decade, nearly one hundred thousand sewing machines had been sold in the United States, and a number of new industries were created which employed thousands of workers. In Troy, New York, during the year 1858, three thousand sewing machines were reported in use in making shirts and collars. Through the invention of the sewing machine and the rapid development in the manufacture of clothing, the Northern states were prepared to clothe their citizens, both soldiers and civilians, throughout the war years that were to follow.

Mechanized Shoemaking. The sewing machine also gave a great stimulus to the shoemaking industry. The sturdy Singer machine was built strong enough to sew leather and was quickly adopted by shoemakers. At first it was used only for sewing uppers, but in 1858 Lyman R. Blake devised a machine to sew the uppers to the soles. This machine was later improved upon by Gordon McKay, who purchased Blake's patent. McKay took out a later patent for "a process of sewing soles of boots and shoes," and proceeded to organize a company for the manufacture of the McKay machine, a machine that sewed one hundred times faster than any other machine. It produced McKay-sewn shoes as contrasted with hand-turned shoes.

Other developments were introduced which greatly increased the production of shoes. The Sturtevant shoe pegging machine invented in 1858 could insert nine hundred pegs a minute, and soon supplanted all other pegging machines in the shoe industry. The New England states alone produced in one year twenty-five million pairs of boots and shoes, all pegged by Sturtevant machines. The shoe industry expanded rapidly. By 1860 more than ninety million dollars worth of boots and shoes were being produced annually in the United States.

Food Preservation and Packaging. The canning industry, though still lacking a scientific basis, made considerable headway before the Civil War. The scientific canning of foods began with Nicholas Appert of France, who discovered that food could be preserved by boiling it in a sealed container. William Underwood of England brought the canning industry to the United States. He arrived in New Orleans in 1817, and traveled to Baltimore, then on to New York, where he set up his business of preserving fruits and pickles in glass jars. He carefully labeled his products "Canned in Britain," however, to ensure acceptance of his product, and established an extensive trade.

But the sterilization process had not yet become a science, for the length of time required for sterilizing had not been determined. Processing foods had always been a matter of conjecture when glass containers were used. In 1825 Thomas Kensett introduced the term "canning," when he preserved oysters and fruits in a tin canister and patented his container. The canning industry spread rapidly and by 1860 had extended as far west as California. Canned goods were to prove a decisive factor in feeding the Union forces during the Civil War.[2]

Glass Industry. Glass production, although making no great technological advances, mounted in value during the second quarter of the nineteenth century. Glassmaking techniques had been kept a trade secret up to that time. On the whole, American manufacturers followed the European techniques, but lagged behind production of France and Belgium. The pressing process, which originated in America in 1826, was improved by introducing better molds and better machinery. A product was manufactured that carried off several prizes at the London Exposition in 1851, and first prize was awarded United States for fine pressed glass at the Paris exhibition in 1867.

Glass began to be used for a number of purposes. It was used in building and for tableware, in addition to its former use for making jars and bottles. The value of glass made in the United States increased over 100 per cent in ten years, from 1850 to 1860, or from four million dollars to over eight million dollars in value. Mason's screw-top jar, patented in 1858, was hailed as the greatest innovation in all the history of fruit jars. In the glass furnaces, coal had replaced wood for fuel by 1860.

American Clocks and Watches. Other technological advances occurred during this period, notably in the production of clocks, watches, locks, lathes, and the manufacture of giant steam engines. America early excelled in clockmaking. By introducing a system of interchangeable parts, clocks were produced that could undersell any on the market. In 1850 machine-made watches appeared. Few watches had been made in this country prior to that date, but after 1854 American watchmakers made most of the parts for their own products. Some watchmakers boasted they could make a perfect watch in 3 hours. Watches and chronometers, made from precision machinery on the system of interchangeable parts as early as 1848, were found to be surprisingly accurate. They were used on the Grinnell Expedition in 1850.

American Locks. Sturdy American-made locks came into worldwide use about the middle of the nineteenth century. Locks, too, were made by precision machines on the principle of interchangeable parts, and defied the most skilled picklocks. In the early 1850's Linus Yale invented the Yale Infallible Bank Lock, known as the "changeable" type. The key for this lock, "made of component parts," could be separated and reassembled to change the combination. Yale made several improvements on his original model, and later brought out the Yale double-dial bank lock, which came into universal use throughout the nation.[3]

Machine Tool and Engine Improvements. Of great significance in the development of machine tools was the turret lathe devised by Stephen Fitch in 1845. From this instrument emanated the long series of automatic tools that have characterized American industry and have made possible the technique of mass production.

In the rise of the factory system, the improved Corliss steam engine, invented by George Henry Corliss in 1849, was of major importance. The chief feature of this invention—rotary valves, separate ones for steam and exhaust parts, and a governor which controlled the valves and the admission of steam to the engine cylinder—set the basic pattern for the reciprocating engine. Its slide valve, valve gear, and automatic release gave to it an evenness of rate of speed that made possible speedier operation of the textile machines. In the history of steam engines the name of Corliss stands with that of James Watt. The principal features of the Corliss engine were copied by engine manufacturers throughout the United States and Europe.[4]

Railroads, Bonds of Unity

Railroads were pushing out in all directions on the eve of the Civil War, binding the nation together. An English traveler, Thomas Sterling, was moved to comment in 1857, "The Railway seems to be the soul of Western Civilization. The Americans are becoming railroad minded."

Increased Road Mileage. In 1850 there were almost 10,000 miles of road in operation. By 1860 this had more than tripled, and railroad mileage in the United States exceeded that of any other country in the world. Also, by 1860, the railroads were carrying more than half of the nation's internal commerce, and 30,600 miles of steel rails were reaching into distant areas. Chicago, which launched its first locomotive, the "Pioneer" in 1848, had by 1860 become America's leading railway center with eleven railroads merging there. By 1852 the railroads had crossed the Mississippi, and by 1855 they were in Iowa. Even California built a short railroad in 1858. In 1857, a southern rail route between the Atlantic seaboard and the Mississippi was completed. Texas had a railroad in 1853, Arkansas in 1858, and by 1859 a system of railroads extended to the Missouri River at St. Joseph.

Pullman Cars, 1858. Not only was the railway system expanding, but it was becoming more efficient. During the 1850's, the transition from wood-burning locomotives to coal-burning locomotives began. Ease and comfort in travel was seen to be desirable, as evidenced by the success of George Pullman's first luxury car in 1858 (see Chapter 26). The light wooden passenger cars were an improvement over those of the earlier generation. They were heated with wood-burning stoves, and lighted by flickering coal-oil lamps. There were still no vestibules, however, or enclosed platforms. Railroads were made safer by the introduction of the telegraph signal system, first used on the Erie in 1851.

Iron Bridges. Some important technological developments in bridge construction were made during this period. Iron began to supplant wood as building material. An all-iron railroad bridge was built for the Reading Railroad in 1845. A tubular girder bridge was constructed at Bolton, Maryland, for the Baltimore and Ohio Railroad in 1847. New construction methods were employed. The Howe truss bridge was erected at Pittsfield, Massachusetts, for the Boston and Albany Railroad. In 1848 John A. Roebling constructed his first suspension bridge to carry the highway over the Monongahela River at Pittsburgh. During the next two years he built four suspension bridges for the Delaware and Hudson Canal and, in the middle of the decade, the celebrated Niagara Falls Bridge. The longest bridge in the world at that time was built at Wheeling, a suspension railroad bridge over the Ohio River. By 1860 there were more iron railroad bridges in the United States than in any other country.

East Meets West. Horace Bushnell, contemporary divine and essayist epitomized in rhetorical language America's progress in transportation in the midcentury. He described Americans as not being content with horses of flesh, but as busy building horses of iron to carry immense loads over mountains, "and soon we may see them extending from one shore to another." "The boundaries of the nation are spanned by bands of iron," he wrote in 1836, "crossed and recrossed many times a day, as freely as the birds of the air fly over them, and as swiftly. The sails of commerce are the wings of truth. Wherever it goes the power of science, and all that belongs to cultivated manhood, is felt."

Technological Future of America

In the two generations since the establishment of the republic, America had made rapid advances in applied science and technology. The Hamilton-Jefferson differences as to the relative importance of agriculture and manufactures had been forgotten. Neither had developed to the exclusion of the other. The great sprawling nation now stretched from sea to sea. Although its citizens on the frontier were still confronted with the danger of the Indian as in pioneer days, there was growing up in this land a technological civilization greater than its people realized.

A contemporary observer, writing in *Living Age* on the scientific and technological developments of the pre-Civil War period, declared that recent improvements in machinery had facilitated production to such a degree that once they became diffused through the world, "only a few hands, comparatively, will be requisite in the mechanic arts. Those in agriculture will be more in a condition of ease." And Dionysius Lardner, eminent British scientist, said in 1858 that young America had become "a theater for the development and encouragement of arts and sciences." This was a significant observation—characterizing this young nation as a theater or a workshop where new skills, new processes, and new tools were first invented and tried out, and then copied by other nations of the world.

Walt Whitman, in 1856, looking at America through the eyes of the poet, pictured it more truly than his contemporaries knew, when he said:

> Colossal foundry, flaming fires, melted metal, pounding trip hammers, surging crowds of workmen shifting from point to point, waste and extravagance of material, mighty castings; such is a symbol of America.

Within a few years the nation became engaged in a bitter struggle which President Lincoln declared would determine "whether this nation or any nation can long endure." In the war that followed neither side could claim an advantage in either patriotism or loyalty to the cause for which each fought. The advantage lay in the superior skills and the diversified technologies that developed in one section as they competed with the limited technologies of the other.

Notes for Chapter 19

1. Quoted in Littell's *Living Age*, XLI (1854), 257.
2. See Chapter 23 for further advances in canning technology.
3. *Dictionary of American Biography*, XX, 591, 592.
4. D.A.B. IV, 441. For a description of the huge Corliss engine that furnished the power for Machinery Hall in Philadelphia Centennial, see Chapter 21.

Chapter 20

TECHNOLOGY IN THE CIVIL WAR

Technology played a major role in America's Civil War. Just as the introduction of gunpowder in the early fourteenth century and the use of armed vessels in the mid-nineteenth century are regarded as turning points in the early history of warfare, the American Civil War marked a turning point in many technological advances. It was the first struggle in which science and machinery played a dominant part, and it was the first time that technological innovations and improvements were applied on a large scale in a major war.

Prewar Status of North and South

Strategic Preplanning. Even though the threat of war had been imminent for years, neither the North nor South had made any special preparation for a prolonged war. President Lincoln, his Cabinet, and his advisers believed that the Northern forces could blockade Southern ports, make a quick dash on Richmond, capture it, and put an early end to the conflict. Less serious, but more vocal, were those who glibly asserted that the inventors of the North would supply the tools of warfare that would cause the South to surrender quickly. Said the *Philadelphia Enquirer,* August 3, 1861, "Take our word for it, these geniuses (Yankee genius) will yet produce some patent Secession-Excavator, some Traitor-Annihilator, some Rebel-Thresher, some Confederate State Milling Machine, which will grind through, shell out, or slice up this war, as if it were a bushel of wheat, or an ear of corn, or a big apple."

The South, pursuing a different line of reasoning, also predicted an early victory. Their leaders planned to hold off the Northern forces until England would be compelled to intervene in order to save her cotton goods industry, and thereby ensure a wartime economy that would triumph over the North. Neither side guessed right. The dash on Richmond by the Union forces was rudely frustrated in the first Battle of Bull Run, and the hope that England would intervene on the side of the South faded in the summer of 1863.

Military Preparation. Nor was either side prepared to take immediate advantage of the military and technological developments that had occurred during the past half-century. A report made in January, 1860, showed that of the 610,000 small arms of all descriptions in government arsenals, most of them (whether in northern or southern arsenals) were old flintlock muskets altered to percussion use, and woefully ineffective. The two armories that the South established at Fayetteville and Richmond did not begin production until late in the year of 1861, and the first shipment of arms from England did not arrive until September, 1861. The South was therefore unable to follow up its smashing victory at the first Battle of Bull Run and attack the badly demoralized forces in Washington, simply because of lack of arms. Even in September, 1861, the South was rejecting volunteers at the rate of five thousand a day because they could not be equipped.[1]

Shortage of Arms in South. The shortage of rifles in the South was desperate. The two factors that saved the Confederate cause from early defeat were the weather and the lack of preparedness on the part of the Union forces. In their desperation the Southern leaders ordered thousands of pikes, the most ancient implements of war, to be hammered out for use by the infantry, just in case the supply of guns and ammunition should completely fail them.

Confederate Plans for Powder. The South had only two small powder mills when the war broke, woefully incapable of producing the needed supply. General Josiah Gorgas, a West Point graduate, (father of William C. Gorgas of Panama Canal fame) became Chief of Ordnance in April, 1861, and hurriedly planned to erect the finest powder plant in the country. Ground was broken in Augusta, Georgia, in September, 1861. Zinc for tanks was torn from old roofs; copper was salvaged from turpentine stills; rollers and other machinery were fabricated at the Tredegor Works in Richmond. A Nitre and Mining Bureau was created to ensure adequate supplies of saltpeter for making gunpowder.[2] But despite all these efforts, the South was never able to produce an adequate supply of powder.

Procrastination in the North. The North was likewise ill-prepared. Although new and improved weapons had been invented, they had not been adopted by the War Department despite the fact that the Sharpe rifle, the first practical American-made breechloader, patented in 1857, had already proved its accuracy and rapidity of fire power in actual combat in the Kansas struggle a few years earlier. The War Department had also refused other breechloaders, such as the Burnside, the Maynard, the Peabody, and the Spencer. Moreover, the army had no effective system for testing any of the new weapons had they been accepted.

Had the Navy been equally reactionary, the "Monitor" would never have been utilized and the South might have broken the blockade, paving the way for a negotiated peace. As a further commentary on the traditional conservatism of the Northern leaders, it should be noted that the Gatling machine gun, invented in 1862, was not officially approved by the army until after the war was over.

But, despite the fumbling and the delaying tactics in the matter of firearms, the North soon began to demonstrate its superior war potential in other lines: notably, in material and productive capacity.

Northern Superiority in Iron and Steel. In every war, iron and steel comprise the basic material for the production of munitions. The total output of pig iron in 1860 was estimated at 859,000 tons; but of this amount only 26,000 tons were produced in the South. Of the total Northern tonnage, Pennsylvania alone produced 580,000 tons—which helps to explain the desperate raids undertaken by the Southern armies to capture the Keystone state in 1863. Of the 11,800 gross tons of steel, more than 9,000 tons came from Pennsylvania. Had the Confederacy succeeded in gaining control of the iron and steel industry in Pennsylvania, the whole course of the war might have been altered.

Production of Arms. There were three cannon factories located in the North in 1861; one each in South Boston, West Point, and in Pittsburgh. Within the next two years, additional factories were established in Providence, Boston, Portland, and Reading. While the Cold Springs Foundry at West Point was the most famous, since it manufactured the Parrot rifled gun, yet the greatest dependence was placed on the Fort Pitt Foundry in Pittsburgh. It was there that the big Rodman smoothbore guns were cast. During the first three years of the War, the Fort Pitt Foundry supplied more than two thousand cannon and mortars, including one hundred eight 15-inch guns; six hundred navy 9-inch shell guns; and three hundred army 8-inch columbiads. It also turned out 10,000,000 pounds of shot and shells.[3]

Armories. Since no one plant can supply an army composed chiefly of infantrymen, armories are as important as foundries. When the war broke, there were two government armories equipped to produce small arms, one at Springfield, Massachusetts, and the other at Harper's Ferry. The latter was burned by Union forces, April 18,

1861, to prevent its capture, but the Confederate forces arrived in time to save some of the machinery and equipment and to take it to the Richmond and Fayetteville arsenals. While some historians have implied that this act by the Confederacy placed the South on an equal footing with the North, the facts prove the reverse.[4] The South lacked the raw materials, the skill, and the technological efficiency needed to produce the rifles and arms for its soldiers.

The North, on the other hand, in addition to having an arsenal at Springfield, was able to call upon a number of private armories. Also, there was the Colt Arms Works at Hartford, Connecticut, the most efficient plant of its kind in the world. Within a few months the North was able to expand greatly its small arms production. Rifle manufacture had by 1860 reached the stage of precision tooling. Approximately one hundred automatic power tools were used in manufacturing the Springfield rifle, a gun that had a range of 1,000 feet. These the South did not have.

Ammunition. The North also had an advantage in manufacturing ammunition. In addition to iron, the needed critical materials were saltpeter for gunpowder, lead for cartridges, and copper for percussion caps. The supply of lead and copper was ample in the North but limited in the South. The lead mines in Virginia, and the small copper mine at Ducktown, Tennessee, were never able to supply the demands. The Confederacy resorted to salvage operations. Turpentine and apple brandy stills were seized and melted down to salvage the copper, and lead window weights and lead pipes were yanked from homes and contributed by patriotic citizens.

In the matter of saltpeter (for gunpowder) both North and South had been dependent on India. It was a product that had to be imported by way of England. The North, as a result of the "Trent Affair," passed through a technological crisis during the first winter of the war. With the supply of saltpeter almost depleted in the fall of 1861, Lamont DuPont made a hurried trip to England, purchased 2,000 tons of the material, had it loaded and was ready to leave port, when England imposed her short-lived munitions embargo.

The imposing of the embargo was in retaliation for the capture of two Confederate agents, Mason and Slidell, who had been passengers on board a British mail ship, "The Trent." After Lincoln, through his Secretary of State Seward, yielded to the British demands in freeing Mason and Slidell, England lifted her embargo, and DuPont returned with his 2,000 tons of saltpeter.

DuPont's Discovery. This incident led to a major technological development in the manufacture of munitions. Determined to become independent of any foreign supply, DuPont perfected a process for using sodium nitrate instead of saltpeter in manufacturing gunpowder. Thus India's monopoly was broken.[5]

Military Communication. Other recent technological developments that aided the North were the installation of the field telegraph and the military balloon employed to improve reconnaissance. The first field-line telegraph was used in Virginia in July, 1861, between McClellan's headquarters and the front line officers. Once installed, the field telegraph was used in many theaters of combat. Of the 5,400 miles of telegraph lines constructed in the Northern states between 1861 and 1866, more than 1,000 miles of wire were used by the military forces.[6]

Balloons in Warfare. The most spectacular technological innovation was the use of military balloons. The use of balloons had been successfully demonstrated as early as 1783, and the French had employed them in military reconnaissance in 1794. But it remained for Professor Thaddeus Sobieski Coulincourt Lowe, a citizen of New Hampshire and a protégé of Joseph Henry, to demonstrate the military value of balloons. Lowe had made a few successful flights before 1861. Soon after the outbreak of the war, he was appointed Chief of the Aeronautic Section and designed a balloon, "The Union." It had 32,000-cubic feet capacity, and cost fifteen hundred dollars. General McClellan, impressed by this newest military weapon, ordered six more to be constructed. Professor Lowe went to Philadelphia, supervised the manufacture of the balloons, and placed them in service early in 1862. Realizing the advantage of making ascensions from water, he designed a special barge as a base of operation and converted it into the "first aircraft carrier in history." [7]

A number of free-lance balloon enthusiasts offered their services to the Union Army. Among these, John Wise of Lancaster, Pennsylvania, gained the distinction of building the only armored balloon used during the war. John LaMountain, of Troy, New York, made several aerial reconnaissances for General Benjamin Butler in the vicinity of Fortress Monroe in 1861, and for the Army of the Potomac in the fall of the same year. His flights caused considerable consternation among the Confederate forces. In December, 1861, General Joseph E. Johnston warned against those "infernal balloons," and General Beauregard ordered General Longstreet to camouflage their dummy guns which they had pointing at the Union lines "so as to conceal their true nature from aerial vision."

Food Supply. Basic in any successful military or naval campaign is the matter of food. Here, paradoxically as it may seem, the South, while predominantly rural, was worse off than the North. "King Cotton" had so completely monopolized southern economy that there was a food shortage. The greatest threat to the South's food supply was the lack of agricultural machinery. Even when the South realized it would have to begin planting grain and cereal crops where cotton had formerly grown, it had no modern machinery to cultivate or harvest the crops. In the North the practice of diversified agriculture, together with improved machines, was producing a surplus. In normal prewar years, the North had shipped millions of bushels of wheat yearly into southern states.

Northern Food Surplus. The introduction of the new labor-saving machines, plus a sizable immigration, enabled the North to produce sufficient food for home consumption, to feed the men at the front, and also to provide a surplus for export. The export of northern wheat jumped from 20,000,000 bushels in 1860 to 60,000,000 bushels in 1862, a factor that greatly influenced the diplomatic developments of the remaining war years. The Northern farmers enjoyed good markets, became prosperous, and thereby helped to strengthen the entire economy of the Union.

The first year of the war also witnessed a marked increase in the use of canned goods and condensed milk throughout the northern states. The entire output of Borden's condensed milk was commandeered by the Union Army early in the war. Also, Borden's condensed berry juices were bought up by the United States Sanitary Commission and served to the wounded.

Northern Clothing Supply. Third on the list of fighting equipment comes clothing. Here again the North, because of new developments in textile technology, had an advantage. The recently invented sewing machine had brought about an extensive ready-made clothing industry. The output of sewing machines doubled between 1860 and 1864. With a sewing machine, shirts could be made twelve times as rapidly as by hand.

Southern Clothing Supply. In the South there was no extensive ready-made clothing industry. One northern city alone (Boston) was producing more ready-made clothing in 1860 than was produced in the entire South. In order to supply the southern armies with uniforms, the Quartermaster Bureau of the Confederacy set up small clothing "factories" in a number of cities late in 1861. They employed several thousand women of high and low estate, and throughout the year 1862 they provided a sufficient supply of cotton uniforms.[8] There were a number of cotton mills in the South, but when the machinery wore out it was impossible to obtain parts for repairs. After the first two years of the war, clothing for both soldier and civilian became scarce throughout the South.

Campaign of 1862

Throughout 1861 and during the early months of 1862, it looked as if the South might win. The Confederate armories were producing one thousand rifles a month during the first part of 1862, and Captain Caleb Huse, purchasing agent for the Confederacy, secured eighty thousand Enfield rifles in London and had them shipped on a steam vessel so as to outrun the blockade of sailing vessels. The Augusta powder mill, under the direction of Col. George Rains began production in April, 1862, with the expectation that powder could be manufactured in unlimited quantities.

New Southern Military Weapons. The South introduced two new weapons early in 1862 which caused consternation among the Northern forces—a land mine, and a rudimentary machine gun. In May, 1862, General Gabriel J. Rains planted a primitive mine beneath the surface of the road to cover his retreat from Williamsburg, Virginia. The pursuing Union cavalry suffered severely from the explosions that followed, and General McClellan issued a bitter denunciation of "this most murderous and barbarous conduct."

A breechloading machine gun, invented by Confederate Captain R. Williams, and manufactured at the Tredgar works in Richmond, was first used at the Battle of Seven Pines, June 1, 1862. This unwieldy implement, weighing 275 pounds, with an ammunition box of 600 pounds, was pulled by one horse and operated by three men. Discharges were made by turning a crank; it fired forty times a minute, and had an effective range of 2,000 yards. When in October, 1863, the Confederates brought six of these new guns into action at the Battle of Blue Springs, Tennessee, and fired two hundred shots a minute, it caused great consternation in the ranks of the Union Army, and one of the Union soldiers exclaimed, "We concluded that the enemy had some new fangled machine for throwing railroad spikes at us."

Early Victories of the Confederate Navy. The greatest alarm that spread throughout the North early in 1862 was caused by the Confederate Navy late in 1861 and during the early months of 1862. The Confederate ironclad "Manasses," propelled by steam, had destroyed a squadron of wooden vessels stationed at the mouth of the Mississippi in October, 1861. Secretary Mallory of the Confederate

Navy was authorized to speed construction of four new iron-plated warships, two at New Orleans and two at Memphis, in addition to the ironclad ram "Merrimac," (Virginia). News of these developments caused alarm throughout the North.

The "Merrimac" (Virginia) was the first of the ironclads to be launched. She was a screw-frigate of 3,200 tons, armed with six 9-inch Dahlgren guns, four Brooks rifles, and was fitted with a heavy cast-iron prow for ramming. Her sloping casemate was covered with two layers of 2½ inch wrought-iron plates, with a 7-inch wooden backing. Her builders believed that this novel construction made the "Merrimac" invulnerable. When on March 8, 1862, she steamed out and broke the blockade at Hampton Roads, sent four large wooden warships to the bottom of the channel, and escaped without injury, a revolution had occurred in naval technology. Her iron-clad covering was not penetrated.

Panic spread throughout the North. The inhabitants of the coastal towns feared this monster of the seas would steam into port and bombard their cities. President Lincoln called a special session of his Cabinet, and Secretary of War Stanton warned of the imminent threat by the Confederates, fearing they might either ascend the Potomac River or shell the White House.

Ericsson's "Monitor." If the North had only known, however, that John Ericsson's "Monitor," " a cheese box on a raft," was moving toward Hampton Roads, fears would have been allayed. The "Monitor" consisted of a small iron hull upon which rested a large raft surmounted by a revolving turret. The raft was 5 feet deep and was protected by a side armor of five 1-inch plates, and its roof was shielded by railroad iron. Two 11-inch Dahlgren shell guns were mounted inside.

This novel floating battery, driven by screw propellers, was intended to aid Admiral Farragut in his campaign against New Orleans, but when her commander heard of the destruction of four Union vessels at Hampton Roads, he hastily changed his plans and headed for that area. Early in the morning, March 9, 1862, he sighted the "Merrimac" (Virginia). A 3-hour naval battle followed, with neither ship able to cripple the other seriously. Finally, after the captain of the "Merrimac" had been wounded, he withdrew from the scene of battle and sailed into the harbor at Norfolk, unable to break the blockade.

The Battle of Hampton Roads marked a turning point in the technology of naval warfare. Sailing vessels and wooden steam vessels were doomed. Iron-clad battleships, propelled by steam power with guns mounted in revolving turrets, introduced a new chapter in the history of sea power. Within one week, Secretary of Navy Gideon Wells ordered work to begin on six iron-clad monitors of 1,330 tons each.

Naval Torpedoes. The Confederacy introduced another new device in naval technology late in 1862. A board known as the Naval Torpedo Service, created by the Confederate Congress in June, 1861, carried on a number of experiments under Commander Matthew Fontaine Maury, relating to a new device in firing torpedoes. On December 2, 1862, the Union iron-clad gunboat "Cairo" was sunk in the Yazoo River by a friction primer and a trigger line, the first warship ever destroyed by torpedo action.

Land Warfare in 1862. While the North was striving desperately to modernize its fleet, enforce a strict blockade along the entire southern coast, and hasten the end of the conflict, things were not going too well in other theatres of the war. The specially trained artillery units upon which the North so largely depended for an early knock-out were not coming up to expectations. Because of poor performance of these units, together with some tactical errors, General McClellan had not been able to crack the Confederate defenses in the Peninsula campaign. His Army of the Potomac was not gaining the victories that he had anticipated. General Joseph E. Johnston of the Confederacy, with a much smaller army, had checked McClellan in spite of vastly superior numbers. Another southern army commanded by General Thomas J. (Stonewall) Jackson swept up the Shenandoah Valley against the Union forces, and was threatening the nation's capitol.

General Lee Leads the South. In the face of these setbacks the North became more alarmed than ever. When in May, 1862, it was learned that a new southern commander, General Robert E. Lee, had succeeded General Joseph E. Johnston, who had been wounded, Northern generals realized they were matched against one of America's greatest military leaders. After a seven-day battle around Richmond, General Lee's army was joined by General (Stonewall) Jackson's army, beat back McClellan's superior forces, and forced President Lincoln to change commanders in the hope of turning the fortunes of war.

It was in the midst of these dark days that the great inventor John Ericsson addressed an historic dispatch to President Lincoln, August 2, 1862, saying,

> The time has come, Mr. President, when our cause will have to be sustained not by numbers, but by superior weapons. By a proper application of mechanical devices alone will you be able with absolute certainty

to destroy the enemies of the Union. Such is the inferiority of the Southern States in a mechanical point of view, that it is susceptible of demonstration that, if you apply our mechanical resources to the fullest extent, you can destroy the enemy without enlisting another man.[9]

Following Captain Ericsson's pleas to President Lincoln, the whole war program was expanded. It included plans for producing not only the latest and most modern weapons and equipment, but, equally important, also the use of the latest developments in transportation technology. Railroads were used at the outbreak of the war to transport supplies, but not troops. They could march on foot.

Military Transportation and Engineer Corps. Proof that railroads could be used advantageously to transport troops came in August, 1862. In four days, August 21 to 25, some two thousand Union troops were transported by rail from Parkersburg, Virginia (West Virginia), to Washington, each contingent moving 400 miles in about 40 hours. They had been ordered to hasten to General Pope's aid in his attack on Richmond, but the Second Battle of Bull Run altered their plans and they were then ordered to Washington; they arrived in time to prevent Lee from capturing the Capitol.

Both sides soon realized the military significance of the railroads, and special raids were ordered with the view of destroying trains, tracks, and bridges. These tactics led to the creation of a special corps in the Union Army in the spring of 1862, headed by Herman Haupt, an experienced railroad engineer. He became Chief of Construction and Transportation Services and was attached to the Department of the Rappahannock. He operated in northern Virginia, in and around Washington. This unit became an indispensable branch of the service, ready for any emergency. By having on hand the standardized, interchangeable parts of bridge spans, track, and trusses, this corps was prepared to move on command, repair torn up tracks, throw up new bridges, or repair those that had been destroyed.

Haupt's corps worked so efficiently that it was said that "The Yankee could build bridges faster than the Rebs can burn them down." At the Battle of Gettysburg, July 1–3, 1863, General Lee's army destroyed most of the main line bridges, but by August 11th, two weeks after the last of Lee's forces had retreated across the Pennsylvania line, the Baltimore and Ohio was reopened for "through" traffic. Specially trained railroad men were indispensable. When General Pope took command of the Rappahannock Division in June, 1862, he ignored the Construction and Transportation Corps and permitted Haupt to return home. Ten days later the

Assistant Secretary of War wired Haupt, "Come back immediately. Cannot get on without you. Not a wheel moving." [10]

The Confederacy at no time was able to build the railroads it needed, or to repair the worn out ones that had been built before the war started. It required almost two years to build a line 48 miles long which was, Lee said, "of vital necessity," connecting Danville, Virginia, and Greensboro, North Carolina; and when it was built, the rails from another road had to be torn up to be used for the new track. In October, 1862, the Confederate Congress voted one million dollars to build a railroad from Rome, Georgia, to Blue Mountain, Alabama, to tap the iron and coal deposits of that region. But because of the difficulty of procuring iron rails, the road was never completed. The lack of adequate rails made it impossible to obtain the iron ore needed to make more rails.

With the approach of winter in 1862, the military campaigns slowed down, and both sides awaited the renewal of activities in the spring of 1863.

Campaign of 1863

An important move toward enlisting the aid of the nation's top scientists was taken early in 1863. On March 3 of that year, President Lincoln signed an Act of Congress creating the National Academy of Science. Senator Henry Wilson of Massachusetts, author of the bill, declared that the need for "a society of scientists had for years been in the mind and on the tongues of the devotees of science." He called attention to the "dark and troubled night" through which the nation was passing, and urged the Congress and the President to call upon men of science with the idea that "they be commissioned to infuse into it [the nation] that vital and animating spirit that shall win this most gigantic civil war and make this bleeding nation one and indivisible."

National Academy of Science. The Academy's charter required that whenever called upon by any department of government, its members were to investigate, examine, experiment, and report upon any subject of science or art. The original membership of fifty included almost every important man of science of that period: Joseph Henry, Alexander Dallas Bache, Louis Agassiz, Asa Gray, Wolcott Gibbs, Benjamin Silliman, Sr., Benjamin Silliman, Jr., and others of equal prominence. During the first eighteen months of the Academy's existence, special committees were appointed to study such matters as magnetic deviation in iron ships; methods for protecting the bottoms of iron vessels; a new weights and measures system; the preparation of more accurate wind and current charts; and experiments on new methods of expansion of steam. While the Academy's contribution toward winning the Civil War were relatively insignificant, its later history included a number of valuable scientific achievements. It helped set the pattern for such wartime organizations as the National Research Council during World War I and the Office of Scientific Research and Development, World War II.

Northern Naval Technology. While the Northern Armies conducted their military campaigns through the entire summer of 1863 without benefit of any major technological developments, the Navy on the other hand came up with a number of new developments. Naval action was directed mainly toward effective blockade and counterblockade measures. The success of the South's blockade runners offered a challenge to the Northern Navy and her naval engineers began a number of new experiments. They introduced the forced draft system in boilers in an attempt to speed up their ships, and the noted naval expert, B. F. Isherwood, made several improvements on the simple steam engine that powered the ships.

The Northern Navy's greatest success, however, was won with her iron-clad monitors of the "Passaic" class. These heavy battle wagons, equipped with thick armor plate, 15-inch Dahlgren shell guns, 440-pound projectiles, and revolving turrets, were next to invulnerable. They gradually closed in on Southern ports, demolished their forts, and bottled up their commerce. By strangling the foreign commerce of the South they inflicted irreparable damage.

Technology Aids Food Production. As the campaign of 1863 dragged on, the South experienced a serious food shortage. The supply of wheat that she expected to receive from England did not arrive. On the contrary, the North's surplus continued to mount. More labor-saving machinery was introduced. In 1863, almost five hundred patents for agricultural implements were granted, as against only two hundred and forty for instruments of war.

Mass Troop Movement by Rail. The rapid advances that the North was making in transportation technology in 1863, particularly in the operation of railroads, became a deciding factor in the military operations of that year. General Rosecrans' Army of the Cumberland, after being routed at Chickamauga, September, 1863, had withdrawn to Chattanooga. General Braxton Bragg was expected to attack, endangering the whole western campaign. Rosecrans' only hope of warding off this expected attack was to secure heavy reinforcements quickly. His request for two full army corps from General Burnside's army in eastern Tennessee could not be met had the troops been compelled to march over land.

Secretary of War Stanton and Daniel C. McCallum, Supervisor of Railroads, proposed an unheard-of plan. They ordered that two army corps from the Army of the Potomac, since they would not be needed during the winter, be dispatched by rail to Rosecrans' aid. President Lincoln was won over to the plan and twenty-three thousand troops, horses, wagons, and equipment were rushed south to join the Army of the Cumberland. When General Grant arrived some days later with additional reinforcements the combined forces again took the offensive. Troop transport by train had turned the tide.

First Armored Cars. The armored railroad car, another contribution to military transportation technology, made its appearance in 1863. These cars, described as "bullet proof and mounting a cannon," were used to patrol the Baltimore and Ohio Railroad, the key supply line for the Army of the Potomac. They were ordinary box cars, covered with iron sheeting outside and inside, and equipped with small cannon. These armored cars succeeded in preventing a number of Southern raids from reaching supply bases. Another advance in military technology was introduced when the Union forces constructed blockhouses near strategic bridges and tunnels. These novel blockhouses were manned by a detail of soldiers. They were the forerunners of the modern pillbox.

Campaigns of 1864–1865

The difference in the supply of guns, breechloading rifles and cannon, between the North and the South became a deciding factor in 1864–1865. After three years of tooling up, the Northern armories swung into mass production.

Mass Production versus Manpower. One hundred thousand breechloaders were in the hands of Union soldiers by the end of the war. Meantime, in 1864, a giant 20-inch Rodman smoothbore cannon, weighing 115,000 pounds, the largest piece of ordnance in the world, was cast by the Fort Pitt Foundry in Pittsburgh.

The Confederacy, on the other hand, had established two new government armories in 1864 which, added to the three already in operation, provided a theoretical production of forty-five hundred small arms a month. The actual output, however, due to shortage of skilled labor and lack of materials, slumped from three thousand a month in 1863 to less than two thousand in 1864.

Trench Warfare and Mine Laying. During the military campaigns of 1864 and the first hundred days of 1865, a little-known technique was introduced, which came to be called "trench" warfare. The use of breastworks for military protection dates back to prehistoric times, but the habitual use of breastworks, thrown up in the presence of the enemy, fortified with wire entanglements, was an innovation in military technology. General Fitz-John Porter's troops of the Union Army had resorted to trench warfare in the Battle of Gaines' Mill, June 27, 1862. A few weeks later, in the Second Battle of Bull Run, General Jackson's Confederate troops had fortified themselves in excavations made for an unfinished railroad, and thus they were impregnable. At the Battle of Chancellorsville, May, 1863, the Confederate troops were so skilled in this new technique as to cause an officer in the Union Army to exclaim, "Why, we never saw any Rebels where we were; only smoke and bushes and lots of our men tumbling about."

The Confederate troops continued to make extensive use of trench warfare throughout the closing months of the war. General Sherman, on his march through Georgia in the summer and fall of 1864, declared that the whole country was one vast fortress, and that, "General Johnston must have had at least fifty miles of connected trenches." General Sherman, however, was the first to master this new technique by outflanking the enemy and then entrenching his own troops for a counterattack.

The planting of mines, a new wartime technique, continued to prove an effective weapon for both armies. The most gigantic display of explosives occurred during the ten months' siege of Petersburg. In a bold effort to terminate the siege, the Union troops worked for months in digging a mine shaft 500 feet long, extending to a point under one of the Confederate forts defending Petersburg, and planted 8,000 pounds of powder in eight chambers at the end of the shaft. When the mine exploded on the morning of July 30, 1864, it blew up the fort and created a crater 170 feet long, 60 feet wide, and 30 feet deep. From a military point of view, this innovation in land warfare on the part of those who planned it was a success. Actually, however, due to an eleventh-hour change in command, the Union forces suffered the heavier casualties. The mistakes were human, not technological.

Naval Warfare. Meantime, the war at sea was marked by two further advances in naval technology—the extended use of the torpedo, evolving into the torpedo boat; and the submarine. The first Southern torpedo boat, the "David," built in 1863 by Doctor S. Julian Ravenel of South Carolina, from a discarded locomotive boiler, was propelled by steam. She carried a torpedo charge at the end of her 30-foot spar. Commander Hunter Davidson built a second steam torpedo boat, "The Torpedo," early in 1864. Northern naval technologists were also busy designing torpedo boats. In October, 1864, a daring Union officer, Captain W. B. Cushing, planted a spar torpedo under the troublesome iron-clad ram "Albemarle," and blew her up off the coast of South Carolina. Torpedo boats continued to be used effectively until the end of the war.

The first ship ever to be destroyed by a submarine was sunk during the Civil War. Captain Horace L. Hunley of New Orleans pioneered in introducing this novel weapon, and built two submarines which, because of their imperfection and inexperienced crews that manned them, sank before they could be put to effective use. The third one, named "The H. L. Hunley," was made of iron and was operated manually by a crankshaft. After being built at Mobile she was shipped to Charleston, where the contact torpedo was replaced by a trigger torpedo containing 90 pounds of explosive. It was operated by a percussion and friction primer.

On February 17, 1864, the "Hunley" managed to move up alongside the "Housatonic" and planted her torpedo. An explosion followed, and the "Housatonic" sank in 3 minutes; but the submarine "Hunley," the first ever to destroy a warship, also capsized and nine crew members were drowned. While the North built at least three submarines, none of which were ever used effectively, the South continued to build iron-clad rams as long as she could secure the necessary materials and machinery. One of her ships, the "Tennessee," equipped with six 7-inch Brooke rifles, and plated with three 2-inch wrought-iron slabs, continued to annoy the Union Navy until she was completely disabled in the historic battle of Mobile Bay.

Battle of Mobile Bay, Aug. 5, 1864. The story of the "Tennessee" is largely the story of that famous naval battle. The only port on the Gulf that was still accessible to blockade runners in the summer of 1864 was Mobile, Alabama. Admiral Farragut planned its destruction, and assembled seventeen vessels, carrying one hundred and ninety-nine guns, and twenty-seven hundred men. The attack began in the early morning of August 5, 1864. Farragut's first battleship, the "Tecumseh," struck a torpedo and sank. The "Brooklyn" slowed down, and as Farragut's flagship the "Hailford" steamed ahead, he was told that "a heavy line of torpedoes lies across the channel." It was then that he uttered his famous message, "Damn the torpedoes: four bells; full speed ahead." His fleet steamed past the fire of Fort Morgan and closed in on the "Tennessee." Unable to pierce her armor, the Union ships rammed her repeatedly until her steering gear was disabled. The "Tennessee" was compelled to surrender. The Union fleet paid dearly for this victory, losing one hundred and sixty-five men, while the Confederates lost only twelve. For the North, however, the capture of Mobile was an important victory. A tightening of the blockade of all Southern ports followed. This meant the end of the naval warfare, while Sherman's march to the sea also meant that the end of land warfare was near at hand.

End of the War, April, 1865. Little of importance in military or naval technology occurred during the winter of 1864–1865. Finally, on April 3, 1865, Petersburg, Virginia, which had been under siege for months, fell to the Union forces; and with it fell Richmond. The fall of Richmond meant the end of the War. Six days later at Appomattox Court House, General Lee and the remnants of his once proud army surrendered to General Grant. The war was over.

Technological Firsts of the Civil War

In summarizing the role of technology in the Civil War the following developments should be noted. This was the first total war in the modern sense of that phrase—a war in which the artisan, the farmer, and the mechanic played a role equal in importance to that of men in uniform. It was the first time that the achievements of an industrial, scientific revolution were put to large-scale military use. Improved breechloading rifles for the infantry, and accurate artillery guns and rifled cannon came into general use. War news was dispatched by telegraph, and balloons signaled the movement of troops. It was the first war to be photographed adequately, and the first to use mass movement of troops by railroad. Steam, rather than sails, supplied power for warships. Iron-clad men-of-war, mounting powerful shell guns and rifled cannon, made the ancient wooden vessels obsolete. The Confederacy used land mines and torpedoes effectively. Both sides pioneered in introducing machine guns, armored cars, surface torpedo boats, and submarines.

Machine Power Predominates. In considering the many advances in military and naval technology that occurred during the Civil War (and there were many), it must be noted that eventually the struggle settled down to a contest between the production preponderance of the North and the industrial weakness of the South. The natural resources of the South early became exhausted. Her farms were laid waste. Her factories were in ruin, and all hope of receiving aid from abroad disappeared with the tightening of the blockade.

Meantime, the North was rapidly growing stronger. By the spring of 1864, war material and military supplies had reached the mass-production stage. The technological revolution in agriculture, the recent developments in food technology, and the introduction of the sewing machine enabled the Northern armies to be better fed and better clothed than the Southerners. Troop transportation by railroads enabled the Union forces to mobilize speedily in any theatre of war, and the field telegraph permitted one central command to dispatch men and supplies where they were most urgently needed.

When the final conference of the high command of the Union forces was held in late March, 1865, to plan the attack on Petersburg and Richmond—the last holdout of the Confederate forces—General Sheridan arrived March 26 via the Virginia Central Railroad, General Sherman arrived the next day by steamer from North Carolina, and the following day President Lincoln arrived by rail from Washington. Such co-ordination and timing could not have occurred in the war-torn South.

Peacetime Industry in Wartime Production. While no one technological achievement brought victory to the North, yet it was the combination of many wartime technologies, plus the industrial power of the North, whose strength rested upon new techniques and new inventions, that finally conquered the South. These recent inventions and technologies represented a remarkable achievement for a young nation that had been devoting its chief energy to subduing a wilderness, opening up farms, building factories, and establishing a network of internal improvements. Most of the inventions and technological achievements had been designed for people who felt themselves secure from all wars. But when the crisis came, those who had pioneered in developing the weapons for the peaceful conquest of a continent reluctantly transformed them into instruments of war. The same was true in a very real sense in World War I and World War II.

Notes for Chapter 20

1. D. H. Hill, "Confederate Ordnance Department," *North Carolina Historical Commission,* Bulletin No. 28 (1921), p. 82.

2. Victor S. Clark, *History of Manufactures in the United States,* II (1928), 43. It should be observed that this Bureau later took over the supervision of iron, lead, copper, and zinc production; thus it was the forerunner of similar war production boards in World War I and World War II.

3. *Scientific American,* New Series, II (1864), 165.

4. See James Ford Rhodes, *History Of The United States from the Compromise of 1850,* III (1899), 410.

5. Otto Eisenschiml, "The Chemist in Three Wars," *Science, New Series,* IVC (Oct. 16, 23, 1942), 347-352, 367-373, 349.

6. William R. Plum, *The Military Telegraph During the Civil War in the United States,* 2 vols., I (1882), 97-98.

7. See F. Stanbury Hayden, *Aeronautics in the Union and Confederate Armies,* Johns Hopkins Press (1941).

8. Charles W. Ramsdell, "The Control of Manufacturing by the Confederate Government," *Mississippi Valley Historical Review* VIII (1921), 233.

9. William Conant Church, *The Life of John Ericsson,* 2 vols., II (1891), 34.

10. Edwin A. Pratt, *The Rise of Rail Power in War and Conquest* (1916), pp. 31, 43.

Three recent studies, Robert C. Black, *The Railroads of the Confederacy* (1952); Thomas Weber, *The Northern Railroads in the Civil War* (1952); and George Edgar Turner, *Victory Rode the Rails* (1953), treat in great detail the part played by railroads in the Civil War.

PART III

RISE OF UNITED STATES AS A WORLD POWER: Civil War to 1900

During the period from the Civil War to 1900, growing America rose from the ranks of a small isolated republic to take her place in the family of nations. She was host to the world at two international expositions when the Old World brought its best to be exhibited on American soil.

In this period machine technology triumphed. The machine invaded the western prairies and harvested the mighty crops that helped to feed America and the world. The great meat packing industry, with assembly-line methods and the introduction of artificial refrigeration, provided the American people with fresh meats the year round.

The machine held sway in giant steel mills and in textile mills where production surpluses sought an outlet beyond our borders. The manufacture of steel rails laid the foundation for a new era in railroad expansion by supplying steel rails to take the place of antiquated streaks of iron rust. American steel made possible the skyscraper, the one distinct contribution of the United States to the world of architecture.

The birth of the petroleum industry, the rise of giant oil companies, and the distillation of gasoline for internal combustion engines ushered in a revolution in transportation technology with the introduction of the automobile.

The harnessing and utilization of electricity gave man his greatest mechanical servant. The incandescent light arrived. Electricity brought a new invention in communication technology—the telephone, all before 1900. The invention of the typewriter, the mimeograph, the multigraph, the developments in photography, and the improvements in high-speed printing brought major changes in the entire system of rapid communication.

Behind all these developments lay the forces of applied science and technology.

Chapter 21

THE NATION'S AWARENESS OF SCIENCE AND TECHNOLOGY TO 1900

Within a few weeks after General Lee's surrender at Appomattox, this country was invited to participate in the Paris Exposition to be held in 1867. Despite the staggering war debt and unstable financial policy, and with one-third of the nation still in ruins, a number of industrial leaders and a small group of scientists urged that the invitation be accepted. President Johnson and the Congress finally agreed. Although the American exhibits arrived late and were assigned to the less desirable quarters, they attracted considerable attention as the weeks and months passed.

American Progress Displayed

Paris Exposition, 1867. This was the first impressive proof to Europe of the advances in science and technology that had occurred in the midst of a civil war. The ambulances and pharmaceutical exhibits displayed by the Sanitary Corps, which captured a Grand prize, won admiration for a nation that gave relief to its sick and wounded victims of war. In many technological exhibits American contributions were recognized as more original and inventive. Though lacking in graceful design, they stood out for being "unadorned and highly useful." McCormick won a Grand prize and the Cross of the Legion of Honor for his reaper. Cyrus Field was honored for his Atlantic cable. Coveted prizes and awards went to the Sharp machine for making screws, the Sellers planing machines, and above all, to the Corliss engine with its rotary valves and governor that controlled the admission of steam into the cylinder.

Gold medals were awarded to the Howe sewing machine, to the American Buttonhole Machine Company for a machine that made buttonholes, to a machine that made six lead pencils in one operation, and to a telegraph printing machine that could record a greater number of messages than any other. Rugs, carpets, textiles, and shoes took various awards. So, also, did machines for peeling apples, beating rugs, brickmaking machines, automatic flour sieves, washing machines, Fairbank's scales, and dozens of others. More than half the American exhibitors carried off awards. The American exhibits were compared favorably with those of the Old World, which seemed to do "little more than perfect old principles, and invent new adaptations for old machines." [1]

Influence of Expositions. During the next three decades the United States participated in more than a dozen international expositions, including two of our own, the Centennial of 1876 and the Chicago World's Fair of 1893. During this period the European attitude toward America and American-made products changed, and the emergence of a new and powerful nation was recognized. Among the factors that brought about this change in attitude was the part played by America in these international expositions. In every one of them the products of applied science and technology were dominant. They presented a striking picture of the development and wide diffusion of technological progress.

In some of these expositions, the United States manufacturers participated half-heartedly and received little government support. But, over the period, they presented a strong advertisement for the power and greatness of a nation that by 1900 had won world recognition. At first, it was the ingenious, the novel inventions that attracted attention. Then came the exhibits of the labor-saving devices and inventive skills. Mass production with accompanying reduction in costs was demonstrated. And, finally, with the Chicago World's Fair, Europe admitted our entrance into the realm of grace and beauty with the superb setting of that exposition and the promise of the new architecture exhibited there.

These expositions were held in a number of European countries —Spain, Austria, Germany, France, and England, and in far off Melbourne (Australia). At Vienna in 1873, it was again American machinery that was praised. The exhibit of an American school with maps, charts, textbooks, and other equipment won admiration. The American school display at Paris in 1878 was even more outstanding; an international jury awarded the American exhibitors twenty-eight more honors than those of any other country, save France.

Paris, 1878. At Paris, also, America's advances in many technologies were recognized. America excelled in machinery exhibits. There was no rival for the Wheelock engine, which supplied power for operating all American machines. In the machine tool section, the greatest technological exhibit was on display. There, the technology of making interchangeable parts by rule and gauge was so unique that it came to be described as "the American system." The awards in all classes granted to America excited surprise and aroused some jealousy. American exhibitors of machinery, however, were flattered to observe other exhibitors had copied many of the original aspects of our machines. Of our 1,200 exhibitors, 854 received awards.

Melbourne, 1888. America participated in two expositions at Melbourne in 1880 and in 1888. In both, there was a feeling of comradeship, because Australia, like America, was a new nation. Despite the long distance of transportation, our exhibits were creditable, though small. Of the 339 exhibits, 214 received awards, with 110 "firsts." Our watches, machine tools, school furniture, and agricultural implements attracted attention at the 1880 fair. The typewriters, sewing machines, carriages, lamps, and musical instruments held the spot light in 1888.

Other European Expositions. At Paris, in 1889, the American agricultural exhibits were recognized as inferior to none. A leading German economist noted that "the American technical know-how will be an important factor in the competition for world markets." Here, America won 53 Grand prizes, 198 gold medals, 266 silver medals, and 233 bronze medals, and received 133 honorable mentions—a good share of the awards. In the electrical exhibits, American products were outstanding.

Specialized international exhibitions of fisheries were held in Berlin in 1880, and in London, 1883, at which the United States ranked first in size of display and received a proportionally large number of awards. Here were exhibited models, maps, and charts of fishing grounds. Examples of the uses of fish and other water products were exhibited. Also, there were exhibits of other government agencies and technical schools interested in promoting scientific projects. The United States geological survey, Naval observatory, and Bureau of Standards won prestige in scientific matters for the United States. This reputation was enhanced by our participation in the Columbia Historical Exhibition at Madrid in 1892, at which the American exhibits in anthropology, archaeology, and ethnology received universal acclaim. Displays were sent from the Smithsonian Institution, the United States National Museum, the University of Pennsylvania, Peabody Museum of American Archaeology and Ethnology, and other private institutions.

American International Expositions. In our two home expositions, international interest progressed from recognition and respect at the 1876 Centennial to admiration and the conviction that Europe had much to learn from America at the Columbian Exposition. By 1893 the achievements in American technology could no longer be ignored. In cultural matters the nation was matching her technological "know-how."

Centennial of 1876. During the early 1870's, in the midst of nationwide financial depression and an orgy of political scandals, plans were announced for observing the one hundredth anniversary of the adoption of the American Declaration of Independence. The proclaimed purpose was to hold an exhibition in Philadelphia under the auspices of the United States government, in which American arts, products, and manufactures would be displayed. The opening of the Centennial, America's first world's fair, was participated in by President and Mrs. Grant, Governor John F. Hartranft of Pennsylvania, and dignitaries from foreign governments. The exhibition was open to all the nations of the world.

President Grant, in officially opening the Centennial Exhibition, reminded his listeners that the chief objective was to bring together exhibits of our attainment in the industrial and fine arts, in science, in agriculture, and commerce. He observed that one hundred years ago the country was new and only partially settled, but the Exhibition of 1876 was a demonstration that the nation now compared favorably with the older nations in science, industry, and business.

Machinery Hall. Machinery Hall, with its jarrying whir of machinery "whose multiplied engines make every man hundred-handed and victorious over difficulties which used to defy human strength," was described as the emblem of America's mighty power. It housed the industrial and mechanical exhibits, and was visited by more people than any other building.

Exhibited here in a setting that dominated the vast hall, was the famous Corliss engine, built by George H. Corliss of Rhode Island. This huge double-acting, duplex vertical engine stood 40 feet high, with 44-inch cylinders of 10-foot stroke. Although built as a 1,500-horsepower engine, it was capable of generating 2,500 horsepower, and weighed 700 tons. Operating as noiselessly as a watch, it communicated its motive power to 10,999 feet of shafting, scattered throughout the 14-acre building, and to some eight thousand mechanisms from carpet looms and huge presses to delicate sewing machines. For six months, "without haste, without rest, and with equal pulse," this giant machine propelled an endless system of belts and wheels. The *Nation* referred to the Corliss engine as "the marvel of all spectators."

Significance of the Centennial. The exhibits in Machinery Hall represented America's advances in the field of invention, applied science, and technology. In commenting on the numerous exhibits, the editor of *Scribner's* said that the developments in building steamships and locomotives, in the cotton gin, steam engines, the telegraph, and in agricultural and mining machinery during the first hundred years of our republic, had revolutionized the world. And the *Atlantic Monthly* (July, 1876) reminded its readers that, "nowhere else are the triumphs of ingenuity, the marvels of skill, and invention so displayed. . . . Surely here, and not in literature, science or art, is the true evidence of man's creative powers. Here is Prometheus unbound."

Foreign observers were impressed by the mechanical exhibits. Among the numerous correspondents from abroad was F. Otto Reuleaux, professor of mechanical engineering, Berlin, who, in writing to his German friends commented on the unparalleled strides the Americans had made in their technical and mechanical development even though, he said, "they were still in the grip of a depression." He observed that it was a well known principle for the Americans to replace with machines the human hands, of which there was a scarcity.

In the manufacture of certain finer steel products, the United States could still learn from the Germans and English, said the writer, but their axes, hatchets, files, and tools for forestry, plantations, and gardening presented such a variety and beauty as to compel the foreign visitors "to stand and gaze in wonderment." Professor Reuleaux was particularly impressed by the plain and circular American-made saws, their size and cutting speed, and the ease with which they were driven. In the art of nickel-plating metals he noted that America surpassed all other nations. The use of nickel plating on hand rails, candelabras, and chandeliers diffused a rare brilliancy throughout the exhibition buildings. In the exhibits of lamps and chandeliers, American designers showed excellent skill in casting, gilding, bronzing, and coloring. Germany, he concluded, "had hardly anything worth mentioning opposed to them."

The exhibits from foreign nations comprised one of the major features of the Centennial. Great Britain and her dependencies sent more than three thousand exhibits; Spain and her colonies sent about the same number, while Portugal came next with twenty-five hundred exhibits. Other countries, such as Germany, France, Russia, Belgium, Norway, Sweden, and a number of South American and Asiatic countries, also sent exhibits. Placed side by side for comparisons, these industrial exhibits of the world, which included every product known to science and art, served as an incentive to bring forth new and useful inventions.

The Centennial of 1876 marked the first time in our nation's history that vast numbers of foreign artists, inventors, scientists, and industrialists flocked to our shores in the belief that they might possibly learn something that could be of value to their own countries.

Close of the Centennial. The Centennial Exhibition closed November 10, 1876, during the nationwide excitement over the disputed presidential election. Interest in the Exhibition had been crowded off the front pages of even the Philadelphia papers by headlines relating to the Hayes-Tilden disputed election. President Grant, in officially closing the Exhibition, declared that it would be of enduring advantage to the American people. He spoke again of the rapid progress in the arts, sciences, and mechanical skills made by our people in a single century. "While the United States may be a little behind some of the older nations in one branch," he concluded, "in others we scarcely have a rival."

The Chicago Exposition. The Columbian Exposition held at Chicago, 1893, served both a national and an international purpose. The tens of thousands of American citizens who visited the fair were awakened to the wonders of the world and to a realization of the civilizations of other lands, as well as an appreciation of what our own nation was achieving. Representatives from other countries witnessed a fuller display of American achievements than was possible in any of the exhibitions held on their soil. They marveled at the American organization in staging and directing such an exposition. For the first time they were brought to a realization of the vastness of America's west, and the growing power of the young nation.

Electrical Exhibits. The dominating feature in the realm of science and technology at the Chicago Exposition was the use of electricity, both as a source of light, and as a source of power (see Chapter 27). The grand opening of the Exposition began with President Cleveland "waving a magic wand, gently pressing a button, which set in motion all the ponderous machinery, the spray of the fountains, and the lighting of the Great White Way."

The marvels of electrcity were demonstrated on every hand. It was used to light streets and buildings, to propel cars and locomotives, and to operate hundreds of machines. It supplied the power for operating the huge fountains, for the elevators used in many buildings, and drove the sewage of the Fair grounds toward the lake.

Electrically operated clocks were displayed in a number of buildings. One central clock controlled all the others. Each day, at the hour of noon in Washington, D. C. (11 o'clock in Chicago), a signal was sounded from the Naval Observatory, and instantly the electric clocks in the Exposition buildings "gave forth their music on a chime of bells." There were program clocks, used for starting trains and for ringing bells at regular intervals—the forerunner of a signaling system that was soon to be installed in school classrooms and in factories.

The electrical feature that attracted most attention of the sightseers was the illumination of the grounds and buildings, the fountains and waterways, and the play of searchlights, "a more striking illustration of the wonders wrought by this science than any mere collection of machinery."

Machine Exhibits. The exhibits of American machines, however, vied with the electrical displays in popular interest. The Hall of Manufactures and Liberal Arts, as well as the Palace of Mechanic Arts (commonly called Machinery Hall) were crowded with the world's most modern machines, with those from the United States predominating. The more than four-score groups and two hundred classes of exhibits represented "almost every mechanical device fashioned by the ingenuity of man." Machines for transmitting every kind of power—electric, steam, hydraulic, pneumatic, wind and water power—were exhibited.

There were also machines and appliances for preparing and analyzing various articles of food. There were machines for printing, typesetting, binding, stamping, embossing, and for other branches of bookmaking and newspaper work. Machines, apparatus, and tools for lithography, color printing, photomechanical processes, and for illustrating, attracted vast numbers of sightseers (see Chapter 28). Machines for working metals, mines, and wood, and machines and implements for miscellaneous purposes, ranging from shaping the head of a pin or the eye of a needle to the construction of a watch were of special interest. Other exhibits that signified recent technological and engineering advances showed an electrical appliance measuring the velocity of missiles fired from selected guns, operated by pressure gauges; also, machines that used oil for fuel (the oil being brought by pipeline from nearby Whiting, Indiana); a novel exhibit of hoisting machines, distinctly American; and machines that showed new developments in refrigeration.

Following the close of the Exposition, editorials, mostly praiseworthy in nature, appeared in the larger daily papers of the country. One writer summed up the effects of the Fair rather well, when he said, "All the contrivances of civilization, ancient and modern, that helped to elevate and ennoble man, refine his tastes, enlarge his ideas, and further his deliverance from the despotism of nature, have been displayed before millions of visitors." That seemed to be the universal note. The gathering of men and their products from all quarters of the globe had quickened their intellectual and creative interests. It had provided a stimulus for greater scientific and technological achievements at home; and for visitors from less developed countries, the Exposition had also pointed the way for improving their economic and social conditions and raising their standards of living.[2]

The Paris Exposition, 1900. The last Exposition of the nineteenth century in which the United States participated was held in Paris in 1900. The major exhibits sent from this country consisted of both light and heavy machinery, the latest models in agricultural machinery, and machines used in mining and in textile manufacturing. While France and her colonies contributed more than one-half of the total exhibits, the United States' exhibits ranked second in number, with a total of more than six thousand.

The American exhibits were so numerous that they could not all be housed in Machinery Hall, erected on Champs de Mars. The overflow exhibits were assigned to space set aside in special buildings that had been erected at the Vincennes Annex. The prize exhibits included model iron bridges, drilling machines, rock drills, tools operated by compressed air, motors, pneumatic riveters, cutting tools that retained their cutting edge at visible red heat "presenting a novel and a remarkable phenomenon," declared the Jury of Awards. Of the total awards given, the United States received six hundred and forty-two, second only to France.

As the nineteenth century drew to a close, the United States was making plans to hold two more expositions—the Pan-American Exposition in Buffalo in 1901, and the Louisiana Purchase Exposition in St. Louis in 1903–1904.

American-European Technology Exchanges

Another evidence of the nation's scientific awakening during the last quarter of the nineteenth century, was the interest of American scientists in the progress of science in Europe, and also the visits to this country by noted European scientists. Americans were especially interested in the great new scientific thought that was developing in Europe—in Thomas Huxley, Herbert Spencer, Clark Maxwell, Lord Kelvin—and in the curricula of the great universities in England, in Paris, Berlin, Göttingen, and Heidelberg. Listed in the group of Americans who went abroad were students and young instructors, who later took up administrative and teaching positions in American colleges and universities.

Exchange of Scientists. Included in this group was Josiah Williard Gibbs who became mathematics professor at Yale, and who later announced his first and second laws of thermodynamics, which laid the basis of a new concept of physical chemistry. There were two others who became university presidents, Andrew Dickson White, first president of Cornell University, and Daniel Coit Gilman, first president of Johns Hopkins. White sounded a new note in American education after returning from an extended trip to the universities in Europe. Speaking at the opening exercises at Cornell University, he declared that, "science will find an asylum here where agriculture, engineering, and the mechanical and technical arts will be taught along with moral philosophy, history, and political economy." Gilman, in his opening address at Johns Hopkins University, proclaimed a new philosophy in American education when he announced that scientific research would be the "soul of the whole institution."

Others sought out the new fields in chemistry that were being developed abroad, especially in its application to agriculture. At Göttingen, Germany, Charles A. Goessman, himself a student of the famed Frederick Wöhler who had attracted an earlier generation of students, was sought by many who became leaders in the American Chemical Society. Among those were Francis H. Storer, who became professor of chemistry at Harvard, and George C. Caldwell, later head of the department of chemistry at Cornell. Nathaniel T. Lupton, likewise a leader in agricultural chemistry, studied under Bunsen at Heidelberg, and became professor of chemistry at Vanderbilt. Eugene W. Hilgard and Wilbur O. Atwater, after completing their studies, took up work in agricultural stations in this country; Hilgard at California Experimental station, and Atwater at Connecticut. Outstanding as students at European centers of learning were

Stephen Babcock, who developed the famous Babcock test for milk in this country (see Chapter 24), and Benjamin Smith Lyman, who later became mining engineer and geologist for the governments of India, Japan, China, and the Philippines.

While American men of science were going abroad, European scientists were coming to this country. English scientists, particularly, began visiting America, where, for the first time they found appreciative audiences.

Tyndall's Lectures, 1872. John Tyndall, noted British physicist arrived here in 1872 to give a series of lectures on the subject of "Light." After lecturing in Boston and Philadelphia, he went to New York, where, long in advance of the opening hour of his first lecture, every one of the fifteen hundred seats in Cooper Institute Hall had been sold, many of them to women. Despite the worst blizzard New York had seen in years, with zero weather and a stoppage of all public transportation (said the New York *Tribune*), enthusiastic crowds turned out to hear every Tyndall lecture.

Interest in Tyndall's lectures on the "Physics of Light" was heightened by the use of extensive illustrative apparatus. "He is a man who can popularize science," said the *Nation.* Tyndall, himself was delighted with the ovation given him. In his closing lecture he told his listeners that their presence during the six nights of inclement weather, "embodying as you do much of the mental force and refinement of this vast city is an answer to my question on the reward for studying science." He urged his listeners to take science "not so much as a servant of mammon," but "as the strengthener and enlightener of the mind of man." He concluded his lectures by declaring that scientific knowledge would soon become available for the needs, comforts, and luxuries of America's civilized life.

The effects of Tyndall's lectures were many. The New York *Tribune* sold over fifty thousand copies of its special supplemental pages containing full reports of Professor Tyndall's lectures, in addition to the large sale of the regular issues in which the lectures were first reported. "The People Have A Wonderful Appetite For Science Just Now," read a banner line. More than one hundred thousand copies of his lectures were sold and read in this country with avidity. Joseph Henry believed they had done more to give precise and definite knowledge of the principles of the sciences they treated than any other series ever published.

Tyndall's profits, beyond his personal expenses, which amounted to 2,500 pounds (British), were assigned to a trust fund to provide science scholarships in Columbia, Harvard, and the University of Pennsylvania for study abroad. The declared purpose was to advance studies in science and to promote original research, particularly in the department of physics. When Tyndall returned to England, he was quoted as saying that "mechanical ingenuity engages a greater number of minds in the United States than in any other nation in the world."

Thomas Huxley's Visit, 1876. Four years after Tyndall's visit another distinguished British scientist, Thomas Huxley, arrived in the United States. Despite his announced aim that he was coming only on a pleasure trip, he was from the start besieged with numerous invitations to lecture, even though some accused him of being an atheist. At Nashville, Tennessee, where he visited his sister, he was amazed by the amount of publicity he received. Crowds gathered about his sister's home and followed him on his visit to the state capitol, and to Vanderbilt University. In a lecture, given from the steps of the capitol, in which he explained his views on the science of evolution, he received a hearty ovation. In order to make himself heard he had to shout so loudly that he injured his voice, an injury which followed him in his later lectures in Baltimore and in New York.

Speaking in Baltimore at the dedication of Johns Hopkins University, Huxley urged the introduction of more courses in science, saying that professors would encounter no difficulty in giving elementary instruction in physics, chemistry, and in the elements of human physiology. He urged the secondary schools and academies also to introduce courses in science. Shortly thereafter, one of Huxley's assistants, Professor H. N. Martin, introduced a department of biology in Johns Hopkins University.

It was in New York, however, that Huxley received his warmest welcome. The *Scientific American* declared that New Yorkers were as familiar with his works as were his own countrymen back in England. He was acclaimed as one whose lectures and essays had placed him among those scientists who were influencing the intellectual thought of the western world. Huxley gave three lectures on "Evolution."

Herbert Spencer's Visit, 1882. The third member of the distinguished scientific trio was Herbert Spencer, regarded by the editor of *Popular Science Monthly* as "the most eminent thinker in the world." Although in poor health, Spencer was determined to come to the United States primarily to observe America's industrial and technological development with the view of writing a book about it. Spencer was more loudly acclaimed in America than in his native

England. His expressed belief in man's material and social progress captured the enthusiasm of the American people. His whole philosophy was summed up in the statement that "science and education would emancipate the human spirit from the thralldom of ignorance and superstition, which had darkened men's minds for ages." His ideas were widely acclaimed by a number of American educators.

In an interview released to the press prepared by Professor E. L. Youmans, Spencer spoke of the unparalleled progress of America, which was ahead of all other nations in mechanical appliances. "Americans," he continued, "have in their very faces a great determination, a kind of 'do-or-die' expression, and this trait of character is combined with a power of work exceeding that of any other people." But, he warned his readers, that was not enough. If along with this material progress Americans would develop a progress of a higher kind (which he neglected to define), there would remain nothing to be desired by the people of this young nation.

This same theme was enlarged upon at the farewell dinner given in his honor at Delmonico's in New York City, attended by several hundred guests. After listening to toasts by William M. Everett, who declared no banquet hall could hold all Spencer's admiring friends; by Carl Schurz, who hailed him as the builder of science, the great apostle of evolution; and other eulogies by William Graham Sumner and Oliver Wendell Holmes, Spencer admonished his friends for working too hard, "wronging themselves and their neighbors into too much work and worry." After reminding his listeners that they had succeeded in educating people in the gospel of work, he declared that it was now time to practice the gospel of relaxation.

Long after Spencer sailed for England the newspapers and journals of science continued to publish articles describing the work not only of Spencer, but also of Huxley, Tyndall, and other men of science whose writings were read by an increasing number of people. A new generation of Americans had arisen, interested in keeping informed about developments other than politics and wars.

Export of American Machines. In addition to the exchange of American and European scientists, Americans skilled in the operation of American machines, together with their machines, were going abroad also. American shoe machinery with skilled supervisors were sought by a number of shoe manufacturers in Germany during the 1890's. A manager of a plant in Breslau induced a skilled American shoe-machine operator and his wife to go there, take charge of his factory, the largest in Germany, and install American methods.

The factory employed more than four hundred workers. It was entirely equipped with American-made machinery—McKay's, Millen's, and Rockingham's—and with Goodyear welt-making machines. Before the American took charge, the factory was producing one hundred forty-two pairs of shoes a day. Within two weeks, the same workers were producing more than four hundred pairs a day, and the pay of the workers was increased by forty-four cents a day.

Also in Breslau, a number of American technicians were being employed in phonograph factories. American engineers were employed in supervisory positions in the ironworks at Laurahütte, and an American firm was engaged (1898) to install electrical equipment in a factory in Kattowitz. The American consul in Breslau reported in September, 1899, that "American machinery of all descriptions is being introduced throughout this district."

Dissemination of Scientific Information

A further recognition of the role of science was evidenced in the periodicals of the country; there was not only an increase in the output of scientific publications, but also in the recognition given to science in the so-called general magazines. The first official cognizance of periodicals and newspapers in United States history appeared in the Census of 1860. Thereafter, each subsequent census reveals the rapid increase in publications.

Popularized Science Periodicals. While, in 1860, magazines devoted to science and technology were not differentiated, newspapers and periodicals were recognized as having become "the popular educators of the day," and as being "vast depositories of discussions and information of all topics that engage the thoughts and enlist the activities of men." Of the 4,051 publications listed as political, religious, literary, or miscellaneous, scientific journals were included in the latter class—an estimated two hundred in all. The Census of 1880 went much farther. Not only were the scientific publications recognized, but they were also classified as science and mechanics, agriculture and horticulture, medicine and surgery, railroads, law, and trade journals. The Census of 1900, retaining the same classifications, revealed an increase in publications during the preceding ten years, varying from 70 to 100 per cent.[3]

Improvements in Scientific Publications. With the great increase in the quantity of scientific literature being diffused, there was likewise a great improvement in the quality. In the earlier publications there was no room for articles describing the advances in pure science. American scientists were forced to turn to European publications for scientific information; and here, too, they found a medium for publishing their own discoveries.

Articles on popular science and applied science continued to occupy a large field in American publications, but with the rise of professional journals and publications of scientific societies, an adequate repository was afforded for American scientific findings. Then too, indexing services developed, enabling the reader to find quickly the information in his field. By 1900 periodical publications had so developed that not only were the popular science magazines being read even by the mechanic's helper, but also many of the journals became the source books for scientific literature. This journal literature, presenting the independent pieces of investigation, became, when classified and verified, the books of literature of the scientific world.[4]

Scientific Articles in Literary Journals. A number of literary magazines increased the popularization of the scientific articles they carried by their wide circulation, due to low cost. Before 1900, *Munsey's, McClure's,* and *Cosmopolitan,* because of improvements in mechanical production, sold at ten cents a copy. Advances in printing technology, in machine composition, in making illustrations, in presses, and in stitching and binding (see Chapter 28) had made it possible for daring editors to put out these excellent magazines at a trifling cost. In 1900 the combined circulation of these three magazines, with the *Ladies Home Journal,* reached almost 2,-500,000 copies per issue.

Harper's began carrying articles bearing such titles as "Making Watches By Machinery" (July, 1869), "A New Theory Of Heat" (August, 1869), and "The Progress of Electricity" (September, 1869). Other magazines such as *Galaxy,* which became the *Atlantic Monthly; Lippincott's,* and *North American Review,* discovering a growing interest in science, devoted space to the new developments in these fields. *Lippincott's,* founded in 1868, informed its readers in its very first number that articles on science would appear regularly in each issue. While this magazine carried a number of articles on Darwin's *Origin of Species,* it also gave generous space to articles on applied science and technology. *Scribner's,* published a report in December, 1880, authorized by Edison, substantiating his revolutionary invention in electric lighting. It was observed that Edison's seemingly inexhaustible store of wonders had made him "the theme of every tongue."

Technical Journals. In addition to the literary magazines that featured scientific developments, there was a phenomenal growth of scientific and technical journals. Listed in this specialized group were the *American Naturalist, Popular Science Monthly, Western Review of Science and Industry, American Chemical Journal, Journal of Applied Chemistry,* the *Botanical Gazette,* and the *American Machinist.* A number of others were devoted to such subjects as ornithology, microscopy, meteorology, archeology, and photography.

The *Engineering and Mining Journal,* established in 1866, in its early issues reveals the unparalleled growth of the mining industry. It pictured profusely the new tools and new machines that were being introduced to supply the fuel needed by the expanding industries throughout the country.

Engineering Journals. More than a dozen magazines relating to new developments in the field of engineering were established during this period, including such publications as *Engraving Magazine, Engineering News, Electrical Engineering, Electrical World, Technologist, Railroad and Engineering Journal,* and *The Street Railway Journal.* These magazines not only described the developments in the different fields of engineering, but also spoke of the growing influence of the engineer himself. "The glowing light of his [the engineer's] usefulness which illuminates the hilltops and the valleys," declared one writer, "will bring comfort and cheer into millions of American homes."

Leading Popular Magazines. Of all the scientific journals, however, the one that enjoyed the largest circulation, and the one that did most to create a public interest was the *Popular Science Monthly,* founded in 1872, by Edward Livingston Youmans. Youmans' primary aim was to popularize science. Technical, scientific treatises were explained in layman's language, in terms which the average reader could understand. This magazine had a phenomenal circulation, and thousands of Americans were for the first time able to keep abreast of the latest discoveries and inventions affecting their everyday life. It was established especially to provide a medium for the publication of the works of Herbert Spencer. The *Scientific American Supplement* was established in 1876 to report the technological and scientific displays at the Centennial Exhibition. It became a veritable "gold mine" for students working in the field of science and technology.

The "House Organs." Another type of magazine, often called the "house organ," sprang up in various parts of the country. Aimed at increasing the sales of special products and promoting good will, these small magazines spread rapidly. The *Fall River Line Journal,* published by the steamship company of that name, has appeared continuously since 1876. Other companies and industries adopted the practice. Some of these magazines contained such valuable historical material that they were indexed, bound, and placed on file in libraries.[5]

The scientific and technical magazines and periodicals introduced a new feature in American journals. They described more vividly than any other publications the major developments in science and technology. In their files could be found the history of iron and steel, of agriculture, of building, of transportation, mining and manufacturing, of coal, oil, and electricity. In short, these magazines and journals of science supplied a major part of the economic, industrial, and technological history of the nation. But they did more than record scientific progress. By the turn of the century, they were instrumental in promoting new ventures in the fields of applied science and technology.

"The Second Industrial Revolution"

The last quarter-century had witnessed a popular, nationwide interest in science. Expositions, both at home and abroad, had drawn millions of Americans to visit and view the products of the world's leading scientists and inventors. American scientists, impelled by an urge to learn about the latest developments in Europe, started going abroad to visit the leading scientists in England and on the Continent. Likewise, noted English scientists began paying visits to America, where they found larger and more enthusiastic audiences eager to hear their lectures than they had at home. Public interest in science was also promoted through the literary magazines, scientific, engineering, and professional journals that had sprung up during the post-Civil War years.

Despite the long years of political apathy and stagnation in progressive government that followed the Reconstruction period, the scientists of the nation, aided by the introduction of new technologies in many of the basic industries, ushered in an era often characterized in American history as the Second Industrial Revolution.[6]

Notes for Chapter 21

1. For a contemporary account of the United States exhibits, see Henry Morford, *Paris in 1867* (1867).
2. Hubert Horne Bancroft, *The Book of the Fair,* 4 vol., illustrated, IV (1893), 400.
3. The classification of the different types of periodicals in the 1880 Census (covering the preceding ten years) revealed one hundred and seventy-three publications devoted to agriculture, sixty-eight to science and mechanics, and a number to other scientific interests. There were two hundred and forty-eight trade journals—with every

major industry in the country having one or more journals devoted to its special interests—reviewing the latest processes and techniques in each special field. Although some of the journals could hardly be called scientific, they served an increasing number of readers who liked to believe they were scientifically minded.

4. See E. J. Crane, *Literature of Chemistry* (1927).

5. The author is indebted to E. H. McClelland, former librarian of the Technology Division of the Carnegie Library, Pittsburgh, Penna., for this thought.

6. It is the author's belief that America's second industrial revolution did not come until the mid-twentieth century: the age marked by the development of atomic power, electronics, and automation (see Chapter 42).

Chapter 22

THE AGE OF STEEL

Iron and steel are basic to the development of all major American industries. The outbreak of the Civil War had found the United States an iron-hungry nation. Uses for iron had increased beyond all proportion to the ability to produce it. Iron was needed for stoves in increasing quantities, for telegraph wires, and for hoops on barrels used in the new petroleum industry. It was also needed in the manufacture of sewing machines, which were being exported to many foreign countries. It was needed for buildings, for facades, and even for structural members. It was in demand for steamships, and some of the iron manufacturers predicted that, with the destruction of the Turkish fleet at Sinope in 1853, iron would be needed for armor plate.

Iron was also taking the place of stone in building bridges that spanned the creeks and rivers of the continent. But, the greatest demand for iron came from the railroads. Not only were heavier rails needed, but iron was also needed to replace the wooden parts of earlier locomotives. Likewise, there was an unlimited demand for rivets and spikes, all to be made of tough iron.

To meet these demands almost one million tons of iron were being produced in the United States in 1860. This amount was being manufactured by more than two hundred and fifty establishments into cast iron, forged iron, and puddled iron. But more iron was needed than the puddler's furnace or the forge and foundry could turn out; moreover, iron had to be produced at a lower cost.

To add to this shortage came the increased demands of the Civil War for weapons, projectiles, and the countless uses a government at war has for iron. Tracks and railway equipment wore out rapidly. The battle of the "Monitor" and "Merrimac" had clinched the argument in favor of iron for vessels and refuted the objection against "unnatural material" raised by the opponents.

Major Developments in Iron Technology

To satisfy the demand for iron and steel created by the war and the postwar period, the next quarter-century witnessed a number of technological developments in the iron industry. The most significant developments were the introduction of a new fuel, coke; the utilization of new ores; the development of new technical processes; and the establishment of a new system of scientific management.

Fuel Requirements. In the process of converting iron ore into pig iron, fuel was a major factor. Charcoal had been used for centuries in the manufacturing of iron, but as the demand for iron and steel increased, forests were becoming depleted. Coal, both hard and soft, had been used with some degree of success, but the search continued for a fuel that would produce even greater heat. The answer was a product called "coke."

Coke was produced by dumping tons of soft coal into brick ovens built like beehives, applying heat, and burning out the impurities, which escaped in the form of gas. The residue was called coke, and when fired it produced far greater heat than the raw coal. As early as the 1840's a limited amount of coke was being produced in the coal fields of southwestern Pennsylvania and was used in blast furnaces for manufacturing pig iron. The practice did not become general, however, until after the close of the Civil War.

Pittsburgh, Center of Iron and Steel Industries. The discovery of an excellent grade of coal in the Connellsville region, near Pittsburgh, brought about the removal of the iron industry from the eastern part of Pennsylvania to Pittsburgh, where rolling mills and puddling furnaces had already been established. With the coming of the blast furnaces, the advantages of assembling all stages of iron manufacture in one community were realized. These developments marked the passing of the isolated iron forges and furnaces and paved the way for the greater concentration and the large scale production that was to follow.

Henry Clay Frick, 1849–1919. Henry Clay Frick entered the Connellsville coal fields of southwestern Pennsylvania in 1871 and began large-scale production in converting coal into coke. He bought up extensive coal lands and built thousands of beehive ovens. He introduced machines for drawing the coke from the ovens, an arduous task when performed by hand labor. From his ovens, trainloads of coke were shipped to Pittsburgh iron and steel mills.

Iron Ore Deposits. The utilization of new ores was likewise effective in bringing about a change in the industry. Swedish iron had been considered superior to American iron, not because of superior technical processes or even to superior ores, but rather to the uniformity in the quality of the iron ores. In the United States the ores were of varying composition, and the quality of the product was unpredictable. Iron ore from Pilot Knob and Iron Mountain in Missouri was considered of excellent quality, but the supply was limited, and by 1890 those mines had been depleted. The opening up of what appeared to be an inexhaustible source of iron deposits in the Lake Superior district, site of the famed Mesabi range, was the second factor that effected the revolution in the iron industry.

The story of the development of the Lake Superior region is one of the sagas typical of American business. The investment of capital in that remote, inaccessible region was evidence of the risks often taken by venturesome Americans. The technological methods devised for handling the freight over its long haul was a challenge to the railroad and shipping magnates. The organization required to control the several operations of mining, shipping, and loading was an example of efficiency of American methods and of the large-scale operations that later characterized American industry.

Mines had been opened up in the rich Marquette range as early as 1845, but because there were no roads the ore was taken out only in winter when sleds were used. A ton and a half of ore was carried on a load, and one trip a day to the sailboat on the lake was the maximum. The ore was then transported on Lake Superior to the falls at Sault Sainte Marie, carried by portage, and reloaded on vessels bound for Erie, Pennsylvania. From there, canal connection brought the ore down into the region of the blast furnaces. The Sharon Iron Company, Sharon, Pennsylvania, placed an order for Lake Superior ore as early as 1853.

The demand for ore became so great that new developments in mining technology and in transportation technology were developed for handling it. In 1856 the federal government constructed a canal at the rapids at Sault Sainte Marie, and a privately owned steam railroad was built to haul the ore from the mines to the lake. Following the end of the Civil War, Robert Wallace devised a hoisting machine capable of unloading a barge of ore in one day's time—a four-day job by the arduous wheelbarrow method. Next, lake transportation was expedited when R. J. Hackett built bulk freighters for use in transporting large quantities of ore by way of the Great Lakes.

New Mining and Transport Methods. With the opening up of the Lake Superior ore fields new developments in mining technology were introduced. The miners, mostly Englishmen and Scandinavians, were lowered to the mining areas, as these became deeper and deeper, in steel cages instead of climbing down and up slippery ladders. Steam and electricity propelled tramcars that supplanted the wheelbarrow in taking the ore to the shaft. Steam shovels took the ores from the stock piles to the shipping cars. Rock drills took the place of the picks of the earlier days; and hoisting machinery and cranes carried the heavy loads.

Ore ships began to ply the Great Lakes in increasing numbers, and by the mid-1880's, three times as much tonnage was passing through the Sault Sainte Marie Canal as through the Suez Canal. The ore ships were built of iron, later of steel, and propelled by steam power. New types of steel freighters were designed. In the late 1880's Alexander MacDougal introduced his "whale backs," long and low in construction, and designed for increased carrying capacity and simplicity in loading. These continued to be made longer, first 300 feet, then 400 feet, and by 1900 they had reached 500 feet in length. The technological problem of transporting the iron ore from mine to mill, however distant, had been solved.

Development of Steel Production

Accompanying this increase in the supply of iron ore was the introduction of the new Kelly-Bessemer process for making iron in larger quantities. This process produced an iron that was different from the type which came from the puddler's furnace or the blacksmith's forge. Also, it differed from the new crucible steel produced in the pots that Joseph Dixon had designed. The new product was called "steel," and to differentiate it from cutlery steel, it was termed "mild" steel.

Kelly's Air-blown "Pneumatic" Metal. The story of this revolutionary development in iron making began with William Kelly. In 1842 Kelly moved from Pittsburgh to Kentucky, where there was an abundant supply of timber which he used for charcoal to make iron sugar kettles. Within a few years, however, the local timber supply was exhausted, and Kelly began experimenting in a new way to make iron without fuel. He knew something of the chemistry of iron and devised a method whereby the burning of the excess carbon in the iron produced the heat to purify it. He did this by making use of the affinity of carbon for oxygen. A blast of air introduced into the pig iron oxidized the carbon, silicon, manganese, and other impurities, which were volatized by the heat developed in this oxidization process. Kelly called this process "pneumatic," or "air-blown." [1] He built six or seven furnaces, or converters, "to convert pig iron into steel without any fuel save cold air." [2]

In England, Sir Henry Bessemer's discovery in 1854 of the same principle resulted from some experiments in trying to produce a superior metal for cannon so that it could fire a projectile he had invented. Bessemer's added contribution, the converter, was made with holes in the bottom for the admission of the blast.

Bessemer Converter. A Bessemer converter has been described as resembling a great iron egg. It was a pear-shaped, open-topped furnace, supported on standards or trunnions, and attached at the top to a pinion that enabled it to swing at an angle of 180 degrees. The bottom of the converter was perforated by ten to twenty circular holes, into which was inserted a conical fire-clay tuyère, or nozzle, for the admission of air to the furnace. Pig iron was poured into this furnace in the molten state. The air blast was driven in through the holes in the bottom, and impurities in the iron ore were burned out in a shower of sparks. Then the converter was tilted to discharge the molten steel.

The blowing of a converter was a spectacular performance. At the beginning, only sparks and a little brown smoke were seen, indicating that the silicon in the iron was burning. Then the oxygen from the air blast turned its attention to the carbon. A flashing flame appeared which grew rapidly until it filled the mouth of the converter and dazzled the spectator. Tiny meteors rained down upon the workers. The temperature rose quickly, and the roar became deafening. Then a cloud of brown smoke arose. The iron continued to burn and the skilled operator, sitting on a platform and wearing goggles of specially prepared glass, knew exactly when to pull the lever to shut off the blast.

Special skill was required to determine the exact second for shutting off the blast. If it were allowed to remain too long the iron was "burned," and the batch was spoiled. In the heyday of Bessemer production, the workman on this job was rated superior to his fellow workers and often received ten dollars a day.

First "Bessemer" Steel. In 1862 the Cambria Iron Works of Johnstown, Pennsylvania, employed Kelly to build a special converter, "the first of its kind ever built for the purpose of making steel," according to Boucher.[3] Beyond demonstrating that this new converter would work, no further use was made of it. Two years later, the Kelly Pneumatic Process Company, headed by William F. Durfee, established a pilot plant in Wyandotte, Michigan, where the Kelly process was introduced, and built a converter of 2½ tons capacity. In September, 1864, the company succeeded in making steel, which, says Swank, "was the first Bessemer steel made in the United States." [4]

Simultaneously, Alexander Holley of Troy, New York, using Bessemer's patent, produced steel in an experimental way in 1865. It was in August, 1867, that the first industrial application of the process was utilized when 2½ tons of steel rails, the first steel rails ever rolled in this country upon order, were made by the Cambria Works in Johnstown, Pennsylvania.

This accomplishment led the Pittsburgh *Chronicle* to state (October 26, 1867): "This country has thrown off another shackle that has hitherto bound it to England. Bessemer steel rails may now be purchased from the Cambria Iron Works." The announcement pointed out that the old iron rails, which lasted about three years, would now be replaced with steel rails that would last twenty years.

Thomas Gilchrist's Method. For a period, America lagged behind England in adopting the Bessemer process. Our ores, due to a high phosphorous content, were not considered suitable for conversion into Bessemer steel, and for a short period the plants using this process imported pig iron from Europe. The adoption of the principle of lining the converter with lime to neutralize the phosphorous, a method introduced by Jacob Reese of Pittsburgh in 1866, partly solved this problem. This method was improved upon a few years later when the Thomas-Gilchrist method, somewhat similar, was brought from England. Later, when John Fritz of the Bethlehem Iron and Steel Company, with the aid of Alexander Holley, built large (for that day) 7-ton converters, the final steps for producing Bessemer steel were completed.

Manufacturing Steel Products

The ingot castings as they came from the Bessemer converter were ready for fabrication into the many needed steel products in the various mills. This was done either in the original steel-making plant or in a separate establishment. They were taken to the foundry where they became castings, or they went to the forge where huge hammers worked them into the desired forms, or they went to the rolling mills where they were flattened into blooms, billets, and slabs, and then rolled into rails, sheet bars, rods for wire, skelp for pipe, or rolled strip from which plate was made. America's steel products broke into the world market during the mid-1870's with steel saws, shovels, frames, and locomotives. A number of later improvements were made in the Bessemer process, but once its efficiency was established, there were no revolutionary innovations.

Improved Blast Furnaces. Accompanying the introduction of the Bessemer process, with its speedy conversion of iron into steel, came a demand for greater blast-furnace production. Blast furnaces are the genii of the steel industry, for it is within these furnaces that the raw materials of iron ore, coke, and limestome are transformed into pig iron. The first improvements over the old hillside furnaces were introduced from England by Thomas Wetherbee of Fletcherville, New York, in 1866. While operating a charcoal blast furnace, he introduced a bell and hopper construction to facilitate charging. A steel jacket with firebrick lining was substituted for the old wall masonry construction. Furnaces of greater diameter and height were built.

Andrew Carnegie, a name later synonymous with steel, had entered the iron industry in 1865 and made his first contribution by improving the blast furnace. He erected two huge blast furnaces in the Pittsburgh area and employed the most talented men he could find to operate them. Carnegie was one of the first to introduce scientific research in blast-furnace production. He brought a German chemist, F. G. Fricke, to "doctor" the operation of his furnaces, observing that while he could not afford to hire a chemist, yet he could not afford to be without one. "Nine-tenths of the uncertainties of pig iron making were dispelled," declared Carnegie, "under the burning sun of chemical knowledge."

Gayley's Dry-air Blast. The aim in blast-furnace improvement was to increase the output without increasing fuel costs. Furnaces were built larger, some of them stood 100 feet high and could produce 100 tons daily. By 1900 an average of 200 tons a day was not unusual. More powerful blowing machines were devised. Steam was introduced to force a terrific blast through the tuyères. Improvements were made for mechanically handling the raw materials, and hoisting machines did the work of human labor. The chemistry of blast furnace operation was further improved when James Gayley, "Father of Modern American Blast Furnace Practice," devised the dry blast, in which the air was dried by refrigeration before it was drawn into the furnace. This was an improvement over Nielson's hot blast of a generation earlier. Bronze cooling plates for the furnace walls, a sheathing of water where the heat was most intense, a compound condensing engine for supplying the blast, and electric power for operating the mechanical equipment, all increased the efficiency, and thereby the output. At the same time operating costs were decreased.

Large-scale Steel Production. While the basic principles that were to develop into America's mighty steel industry had been established by the early 1870's, production figures had not yet begun to soar. In 1872 only 2 per cent of the pig iron made in the country was converted into Bessemer steel. As late as 1877 more rails were made of wrought iron in the old way than were made of Bessemer steel. Moreover, railroads continued to import steel rails. A number of improvements were still needed before the nation could hope to become a leader in the steel industry. Plant mechanization and plant management remained to be effected. More efficient manufacturing techniques in hundreds of new products had to be established, and the application of chemistry as a factor in steel manufacture was yet to be fully recognized. A new generation of industrialists, engineers, and scientists was about to usher in a new era in the nation's industrial history.

Carnegie Revolutionizes Management. Andrew Carnegie, more than any other individual laid the foundation for efficient organization in the iron and steel industry. Carnegie brought together the vast mineral resources of the Lake Superior region and the new technological processes by which iron was made into steel, and then transformed them into finished products. He bought up ore mines, coal fields, lake steamers, the whole of Conneaut Harbor in Ohio, and the railroad line connecting that harbor with his Pittsburgh plants. He authorized old plants to be torn down and new ones built as improvements were introduced. Although not always a pioneer in new manufacturing techniques, yet when the value of a process or a piece of new equipment was proved to be superior, he spared nothing to secure it.

Carnegie declared that the nation that made the cheapest steel would, within a few years, "have all other nations at its feet." That was his creed—produce steel cheaper than any one else in the world. As prices fell, profits grew. Carnegie had little direct participation in the daily management of his own companies, but he knew how to pick men as executives, superintendents of mills, operators, and managers, to whom he delegated all special tasks. Equally significant was Carnegie's decision to employ expert chemists and metallurgists who, because of their scientific research and laboratory techniques, were able to improve the quality of steel and at the same time produce it more cheaply. When all these factors were brought together in the Carnegie Works in Pittsburgh, that city became the nation's steel capital. Here was the largest Bessemer plant, the largest crucible plant, the largest steel-rail mill, and the largest steel-freightcar works in the world.

Chemistry and Metallurgy in Steelmaking. While new machines, new technological processes, and efficient organization contributed greatly to America's supremacy in the steel industry, the contributions of the chemists and metallurgists were equally important, since the making of iron and steel is a chemical process. The steel furnaces and converters are in reality huge test tubes where the ingredients, according to formula, are transformed by chemical changes into a new metal product. In the early days, the making of steel was an empirical process, and determining its composition was a craft, not a science. Gradually the rule-of-thumb practices were eliminated, and skilled artisans relinquished their positions to research scientists in the laboratory.

Industrial and Educational Research. Scientific research became the basis for steelmaking. Colleges and universities introduced courses in metallurgy and chemical engineering. Young graduates trained in science found good positions in the steel industry. Fred Wood, a graduate of M.I.T., made a name for himself at the Pennsylvania Steel Company at Steelton, and built the Sparrows Point plant in 1887. Robert Hunt, a pioneer chemist in the Cambria Steel Company, opened up a consulting laboratory in 1888. One of the earliest commercial laboratories was established in Pittsburgh by a German scientist, Dr. Otto Wurth, and his associate, S. G. Stafford. Homer D. Williams, a metallurgical student at Lehigh University, began as a chemist at the Joliet works of the Illinois Steel Company and later moved up to the presidency of the Carnegie Steel Company. William E. Corey, employed in the metallurgical laboratory of the Edgar Thompson works in Pittsburgh, came up through the ranks to become president of the giant United States Steel Corporation. Albert Sauver, hailed as the dean of American metallurgists, pioneered in the study of steel under the microscope. He tried, unsuccessfully, to persuade President William Rainey Harper of the University of Chicago to inaugurate a course in metallurgy, nor could he interest William E. Corey of the United States Steel. As a result he established an independent laboratory and was soon training a number of metallurgists for the steel industry.

Organization and Equipment Growth. It was in the organization of the steel plant itself, however, that the greatest technological improvements were made. By 1900 a modern Bessemer plant was marked by a vast array of machinery and relatively few men. Mechanical workers rose from the floor, reached out from the walls, and clung to the ceiling. Some were stationary, while others traveled at great velocity. Steam engines operated belts and pulleys. Locomotives transported truck cars. Hydraulic apparatus lifted furnace doors. Electrically operated tongs put pieces of iron into the fire and drew them out again. Compressed air drove the tools. At the close of the century the plant in Homestead, Pennsylvania, was the largest single iron- and steel-making plant in the world. The buildings extended 1½ miles along the Monongahela River front. The plant operated a private railroad, using seventy locomotives, and employed seven thousand men.

Equipment Replacement. New machines were ever replacing the old. Captain William Jones of Pittsburgh, one of the great steelmen of all time, became known as "scrap pile Bill," because he was always ready to throw out any piece of machinery as soon as a new and better one was introduced. He told a delegation of British producers that while they, the Britishers, tried to increase production by enlarging their machinery, in America production was increased by bringing in new machinery.[5] Andrew Carnegie declared in his *Autobiography* that he was able to increase the output of his mills by 60 per cent by introducing "the newest and latest machine." When asked about the indispensability of scientists in manufacturing steel, Carnegie replied, "If faced with the loss of buildings and plants, or the services of scientists, I would prefer to lose the former. Plants would be less difficult to replace than scientists."

Electric Power. Almost as important as the machinery of steelmills was the introduction of electrical power, used to perform the arduous tasks that had been done by human brawn. Alva Dinkey revolutionized rolling-mill practice by using electricity to drive feed tables, thus displacing the hooks and tongs operated by hand labor. In 1895 he invented a reversing controller for electric motors which permitted flexibility in operation and greater ease in manipulation. Electric lights were introduced in the steelmills in the 1890's, which not only brought about greater efficiency in operating the machines but were an added safety factor.

Continuous Production System. By 1900 the trend in blast-furnace production was toward fewer furnaces and greater output per furnace. The blast-furnace operation was no longer a separate industry, but a part of a large-scale production of iron and steel. These great furnaces consumed three to four trainloads of fuel and iron ore daily, and had to be fed every 15 minutes day and night. Production in blast furnaces was a continuous process while the furnace was "in blast," which at times extended over a period of several decades. Feeding required little manual labor. Small cars traveled up and down an incline and automatically dumped their contents into the furnace. Furnace gas was converted into electric power to do the work of many hands. In former years one ton of pig iron was produced for each man employed. By 1900, the output had been increased to 6 tons per man.

The American Steelworker. While new machines, improved technology, efficient organization, and vast capital contributed greatly to the success of the steel industry, there was still another factor to be considered. Behind these machines stood the brawny steelworker whose heart was in his job. The men of the mill were a hardy lot. Stripped to the waist, with muscles flexed, and with their eyes protected by heavy goggles from the glare of molten metal and flying sparks, they tossed steel ingots about with heavy tongs as if they were toys. One of their spokesmen claimed that the reason for cheap steel production in the United States was not due solely to the bigness of the industry, but also to the greater machine energy possessed by the American workmen. "On the other side [of the Atlantic] when you see a man trundling his wheelbarrow he tires and takes his time. Here we load the article on a truck and have an engine pull it." The American steelworkers developed a pride in their profession and were regarded by many as the aristocrats of the labor world. The mythical hero of the steel industry, Joe Magarac, symbolized in iron and steel what Paul Bunyan symbolizes in the lumber industry.

Increased Uses of Steel. Rails made the first great demand upon the steel industry, and in order to supply this demand rolls were so perfected that in the largest steelmills, rails could be made 184 feet in length. While still hot they were cut by circular saws to the required 30-foot lengths, allowing two "crop ends." Rails increased in weight from 56 to 85 pounds.

The demand for wire nails stimulated a demand for another new industry. In making nails, the wire was drawn through a die by a gear-driven drum. (Cut nails were made of steel instead of iron.) Wire nails were not generally accepted at first by the trade, and the development was slow. During the 1890's, however, the production of wire nails had overtaken the cut-nail output and ranked fourth in steel output.

The growing demands for wire, in turn, made heavy demands upon the steel industry. One of the greatest demands came from the farmers and ranchers of the West and Southwest. John W. Gates, interested in wire and steel, and a high-pressure salesman for wire fencing, sold the Texas cattlemen the idea that barbed wire, designed by Joseph Glidden of Illinois, would fence out the roving cattle and fence in their own. Wire fencing became popular with farmers, due partly to the increasing shortage (in certain sections, complete absence) of timber for fences. While a board and post fence cost one dollar a rod, a barbed-wire fence made of Bessemer steel cost less than fifty cents.[6]

Tin-plate Industry Introduced. Another challenge to American ingenuity resulting from the manufacture of steel was the tin-plate mill. England had long had a monopoly on the tinning art, and it was thought that none but Welsh tinners, with centuries of experience in making tin by a slow, laborious, hard process, could produce satisfactory tinware.

Tariff on Tin, McKinley Act, 1890. Meantime, the United States was becoming the world's largest consumer of tin plate, but there were no mills in this country to produce it. The rapidly growing canning industry had increased the demand for tin plate, and pressure was brought to bear upon Congress to provide protection for the manufacturer of tin plate. This was done in the McKinley Tariff Act of 1890.

Importation of Skilled Labor. There was not a tin-plate mill in the country when this Act was passed, but by September, 1891, five concerns had entered the field; early in 1892, nineteen plants were producing tin plate. Skilled workmen were induced to come from Wales to establish the business. By the summer of 1892 one of the large tin-plate industries in Wales moved to the United States and established a plant at Norristown, Pennsylvania. Just as Coates was forced to bring his thread, Gillet, his steel pens; Faber, his pencils; Sheffield, its razors; and Belgium, its glass; so did Wales bring its tin-plate industry to American shores.[7]

Americans Improve Tinmaking Methods. In this country, however, the industry took on a new form. Nearly all the old Welsh ideas for hot rolling, shearing, annealing, and tinning were discarded by American manufacturers. America substituted machinery for hand labor. Improvements were made both in the plate and in the tinning process. Improved steel, malleable and ductile, and of the desired low-carbon content, was manufactured by the Bessemer process, supplanting the original rolled or puddled iron. The black plate for tinning, however, could not be made on ordinary rolling-mill machinery, so the old hand process was continued, passing the metal back and forth through a single stand of rolls.

New Machines Speed Production. After coming from the rolling mill the refining process—which involved annealing, treating with palm oil, and cold rolling—was repeated until the desired smoothness and quality was obtained. Then the plate was given its coating of tin, lead, or zinc as desired. Machines were introduced to carry on many of these operations. Samuel Buckman of Philadelphia in the early 1890's brought out a machine that would clean, scour, and pickle the sheets, pass them into the flux box, and finally into the tin pot, all by means of a series of rolls. The Norton Brothers of Chicago developed a tinning machine that did the work as quickly as the rapid-process acid-flux machines used in Wales, and permitted the use of palm oil as a flux. A large tin-plate manufacturer in Swansea, Wales, was one of the first to purchase the American-made machine. By 1900 the Welsh tin product had been driven from the market in the United States.[8]

Steel Becomes Leading Industry

Steel for Bridges. The advantage of iron over wood in bridge construction had been recognized long before the days of steel when cast iron and wrought iron had been used. With the coming of the steel age, however, steel began to replace iron. The world's first steel bridge was erected in Holland in 1865. In America, steel was used in the Eads' bridge across the Mississippi River at St. Louis in 1873, in numerous railroad bridges, and as a crowning glory, in the Brooklyn Bridge, the first suspension bridge to use *steel* wire.[9] By 1900 wooden bridges had generally given way to steel.

Economic and Political Effects. The economic consequences of the new technological processes produced a number of significant changes in the steel industry. By 1900 steel was being produced and sold at 2 pounds for three cents. Steel rails that cost ninety-three dollars a ton in 1870, could be bought for twenty-four dollars in 1900. American railways and many foreign railroads were being built with Bessemer steel. Rails, axles, springs, and boiler plates of Bessemer steel replaced the wrought iron formerly used. Spades, forks, hammers—all conceivable tools, save cutting tools (made from crucible steel)—were made from Bessemer steel. The technological revolution in steel production gave America her first position as a world power.

Manufacture of New Type of Steel. As the uses for steel broadened, another type of steel known as the "open-hearth" product, began to be manufactured. The open-hearth process had been developed in England in 1858 by the Siemens brothers when they patented their regenerative gas-fired furnace. This became the basis for the open-hearth steelmaking process.

First American Open-hearth Steel. In 1868 an open-hearth furnace was introduced in America at the Cooper Hewitt Company. At first there was little demand for open-hearth steel. A little over 1,000 tons were manufactured in 1870, and only 100,000 tons in 1880—still a negligible amount. The Carnegie-Phipps works at Homestead introduced the open-hearth in 1888, and by 1900 the open-hearth was producing more than half the tonnage of Bessemer steel. In that year, Carnegie recognized the open-hearth as the future method of steel production.

Superiority of Open-hearth Steel. The open-hearth furnace was actually neither open nor a hearth. The furnace consisted of a huge enclosed shallow basin in which were thrown pig iron, scrap, lime, and whatever alloying metals were needed. As these were melted, the impurities in the iron were carried off with the slag. The resultant steel was considered tougher than Bessemer steel, and softer, more nearly resembling puddled iron. Once it was realized that steelmaking was a chemical operation, the advantages of the basic open-hearth process became apparent, for it was then possible to determine the chemical composition of the product.

Open-hearth steel was not made in the same spasmodic manner as was Bessemer steel, for the process (requiring from 31 to 32 hours) could be interrupted to take time for testing. As tailor-made steel came into demand, exact composition was determined by the addition of alloys and the proportioning of ingredients, a factor not possible in the Bessemer process. The open-hearth process was more reliable than the Bessemer process. It was often referred to as "the scavenger of the steel industry," because scrap and waste steel could be utilized by this method.

United States Leads in Worldwide Steel Production. The consumption of iron had been long recognized as a measure of a nation's progress, but with the coming of the steel age, steel came to be regarded as a barometer of trade and the nation's economic weathervane. The industrial progress of any nation in western civilization is measured by the quantity of iron and steel that it consumes. It is reflected in the extent of its railroads, its steamships, and war vessels, in its trolley lines, in its iron piping for gas and oil transmission, in its immense structures of skeleton steel for buildings and bridges, and in its enormous shafts and subway excavations requiring steel linings and braces.

An Economy Based on Steel Consumption. In a little more than a quarter of a century of steel production, America's output had increased to twice that of England or Germany, nine times that of France, Austria or Hungary, and ten to twenty times that of other steel-producing nations. In 1897 Senator Wheeler of Alabama declared that his state alone, which had just recently expanded its iron and steel industry, was producing twice the amount of steel produced by Sweden and Norway, and an amount equal to that of Russia or France.[10]

Steel found its way into a variety of uses, supplanting wrought iron, cast iron, and wood. Railroads, which originally wanted steel to replace rapidly wearing iron rails, began demanding heavier rails, rails weighing up to 100 pounds. Steel replaced the old wooden, cast-iron, and wrought-iron parts in the locomotives. Boilers, engine frames, cabs, tires, tanks, all were made of steel. Locomotive construction became largely an American industry because of steel. In 1900 there was only one foreign-made locomotive in the country.[11]

Skyscrapers used beams and girders of steel with reinforced concrete in such quantities that mills were hard pressed to meet the demand. The canning industry consumed huge quantities of tin plate. Woven-wire fences and barbed-wire fences were in demand. Wire nails were used by the building industries. Shipbuilding became a major industry in the 1890's, and of the sixty-one vessels built in the twenty shipyards of the country in 1900, only six were of materials other than steel.[12] New naval vessels of steel construction began to supplant the antiquated wooden hulls of former years.

Steel Industry Sets a Pattern. The technique of manufacturing steel set the pattern for large-scale production in many other industries. Economy of heat, power, and time was effected in the large plants by making steel production continuous. The manufacture of steel started with ore, coke, and lime in the blast furnace, from which with one heat it ran into the converter, thence to the ingot, from there to the bloom, and was finished into a steel rail without being reheated. This whole operation could be accomplished in 24 hours. Continuous steel production marked the beginning of the modern assembly line.

Impact of the Steel Age

The significance of the steel industry on the nation's history is not to be measured merely by the production of materials made from iron and steel, however, but rather by the impact of the steel age upon all phases of American life. History is affected as much by technology as by politics, and the human race is affected as much by a giant industry as by a dynasty. A contributor to the *Dublin Review,* in commenting on America's steel age, wrote near the end of the century that the great industrial revolution that was occurring in the United States was as far reaching in its effects as were "those more clamorous social upheavals that reverse thrones and dislocate empires." "The fate of nations," declared the writer, "can be as truly altered by an invention as by a conquering horde or by a defeat in battle." [13]

Notes for Chapter 22

1. See John N. Boucher, *William Kelly: A True History of the So-Called Bessemer Process* (1924). See also James M. Swank, *History of Iron in All Ages* (1892).
2. Boucher, p. 12.
3. *Ibid.,* p. 80.
4. James M. Swank, *History of the Manufacture of Iron in All Ages* (1892), p. 409.
5. *Iron and Steel Institute Journal,* XVIII (1887), 129.
6. The wars that occurred between the cattle drivers and the ranchers over the introduction of the barbed-wire fence were debated at length in the sessions of the Forty-eighth Congress, 1883–1884. Pages of testimony and statistics were introduced, all of which supply excellent material for the history of the barbed-wire fence. See also, Walter P. Webb, *The Great Plains* (1931); and Earl W. Hayter, "Barbed Wire Fencing," *Agricultural History,* XIII (October, 1939), 189-207.
7. Rep. Broderick (Kan.), *Congressional Record:* 52nd Congress, 1st Sess. (June 18, 1892), p. 5401.
8. Victor S. Clark, *History of Manufactures in the United States,* III (1928), 133.
9. See David B. Steinman and Sara Ruth Watson, *Bridges and Their Builders* (1941).
10. *Congressional Record:* 55th Congress, 1st Sess. (Mar. 22, 1897), p. 125.
11. *U. S. Industrial Commission Report,* V (1900), 232.
12. *Ibid.,* XIX, 479.
13. E. M. Clarke, "Age of Steel," *Dublin Review* (July, 1897), p. 54.

See also Thomas Southcliffe Ashton, *Iron and Steel in the Industrial Revolution* (1951); Herman Hollis Chapman, *Iron and Steel Companies in Years of Prosperity and Depression* (1935); Edwin Clarence Eckel, *Coal, Iron and War; A Study in Industrial-*

ism, Past and Future (1920); Henry Oliver Evans, *Henry W. Oliver, Iron Pioneer, 1840–1904* (1942); Douglas Alan Fisher, *Steel Serves the Nation, 1901–1951; The Fifty Year Story of United States Steel* (1951); Stephen Lincoln Goodale, *Chronology of Iron and Steel* (1931); Maud and Miska Petersham, *The Story Book of Iron and Steel* (1935); James Moore Swank, *History of the Manufacture of Iron in All Ages, and Particularly in the United States from Colonial Times to 1891* (1892).

Chapter 23

THE BIRTH OF THE PETROLEUM INDUSTRY

When "Colonel" Edwin L. Drake developed the technique of drilling for oil in 1859, rather than waiting for it to seep through the ground and accumulate in pools, he opened up not only the hidden reservoirs of oil, but also a new industry and vast new fields for science and technology. Following in rapid succession came improvements in drilling operations, the invention of many new tools, new developments in transportation technology, and with the co-operation of science, the discovery of vast new uses for petroleum products.

Pioneer Methods and Discoveries

A mere two thousand dollars had been doled out to Drakc for his initial experiment, but within ten years more than two hundred million dollars had been invested in the petroleum industry. This new development came during the postwar Civil War depression years, and was the one major industry least affected by the panic of 1873.

Petroleum Creates Companion Industries. The oil industry necessitated the building of new lines of railroads. It increased the production in iron and steel. It provided employment for coopers on the Atlantic seaboard who were out of work. It stimulated barge- and shipbuilding, and in an ever widening circle, developed the wealth of the country. The tax levied on the production of petroleum alone from 1862 to its repeal in 1866 netted the United States government more than twelve million dollars in revenue. This abundant supply of petroleum developed an industry which had such a phenomenal growth that in ten years oil became the country's second greatest export, surpassed only by cotton. In its far reaching effect, the oil industry has changed our whole civilization by making oil abundant, useful, and inexpensive.

Demands for Oil. A number of factors were responsible for the birth of the oil industry. There was an urgent need for a new illuminant, because of the decreasing supply of whale oil. There was the need for a better illuminant for the manufacturing plants that were springing up throughout the country. There was need for lubricating oil for the tools and machines that were being introduced in the fields and factories across the country. Previous efforts to meet these needs had failed. The production of lard oil had increased, but not in sufficient quantity to satisfy the demands. Coal oil was extracted from shale and from bituminous coal, but at great labor and expense. Just when the artificial sources of oil seemed about to fail, a substitute was found flowing in inexhaustible streams from the carboniferous strata.[1]

Search for Oil Begins. The search for a greater supply of oil was greatly stimulated in the mid-nineteenth century. The region of northwest Pennsylvania, where quantities of oil oozed from the ground and collected in springs and creeks, became the center of that search.[2] The scientific discoveries of Professor Benjamin Silliman, Jr., of Yale, who predicted the many possible uses of petroleum, and the business acumen of such men as George H. Bissell, who worked under Silliman's guidance, provided the impetus for this new development. The persistence of the Connecticut Yankee, Edwin L. Drake, and the technical skill that had been developed in drilling salt wells, initiated the industry. The new product was hailed as an effective medicine, a solvent, a fuel, a lubricant, and an excellent illuminant.

"Colonel" Drake's Discovery, 1859. A one-time railroad conductor, Edwin L. Drake, was employed to go into the oil region around Titusville, Pennsylvania, and experiment with new methods for increasing the flow of oil. "Colonel" Drake,[3] as he was called, decided to use the same technique in drilling for oil that was employed in drilling salt wells. He persuaded "Uncle Billy" Smith, a Tarentum blacksmith, who was familiar with salt-well drilling to join him in Titusville and drill for oil.

A site was selected where the oil oozed through the ground. Drake's helpers constructed a wooden derrick, 40 feet high to house the drilling rig, secured a 6-horsepower steam engine, and a "Long John" stationary tubular boiler of the type used on steamers on the Ohio.

Drake Initiates New Technique. They began by digging a pit through the surface, as was the custom in the salt-well areas, but after repeated failures, due to the cribbing caving in, they abandoned this method and introduced a new technique—the first in drilling history. They drove a pipe down to bedrock, a distance of 36 feet, and then commenced a drilling operation through the pipe. They began drilling about the middle of May, and on August 28, 1859, a vein of oil was struck at 69½ feet. The oil arose to within 10 feet of the surface. Next, the well was equipped for pumping and it began to produce about forty barrels a day. Gradually, production decreased, and within a few years it was producing only fifteen barrels a day. Regardless of the flow, a new industry was born.

The Big Oil Rush. When it was established that the supply of oil could be produced by merely sinking drills in the region where it was known to exist, a rush to the oil fields comparable to the California gold rush of 1849 ensued. Land values mounted. Promoters talked in terms of hundreds, and then thousands of dollars of profits; and in a few years, million-dollar deals were common. Pithole, a village of a few houses, within three months boasted of having fifty hotels, three daily newspapers, an opera house, a waterworks system, and a population of fifteen thousand. Its postmaster handled more mail than any other city in Pennsylvania, except Philadelphia and Pittsburgh. Living costs soared. Banks were organized. Farmers sold their oak trees for barrel staves, and gave up farming to work for oil drillers. Machine shops and blacksmith shops worked 24 hours a day repairing boilers, pumps, and drills, and designing tools for any one who wanted to introduce a new method of drilling.

Amazing Production Records. By the close of 1860, seventy-five wells had been drilled and five hundred thousand barrels of oil produced. In the first ten years of the industry, twenty-five million barrels of oil were produced. The market was highly speculative. Oil sold as high as twenty dollars a barrel and dropped as low as ten cents a barrel. Seventy-five per cent of all wells drilled were either dry or abandoned within five years. On the other hand, great producers like the Empire Well, yielded three thousand barrels a day, and the Noble and Delemeter Well yielded four thousand barrels daily. A few gushers reported eight thousand barrels a day. Some of these had no facilities for containing the oil as it poured over the earth. People scooped it up in buckets, making from thirty to fifty dollars a day from their find. In most cases drilling involved an outlay of time and money; in others, operators "kicked down" a well with a spring pole and found themselves lucky in the gamble.

"Kicked-in" Wells. In this process, a 40- to 50-foot pole was used. It was supported over a forked upright, with the heavy end of the pole secured in the ground, and the smaller end, to which a drilling

bit was attached, suspended over the hole. By means of loops of rope the "kicker" pulled the pole down, and this motion lowered the tools, causing the bit to cut into the rock. One lucky prospector brought in a five hundred barrel well by this method in the summer of 1861. Some wells flowed oil for years; some flowed for a period, then ceased. Some dripped, or were "weeping wells," while others refused to yield without technical persuasion.

Oil Pumps. To control the output and to ensure a constant supply of oil, pumps were introduced. The first pumps used in the oil industry were made of sheet iron, riveted and soldered together; a piston, with a flat piece of leather at the bottom, to act as a valve, worked up and down inside the casing when power was applied. These pumps came to be operated by steam. Later they were operated by gas engines.

At first, each oil well was pumped as a single unit with a boiler and engine of its own. Then several wells were pumped by one boiler with but one pumper in charge. A further economy was effected when several wells were coupled to one engine by means of shackle rods. In this arrangement the down stroke of one well coincided with the upward lift of the other.

The early production of oil was always a gamble and a challenge. As one writer described it, "The death of every oil field begins the day it is opened." Many of the early wells produced abundantly for a few weeks or months, and then dwindled to almost a trickle. Here, again, a new technology was introduced that greatly affected the oil industry.

"Shooting" Wells. Colonel E. A. L. Roberts of New Jersey, while in military service during the Civil War, had witnessed a shell explosion in a millrace at Fredericksburg. His inventive mind conceived the idea of "opening up veins and crevices in oil bearing rock by exploding an elongated shell or torpedo therein."[4] Roberts' method differed from the old-fashioned method of blasting with gunpowder, which always shot upward. A device was needed to cause a sideward and downward explosion. Roberts, profiting by the newly discovered nitroglycerin, designed a torpedo, which he patented in 1864. The following year he tried his new torpedo in a 600-foot well in Pithole. A new technological device of "shooting" wells, which continues to this day, was introduced.

"Fishing" for Lost Tools. One of the early difficulties in drilling was that the rods broke frequently, resulting in the loss of bits and tools. To recover these losses, "fishing tools" were devised, and "fishing" became a trade in itself. One such tool known as the "Luther socket" was devised by a local blacksmith, L. H. Luther. It con-

sisted of a hollow rim in which a pair of clamps was inserted. The rim slipped down over the tools at the bottom of the well, and when pulled up, the clamp lips on the jar caught the shoulder of the bit and hauled it to the surface.

Cable Drilling. The first established method of drilling was the cable-tool, or percussion method. This was Drake's method, and it is still employed today in many cases, with a number of improvements. The derrick, which marks the oil field to the observer, had before 1900 risen to an accepted height of 84 feet. From the crown block at its top, the cable with the drilling tools extended many times its height into the bowels of the earth. Three pulleys mounted on this crown block held the casing line, the sand line, and the drilling line. Housed in the derrick at the base were the bull wheel, around which the drill line was coiled, and wheels to take care of the other lines. From an engine nearby, power was transferred by a band wheel to a walking beam. As the walking beam vibrated, the drilling tools struck a heavy blow on the rock in the well. The sand pump, attached to the sand line, was used to remove debris from the well, and the casing line drew the casing that lined the hole. The drilling tools were a string 60 feet long and 5 inches in diameter, weighing 2 tons (including a bit), attached to an iron stem, drilling jars to give play to the strokes and jar the tools loose when they became stuck, a sinker bar, and rope socket.

Tool Innovations. Tools that were invented on the spur of the occasion to meet an unexpected difficulty were numbered by the hundreds. The copper tubing, required by the salt drillers to prevent corrosion, was replaced by iron. Casing was used to line the hole to exclude the water and sand, and to confine the oil to its immediate horizon. This was at first a 3¼-inch pipe, cut in 20-foot lengths which were threaded at both ends so that they could be taken apart by a tool wrench while still in the hole. Later a 5⅜-inch pipe was used. The patented Forgie wrench was said to save the operators thousands of dollars. Steam engines were made more powerful and more efficient. Gas engines, utilizing the waste gas coming from the wells, were used later.

The production of petroleum was a new industry, but it was introduced at a time when technology had so far advanced that there were materials suitable for the construction of the necessary tools, and the technique ready to devise them. Electric power, the telegraph, and the telephone, all played a part in the later development of petroleum technology. As observed by a visitor to the oil fields, "How fortunate that the oil that has been contained inside the earth for millions of years remained there untouched until methods for drilling had been sufficiently advanced to tap it and put it to man's use!" [5]

Discovery of New Fields. The Oil Creek area did not long monopolize the oil industry. Within a few years all Western Pennsylvania was alive with prospectors. The area, extending down into West Virginia, became known as the Appalachian oil field. The McDonald pool, opened up in 1891, had an amazing production of ninety thousand barrels a day. It covered a small area only 9 miles long and 2 miles wide. The Sistersville pool in West Virginia, tapped the next year, was also highly productive. In 1885, the Lima field in northwestern Ohio had been opened up where a different grade oil was found. In that field twenty thousand productive wells had been drilled by 1890. In 1891, the Indiana field was in commercial operation and the oil excitement spread to that state. California oil was discovered as early as 1876, and by 1900 that state was one of the major producers of oil. Pennsylvania production reached its peak in the early 1890's.

Early Petroleum Geologists. Efforts to employ scientific methods aimed at taking the guesswork out of drilling, and to aid in the location of oil deposits, were made by geologists at an early date. One of the theories, the "anticlinal theory," was advanced separately by F. W. Minshall and Dr. C. C. White. The Census of 1880 published a drawing of the measurements and the mapping made in support of the anticlinal theory, which was said to have been responsible for opening up the successful West Virginia field. But as long as "wildcatters" brought in gushers where the theorists failed to locate oil, and drillers got "dusters" where oil was supposed to be, the practical driller paid little attention to science. Many wells were produced, however, in the strata that geologists had specified as oil-bearing rocks.

Rotary Drill in Texas Fields. For the more sandy soil in the Southwest, rotary drilling was devised to bring in the famed Spindletop Well in Texas in 1901. By this method the earth's crust was ground away by constant rotation of a drill, as contrasted with the intermittent percussion of the drill bit in the cable-tool method. The rotary system proved to be better adapted to deep wells and to regions where the earth's crust presented a soft formation. For rock drilling, the percussion instrument was necessary. The great advantage of the rotary system was in its speed.

Petroleum Transportation

The shipment of vast quantities of oil brought about a novel development in transportation technology. Pittsburgh, the nearest city of any size to the earliest oil region, was some 90 miles away, with no good connection either by road, railroad, or canal. Shipments could be made by water by way of Oil Creek and the Allegheny River, but these streams were ice-jammed in winter and too low in summer. The nearest shipping points by rail were 30 and 35 miles from the oil fields.

Early Wagon Shipments. The only available method was to haul the oil in barrels by wagon. Some six thousand two-horse teams and wagons were engaged in hauling oil to shipping points during 1862 and 1863. The district roads became a bottomless quagmire. Solid lines of teams extended miles in length, and the horses suffered outrageously from the whips of the drivers who tried to spur them on. Wagons dropped into mud up to the axles, horses sank to their bellies. Teamsters tore down fences, drove through fields in an effort to avoid rivers of mud. The customary load was five, sometimes six, barrels. Moreover, the shortage of barrels added to the difficulties.

Railroads Expand into Oil Fields. Realizing the potential bonanza that lay ahead in the oil industry, the railroads began early to extend their lines into the oil region, and by 1865, four separate, independent companies had built roads into the oil-producing areas. The problem still remained, however, of transporting the oil from each new well to railroad sidings.

"Pond Freshet" Transport. One alternative, described as the "pond freshet," provided a method for shipping great quantities of oil by boat down the Allegheny River to Pittsburgh. This ancient method, borrowed from lumbermen, provided for creating artificial floods caused by releasing dams erected on upper Oil Creek and its tributaries. When the barges loaded with barrels of oil had been assembled, the dams were opened and the rush was on. During busy seasons, the opening of the dams occurred twice a week. The boats were of all kinds and sizes, and carried the oil either in bulk or in barrels. The cost of a "freshet" varied, running up as high as four hundred dollars, the sum being paid as toll by the boatmen.

When the boats were loaded, the word was sent to "pull the shoats." With this, the waters started and the series of dams were cut on schedule as the waters advanced. Giddens described these pond freshets as affording an unusual sight. "Flat boats, little and big, loaded with 10,000, 20,000, 30,000 barrels of oil, either barreled or in bulk, floating alone endways and sideways on a rushing flood, and wildly fighting their way down Oil Creek, which was only twelve rods wide, and very crooked as it wormed its way through steep hills."[6] It required the skill and strength of hundreds of boatmen to avoid collisions. If a jam occurred, the oil was spilled into the creek. When the boats reached their destination safely, the rewards were fantastic. But, there were many disasters, too, which resulted in lining the streams with abandoned boats and burst barrels, pouring thousands of gallons of oil into the Allegheny River.

Watercraft Transport. A more orderly means of transportation was introduced with the steamboat. A young river captain, Jacob J. Vandergrift, adapted a coal boat (the "Red Fox") and barge to ply between Pittsburgh and Oil City, with bottoms filled with empty barrels consigned to oil operators up river. He also built twelve bulk boats with a carrying capacity of four hundred barrels. The boats were lined with tar and caulked with oakum for waterproofing. When the Allegheny waters were low or frozen over, which happened during a good part of the year, the river was not navigable.

Storage Methods. To take care of the immediate needs, either that of saving the oil from a gusher or holding it for transportation, tanks were constructed. The first tanks were wooden structures used merely to store the oil. An improvement introduced in 1863, called the "Christmas tree" tank, conducted the oil from the well to the storage tank by arranging a discharge pipe which was connected with the tubing at the top of the well. The oil could be conducted in any direction to the tank. Because of the danger of fire, and because of possible leakage, tanks made of iron appeared at an early date. Three such tanks were erected in 1864 at Oil City, each with a capacity of eight thousand barrels. They were covered with a gravel roof, "fire proof and leakage tight." The demand for larger tanks resulted in erecting structures of thirty thousand barrel capacity. Tank construction became a regular business, and as the industry developed, tanks of enormous size were erected at numerous points throughout the country.

Development of Railroad Tankcars. Shipping by rail marked another development in oil transportation technology. The first oil tankcars were wooden barrels set on a flatcar and were tried out as early as 1864. A novel tankcar was designed by placing oil tanks inside a boxcar, but the tanks leaked and never proved satisfactory. Next, iron cars of a boiler type were introduced. These cars were leakproof and theftproof. Since oil expands with the change in temperature, allowance was made for this expansion, and a dome was erected on the top of the tank. The tankcar, originally designed by

Charles P. Hatch, who moved to Pittsburgh in 1865, has continued to this day.

Tank Ships. Tank ships early made their appearance for transporting oil abroad. As barges were used for bulk transport, so wooden sailing vessels became tanks and were used for a number of years. The iron ship "June," built in 1863, proved successful as an oil transport in carrying petroleum from New York to Liverpool. The Belgian ship "Charles," fitted with a number of separate tanks was constructed to carry oil in bulk, loading from New York in 1870. Its capacity was seven thousand barrels and it was supplied with pumps to load and unload its cargo of oil. Tank steamers continued to be a success from the start.

Early Pipe Lines. There still remained the problem of getting the oil from the tanks at the wells to the barges and railroads. Also, there was always a shortage of barrels, and, too, the demands of the teamsters were exorbitant. As a result of these obstacles, oil operators turned to a new device in transportation technology—the pipe line. The first pipe lines were gravity lines. They extended from the oil well down to the river, or to a tank on a railroad line, and could be used only when the elevation of the tanks at the well was higher than the tanks into which the oil flowed. This handicap was overcome by a system of pumps.

The first successful pipe line operated by pumps was introduced in 1865 by Samuel Van Sickle. It extended a distance of 5 miles, over ridges and hills, and was described as, "the most wonderful of the many wonders produced at Pithole." The pipes, each 2 inches in diameter and 15 feet in length, lay for the most part on top of the ground. Teamsters, seeing a threat to their business, tore up the pipe line and smashed the pumps. But after a few of Van Sickle's brawny "pipe protectors" had tusseled with some teamsters, the trouble quieted down and the pumps and pipes continued to deliver the oil to the waiting barges and railroad lines.

Long-distance Pumping. It was first believed that pumps would have to be installed every mile or two in order to force the oil up hill. An ingenious operator for the Titusville Pipe Line Company disproved this theory when he laid a 2-inch line 10 miles long. It required eighteen barrels of oil to fill one mile of the pipe line, and although he installed intermediate pumps, he used only the initial pump. After 3 hours of slow, even pumping, the oil began to empty into the tank 10 miles away. Long distance pumping marked another advance in petroleum transportation technology.

By 1872 the pipe-line system was fairly well established. There remained the consolidation of the lines, the building of trunk lines, the extension of the pipe lines to the seaboard and to refineries outside the producing district, and the extension of the lines into new fields. These vast developments, however, required the aid of capitalists and petroleum engineers.

Extension of Pipeline Network. A pipe line from the oil fields to the Atlantic Coast was laid in 1879 through New York state by way of Olean to New York City. Five years later a line was thrown across the mountains of Pennsylvania by way of Wilkes Barre to the coast. With the opening up of oil fields in southwestern Pennsylvania and West Virginia in the late 1880's, and in the region around Lima, Ohio, trunk lines to Toledo and Chicago were built and connected with the lines to the East. Bear Creek Station became a great pipe-line center for all East–West transit. A perfect network of pipe lines was laid out, including Chicago and New York, Cleveland, Pittsburgh, Philadelphia, and Baltimore.

"Underground Railroads." By 1900 the eastern fields from Chicago to the seaboard were crisscrossed with "underground railroads," carrying both crude and refined oil. The system consisted of lines carrying the oil from the wells to the tanks, trunk lines extending from refineries to distributing centers, pumping stations, storage centers or terminals, switch systems, dispatchers, and communication systems. The mileage of this underground system increased from a little less than 700 miles in 1879 to almost 18,000 miles in 1900, of which 6,782 were trunk lines.

Pipe Lines versus Railroads. Pipe-line transportation became a law unto itself. Railroads, particularly the Erie Railroad, controlled by Jim Fisk and Jay Gould, waged a cutthroat battle with the Pennsylvania Railroad in an attempt to capture the oil shipments from Pennsylvania, only to discover that a new transportation system —the pipe line—was the competitor of both. John D. Rockefeller, in his testimony before the United States Industrial Commission hearings in 1900, declared that the entire oil industry depended on the pipe-line system, "without it every well would shut down, every foreign market would be closed."

Petroleum Refining

Technology's greatest role in the petroleum industry relates to the refining process. The principle of refining had been established long before Drake brought in his first well. In England, James Young, in the early 1850's, had produced paraffin oil from coal. In this country Abraham Gesner had produced illuminating oil from coal, which he called "kerosene." In Boston, two chemists, the Atwood brothers, made a lubricant from coal tar in 1852, believed by some to be the first produced for sale in the United States. Their product was a mixture of coal oil, cheap animal fats, and vegetable oils. Since they put their product on the market at the time of Napoleon III's coup d'état, they called it "coup oil." A company was organized to manufacture the product, and in spite of its disagreeable odor, it found a ready sale to cotton-mill owners and railroad companies.

Coal Oil Production. Improved refining methods, encouraged by the manufacture of lamps, helped to increase the demand for a satisfactory illuminating oil. In New York, Pennsylvania, and Ohio some forty firms were engaged in producing coal oil during the late 1850's. The manufacture of lamp oil from cannel coal (coal oil) had become a sizable industry by this time. Almost two million dozen coal-oil lamps, each consuming an average of 4 gallons of oil annually, were in use.

Kier's "Seneca Oil." It was during this same period that Samuel M. Kier of Pittsburgh was carrying on some experiments that made him the world's first refiner of petroleum. Kier, a salt-well operator, utilized the oil that was a nuisance in the salt wells. He bottled it, pasted some fancy labels on the bottles, and called it "Seneca Oil," a cure-all for every known ailment.

Kier believed that the oil could be used for something other than a liniment. He sought the services of a Philadelphia chemist, J. C. Booth, who advised him to distill it. After observing the technique used by a local distiller in manufacturing raw whiskey, Kier built a model still. He used a huge iron kettle as a container, to which (with the aid of a plumber) he attached a worm pipe to drain off the distillate. He filled the kettle with crude oil, built a wood fire under the kettle, and as the oil "boiled and bubbled," a thin trickle of fluid drained off into a pan. From this crude "refinery" a new petroleum product, lamp oil, was developed. Lamp oil found a ready market.

Another pioneer refiner was A. C. Ferris of New York, who observed one of Kier's lamps in Pittsburgh, returned home, distilled some oil of his own, and introduced the product there. He eliminated the objectionable odor that was a deterring factor by adding caustic soda and sulfuric acid, a method borrowed from the coal-oil industry. In New York, as in Pittsburgh, the demand far outran the supply.

Refining Technology. When the Drake Well came in, followed by hundreds of others, refineries sprang up in the oil fields of northwestern Pennsylvania and in nearby cities. Erie, Cleveland, and Pittsburgh became refining centers. The same refining methods as those employed for coal oil were used and the product was practically identical. Most of the early refineries were small and many of them were crude. The coal-oil industries were soon forced to go out of business or convert to petroleum refineries. In 1873 there were more than one hundred refineries in the country, with Cleveland leading, New York second, and Pittsburgh third.

Early Processes. The refining process was simple. Refiners ran crude oil into retorts and subjected it to excessive heat, which caused evaporation. The vapor passed through a worm in which it was condensed, and from this the product ran into distillate tanks. The first product was naphtha, the next was illuminating oil, and the last was a heavy oil containing paraffin. The naphtha was discarded. Illuminating oil was the desired product and it was graded by color, with astral white the superior quality. It was treated with sulfuric acid, caustic soda, and other chemicals to clarify and deodorize it. In this way 75 to 80 per cent of the crude was converted into illuminating oil.

Refining Improvements. Refineries underwent the same development characteristic of all industry in the last two decades of the nineteenth century. Efficiency was increased, organization was instituted, and the small independent refinery gave way to the large corporate establishment. By 1900 there were seventy-five refineries in the country, located in twelve states, with Pennsylvania being the greatest producer. With the introduction of new oils, however, refining technique underwent some technological changes. The groundwork was laid for the introduction of scientific methods into refining. The product from the Lima field (Ohio) was not an acceptable illuminant, since it had a foul odor due to the quantities of sulfur in the oil. In 1885 at the Whiting plant in Indiana, research methods were applied to find a way to develop a product that would have uses other than as furnace oil. The initial research program not only made the Whiting plant one of the world's greatest, but also pointed the way to a new era in refining technology. This—together with the advent of the automobile with its demand for the former waste product, gasoline, and with gas and electricity challenging the oil as an illuminant—forced the petroleum industry to venture into greater technological improvements in oil refining.

Lubricants. Although there had been a great need for a better lubricant in the new high-speed machines, manufacturers were skeptical about using the new oil. It was the oil industry itself, rather than public demand, that developed the manufacture of lubricating oil from petroleum. The industry did this in an effort to utilize the waste product from the refining of illuminating oil.

In the early development of the oil fields, some grades of crude oil were found to be applicable for use as lubricants in their natural state, but the amount was small in comparison with the quantity needed. Hiram B. Everest, a leather manufacturer, is given credit for laying the foundation of the lubricating oil industry, since he discovered a method for utilizing the residuum from the refining of kerosene, and made a cylinder lubricant from it. He first used the waste oil from a small refinery to process his leather. While working with the product in this field, he discovered its qualities as a lubricant.

Lubricating Technology. The range of lubricating oils extended from the light oils to the solid greases. The methods of manufacture involved filtration without the application of heat; refrigeration which made it possible to beat out the undesirable waxes; and steam distillation which guarded against overheating. Treatment with sulfuric acid and lye aided in removing the impurities and neutralized the acidity. Mixtures of lime and soap provided a medium for holding the oil.

While the public was slow to accept mineral oils as lubricants, using them only in conjunction with other oils, the railroads found them indispensable. Within a few years the rolling stock of a majority of the nation's railroads was lubricated by petroleum products. "Without it [petroleum oil]," declared a superintendent on the Baltimore and Ohio Railroad in 1882, "every wheel would grind to a stop within twenty-four hours."

Rockefeller, Master Organizer

The early years of the oil industry were marked by profligate waste and inefficiency in production, refining, and transportation. Out of this wild chaos and confusion emerged John D. Rockefeller of Cleveland, who became the master organizer of the petroleum industry. He brought about orderliness, stability, and ultimate virtual monopoly. Rockefeller was not interested in oil production when he entered the industry. He confined his activities to conserving, regulating, and controlling the supply as it flowed from the earth.

His aim was to profit by eliminating waste in handling the product. With businesslike acumen he developed a system of aggregation and integration that resulted in greater efficiency and in the establishment of a centralization of the industry.

Rockefeller began with the acquisition of refineries. Realizing the advantages of the pipe-line system, he brought about the organization of a system for transporting the crude oil from the wells to the refineries by means of pipe lines. He organized a system for distributing the refined oil through bulk shipments and storage facilities. He acquired timberlands for material for cooperage plants. He built tanks and tank cars, and he pioneered in scientific research in petroleum technology. He standardized the many products of oil until his organization reached from the oil pumps in the fields to the consumer in his home or factory.

All these developments in the oil industry had occurred by the close of the century.

Petroleum Creates a New Way of Life

The need for a new illuminant, which led to the birth of the oil industry, released a product from the earth's interior that was soon applied to thousands of uses. Petroleum was first sought as a product for supplying artificial light, but its utilization as an illuminant soon became negligible in comparison with its many other uses. President Grant, on a visit to Titusville (birthplace of the oil industry), in September, 1871, spoke of the part played by petroleum in bringing about a new industrial age. "By utilizing her vast oil deposits," declared the President, "America may soon become one of the great industrial nations of the world."

Through petroleum technology, earth's oil soon began to supply the country with a greater abundance of power than light. It provided lubrication for the wheels of industry. It made possible the internal combustion engine, which came to be a major source of the nation's expended energy. The discovery of petroleum, compared with the discovery of gold ten years earlier, changed the American way of life to an immeasurably greater degree.

Notes for Chapter 23

1. *House Exec. Documents,* 37th Congress, 2nd Session, No. 116 (May 20, 1862), p. 70.
2. The standard work on the early history of petroleum is Paul H. Giddens, *Birth of the Oil Industry* (1938).
3. When Drake left for the oil fields he was told, "Your mail will be addressed to 'Colonel' Drake"; thus the title, wholly fictitious, followed him the rest of his days.

4. Paul H. Giddens, *Early Days of Oil* (1948) p. 72. Oil City, *Derrick Handbook of Petroleum* (1898).

5. *Ibid.*, p. 106.

6. See Giddens, *Pennsylvania Petroleum 1750–1872, A Documentary History,* Pennsylvania Historical and Museum Commission (1947), pp. 227-228; The Oil City *Derrick Handbook,* I, II, III, offers the best material on the early years of the oil industry. See also, Allan Nevins, *Study in Power: John D. Rockefeller, Industrialist and Philanthropist,* 2 vol. (1953) for an excellent study of this pioneer oil magnate; Giddens, *The Beginnings of the Petroleum Industry,* Historical and Museum Commission (Pennsylvania, 1941); *Early Days of Oil, A Pictorial History* (1948); "History Looks at Oil" (1945), a paper delivered at American Petroleum Institute, Chicago; Benjamin Crew, *A Practical Treatise on Petroleum* (1887); William T. Brant, *Petroleum* (1895); Gustov Egloff, *Earth Oil* (1933); Lawrence Babcock, *The First Fifty Years of the Standard Oil Company* (1939); American Petroleum Institute publications are valuable for many studies on various phases of the petroleum industry.

Chapter 24

THE COMING OF THE ELECTRICAL AGE

"The belief is a growing one that electricity in one or more of its manifestations is ordained to effect the mightiest revolutions in human affairs . . . ," wrote the United States Commissioner of Patents, Thomas Ewbank, in his annual report of 1849. After pointing out the uses of electricity in sending messages by telegraphy, the Commissioner prophesied that, "it [electricity] can be drawn rapidly from its hiding place, and made to propel land and water chariots, animate manufacturing mechanisms, become an agricultural laborer, and a household drudge of all our work." Impressed by the current nationwide excitement caused by the California gold rush of that same year (1849), Ewbank suggested that those who were seeking permanent wealth, and who desired to enrich their country, "should seek placers inexpressibly more precious than any to be found on the Sacramento by digging in the mines of the motors."

Electricity Lights America

The next half-century saw the fulfillment of Commissioner Ewbank's prophecy. Electric currents from batteries were already carrying messages by telegraph over hundreds of miles of wire, and electromagnetic motors had been developed. Joseph Henry, first secretary of the Smithsonian Institution, Thomas Davenport, the Brandon blacksmith of Vermont, and Charles G. Page of Massachusetts had pioneered in building a number of electric motors and dynamos that were in limited use by 1849. But these new sources of energy were not yet geared to fulfill the Commissioner's prophecies.

It was not in the application of power, but in the production of light that electricity was first adapted to man's use. Experiments in electric lighting got under way soon after the close of the Civil War.

The first electric light to be commercially applicable was the arc light.

The Arc Light. The arc light was produced by passing an electric current between two carbon electrodes placed in a vertical position. When heated to incandescence, the electrodes emitted a brilliant arc as they were drawn apart. The arc light had been discovered in the laboratory and remained a laboratory instrument for many years before the inventors saw a practical use for its application. It was not until the steam-driven electric generator was developed that electric power could be produced cheap enough to make electric lighting an economic possibility. The battery was effective in the laboratory, but too costly for commercial use.

Brush's Arc Lamp. Scores of inventors in both Europe and the United States entered the field of electric lighting in the 1860's and 1870's, and many different types of arc lamps were developed. In the United States Charles F. Brush of Cleveland led the way. He produced a simple mechanism for the lamp—a clutch and two electromagnets by means of which the arc was struck when the voltage was applied. He constructed a generator that enabled him to connect a large number of lamps in series instead of having a dynamo for each individual lamp. He also devised a "shunt coil" which enabled the current to by-pass a burned out lamp, allowing the other lamps to operate. Brush's initial apparatus, a single-light generator and arc lamp, installed in the home of Dr. Nicholas Longworth in Cincinnati, in 1877, attracted citywide attention.

John Wanamaker of Philadelphia employed Brush to install a series of arc lights in his store, and on the night of December 26, 1878, he startled the business world by turning on the lights. Crowds gathered to witness "twenty miniature moons on carbon points held captive in glass globes." Next, a Boston clothier bought a number of Brush lamps and installed them over the sidewalk in front of his store. New England factories, realizing that this invention made night work possible, began installing electric lights. New York and Boston displayed their merchandise under arc lights, and night window shopping became a fashionable pastime. Hotels as far west as San Francisco hung arc lights in their lobbies. "City cellars everywhere," declared one observer, "were filling up with grunting little steam engines and whirring dynamos," to produce this new kind of light.

Many other arc systems were developed besides the Brush system, each claiming some special advantages. Inventors hastened to machine shops with their special designs, and bombarded the patent office with applications. City governments, department stores, and other business establishments were urged to install the new lighting systems.

Street Lighting. Street lighting by arc lights on a large scale was introduced in Cleveland in 1879. The same year, in San Francisco, the California Electric Light Company was incorporated, and supplied service from an arc-lighting central station, claimed to be the first in the world to produce and sell electric service to the public.[1] In 1880, when the little town of Wabash, Indiana, discarded its old-fashioned gas lighting and installed the Brush system of electric arc lighting, the opening night was a gala affair. Thousands of people came to witness the event; some had even come from distant cities. "Men fell on their knees, groans were uttered at the sight, and many were dumb with amazement." [2]

Among the larger cities to install electric arc lighting systems were New York, Philadelphia, Baltimore, Cincinnati, Buffalo, and Rochester. Washington, D. C., experimented with the system in 1884 by placing large arc lamps, of 4,000-candlepower, each fitted with conical reflectors, around the dome of the Capitol, and "provided illumination for the various avenues radiating from the Capitol." When the Brooklyn Bridge was opened in 1883, the only system of lighting considered was the electric arc light.

Limitations of the Arc Lamp. While the arc light had its limitations, it remained the accepted system for street lighting for many years. In 1900 more than one hundred and fifty thousand arc lamps were produced. They were used chiefly as searchlights, stage lights, ship lights, and as the beacon in lighthouses. There was scarcely a steamboat on the American rivers, or an ocean-going steamship without electric arc lights. With their huge reflecting projectors the lights could be seen a distance of several miles.

Incandescent Lamps. While scientists, inventors, and engineers were developing arc lamps, others were interested in the incandescent lamp. The two great problems were to find a satisfactory filament of a high resistance substance that would glow with a white heat and not be consumed; and to discover a method to "subdivide the current," that is, to tap the electric current for a single lamp as well as for a series. Again, researchers in the laboratory pointed the way, while the applied scientists were devising a practical, feasible method.

The incandescent lamp was everybody's field. Laboratory scientists discovered that platinum and carbon were high-resistant substances, but the applied scientists found that they were not sufficiently durable to be practical. As early as 1845, the youthful J. W. Starr took out patents for lamps with platinum-foil filament, and also for one having a thin plate of carbon. His untimely death terminated his achievements. When Sprengel invented his mercury vacuum pump in 1865, a means was developed for obtaining a reasonable protection for a filament by enclosing it in a vacuum. Others in this country and in Europe were testing new materials. Some nineteen electric lamps had been invented before 1880.

Meantime, other inventors, including William E. Sawyer, Hiram S. Maxim, and Moses G. Farmer, were attempting to find a method for "subdividing" the current, or more accurately, to develop a method for "subdividing" the light produced by a single current. Professor Farmer in this country, and Richard Werdernmann in London, at about the same time showed the way by connecting the lamps in parallel across a constant potential circuit, instead of through a continuous circuit. Farmer was the electrical superintendent of the United States Naval Station at Newport, Rhode Island. In 1875 he successfully subdivided the current produced by a small electrical machine into forty-two branches, putting a light in each branch.

Thomas A. Edison, 1847–1931. During these same years, Thomas A. Edison, with the aid of some electrical scientists, was carrying on a number of lighting experiments in his laboratory at Menlo Park, New Jersey. Edison, whose name will always be synonymous with the electric light, was not only a colorful genius, but also an inspiration to the members of his staff in their constant search for new devices. He became interested early in the problem of subdividing the electric current and persisted in an effort to discover a satisfactory filament. The search which he and his assistants conducted comprises one of the dramas of the electrical age.

After months of unsuccessful experiments with various metallic filaments, Edison and his staff concentrated on an organic filament. Edison was quoted as saying, "Somewhere in God Almighty's workshop there is a dense, woody growth with fibers almost geometrically parallel and with practically no pith, from which excellent strands can be cut." Despite the doubts of his contemporaries, Edison began searching for such a fiber. Thousands of different plants and grasses, and hundreds of non-vegetable substances were tested. Even hairs clipped from the beards of men on his staff were tried. On one occasion, Edison carbonized a bamboo strip from a palm leaf fan, and discovering that it made an excellent filament, dispatched men to the jungles of Brazil, Java, Borneo, India, China, and Japan to search for better fibrous materials. A particular species of Japanese bamboo that proved to be the most satisfactory was adopted, and was used to provide filaments for his lamps for some six or eight years.

Edison's Patent, 1880. Next, Edison produced a "squirter" filament made from cellulose. In late October, 1879, one of his bulbs equipped with a carbonized cotton filament burned for 40 hours—long enough to prove that such a lamp was practical. Edison's patent on this invention, dated January 27, 1880, stands as the basic patent in the birth of the incandescent lamp.

But Edison's lamp was successful in more features than the filament. His aim was to have an electric light that could be turned on or off like gas. To do so, he employed the arrangement whereby the current flowed in two parallel conductors, resembling the sides of a ladder, and the lamps were connected to cross conductors as the rungs of a ladder. Each light could be controlled separately. A three-wire system, another of his inventions was installed in Sunbury, Pennsylvania, July 4, 1883, and Edison himself personally pulled the switch opening this pioneer station. This system superseded the ladderlike parallel distribution arrangement. He adopted 110 volts for his system, thereby necessitating a much higher resistance in the filament. His was a 110-watt lamp, with 110-volt circuit, and with a resistance of 110 ohms as compared with one ohm in the existing lamps. His lamp had a high-vacuum chamber, hermetically sealed, and was fitted with platinum intake wires.

Edison's aim was for endurance and low cost. When it came to constructing his lamps commercially, he found the cost was $1.25 per lamp, but he had set the selling price as forty cents. He continued to sell the bulbs at a loss until costs were reduced, after three years, to thirty-seven cents. Then in one year the sales were so great that he made up the deficit of the preceding years.

Edison's Generator. Edison's lighting system included the generation of electric energy, and its distribution throughout an extended area. Edison improved the dynamo to carry the load with a constant voltage. He devised what was termed the "Jumbo" dynamo, with a drum-wound armature that maintained 110 volts. He sent this type of dynamo to the Paris Exposition, 1881, where it received great acclaim.

Centralized Electric Power. But even more far reaching than Edison's other contributions to lighting was his establishment of a central power station. Up to this time, wherever electric power was installed, the purchaser operated his own plant. Basements were chugging with power plants, and it became evident that it would be more economical if electric power could be generated at one central station and distributed to nearby consumers.

The opening of the Pearl Street Central Station in 1882 in New York City marked the beginning of the electrical age. Within four months the Edison Company was lighting more than five thousand lamps for two hundred and thirty customers, and the demand for lamps continued far beyond the company's ability to supply them. By 1890 it was supplying current to approximately twenty thousand lamps in many of the major office buildings in New York City.

Distribution of Electrical Energy. The Pearl Street station embodied the world's first system of extensive distribution of electric energy. Besides providing energy for light, the station also provided power for electric motors, for fans, printing presses, and heating appliances. Other cities soon followed. In Appleton, Wisconsin, water-power driven Edison dynamos supplied electric current for a paper mill, a hotel, and some private homes. The Elgin watch factory in Illinois claimed the distinction of having an Edison station in 1885. The town was almost completely lighted by electricity, including stores, streets, and private homes.

Small isolated towns in the far West fell into line. Sheridan, Wyoming, located 200 miles from the nearest railroad, with a population of six hundred, installed a central electric power plant. Separate units of the plant were carted on horseback, piece by piece, and when the plant was installed the inhabitants boasted of a "lit-up" town. Dozens of other towns in Montana, Idaho, and Utah had electric power plants by the 1890's.

Other companies entered the field. Westinghouse, also, installed a number of central stations in eastern cities for supplying direct current to hotels, office buildings, and a few factories. In 1888 the Westinghouse Company was awarded a contract for building a central station plant in London for twenty-five thousand incandescent lamps. The development of the central station was one of the major improvements in the use of electric power.

Electricity Moves Machines

Motors. The central power stations gave great impetus to the development of electric motors. With the system of privately operated power plants, there was little incentive to use electric power. If a factory installed its own steam engine, it utilized the mechanical power directly instead of converting it into electric power first. In the smaller factories, the establishment of a steam engine was not warranted. For them, a small electric motor was the answer.

The first motors were small, with 5-, 10-, or 15-horsepower capacity, and were used to operate sewing machines in tailor shops, small printing presses, ventilating fans, and coffee grinders. Among the early inventors of motors were Frank J. Sprague, pioneer in electric railways, Edward Weston, and Thomas A. Edison. By 1890, when a score or more of motor manufacturers were producing some ten thousand motors, the stage was set for installing electric motors on street railways, elevators, and in factories.

Street Railways. The application of electricity to transportation technology marked a new era. As the cities expanded, speedier and better methods of transportation were required than the horse-drawn streetcars. Electric power, properly geared to railway cars, was the answer. The first electrically operated cars were converted horsecars fitted with small electric motors. The problem was the transmission of the electric current from the power station to the motor on the car. The danger resulting from "live" rails, the first system to be used, caused many cities to prohibit their use. Underground conduits, called "third rails," and overhead trolleys were attempted in some cities. They were often accompanied by the cry of the "deadly trolley" after several accidents were reported.

Overhead Trolley System. A number of inventors, however, including Joseph R. Finney of Allegheny, Pennsylvania, Charles J. Van Depoele of Chicago, J. C. Henry of Kansas City, Thomas A. Edison, and Frank J. Sprague were busy demonstrating both the efficiency and safety of the overhead trolley. When Sprague began operating trolleys in Richmond, Virginia, in 1888, claimed to be the first extensive trolley system in the United States, the surprised inhabitants stood in awe as they watched the steep grades conquered by an unseen power. Other cities soon voted to adopt this newest system of transportation. New Orleans rallied her citizens by issuing a handbill which read, "Lincoln set the slaves free; Sprague set the mule free; the long-haired mule shall no longer adorn our streets."

Lengthening Travel of Electricity. The pioneers in electricity realized that the inability to transmit electric current great distances was the greatest deterrent to the industry. The need for some device that would enable central stations and trolley lines to cover large areas became urgent. When the use of electricity was first introduced it was impossible to transmit electric power for distances greater than a mile or two. Moreover, an increase in the amount of current necessitated an increase in the size and number of wires. To light Fifth Avenue, New York, from Fourteenth Street to Fifty-ninth Street, it was estimated that wires as thick as a man's arm would be needed, and a score or more of power plants would be required to supply the entire city. This limitation was due to the fact that the early application of electricity employed direct current, a one-way current that flows in one direction like water in a pipe.

Alternating Current. The answer to the problem was found in the introduction of alternating current, a pulsating flow of current that could be stepped up to high voltage or lowered to low voltage by the employment of a new device called a "transformer." George Westinghouse of Pittsburgh and his associates pioneered in developing this system. Westinghouse, after achieving fame and wealth from his air-brake invention, became interested in electricity. His early experiments in the direct-current field convinced him of the need for developing another system. He was attracted by an invention of two Europeans, Gaulard and Gibbs, called "a secondary generator," for which he secured the patent rights.

By means of this device, a transformer, (which was really an induction coil), electric power in the form of alternating current was sent over high-voltage transmission lines and then "stepped down" at the various consumer plants to the voltage required for the service.

William Stanley, 1858–1896. After securing the patent rights to the Gaulard and Gibbs system, Westinghouse assigned William Stanley and other members of his engineering staff to work on alternating-current problems. Stanley, working in a laboratory equipped for him at Great Barrington, Massachusetts, in March 1886, gave a public demonstration of an alternating-current lighting system. Within a few months, the cities of Pittsburgh and Buffalo had installed Westinghouse stations for generating alternating current. Among other Westinghouse engineers who made contributions to the system were Oliver Shallenberger, Benjamin Lamme, and Franklin L. Pope. Elihu Thomson of the Thomson-Houston Company secured Westinghouse rights and made a number of contributions of his own.

The "War of Currents." The introduction of the alternating current was bitterly opposed by Edison and other advocates of the direct-current system, and led to "The War of the Currents," the most bitter feud in the early years of the electrical age. The Edison interests charged that alternating current could not be safely regulated and that it was a menace to life and limb, to property, and to livestock.

Bills were introduced in some of the state legislatures to prohibit its use. Edison, in an article in the *North American Review* (November 1889), blasted the users of the "high tension alternating-current system," and charged a number of deaths to its use. He cited the recent installation of the alternating-current system by New York State prison authorities to execute criminals.

The Westinghouse interests refuted the accusation in the same magazine one month later, and explained that in alternating current, the low-tension wires leading into a house were wholly separate and independent from the voltage on the high-tension transmission wires. This was made possible by converters and transformers placed where the house wires led off from the power lines. Each transformer was composed of two separately wound coils, one of thin wire connected with the mains, the other of thick wire connected with the building. The voltage was transformed from 2,000 volts in the core to 50 volts in the thick wire.

Alternating Current Accepted. Incandescent electric lighting by means of the alternating current became standard. Incandescent lamps arranged in every conceivable manner were used for advertising and decorating. In 1898 the Capitol building in Washington was lighted by incandescent lamps—fourteen hundred lamps of 16-candlepower each were installed in the House of Representatives, and twelve hundred were suspended from the ceiling of the Senate chamber. Factories, mines, and mills took up this new system of illumination. By 1900 more than twenty-five million incandescent lamps were in use in the United States, and almost one billion dollars had been invested in the electrical lighting industry.

Alternating-Current Motors. Westinghouse, believing that the alternating current could be used as a source of power as well as for illumination, secured the services of Nikola Tesla, a brilliant scientist born in Croatia who had emigrated to America. Here he began building motors. Tesla succeeded in developing a "polyphase" alternating-current system, with a unique induction motor suitable for the alternating current. Because of the pulsating character of the energy supplied by alternating current, the existing motors would not operate satisfactorily, but would stick on "dead center." Tesla's invention produced a condition favorable for motor operation. It was called "polyphase," because two or three alternating currents were combined to produce a magnetic field, practically constant, and rotating at a constant speed. Induction motors were designed to use the polyphase system. It supplied twenty times more power than the largest direct-current apparatus.

William Stanley, of the Westinghouse laboratory in Great Barrington, also designed induction motors of the alternating-current type, as did Benjamin Lamme, Charles Bradley, and Elihu Thompson, whose inventions became a part of the General Electric Company. By 1890 the alternating-current induction motors were being used to operate printing presses, mining machines, carpet looms, lathes, grinding machines—in short, they were used in almost every existing industry.

Alternating-Current Power Applications. With the development of the alternating current, great impetus was given to the utilization of electricity. It was found to be especially adapted to operating trolley cars (for the old storage battery service was becoming obsolete), and rapid improvements were being made in producing generators and motors for street railways. An electric snowplow was used in St. Paul in 1891. St. Louis in 1892 claimed to be the first city in the world to install special electric cars for mail and express suburban service. In a single year, 1892, some thirty thousand horses and mules used in cities throughout the United States were replaced by electric power. By 1900, the United States had more than 15,000 miles of electrified railways, whereas Canada, England, France, and Russia combined had approximately 1,000 miles. Electrically operated trolley cars were carrying nearly six times as many passengers as the steam railroads. A few electric locomotives were introduced during the 1890's, chiefly to operate trains through tunnels.

Steam Turbines. The coming of the steam turbine, and a corresponding decrease in cost in producing electricity, ushered in a new era of great expansion. It brought about a significant technological development in the electrical age. Factory owners discovered they could now purchase electric power from a central station cheaper than they could produce it in small isolated plants. Home owners also profited, since they could obtain electric power from distant central stations at greatly reduced costs.

Gas Engines. Another source of power for dynamos, the gas engine, was beginning to attract the attention of engineers and scientists by 1900. Already gas engines were being designed for use in both central stations and in isolated electrical plants. With the opening of the twentieth century the groundwork was being laid for internal combustion engines which could be applied to generating electric power. The birth of a new industry, that of public utilities, was at hand.

ELECTRICITY INVADES MANY FIELDS

Chicago Exposition, 1893. At the Chicago World's Fair in 1893, the new electrical displays attracted more attention than all the other exhibits. Edison and Westinghouse were rival contenders for the contract, since the display would be of incalculable advertising value. The contract finally went to Westinghouse, not only for lighting the Fair, but also for supplying the electric power for the numerous exhibits and industries. Westinghouse had just lost a patent suit to Edison which prohibited him from manufacturing a lamp made "entirely of glass," and was forced to devise another type. He came up with his famous "stopper lamp," which proved satisfactory.

When the Fair opened, Westinghouse had provided two hundred and fifty thousand lamps, the necessary transformers, generators, switchboards, underground cables, power stations and dynamos, and hundreds of miles of wiring. While the Chicago Fair had not been designed as an electrical exhibition, it turned out to be such. For the first time in the history of international expositions, separate buildings were devoted solely to electrical exhibits, and electricity was used almost exclusively for the transmission of power to the various buildings. The electrical fountain, admired by millions of visitors, became the chief attraction.[3]

Harnessing Niagara Falls. The possibility of utilizing the great thundering power of Niagara Falls that was going to waste had challenged engineers for a half-century, but it was not until the 1890's that a plan was formulated to harness some of the potential horsepower exerted by the Falls, estimated at fifty million. Several attempts, some of which were successful, had been made to utilize the water power for local purposes, but no one had succeeded in transmitting power any distance from the immediate vicinity of the Falls.

In 1889 an International Niagara Commission of five distinguished scientists of Europe and America was organized, with Lord Kelvin (William Thomson) as chairman. Prizes were offered to attract the best engineering knowledge and experience "to propose plans for producing and transmitting power from Niagara Falls." Proposals came from all over Europe and America, none of which was acceptable. In the summer of 1893, the Cataract Company (organized by the Niagara Commission) announced that it had decided upon a polyphase alternating-current system for the Niagara project. Since Westinghouse had just installed the only complete polyphase alternating-current system of generating, transmitting, and utilizing electric power on a large scale, he was awarded the contract.

Niagara Attracts Industry. After two years of work the first turbo-generator of 5,000 horsepower was put into operation April 1, 1895. The immediate demand for more power led to the installation of two additional turbo-generators. Industries began to locate in the vicinity of Niagara. One of the first was the Pittsburgh Reduction Company (today the Aluminum Company of America), which needed a great amount of electricity to manufacture the new product, aluminum. Other industries soon followed, and Niagara became America's first center of industry powered by electricity.

Long-distance Transmission. The inhabitants of neighboring towns sought this new source of power both for lighting their homes and for operating their industries. In 1896 two transmission lines were extended to Buffalo, 22 miles away. When a three-phase, 11,000-volt electric current was sent that distance a new technological achievement had been reached. Engineers and electricians were attempting to determine just how far power could be transmitted by electrical means. Lord Kelvin observed, after inspecting the Niagara Falls installation in 1897, that he believed 300 miles would be about the limit, a limit rarely exceeded even today.

Factories Electrified. The transmitting of electric power to factories marked an important development in the history of American manufacturing. The *Census Report* for 1880 had not mentioned the use of electric power in industrial use, but by 1900 the capacity of industrial electricity was rated at almost 200,000 horsepower. Factories no longer had to be cluttered up with long lines of shafting, pulleys, and belts, powered by steam engines or water wheels. Nor were manufacturers any longer forced to locate their plants with regard to water power or raw materials.

Home and Farm Uses. The first use of electricity in the home was for supplying light. After central stations became popular an increasing number of homes installed electric lights, and other uses developed rapidly. Heating devices, such as hot plates, flat irons, and cooling devices like electric fans were found in a number of homes by 1900. Electric heating also became available before the end of the century. At the Chicago World's Fair many varieties of electric cooking utensils were displayed. Electric pressing irons, bed warmers, and radiators were on exhibit, and many of the Fair's visitors returned home with orders for installing these new devices in their homes. Another new household device, the electric heater, was introduced—a forerunner of the great changes in heating technology that were to take place in the early decades of the twentieth century.

Rural electrification as known today did not come about until the second quarter of the present century. However, many attempts were made to introduce electricity into some farm operations during the latter part of the nineteenth century. L. H. Bailey, eminent horticulturist of Cornell University, discovered that plants, when placed in glass houses under arc lamps, matured faster than plants grown under ordinary conditions.[4] A vegetable farmer near Boston claimed that the use of electric lights matured his lettuce crops five to six days earlier than usual. The experimental farm of the Alabama State Agricultural and Mechanical School substituted electricity for steam power to gin and press cotton as early as 1892.[5] In California, during the late 1890's, some progressive farmers utilized a running stream to generate electricity for lighting their homes and barns, and to power some of their lighter machines. Storage batteries were used as auxiliaries in emergencies.

Electricity in Naval Warfare. In the 1880's electricity was used to provide illumination for naval vessels, to propel small boats, to fire guns, to operate foglights and searchlights, to run motors for repair shops, and to explode submarine mines and torpedoes. Captain Bradley Fiske in 1891 invented a successful range finder that was used on American vessels and was later adopted by the French, Italian, and Russian navies.

During the Spanish-American War all regular vessels of the United States Navy were equipped with electric lighting plants. On the battleship, "Brooklyn," two turrets were controlled by electric motors and were declared to be superior to steam-operated turrets. A few of the hospital ships were equipped with electric cooking outfits. Electric plants were installed in coast fortifications, storerooms, and magazines, and electric signals dispatched messages both on shore and on naval vessels. At Fort Monroe a school was established for instructing noncommissioned officers in the use of electrical instruments, a forerunner of the specialized training schools for radio, radar, and electronics that were organized during World War II.

The Electric Motor Serves the Public. "The electric motor has won its place," said a writer in the *North American Review* (April,

1888), "and laid hold on the future; it has begun a tremendous revolution in the mechanical world." Electric motors were used to operate elevators and escalators in office buildings, hotels, and apartment houses. By 1900 there were more than three thousand elevators of various kinds in daily operation. Even the quiet dignity of the Washington Monument had been invaded by the installation of an elevator to transport visitors to a height of 555 feet.

By 1900 there were some three hundred electrically powered taxicabs in New York City. Each vehicle was equipped with two motors, a storage battery, and a lever controller to regulate the speed. At the New York auto show that year, there were more electric vehicles on display than gasoline cars. In Washington, D. C., electric buses were operating between hotels and railroad stations in 1900.

Electricity in Medicine. Electricity was used by the medical profession, and despite the claims made about its miraculous curative power by certain "quacks," it proved to be a scientific wonder when properly used. The incandescent lamp made it possible to explore interior parts of the human body. The electrolytic needle was being used before 1900 to eradicate birthmarks, remove surplus hair, and to decompose tumors. Electromagnets were used to extract small pieces of metal from the eye. High-voltage currents were used to anaesthetize any portion of the body to lessen pain and to excite paralyzed muscles.[6] The electric motor was being used by dentists to operate their drills, and Roentgen's X-ray (1895) was hailed as one of the greatest contributions in all medical history.

Electricity Marks a New Power Era. The electrical age introduced the second great power age in our nation's history, steam being the first. By 1900 the basic types of motors had been invented. The alternating current had become the system for generating, transmitting, and utilizing electric power. The transformer, high voltages, the induction motor, and long distance transmission made possible the rapid spread of electric light and power. The electric railway had arrived. American electrical engineers were designing and building electric plants for European countries, and electrical equipment valued in excess of five million dollars annually was being sent abroad by 1900.[7]

Wireless Communication. Electricity was performing many heretofore unknown services by the close of the century. It was no longer a toy used merely to amuse the public. The discoveries of Faraday, Galvani, and Henry had been developed to the point where they were being used as man's servants. One of the most amazing uses of this new phenomena was in the field of communication. Electrical communication, without wires, had moved out of the experimental stage by 1898. In that year Reginald G. Fessenden of the University of Pittsburgh, was giving demonstrations of a system of wireless transmission for the United States Weather Bureau, a portent of a new development in electronics.

Theory and Practice Unite

The pioneers in the electrical field also pioneered in another field —that of uniting the forces of pure science and applied science. Again, it was Edison who led the way in breaking down the barriers between the scientific theorists and the practical scientists. He demonstrated through his many electrical discoveries that the role of the applied scientist, the technologist, was as important as the work of the pure scientist. While he respected the work of the theoretical scientist in his search for truth, yet he demonstrated through his many electrical discoveries that the laboratory technician, the inventor, was likewise working as a pure scientist. He not only contributed to the storehouse of pure knowledge, but he also showed how the knowledge thus acquired could be put to man's use.

Professor Elihu Thomson. Another outstanding example of this liaison between applied science and pure science was the role played by Professor Elihu Thomson, who moved from an academic background into the commercial field. A great teacher and a respected scientist, he became a great inventor and a leader in the electrical world. Proud of his professorial title, he carried his teaching technique of experimenting and testing into the Thomson-Houston Company, makers of dynamos and other electrical apparatus, and later into the giant General Electric Company. His experimental work had, by the turn of the century, led to the establishment of a pioneer research laboratory, claimed to be the first in the industrial world.

Schools of Electrical Engineering. The harnessing of this new source of power, electricity, and the many uses to which it could be applied, brought about changes in the industrial and social world even more important than those that accompanied the age of steam a century earlier. The most immediate problem was that of training men for careers in a new science. Electrical engineers were in demand, and this demand in turn led to the introduction of courses in electrical engineering in colleges and universities, and later to schools of engineering throughout the country. The rapid rise of engineering schools caused Senator George Hoar of Massachusetts to observe (1896), "that every technical school in the country has established, or is seeking to establish, a department of electrical engineering." He noted that the numerous establishments engaged in the manufacture of electrical equipment and machinery, "were greedily seeking inventive men with the best scientific training. Youth, possessing these qualities," he added, "command large salaries instantly."

Public Acclamation. The new science of electricity was a challenging one. A contributor to the *Scientific American Supplement,* March 7, 1900, noted that, "The public has the notion that if there is anything under the sun that electricity can not do, it is not worth doing." Marvelous as these services were in 1900, they were only a beginning of the revolutionary developments in the electrical field that were to occur within the next half-century.

Notes for Chapter 24

1. *U. S. Census,* "Report on Manufacturers" XVI (1890), 239 *ff.*
2. Franklin L. Pope, "The Illuminating of Streets and Public Places," *The Engineering Magazine,* IX (May, 1895) 263.
3. Charles E. Skinner, "Lighting the World's Exposition," *Western Pennsylvania Historical Magazine* XVII (1934), 13-22.
4. "Electricity in Agriculture," *Cassiers,* XI (February, 1897), 266 *ff.*
5. A. F. McKissock, "Electricity on a Southern Farm," *The Electrical Engineer,* XIII (June, 1892), 549-550.
6. George H. Guy, "The Uses of Electricity in Sanitariums," *The Chautauquan,* XXV (August, 1897), 476-479.
7. *U. S. Census Report,* X (1900), 155.

For a recent excellent review of the developments in electricity, see Harold C. Posser, *The Electrical Manufacturer: 1875–1900* (1953); Haig Gordon Garvidian, *Thomas Alva Edison, Builder of Civilization* (1947); Floyd Albert Lewis, *The Incandescent Light* (1949); *Diary and Sundry Observations,* edited by Dagabert D. Runes (1948); Malcolm MacLaren, *The Rise of the Electrical Industry during the Nineteenth Century* (1943); Henry G. Prout, *Life of George Westinghouse* (1926); Walt Bowden, *Bulletin No. 539, U. S. Department of Labor: Technological Changes and Employment in the Electric Lamp Industry* (1923).

Chapter 25

AGRICULTURAL TECHNOLOGY

The age of "big business," marked by the birth of the steel industry, the petroleum industry, and the coming of the electrical age, also witnessed a number of major advances in agricultural technology. Despite the fact that the manufacturing centers had caused an exodus from the farms to the factories, the United States was still a predominantly agricultural nation at the close of the nineteenth century. Its farmers were producing not only enough food for home consumption, but also great quantities for export.

Agricultural Implements

Agricultural Scientists Abroad. In the midst of the Civil War, Isaac Newton, first United States Commissioner of Agriculture, had taken an important step in introducing principles of science in agriculture. He had dispatched a botanist and an entomologist to Europe and Asia in the summer of 1862 to study the habitat of plants and various crops with the view of introducing new varieties into this country. The agents returned with bags of grass and grain seeds, with cuttings of plants and vines for horticultural purposes, and with copious notes on methods for combating insects that were injurious to plant life.

Commissioner Newton, after making a two-year study of the nation's agricultural needs, advocated greater diversity of crops and the introduction of new grains. In his annual report of 1864 he said, "that of the two requisites for the successful cultivation of the nation's crops—fertilization and pulverization—the latter could be achieved only by improved implements and machinery." He called attention to the many new tools and machines designed to perform most of the farmer's tasks, including improved plows, harrows, drills, cultivators, and reapers. "Only the picking of cotton and tobacco," he said, "has defied our mechanicians."

Need for Improved Plows. The most immediate need in expanding and diversifying the nation's agriculture was an improved plow. Although John Deer had undertaken the manufacture of other plows in the 1840's, most of the plows were still being made from cast iron. They were never satisfactory, particularly in tearing up the tough prairie sod of the Midwest. Farmers were calling for new, improved plows throughout the war years, and inventors were attempting to meet those demands. The demands became greater following the close of the war and the opening up of new farm lands. The two hundred fifty-five patents issued on plows in 1869 (the largest number issued for any farm implements in that year) indicated the desire of both farmers and manufacturers to introduce improved plows.

James Oliver, 1823–1908. The old cast-iron plows had one major defect—their moldboards would not scour well. James Oliver of Indiana who "spent all of his money and lost most of his friends" received a patent in 1868 for producing a plow from chilled iron. The formula for producing the chilled iron desired by Oliver prescribed a method for properly cooling the metal by passing a stream of warm water over it. After Bessemer steel became available, Oliver substituted steel for iron. This enabled him to make further improvements. He cut small grooves along the inner surface of the mold, thereby allowing the gases to escape. He invented a method of annealing that made the metal tougher, and produced a mirror-like finish that made the metal rust-resistant.

Interchangeable Parts. Oliver also introduced the "slip-nose share," which permitted the owner to reverse or replace the worn-out part. The plow was so designed that it reduced the draft on the team from 25 to 40 per cent, depending upon the type of soil. Because of its careful design, efficiency of the chilled-steel moldboard, and its lightness in weight, few changes have been added to this model of walking plow during the past eighty years.[1]

Riding (Gang) Plows. Other improvements, which the farmers of the prairie region welcomed, were the introduction of the sulky plow, carried on wheels; and gang plows, which had a seat attached, permitting the operator to ride. J. C. Pfeil of Illinois invented a gang plow in 1865, which the Commissioner of Agriculture declared, "brought forth incredible stories of its efficiency in the prairie fields of Illinois and other western states." Following a plowing exhibition, a committee tested the draught of the plow with a dynanometer and found that, "it ran lighter by one hundred forty pounds than other plows running at the same depth that were held by the ploughman while on foot." Such wheel plows became popular in the western country. In 1869 the Commissioner of Agriculture wrote: "It is worthy of note that the patents granted on wheel plows . . . to residents of California and Oregon largely exceed in number those granted for inventions of a like character from all the other states of the Union." [2] The wheel plow won great popularity in the prairie country, especially in the Red River Valley wheat region. Manufacturers could not keep pace with the demand. In many sections of the country the old-fashioned walking plow became obsolete.

Disc Plows. In 1868 J. S. Godfrey introduced a variation in the construction of disc gang plows by using a concave moldboard with revolving scrapers. J. M. Cravath of Bloomington, Illinois, patented a plow that had gangs of three solid discs. Although the side thrust was not practicable in principle, it was basic to the future disc plow.

Steam Plows. A number of inventors were experimenting with steam plows during the post-Civil War decade. In 1868 P. H. Standish of California was granted a patent for "a steam-plowing outfit." The following year, five patents for improved steam-operated plows were issued, "which if workable, would be advantageous to prairie farming." [3] The limitations of the steam plow were never entirely overcome; the introduction of the internal combustion engine and gasoline tractors curtailed their further development.

Spring-toothed Harrow. The old spike-toothed harrow was still the most commonly used harrow until after the Civil War. It crushed the clods of earth, pulverized the lumpy soil, and smoothed the surface to a depth of 2 to 3 inches. An improved spring-toothed harrow that came into use in the late 1860's was hailed as a great improvement. Long, curved, spring teeth made of steel, shaped like the teeth of a hayrake, only shorter, were substituted for the iron spikes. The chief advantage of the spring-toothed harrow was that by means of a lever it could be adjusted to any angle or depth required. The teeth would spring when striking roots or rocks and automatically dislodge the accumulated debris. The spring-toothed harrows were made in different sizes, ranging from one section to four sections. With the larger harrows farmers could pulverize from 20 to 25 acres per day.

Disc Harrow. Another improved harrow, known as the "disc" harrow, came into general use in the 1870's. The discs were made concave, convex, or straight, and varied in diameter. Each section of the harrow could be controlled by an independent lever. An improvement on the disc harrow was an automatic scraper for cleaning the discs as they revolved, preventing them from clogging with mud, weeds, and debris. Since the addition of these improvements, no major technological improvement has occurred in the construction or operation of this implement. Heavy rollers, some equipped with spikes, were later introduced, but they involved no major technological changes.

Corn Planters. Improved corn planters and grain drills during the post-Civil War period enabled the farmers to increase their acreage and meet the growing demand for foodstuffs. The improved corn planter of the 1880's consisted of a furrow opener and a seed box with a special metering device which counted out the number of grains of corn to be dropped. Next came the corn "checkers," often called the "check-rower," which was operated by a trip wire and dropped the grain at regular intervals. With these improved machines, farmers could plant two rows at a time and from 12 to 15 acres a day. The days had passed when the farmer with a sack of grain about his neck, slowly plodded down a furrow and measured out three grains of seed corn to the hill.

Grain Drills. Improved grain drills were in even greater demand than corn planters. Scattering grain broadcast was a slow, laborious, wasteful task, and resulted in irregular seeding. Drills, mounted on two wheels, equipped with force feeding devices, and fluted hoes that opened the furrows for the seed, were in use in many farming areas during the 1870's. American-made drills made an excellent showing in the Centennial of 1876. John Coleman, chairman of the section on farm implements, declared that the American drills led the field and had caused the British to refrain from exhibiting their drills. In 1878, at the Paris Exposition, the "Farmer's Friend," an American-made drill, was awarded a gold medal because of the efficiency effected in regulating the amount of grain to be sown, the ease of operating the shovels, and the facility for using fertilizer and seed attachments. By the mid-1870's, most of the winter wheat grown in the upper Mississippi and the Missouri Valley areas was sown by drills. Few fundamental changes have occurred in grain seed drills since this period.[4]

The Cultivator. The straddle-row cultivator, which came into general use during the 1870's, proved to be a great labor-saving machine. With this new two-wheeled, double-shovel machine drawn by two horses, a farmer was able to cultivate twice as many acres as with a one-horse plow or a double shovel. By placing a seat on the cultivator, he could ride and operate the shovels by foot and save himself many weary steps. Other cultivators, equipped with discs which opposed each other, were introduced about this time; they pulverized the ground even more thoroughly. The cultivator could be used on any row crops—corn, cotton, beans, peas, potatoes. The age-old hoe was becoming obsolete.

Harvesting Machines

The Reaper. The greatest of all labor-saving devices in agricultural history was the reaper which, following a number of technological improvements during the 1870's, came to be called the "binder." When cutting grain with the earliest reapers, several extra helpers were needed: one to operate the machine and drive the team; another to rake the straw into bundles; another to bind the bundles; and still another to set the bundles upright in shocks.

The development of an efficient, automatic binder was the aim of scores of inventors, including the Marsh brothers, the Gordon brothers, Jacob Behel, S. D. Locke, and many others. One of these inventions, the Locke wire binder, found wide favor among farmers, but only for a season or two. The machine itself was satisfactory, but not the wire. Bits of wire proved to be an annoyance to threshing crews; they became a menace in straw piles and when swallowed by cattle often caused death. Occasionally, tiny bits of wire were even found in flour.

The Twine Binder. A new technological development, a device which made bands from twine, and the invention of a knotter which tied the twine around the bundle of the grain and then cut the twine, drove the wire binder off the market. The twine binder brought about the final evolution of the harvesting machine. John F. Appleby, Jacob Behel, and Marquis L. Gorham were the pioneers in developing the twine binder and knotter. Imported Manila jute and sisal were woven into balls of binder twine and sold to every farmer who owned a binder. The twine binder, called a "self-binder," more than any other single machine enabled the farmers to expand their wheat crops.

In 1882 the McCormick Company, having turned from wire binders to twine binders, sold over fifteen thousand twine binders. The twine binders with their automatic knotters made possible the rapid extension of the wheat belt into the West and Northwest; and large scale farming became common practice in those areas. Schafer cites one farm near Casselton, North Dakota, on which sixty self-binders were employed as early as 1882.[5]

Threshing Machines. During the 1880's a number of developments occurred in threshing technology. Straw stackers, which carried the straw up a light frame by means of an apron, were added to the threshing machines. They saved the labor of two men. Automatic band cutters, self-feeders, weighers, and blowers were introduced during this time, all of which did the work of five to six men.

Steam Power versus Horsepower. The most important advance in threshing technology prior to 1900 however, was the substitution of steam power for horsepower. Steam engines brought about larger machines, provided greater speed in operation, and greatly increased the threshing capacity. Steam threshers could easily thresh from 700 to 1,000 bushels of wheat in one day. Rogin, quoting from an article in the *Country Gentleman* (1882), describes the threshing procedure on a Dakota farm of 800 acres. Immediately after the harvesting had been completed, an engine was set in position and was "steamed up."

> With six teams to haul bundles, four men to pitch on to the wagons, one team to haul water for the engine, three more to haul the grain away, we commence threshing. Feeding from two sides we have two band cutters, two feeders, three men to sack the grain, one engineer and one fireman. With this force we thresh about fifty acres of wheat per day, or from one thousand to twelve hundred bushels. It takes us but thirty minutes to move, set up, and commence work.[6]

Combine. The next major technological development in the history of grain harvesting was the invention of the combine: a combined harvester and thresher. The combine, when operated by a large steam-engine tractor (or more often by a "thirty-mule team") and drawn across the fields, cut, threshed, cleaned, and bagged the grain in one operation. Combines, drawn by horses or mules were best adapted to large level fields and were first used extensively in California in the 1880's, where the scarcity of labor made them popular. Some of the larger tractors could cut and thresh 30 acres of grain in one day.

Milling Processes

Flour Milling. The term, "a revolution in milling," crept into the industrial journals and magazines of the 1880's. Prior to that era, small gristmills dotted the countryside and supplied each local community with flour and corn meal from grain grown in the neighborhood. The local miller seldom contemplated any change in either milling technology or plant expansion. The "low-grinding" process, in which the millstones operated close together, produced the maximum amount of flour in the first grinding, and that seemed to satisfy everyone. But this ancient method became obsolete during the post-Civil War period.

The rise of cities, with the increase in city bakeries, created a demand for a greater production of flour. The expansion of vast wheat areas of the West, and more important, the introduction of hard spring wheat in Minnesota and the Dakotas which defied millstone grinding, necessitated the development of a new, large scale milling technology. The antiquated gristmills gave way to high-speed mills equipped to handle the spring wheat.[7]

There had been no major developments in milling technology from the time of Oliver Evans until the 1870's, when two innovations, the middlings purifier and the roller mill were introduced. With these two inventions the successful grinding of hard spring wheat was accomplished.

The "New Process." A number of millers in Minnesota during the late 1880's were experimenting with various milling methods for purifying their flour, which was dark in color and could not compete with the product made from winter wheat. They came up with three major discoveries. By operating the buhrstones slowly, which kept down the heat, they prevented discoloration of the flour; and by using a high grinding method, they cracked the wheat grain, rolling it into coarse particles rather than grinding it thoroughly. From this method they obtained about equal amounts of flour and middlings from the first grinding. They then reground the middlings to secure the choicest flour.[8] These country millers had discovered the basic secret for purifying the middlings and using them in flour. Their method later became known as the "New Process."

La Croix's Purifying Process. While the Minnesota millers were experimenting with these new developments, a Frenchman, Edmond L. La Croix, who had observed a "purifying" process that had been tried out in a flour mill in Paris, arrived in Minnesota, accompanied by his brother. Alexander Faribault, a miller, engaged the La Croix brothers to build a mill on the Straight River in that state. The "purifier" consisted of a series of silk sieves, designed to separate the starchy flour from other materials, while a stream of air was forced up under the sieves to carry off the bran particles, thus completing the purifying process. Neighboring millers also engaged the La Croix brothers to build purifiers for their mills. They were "too busy" they later testified in a patent suit, "to bother with taking out patent rights."

The "Middlings Purifier." In 1870, George H. Christian, a Minneapolis flour broker, C. C. Washburn, a flour manufacturer and one time governor of Minnesota, and George H. Smith, an expert head miller, organized a partnership and engaged Edmond La Croix to build a purifier for their mill. When Smith, the head miller, added a series of brushes that traveled back and forth and automatically swept the flour from the clogged sieves, a task that required the services of one man, the famous middlings purifier was perfected. The first major revolution in milling in more than eighty years had been achieved.

The hard spring wheat, with its abundant glutinous cells, was ground and passed through the purifier again, and the wheat berry was converted into a clean flour of fine texture. This flour brought from $1.50 to $2.00 a barrel more in eastern markets than the flour milled by the old method.[9] Handsome profits led to a rapid expansion in mill construction aimed at producing the New Process flour. This was especially true in Minnesota and the Northwest. "Its progress," declared Edgar, "amazed and troubled flourmakers everywhere, stirring the whole milling world at home and abroad with a vague spirit of unrest and uneasiness." [10]

Northwestern Wheat Fields. The Northwest became the center of the wheat and flourmilling industry, for it was the center of the spring wheat country. Settlers swarmed into the Red River Valley in the 1870's and 1880's and began to produce wheat on a scale never before known. During the decade 1870–1880, in Minnesota alone wheat production more than doubled; it jumped from 14,000,000 to 34,000,000 bushels. The number of mills also doubled, from 216 to 436; and the value of the flour produced rose from $7,500,000 to over $41,000,000. The mills in Minneapolis alone had an annual output of 2,000,000 barrels. Minneapolis flour began to invade the foreign markets. Governor Washburn sent a representative abroad to introduce the New Process flour in England and on the Continent. For years Minneapolis and Hungarian flour, says Edgar, "were battling it out in Europe for world leadership." The middlings purifier was to milling what the reaper was to agriculture.[11]

The Roller. The second revolution in flour milling was the introduction of the roller. This process of crushing the grain, rather than grinding it, was said to be "the most profound in its results of any device ever invented in the mechanics of flourmilling." [12] Credit for this invention was given to John Stevens, of Neenah, Wisconsin, whose pioneer work was as important in milling technology as was

La Croix's middlings purifier. Stevens began in the early 1870's to experiment with a device that cracked the wheat grain open, but left the bran intact. He crushed the wheat between rollers made of wood. Later he experimented with porcelain and with chilled-steel rolls. He achieved his greatest success when he finally introduced corrugated rolls. With this equipment, he erected in 1874 a complete new mill which could be operated by either water power or steam power. The results were astonishing.

The use of rollers raised the output from a daily capacity of two hundred barrels of flour to five hundred, yielding 90 per cent good flour in contrast to 25 per cent produced by the old method. Rival millers broke into Stevens' building to copy the plan, then improvised their own rollers, and soon the entire milling industry began shifting to the roller-mill process. "Nearly every mill which makes any pretense to do nice work is now using rollers," noted the *Northwestern Miller* in September, 1879; and then Edgar, editorially declared, "The long reign of King Millstone over the Flouring Kingdom was approaching its close in a milling revolution which was to depose him and establish in his stead the Republic of the Roll."

"Patent Flour." The introduction of the roller mill, and the gradual reduction process gave birth to the term "patent flour," used to designate the best grades, irrespective of the milling system used. While the middlings purifier changed the quality of the flour, the roller mill greatly increased the quantity of flour that could be manufactured.

Minneapolis Becomes Milling Center. The results of the roller upon the flourmilling industry were many. Minneapolis became the flour-mill center of America. The introduction of the roller system, declared Lawson,

> . . . added 100,000 people to the citizenship of that place in five years, and made it at a single bound the flour milling emporium of America; by 1886, sweeping into its mills annually thirty-three million bushels of wheat, that ten years before was almost worthless; and settling the bleak northwestern prairies with several millions of hardy pioneers, raising wheat.[13]

Milling Monopoly. The period of the 1880's saw the introduction of several technical devices that led to mass production in milling. A belt drive for rolls displaced the gear drive. Steam power displaced the less dependable water power. Safety devices, such as dust collecting machines, were installed, aimed at keeping the mills free from the dangers of flour-dust explosion. The larger mills, backed by greater capital, began mass production on such a scale that they drove smaller mills out of business. "How could a crossroads miller," asked one observer, "successfully compete with a Charles A. Pillsbury?"

By 1900 the flourmilling industry had become almost wholly mechanized. The middlings purifier and the roller mill, geared to mass production, greatly increased the annual production of flour. Significant as was this technological revolution in the flourmilling industry, however, the food technologists were not satisfied. They pointed out that this supermilling process was destroying the nutritional value of the wheat grain. The white dainty flour delighted the housewife, but it was deficient in food values. The more progressive millers then turned to the chemists in an effort to retain the food values, and at the same time preserve the technological triumphs of the middlings purifier and the roller mill.

Grain Elevator. The grain elevator (see Chapter 16), originally designed before the Civil War and used to provide storage space for the country's surplus crops, came into universal use throughout the vast wheat-growing areas. Chicago, where agricultural and manufacturing technology joined forces, became the world's largest grain market and greatest shipping center. The need for storing grain as it arrived by trainloads, and holding it for later shipping, led to the erection of these huge structures. They were built somewhat like skyscrapers and held from 500,000 to 5,000,000 bushels of wheat.[14] These tall, slender, perpendicular bins were equipped with an endless, power-driven belt, fitted with buckets or cups to carry the grain from the bottom to the top of the building. The grain was then distributed to the proper bins, from which cars and boats could be loaded by the gravity flow of grain. The grain elevator made possible the rapid handling of the nation's largest agricultural commodity in both domestic and foreign commerce.

Learning, Research, and Experimentation

It was during the last quarter of the nineteenth century that a new generation of farmers began to turn from the old traditional methods of farming and adopt more scientific methods. They became interested in all the latest improvements. They indicated a willingness to "go to school" to study recent developments in agriculture and to learn about the operation of the latest farm machinery. The Morrill Act of 1862 had specifically provided for the endowment, support and maintenance of colleges where the leading objective was "to teach such branches of learning as agriculture and the mechanical arts. . . ." By 1880 twenty-eight agricultural colleges, in as many states, had a total enrollment of approximately five thousand students.

Federal Aid for Agriculture. A move aimed at aiding farmers to improve their methods and to further increase their production gained congressional support. Senator John T. Morgan of Alabama, addressing his colleagues in the forty-sixth Congress (1880) on "The Glory of Agriculture," called for a more generous appropriation, "so that we may lay broad and deep those foundations (agricultural education) in order that our magnificent production may be carried to all parts of the earth." He observed that recent improvements in agricultural implements and "our agricultural skills" had driven out the productions of foreign markets, and that the United States had a monopoly of the markets in most countries of the world.

Agricultural Research. Congressmen from the farming states (which included most of the nation) were urged by their constituents to support the Commissioner of Agriculture in obtaining a special appropriation to aid him in diffusing useful information on subjects related to agriculture, the invention of new farm machinery, methods of acquiring, propagating, and distributing "new and valuable seeds and plants." (Annual Report, 1882.) A number of studies were under way covering other subjects, such as the best methods for eradicating harmful insects, the use of new fertilizing materials, methods of vaccinating livestock against deadly diseases, information on new cotton ginning machines, and methods of sinking artesian wells throughout the plain states.

Sugar Beet Industry. The sugar beet industry was one of many new projects in which the department of agriculture was particularly interested. Members of the Commissioner's staff had conducted experiments on the department's grounds in Washington during the 1870's, and in 1880 the commissioner reported on the improved English and French implements of tillage that had been designed for the special cultivation of beets. This information was passed on to sugar beet growers throughout the country, with the request that they submit reports on their production. California growers reported success in producing sugar beets that were used in sugar manufacture in that state.

Department of Agriculture, 1889. The nationwide interest in agricultural development led to raising the Bureau of Agriculture to the status of a Department of Agriculture in 1889, and the appointment of a Secretary of Agriculture as a member of the President's cabinet. One of the first acts undertaken by the newly created Department of Agriculture was a program of scientific research. A special report issued by Assistant Secretary Edwin Willits, in 1890, declared that successful agricultural practices must be conducted on scientific principles. "Nature cannot be cheated," he added, "and her implacable laws will surely find out their transgressor." American farmers were told that plagues, famines, stricken soil, and poor vegetation were not accidental. They all could be prevented, if only the farmers would apply the knowledge resulting from scientific experiments.

Agricultural Experimental Stations. A number of state agricultural experimental stations were established during the 1880's, the most noted being the experimental station at Wesleyan University in Connecticut. Norman J. Coleman, able Commissioner of Agriculture, after years of agitation succeeded in 1887 in having Congress pass the Hatch Act, which provided for the establishment of an agricultural experimental station in every state in the union. The United States was the first nation in the world to take such a step. The experimental stations rendered valuable service by acquainting farmers with the value of fertilizers, proper cultivation of the soil, the latest methods of fighting insects (such as boll weevil), and the most scientific methods for breeding and feeding live stock.

U.S. Weather Bureau. The department was given twenty thousand dollars in 1890 to investigate "the proper location for artesian wells and their use in irrigation in the semi-arid regions lying between 97th degree west longitude and the eastern foothills of the Rocky Mountains." Also, in 1890 the United States Weather Bureau was transferred from the Signal Corps of the United States Army to the Department of Agriculture. The Bureau began issuing daily wind, weather, and temperature forecasts from 8:00 A.M. to 8:00 P.M. It gave out weather information to press associations, telegraph companies, and newspapers throughout the entire United States. "Very few disastrous storms occur now," declared the Secretary in 1894, "of which the people are not warned twenty-four to thirty-six hours in advance of their culmination." That same year the Department announced that it had allocated more than 51 per cent of its annual appropriation to scientific research and investigation, and the range of topics covered every branch of agriculture.

Approximately one thousand of the two thousand and two hundred persons employed by the Department of Agriculture in 1896 were engaged upon "purely technical and scientific work," and more than 70 per cent of the total appropriation was allotted to scientific and technical research.

Teacher Training in Agriculture. Schools of agriculture began introducing teacher-training courses in agricultural subjects during the 1890's, with Cornell University and the University of Missouri taking the lead. An Act passed by the Missouri Legislature provided for instruction in agriculture and horticulture, "in the common schools of the State." By 1890 every state had established an agricultural school. In some of the southern states, agricultural and mechanical training was provided for both white and negro students, with Hampton and Tuskegee Institutes taking the lead in training negro students.

Graduate School of Agriculture. In his annual report for 1902, the Secretary described a new enterprise in agricultural education that had been inaugurated with the opening of a graduate school of agriculture at Ohio State University, where advanced students of agriculture, teachers and investigators from the experimental stations, and the United States Department of Agriculture would present summaries of the latest discoveries in scientific research. Seventy-five students, twenty-seven of whom were instructors in agricultural colleges, including teachers from Tuskegee Institute, enrolled in the nation's first graduate school of agriculture.

Chemistry in Agriculture. The program of chemical research in the department of agriculture grew rapidly. The Secretary reported in 1899 that American agricultural chemists had, during the past few years, taken part in international conferences in London, Paris, and Vienna, and that the chemical services of the Department of Agriculture were the most important of all the chemical work performed for the federal government. During the year 1898 the department had analyzed over thirty thousand samples of seeds and soils of all kinds, from all parts of the country.

Agriculture Dominates National Economy

At the close of the nineteenth century, agriculture was still the nation's biggest business. The center of agriculture had shifted from Pennsylvania, New York, and Ohio to the midwestern states. Farms of over 1,000 acres in size had doubled between 1880 and 1900, due chiefly to the introduction of mechanized farming. The fixed capital investment in farms, buildings, machinery and livestock amounted to more than twenty billion dollars—a sum far larger than that invested in manufacturing.

Beginning of Farm Mechanization. While farming continued to be a drudgery for those without the latest machines, yet there were labor-saving machines available for those who could afford them. In an article entitled, "Farming By Machinery" (*Scientific American Supplement,* July 21, 1900), the writer, while presenting an over-drawn picture, nevertheless described the many farm tasks that could be performed by machines. "Machinery was supplementing manual labor in all agricultural processes from the turning up of the soil to the harvesting of the crops," said the writer. "The plowman now uses a sulky plow," he continued, "upon which he has a comfortable seat, from which to guide a pair of horses. The machine does the rest." Corn planters and seed drills were making obsolete the ancient method of planting corn by hand and broadcasting grain from a bag tied around the farmer's shoulders.

The farmer was made to believe that, because of improved machinery, he would soon become a man of leisure. He was harvesting his crops with mowers and binders, rather than with scythes, sickles, and hand cradles. With a haytedder he was turning and spreading 4 acres of cut grass an hour. With a machine hayloader "he was able to pile on a load of hay within ten minutes."

Another contributor, writing at about the same date, to the same journal, declared, "Indeed there is scarcely a thing done on the farm today in which patented machinery does not perform the greater part of the labor." Describing some of the latest machines as, "mechanical wonders," he observed that, "grain is sowed, cut, bound, threshed, cleaned, sacked, stored and transported by machines." He noted also that potatoes were being planted and dug "by machines or implements," and even hogs were being slaughtered and chickens were being hatched by machines.

Farm Population at Peak. While there had been many technological advances pointing toward the mechanization of agriculture by the opening of the twentieth century, nevertheless, approximately forty million people, more than one-half of the nation's population, continued to live on farms. Those who were employed on the farms outnumbered by more than three million those employed in manufacturing and mechanical pursuits. The products of the American farms, including all crops and all livestock, totaled almost fifteen billion dollars, and the volume of farm products exported comprised approximately two-thirds of the nation's total exports.[15]

This remarkable achievement had been made possible, mainly, because of the recent advances in agricultural technology.

Notes for Chapter 25

1. For the record, let it be noted that the author is not related to the distinguished Olivers of manufacturing fame.
2. *Report of the Commissioner of Agriculture* (1869), p. 320.
3. *Report of the Commissioner of Agriculture* (1869), pp. 316-318.
4. See Russell H. Anderson, "Grain Drills through Thirty-nine Centuries," *Agricultural History,* X (October, 1936) 185 *ff,* for an excellent summary of grain drills.
5. Jacob Schafer, *The Social History of American Agriculture* (1936), p. 85.
6. See Leo Rogin, *The Introduction of Farm Machinery in Its Relation to the Productivity of Labor in the Agriculture of the United States during the Nineteenth Century* (1931). By permission of the University of California Press.
7. See Henry Adams Bellows, *A Short History of Flour Milling* (1924).
8. Charles B. Kuhlman, *The Development of the Flour Milling Industry in the United States, with Special References to the Industry in Minneapolis* (1929), pp. 115-116.
9. George D. Rogers, "History of Flour Manufacture in Minnesota," Minnesota Historical Society, *Collections,* X (1905).
10. William C. Edgar, *The Medal of Gold* (1925), p. 163.
11. *Ibid.,* p. 155.
12. Publius Virgilius Lawson, "The Inventor of the Roller Flour Mill," State Historical Society of Wisconsin, *Proceedings* (1908), p. 250.
13. *Ibid.,* p. 256.
14. Guy A. Lee, "The Historical Significance of the Chicago Grain Elevator System," *Agricultural History,* II (1937), gives an excellent account of this American invention.
15. "Annual Report of the Secretary of Agriculture, Ending June 30," XXVII (1902), *House Document No. 6,* 124.

For a recent excellent study of flourmilling, see John Storck and Walter Dorwin Teague, *Flour for Man's Bread: A History of Milling* (1952). For a discussion of the effects of barbed wire, see Arthur G. Warren, "Barbed Wire, Who Invented It," *The Iron Age* (1926).

Chapter 26

FOOD TECHNOLOGY

Food technology comprises the application of the principles of science to the preparation and preservation of food. Because of the advances in food technology, the American people are the best fed people on earth. Secretary of War John C. Calhoun had declared, near the end of his political career, "Our people, even the poorest, are accustomed to a plentiful mode of living. . . . An American would starve on what a Tartar would live on with comfort."

The Food Container Industry

Improved food technology in the United States during the late nineteenth century was brought about by two main forces. First, was the necessity for preparing for the lean years and the unproductive seasons. Second, was the need to supply the rapidly growing urban population. To meet these demands, new technological processes to prevent spoilage were devised. Methods for supplementing the age-old practice of preserving food by cold storage, by pickling in brine, or by using spices had already been introduced during the first half of the nineteenth century. The last half of the century witnessed the introduction of a number of mechanical devices used in preparing the food for processing, and machines for making cans used as containers.

Food Canning. In the earlier period, food preservation in tin-coated cans had been attempted with some success, and the manufacture of cans had developed into a separate industry. During the 1850's a machine press had been invented for cutting tops and bottoms for the cans, and a die was introduced that produced a can bottom in one operation. Improved tops were made by cutting a hole in the top large enough to permit the food to be put in after the top had been soldered on. This was called the "hole-in-top can." The round disc, or "cap," was soldered over the hole after the processing had been accomplished.

Tillery Copper. Following the Civil War, several improvements in this type were made. In 1876 the Howe floater eliminated the tinner's soldering iron, since a machine rolled the cans at an angle in a bath of solder. The Tillery copper was even more efficient. It spread solder around the edge of the cap in one operation.[1] In 1887 the Cox capper transformed the process into mass production by operating on six cans at once, completing a tray of twenty-four cans automatically.

Canmaking had become a machine production, and the industry had become largely an independent one with the canners purchasing their cans instead of making them themselves.

Automatic Machines. More improvements were added, so that by 1900 the entire operation of canmaking was one continuous line. In 1880, with the introduction of canmaking machines, two men and a couple of boys could turn out fifteen hundred cans a day. By 1900 automatic machinery was turning out cans at the rate of fifty a minute.

Sanitary Can. But canmaking had not yet reached the perfection the trade demanded. The soldering operation was objectionable, because of the popular prejudice against solder and because the canners resented destroying the attractiveness of their fancy fruits by crushing them into the small hole-in-the-top cans. The problem was solved near the close of the century when a machine was devised that mechanically crimped the top, fitted it over the flanged edge of the body, and thereby interlocked it. With a grooved roll operating at great speed, the layers of tin were pressed together. A closing machine completed the operation. The fruit could be inserted into the can without breaking, and the can was solderless. The Sanitary Can Company was organized soon after 1900 to produce this type of container. It was not until around 1920, however, that the sanitary can was generally received and the old hole-in-top type was discarded.

Machines Prepare Food for Canning. Machines for the preparation of food for canning were equally important as the machines for making cans. When it was discovered that fruits and vegetables should be processed as soon as possible after they were picked, this fact, together with the shortage of labor, led to the invention of machines to expedite the process. The local corn-husking and pea-shelling "bees" of former years were supplemented by itinerant laborers, not always dependable or skilled in the art. As new machinery was devised, canneries were often moved to the centers of crop production, which obviated the necessity for the long haul. But the employment of the itinerant workers, together with their housing problem, had become a great social problem. The labor-saving machinery in all phases of canning became more necessary as canneries came to be located in areas remote from centers of population.

Barkley's Machine. With canned corn always in demand—America has often been described as a corn-fed nation—the preparation of corn brought forth a number of inventions. Isaac Winslow's corn knife of the earlier period had given way to Volney Barkley's machine by 1875. This was a crude contrivance, consisting of a cast-iron frame about 4 feet long, attached to a trough. Four spiral knives were fitted in one end. An ear of corn was placed in the trough and forced against the knives by applying a plunger. The machine, operated by one man and a boy, could cut the kernels from 3,000 pounds of corn in 10 hours.

Warfield and Sprague Cutters. The process was further improved by the Warfield cutter, which cut a bushel of corn a minute. Streamlined production was next inaugurated with the continuous chain-feed cutter invented by Welcome Sprague, which came into use in the early 1890's. Resilient wire springs were needed to hold the ears, but because wire mills were not at that time turning out the quality of wire needed, Sprague substituted rubber teething rings for the purpose. A Sprague machine could cut sufficient corn in one day to pack fifteen thousand cans. An automatic corn husker completed the mechanization of the preparation of corn. This was fashioned somewhat like a washing machine wringer. It automatically grasped six ears at a time and ripped the husks off at the rate of three thousand ears a minute, and it was practically independent of human hands.

Pea Sheller. Canned peas, often called the "foreign aristocrat" among our vegetables, were for many years imported chiefly from France. In preparing peas for canning, hand picking and hand shelling were the chief handicaps, until the Scott and Chisholm pea sheller came into use in the 1880's. This machine removed peas from the pods as fast as six hundred hand workers could shell them. The task of removing the pods from the vines, another handicap to rapid canning, was solved during the 1890's when a combination podder and viner was invented.

Pea Podder and Viner. Meantime, horticulturists had produced a pea that matured more uniformly, and thus permitted simultaneous picking of the entire crop. Harvesting machines, loaded with racks similar to hayracks, moved into the fields, hauled the peas to a nearby factory equipped with a combined podder and viner, into which the peas were pitched like so much hay. Here they were shelled, sorted and sized, and the pods and vines were discarded for the silo. The podder and viner were major technological contributions. Within a few years they were supplanting the hand work of thousands of persons; [2] and by 1900, some two hundred and fifty mechanized viners were performing the work that formerly would have required the services of fifty thousand laborers.

Tomatoes. Tomatoes and tomato products led the list of canned vegetables. Although Harrison Woodhull Crosby, chief gardener at Lafayette College, first successfully processed tomatoes commercially in the 1840's, the superstitious belief that tomatoes (love apples) were poisonous kept the public from accepting the product. It was not until the 1870's that tomatoes became one of the "big three" in canned vegetables. Tomato juice, however, continued to be a nuisance in the trade until the 1920's.

Fruits. Other fruits and vegetables soon became an important part of the canning industry. Peaches, which had been a product of the eastern states, were introduced into California, where the cling variety became an important article for canners. The lye peeling method, borrowed from the technology of the Indians, was so adaptable to the preparation of cling peaches that it long kept the canning of the free-stone variety out of the industry. Mechanical peach pitters and cherry pitters were introduced and became great labor-saving devices. Pineapples came into the market with the establishment of a cannery in Hawaii in 1892, but only on a small scale.

Canned Salmon "Iron Chink." One of the most remarkable mechanical aids in preparing another food product for canning was the invention of a small machine called the "iron chink." It was used in the salmon industry, which did the work of butchering the salmon and preparing them for the canneries, formerly done by Chinese on the west coast. An endless belt conveyed the salmon to the machine. The machine adjusted itself to the size and dimensions of each salmon, removed the head, tail, and fins, slit open the body, discarded the entrails, and conveyed the edible portions to the canner. It could process seventy to eighty salmon a minute.

Assembly Line Production. More important, however, in the history of food technology than any of these labor-saving devices, was the organization of the continuous-line system, whereby the canning process was carried on in one automatic mechanical operation. The factory operations in canning include preparing the product for canning, washing, peeling, grading, filling, cooking, cooling, packaging, and warehousing. Each process was a separate time-consuming task until a method was developed for streamlining all the operations. Daniel Trench of Chicago combined a corn cutter, a corn filler, a rotary capping machine, pressure kettles, hoists, and conveyors in one continuous production line.

Automatic Canning. As early as 1883 the Norton brothers began the establishment of a complete automatic line for the manufacture of containers. A factory in Indiana, a canning-center state, perfected an endless chain in a processing method whereby the cans were filled and passed on a belt through a temperature of 250 degrees, then to a water cooler, and then to the warehouse without any stoppage in the process. During the operation, cans were not touched by hand.

Mass Production Lowered Cost. A number of other inventions aided in improving and refining the continuous-line process, which brought mass production in the nation's food industry, with resulting lower costs. Senator George Vest (Missouri), in noting the advances in the canning of foods, declared that, "Years ago canned foods were delicacies, but today (1890) they are the poor man's food, being cheaper in the cans than in the green articles." He noted that the canning industry had become one of the largest industries in the country, and that our canned fruits, meats, and vegetables were supplying the French, German, and English armies.

Food Processing

The same strides had not been made in food processing, however, which was a scientific operation, as had occurred in the mechanical fields. Food processing was as much an enigma near the close of the century as it had been in the beginning.

Food Spoilage Theories. Nicholas Appert had originally advanced the theory that food spoilage was due to a bacteria within the food, but later theories had supplanted this idea and for many years it was held that food spoilage could be prevented if the air were excluded. All methods of processing proceeded on this erroneous idea until the late 1890's. It was the discovery of Louis Pasteur that fermentation is caused by minute organisms—yeasts or bacteria. This discovery led to the scientific method of food processing. Proceeding from this basis, science disclosed that not air, but bacteria, produced food spoilage.

Variety of Methods. Before this theory was advanced canners had developed various new methods of food processing, all proceeding on the "witch doctor" principle, with varying and uncertain success. The early water bath processing system in which the food, packed in containers, was immersed in water and boiled for a prescribed time, was improved by raising the temperature of the water by the addition of calcium chloride.

Pressure Cooker Method. Later the pressure-cooker method was instituted, and food was processed in a retort by steam. Andrew K. Shriver, introduced a closed steam-kettle cooker in 1874. The kettle was an iron vessel with a steam supply pipe passing through a true bottom, diffusing steam through perforations in a false bottom, around which the cans were held mechanically. Steam pressure was regulated by a steamcock and was registered on a steam gauge, as in a locomotive boiler. A removable iron top made the kettle steam-tight. The resultant high pressure shortened the time of processing and eliminated much of the breakage during the operation. With the introduction of the retort, mechanics and machine shops were busy making new apparatus to facilitate the process further.

Cold and Hot Pack Method. Other improvements were made in the treatment of the food in preparing it for the processing. One was the cold-pack method, in which the raw food was packed cold in the container and then immersed in the water bath for processing. Then, to save time in coming to the boiling point, and to save breakage when glass containers were used, the hot-pack method was introduced. In this method, Merrell and Soule, canners at Syracuse, New York, introduced what was called the "gun cooker," patented in 1873, wherein the corn was heated before it went into the can. This was perfected into the "improved corn steamer and can filler method," whereby corn cut from the cob was thrown into a hopper, carried on a spiral conveyor through a long horizontal tube where steam was applied until the corn emerged through a funnel at the opposite end. There it was forced into cans by steam pressure. The canned corn, still hot, was rushed to the retort, where 15 pounds of pressure completed the operation. The entire time consumed was one hour.

Henry L. Russell. None of these systems, however, had proved satisfactory. Meantime, demands for canned foods were mounting, and manufacturers and dealers alike became desperate. A canning company that had been having trouble with exploding cans with resulting heavy losses, sought the services of a young scientist, Henry

L. Russell of the University of Wisconsin. Professor Russell, after a number of experiments, concluded that the swelling and unpleasant odors were caused by the bacteriological action of fermentation. He recommended that the heat be increased to 242 degrees in processing, and that the length of treatment be extended 15 minutes beyond the period allowed by the cannery. The remedy proved effective, and Russell's report, "Gaseous Fermentation in the Canning Industry," published in 1895, brought a new era in the history of food technology.[3] In 1902, in Pittsburgh, a canners' research laboratory was organized, consisting of an association of canners who paid a fee for its maintenance, and was directed by E. W. Duckwell of the H. J. Heinz Company.

Refrigeration

All the while, other methods of food preservation were improved upon. During the last half of the nineteenth century, refrigeration was developed in both mechanical and scientific fields.

Ice Box on Wheels. Following the close of the Civil War, special refrigerator cars were designed. With the extension of railroads, the distribution of natural ice increased throughout the Midwest. In the early 1870's cars roughly insulated and built like "ice boxes on wheels" were introduced. The ice was stored in a central box, allowing the water to drain through a hole in the floor.

Ice Famine, 1889–1890. But the supply of natural ice was never equal to the demand. When the winter seasons were long and sufficiently cold to freeze the ponds and rivers to a depth of 10 to 15 inches, a good supply of ice was assured, but mild winters produced an ice shortage. Such a winter occurred in 1889–1890 and brought on "the great ice famine." It was the warmest winter on record, followed by a sizzling summer. By July, the amount of ice that had been stored was gone. Newspapers in the northern cities reported in lurid detail the prevalence of "summer complaint," an excessive number of cases of typhus, and the spoilage of tons of food. From the South came reports of milk souring in wholesale quantities, of meat spoilage, and the rotting of vegetables.

Dr. John Gorrie's Artificial Ice. Meantime, a number of inventors had been experimenting with methods of producing ice by mechanical methods. As early as 1850, Dr. John Gorrie of Florida had successfully designed a compressed-air icemaking machine, and was supplying ice to cool the rooms for his patients suffering from burning fever. While Dr. Gorrie's invention won for him a statue in the Hall of Fame in the national capital, he was about a half-century ahead of his time.

First Mechanical Refrigerators. Mechanical refrigeration owes its origin to the quiet, patient research of a handful of scientists. A number of basic discoveries and inventions needed for mechanical refrigeration had been developed before the nineteenth century. Galileo had invented an accurate thermometer as early as 1595. Robert Boyle had discovered the law of compression of gases in 1622. In the mid-eighteenth century, Joseph Priestly added the scientific discoveries of ammonia, oxygen, and carbon dioxide gas. When Michael Faraday demonstrated that gases could be changed into liquids under pressure, and Joule proved that heat and energy are mutually convertible, the stage was set for Lord Kelvin, often called the "Father of Mechanical Refrigeration," to discover the chemistry of refrigeration.

Types of Refrigeration. Three types of mechanical refrigeration were evolved. One was the compression system, which involves repeated cycles of compression, condensation, expansion, and evaporation. A second was the absorption system, which utilized heat to bring about an action similar to compression, and cold to correspond to suction. The third was the air system in which cooled compressed air operated an air engine that expanded and cooled the air, which was piped off for use in refrigerating rooms. When, in 1875, Dr. Carl Linde produced a successful compression machine, and ten years later, E. E. Hendrick patented an exhaust-steam absorption machine, the basic principles for mechanical refrigeration had been developed.

Early Attempts to Freeze Foods. Refrigeration had been used early as a means of food preservation, a method known since the time of Bacon, when he stuffed a fowl with ice to preserve it. No thought was given, however, to packaging in order to prevent dehydration. Nor had any method been devised to keep the product frozen until it reached the consumer. Foods were preserved by freezing as early as 1861, and beginning around 1900, much of the poultry and fish was kept frozen in warehouses for shipment to market, but the frozen-food era did not come until after World War I.

The Meat-Packing Industry

Of all the developments in food technology none has been more important than the meat-packing industry. At first the industry was decentralized, with slaughter houses in every community and on many farms, much as had been the custom in Europe.

Centers of Meat Packing. Cincinnati early became a meat-packing center and was given the name "Porkopolis." But, in the 1860's, Chicago became the world's largest meat-producing center. There, mass production was first developed and became the major factor in the meat-packing industry.

The "Long Drive." Raising cattle on the western plains involved an expanse of territory, with trails following the pasture season northward. To drive cattle up the Chisholm Trail from Texas to Abiline, Kansas (a famous cattle center), required two to three months. The Dodge Trail, from Texas through Oklahoma to Montana, required from four to five months. When the railroads were built into the region, they helped to solve the problem of livestock transportation to the packing-house centers.

Refrigerator Cars. Centralizing the packing industry in Chicago involved the shipment of dressed meat to the seaboard, not the livestock in cattle cars. Packing firms in Kansas City began shipping dressed beef to Philadelphia, New York, and Boston. Beef, mutton, and pork from more than a half-dozen packing houses were being shipped to England during the 1870's. These products were stored in insulated compartments between decks, with an ice bunker and a fan for circulating the air. In one year, 1876, approximately 10,000 tons of American beef were shipped to England. Cummings quotes a Scotsman as saying, "What the repeal of the corn laws has been to the cultivation of grain in this country, the invention of refrigerators is likely to be to the raising of cattle." [4]

A number of improvements were made in the design and construction of refrigerator cars. It was discovered that meat while in transit must not come in contact with the ice, and that a circulation of air was necessary. To accomplish this, the meat was hung from the ceiling of the cars, but the swaying of the meat as the train rounded curves resulted in train wrecks. New methods for icing and for containing the meat were devised.

Pioneers in Meat Packing. Leadership in the meat-packing business was provided by a few pioneers who established the industry on a family basis, and these families continued in control to the third and fourth generations. Among the first to enter the industry were the Cudahy's, the Swift's, the Armour's, the Hormel's, and the Morrell's. Philip Armour moved from Milwaukee to Chicago in 1867. Gustavus F. Swift came from New England to buy cattle, and seeing the advantages of the Chicago location, settled in business there in 1877.

Assembly Line Production. The larger packing houses introduced mass production and the assembly-line principle, the two major technological developments that revolutionized the meat-packing industry. (Henry Ford is said to have gotten his idea of assembly-line production in the automobile business from visiting a packing-house plant.) As there were no standards established, the early packers worked out their own system. A large amount of skilled hand labor was needed, and plant mechanization involved the establishment of facilities for specialization in skilled work. This led to the assembly line. Power-driven machines moved the carcasses. A hog-scraping machine was devised. Tanks and digesters were introduced for making lard. Sausage was made by machinery. All the equipment soon became standardized.

Since plant refrigeration was essential, it became the determining factor in all later developments. Rooms in the plant were refrigerated by building chambers with communicating air spaces over the rooms to be chilled. These chambers were filled with cakes of ice. After other refrigerants were introduced, packing plants enlarged their field. The space that had been used for storage of ice was made available for other uses. Besides the production of fresh meats and cured meats, the manufacture of numerous by-products was introduced, and meat canning became a major technological development. By 1900, the meat-packing industry was supplying approximately 30 per cent of the nation's food.

Dairy Products

The dairy industry, which includes milk, cheese, and butter products, became a major development in the history of food technology before 1900. Like other major industries, it owed its existence to a few simple scientific discoveries and inventions that were made in the half-century following the close of the Civil War.

Milk Supply for Cities. As cities increased in population the demand for milk and dairy products also increased, and the efforts on the part of farmers to meet these demands resulted in a number of scientific and technological improvements. These included methods for improved feeding and breeding of herds of milk cows, improved facilities for collecting and distributing milk and dairy products, new methods for scientifically testing milk, and the development of new and greater varieties of cheeses.

Silos, Dairy Feed. During the late 1870's farmers discovered that when they fed green feed (ensilage) to their milk cows they produced more milk. This led to the erection of all kinds of storage bins, even to the use of cellars and cisterns for storing green grain and green stalks of corn. These experiments resulted in the erection of large cylindrical buildings, constructed like a huge hogshead. This storage structure was called the "silo." Before the close of the century, silos were common sights on all large dairy farms.

Milk Trains. Important as were the methods of increasing milk production, even more important were the transportation techniques used in moving the bulk milk products from the farm to the factory or city. As the urban centers grew, the demand for fresh milk increased, which in turn meant that the "milk shed" (the outlying districts that supplied the milk) expanded. Regular dairy routes were established and daily shipments were made, often from a distance of 200 miles. Milk trains frequently had the right of way over passenger trains. The dairy enterprise had moved into the ranks of big business during the 1890's, and before the turn of the century the Dairy Division in the Bureau of Animal Husbandry of the United States Department of Agriculture was established.

The Babcock Test. As the dairy business expanded, new methods were devised for handling the milk and disposing of it. Milk depots were set up and plans were developed for evaluating the quality of the milk in relation to its butter fat content. Chemists worked for years on the problems of devising a practical milk test. In 1890 Dr. Stephen M. Babcock of the University of Wisconsin invented a simple machine known as the "Babcock Centrifuge," commonly called the "Babcock test," one of the most significant inventions in the history of dairying. Dean Henry L. Russell of the School of Agriculture, University of Wisconsin, declared it to be, "the foundation stone on which modern dairy education has been built."

Butter Fat Measurements. The machine consisted of a small wheel mounted horizontally on a driveshaft and holding a number of bottles on its periphery. The bottles were suspended in an inclined position and swung outward when the wheel rotated at an excessive speed. The mechanism was enclosed in a suitable steel or iron protector and was equipped with a heating device to regulate the temperature of the test. As the machine rotated, the fat globules collected in the neck of the test bottles, where the amount was measured on a graded scale. The test could be performed either on individual farms or in a factory or creamery to determine the butter-fat value.

Milk Tests. The Babcock test made the manufacture of butter and cheese more profitable and stimulated the improvement of dairy herds. This simple accurate device provided a standard by which a fair price could be paid for whole milk; "watered down" milk could be instantly detected. One observer noted that the Babcock test "had more influence than the Bible in making dairymen honest." It furnished the dairy farmer with the means of determining the individual performance of each animal in his herd. The Babcock test was the prime factor in transforming dairying from a routine daily performance into a science.[5]

Cream Separators. The cream separator was another important technical device that greatly affected the dairy industry. The age-old method of pouring milk into cans, waiting for the cream to rise, and then skimming it off, was both time consuming and inefficient. A number of inventions were designed to speed up the process of separating the cream. In the late 1870's a machine known as the "DeLaval separator" (Swedish invention) was introduced into the United States. The first separators were hollow-bowl machines and could be operated at rapid speed. They forced the thinned milk to the outer edge of the bowl, while the cream was forced toward the center. The first machines were of the larger capacity power-type, but around the turn of the century a smaller bowl separator was invented and widely adopted by dairy farmers.

Factory-made Butter. Butter was the traditional outlet for excess milk. Although homemade butter always varied in quality and seldom kept its flavor for any extended period of time, the American people were slow to give it up and accept factory (creamery) butter in its place. One of the first factories to begin producing butter and cheese was opened in Orange County, New York, in the early 1860's. It was in the Midwest, however, that factory-made butter had its greatest development. Beginning with the Stewart Creamery in Manchester, Iowa, in 1871, creameries sprang up in Wisconsin, Illinois, and Minnesota.

Improved Churns. Accompanying these developments came an improvement in churning methods. In the early 1890's a combined churning and buttermaking machine was introduced which eliminated the manual handling and working of the butter. It consisted of a wooden drum and a cylindrical-shaped churn. The cream was churned, the buttermilk drawn off, and the butter was rinsed by a machine. When the process was completed, the machine was opened and the butter was removed with a sterilized wooden paddle and packed in tubs.

Butterprint Machines. The demand of the city housewives for small, easily handled packages led to one-pound printing devices, and woodcarving artists came up with an assorted variety of prints. The Lafayette butter printer, patented in 1892, was operated by hand and printed one pound at a time. A number of butter-printing machines followed, with capacities varying from a few dozen prints to thousands of pounds daily.

Factory-made Cheese. The factory production of cheese, undertaken as early as the 1850's by some farmers in Herkimer County, New York, developed rapidly following the close of the Civil War. Commercial rennet, which had been developed in Denmark, was brought to the United States and stimulated the factory production of cheese in a number of cities. Cheesemaking, like a number of other home industries, had been taken over by the factory by 1880. Although mechanical equipment hastened the mass production of cheese, the science of chemistry was also important. Formulas based upon chemical knowledge and bacteriological research removed the art of cheesemaking from the old "dip and taste" methods, and placed it in the laboratory where dozens of different brands were produced to satisfy as many different tastes. By the turn of the century, numerous cheeses were being processed by blending, pasteurizing, and packaging in tin foil or wood containers. This resulted in increased production. Although the Americans have never eaten cheese in such quantities as the Europeans, the per capita consumption has been constantly increasing since 1900.

Nationwide Variety and Quality

The progress in food technology by the turn of the century was marked not only by mass production, efficiency, and labor-saving methods, but also by the development of an improved quality of food. Better foods were being produced than by the old hand-labor method. A greater variety of foods began to be produced, under more sanitary conditions. Firms advertised, "our food is not touched by human hands." Meats were being shipped hundreds of miles from "cattle range to gas range." Home-grown fruits and vegetables were carted to huge processing plants and returned to the housewife's kitchen cupboard, where the can opener became the indispensable utensil. Homemade butter and cheese passed from the farmer's springhouse to a refrigerating plant, where it was molded into fancy prints from large tubs and delivered to the local grocery. By 1900, many farmers were buying butter instead of making their own. Food habits, always hard for a people to change, were developing not only a better-fed nation, but also a better-nourished people.

NOTES FOR CHAPTER 26

1. See Earl Chapin May, *The Canning Clan, A Pageant of Pioneering Americans* (1937).

2. *Ibid.*, p. 281. By permission of The Macmillan Company.

3. *Annual Report of the Experimental Station,* University of Wisconsin (1895).

4. For an excellent study of the technology employed in providing natural ice, see Richard O. Cummings, *The American Ice Harvests, A Historical Study on Technology 1800–1918* (1949).

5. For a short, simple explanation of the Babcock test, see brief biographical sketch of Stephen M. Babcock in *Dictionary of American Biography* XXI, *Supplement I,* 37.

See Richard Osborn Cummings, *The American and His Food Habits: A History of Food Habits in the United States* (1941).

Chapter 27

CLOTHING TECHNOLOGY: BOOTS AND SHOES

In no branch of industry was America's aim to mechanize her industries more marked than in the clothing and textile trades. Before the close of the nineteenth century, the time had passed when the coarse jeans and the imported broadcloths differentiated the farmer from the "gentleman." Machine production had made it possible for the farmer and the small town merchant to dress as well as the city banker. His suit was made of material of the same pattern, and was cut in the same style, and only the closest scrutiny could reveal the difference in texture or in the refinements of tailoring. In America, where the road to the White House was open to everyone, the clothes one wore were no barrier to successful travel of that road.

Machinemade Textiles

Clothing technology became mechanized in all its phases. Machines made the tools for the textile factories, and machinery within these factories produced the fabrics. Other machines fashioned them into ready-made garments. Machines, containing many intricate details, made the boots and shoes for which the leather was prepared by factory tanning. Hats were machine made. When the twentieth century dawned, clothing and textile technology were synonymous with machine production.

Textiles: Big Business. This era of machine production in textile technology, which had its beginning with Samuel Slater, the "Father of American Manufactures," and which had developed into the factory system in the Lowell Mills, had passed through the formative period by the outbreak of the Civil War. By the mid-1870's the manufacture of cloth had become one of the nation's leading industries. But it was during the last quarter of the nineteenth century that the textile industry—along with steel, petroleum, electricity, and meat packing—placed the United States among the top industrial nations of the world. Not only were new industries created at a pace never before known, but also the production of industrial equipment, says Gibb, increased more rapidly than ever, "giving America more, faster, and better machines than any other nation."[1] Textile technology felt the effects of these faster and better machines.

Migration of Textile Industry. This was also the period that marked the beginning of the geographical distribution of the textile industry. While Lowell remained the Manchester of America with its twenty-five thousand looms in operation, and its more than a million spindles, a new regional competitor loomed on the horizon. After the Civil War, textile industries began to spring up in the South and their development played a part in advancing textile technology, as well as affecting the economy of that region. In the Piedmont country and the Carolina's, a textile industry developed which was eventually to surpass New England's. Before 1900, however, the southern mills, rather than being competitors, proved an outlet for northern capital and northern skills. The expansion of the southern industry began in the 1880's, when some of the confusion of the Reconstruction period had abated. Building textile mills in the South proceeded at a pace unequalled in America since the cotton factory days in New England during the early nineteenth century.

The textile industry also moved west, but not on such a large scale. Small woolen mills were established in the areas of Cincinnati, Milwaukee, and Chicago, and a few as far west as California. In a convention of the woolen manufacturers of the Northwest held in Chicago, 1870, fifteen hundred different lots of goods were on display. Other cities held similar expositions. But textile mills in the West were small and were operated on a principle similar to the flour mills, where the custom of bartering raw material for milled products prevailed. With plant consolidation and the development of big industry in the Northeast, and with a marked decline in the wool-raising areas of the Midwest, it became more difficult for these small industries to survive. New England continued to contribute the bulk of the textiles produced in the country, with Pennsylvania leading in specialized fields.

Bulk Production. The flow of inventions in textile production was continuous and provided many technological improvements. In all but a few cases these were not individually important or revolutionary, but they were designed to accomplish the objective of speeding up production. The conflict that arose between speed and quality was resolved in favor of speed. It became an American technique to prepare and handle bulk quantities with larger production at less cost per unit. Once a minimum standard of quality was attained, speed became the goal.

Major Advances in Textiles. Three important advances in textile technology during this period included the introduction of the revolving flat card, the evolution of the spindle, and the invention of the automatic loom.

The Flat Revolving Card. The major improvement in carding technology in cotton goods was the invention of flat cards of steel construction. These replaced the former stationary wooden type. They were more rigid and more effective in tearing apart the raw cotton, and permitted greater speed of operation. Cotton combing machines, used to draw the fibers into strands, were also improved, and larger cotton drawing frames of sturdier construction were introduced.

The Spindle. The evolution of the spindle, while not spectacular, was an important factor. The ring spinners were designed to draw and twist the thread consecutively, and provided a system of continuous spinning. These were geared to operate more speedily. Spindles moved so rapidly that they seemed not to be moving at all.

The ring spindle underwent a number of improvements, one of which made possible the spinning of finer threads. An English textile manufacturer observed in 1894 that the advantages derived from this latest improvement were "increased production, a saving in power, a saving in wages, and a saving in wear and tear." While in the 1880's the ring spindle had been capable of 4,500 rpm, by the mid-1890's its speed was doubled. The mule spindle had disappeared from the American cotton mills.

The Automatic Loom. The introduction of Northrup's "triumph loom" also marked a major advance in textile technology. James H. Northrup, for George Draper and Sons, had developed a completely automatic loom by the mid-1890's. His first invention eliminated reloading the shuttle by hand. It consisted of a magazine similar to that in an automatic pistol which mechanically pushed a fresh bobbin into the shuttle when the weft was depleted. A weft detector determined when this action was necessary. With this loom, one hundred stops a day were eliminated. Northrup then added an automatic threading device, eliminating the antiquated hand operation. One weaver could attend from twenty-four to thirty of these looms. Low-cost production in simple fabrics followed. A "loom for every fabric" became the slogan of textile manufacturers. Competition among factories led to the scrapping of thousands of workable looms in order to install the automatic machinery.[2] By 1900, hand operations were almost extinct.

"Fancy" Pattern Chain. Another development in textile technology during the 1890's was the introduction of the pattern chain. Prior to this period the American cotton manufacturers produced mostly the coarser cottons, without devoting much attention to fancy patterns. With the calico fad in the 1880's, a demand arose for domestic goods in calico patterns. To produce these, a special "fancy" loom was necessary, a loom that used two or more shuttles. Plain looms used only one shuttle. Years before the Civil War, William Crompton, an American, had invented a pattern chain, designed to control the warp threads. Unable to obtain financial backing in this country, he introduced it in England. Years later, American textile mills adopted Crompton's chain after E. H. Ryan introduced a shuttle-changing feature, and by 1900 American textile mills were producing any desired pattern or quality in unlimited quantities.

Materials for Clothing

Woolen Textiles. The demand created by the Civil War and the recession in the cotton trade had given impetus to the woolen industry. Sheep raising, aimed at producing greater quantities of wool, and a superior quality, became a specialized branch of farming. New fabrics and new patterns came into demand, and a new type of material was produced. Since there was not enough new wool available, manufacturers resorted to mixing old wool and hitherto discarded materials with a new material. This substitute material was termed "shoddy," and the custom of manufacturing wool mixtures made it possible for cheap clothing to be made in imitation of the expensive weaves.

New Woolen Products. Demands arose for novel and finer fabrics. Brilliantines and alpacas, materials with a luster and a sheen, were made. A demand for soft clinging woolens developed the delaines in imitation of the French. Knitted goods were desired, and knitting yarns became an important product. Fine mohairs were made. The manufacture of worsted materials, those made with well-twisted yarns of long staple wool which took the lead for men's suits, more than doubled in the decades from 1880 to 1900.

Machines Replace Hand Labor. The period from the early 1880's to the turn of the century outranked all previous decades in the number and variety of technological advances. Mechanical cutters replaced hand labor; buttonhole machines, power-driven sewing machines, and mechanical pressers were introduced in most of the larger establishments. In 1887 the Fenn patent cutting wheel, operating with a circular rotary knife, was introduced. In 1897 George Eastman's straight-knife cutting machine marked another notable improvement, since it could cut one hundred and twenty layers of material in one cutting. The overlock machine was capable of trimming, sewing, and "overedging" the cloth in one operation, along with stitching at the rate of three thousand stitches a minute. Many of the machines used in clothing production were electrically operated before 1900.

Other mechanical devices were introduced to speed the work of preparing the raw wool. Wool-burring machines cleaned the fleece more quickly and more effectively than could be done by hand. A machine was invented that burred 600 pounds of wool an hour. For cleansing the wool, improvements over the old potash and soda methods were made; and mechanical rakes and squeezers were introduced. William Bromwell's automatic wool-weighing and feeding mechanism mixed and weighed the wool. One worker could operate twenty to thirty feeders.

Textile Training Schools. The necessity for technically trained young talent was recognized as being as important as the need for new tools and machines. The textile industry turned to special schools to provide this. The Philadelphia Textile School, founded in 1884, set a pattern for textile education. In Massachusetts a law passed in 1895 made possible the establishment of the Lowell School in 1897, and one at New Bedford in 1899. Here, as at Philadelphia, courses were offered in cotton manufacture, chemistry, dyeing, and textile engineering, as well as designing. The South opened schools to train expert operators and others for managerial positions.

Ready-made Clothing. The improvements in textile technology brought forth a number of social changes, one of which was in men's wearing apparel. The ready-made clothing industry rose rapidly during the last decades of the century. Although the manufacture of ready-made clothing had its beginnings before the Civil War, it did not become one of the nation's big businesses until the last quarter of the nineteenth century. During twenty years, 1870 to 1890, the increased production was estimated at 600 per cent. While no satisfactory statistics were compiled until 1890, the industry's output in that year was valued at two hundred fifty-one million dollars. The great majority of the male population of the country was attired in ready-made clothing before the close of the century; and an estimated 50 per cent of women's and children's wearing apparel was being produced in central shops and factories equipped with the latest machines.

Knitted Goods. Knitted goods were also affected by advances in textile technology. The knitting industry combined garment manufacture with textile making. As knitting machinery was developed, the objective was to fashion and to shape the garment in one process. This made it possible to shape and seam goods automatically. Isaac W. Lamb, a Michigan clergyman, invented a knitting frame (1866) consisting of two horizontal flat beds, which speeded the production of knit goods. Lamb was said to have developed his idea from the technique employed in making whip lashes and in braiding leather thongs. His machine was developed to do both flat work and tubular, and produced many different kinds of knitted goods. Knitted products began to appear in quantity during the next two decades.

Seamless Hosiery. Benjamin Shaw's automatic device for seamless goods, invented in 1867, became popular in both America and Europe. By 1870 a circular knitting machine was making complete seamless hosiery, and in the year 1882 some twenty patents were granted for improvements in knitting machines. Higher priced articles were still imported, however, until an invention of Charles Fletcher of the Providence mills in 1883 made possible the machine production of ladies full-fashioned hose. By 1890 the circular hosiery machine operated automatically. The centers of the knit-goods industry were the Mohawk Valley of New York for knit underwear, and Philadelphia for stockings. Besides underwear and stockings, knitted goods included jersey, a knitted fabric. Yarn became an industry in itself, with the Germantown yarn being most in demand.

Weaving Silk Threads. The silk industry also profited by a number of inventions in textile technology, and the improvements covered virtually all phases of production. Reeling the silk from the cocoon, always done in the country where the silk is produced, was now done in Italy on American-made reeling machines. The Italians supplied American manufacturers with a better grade of materials. The manufacturing processes in this country included that of converting the raw silk into threads and yarns for textile purposes and weaving, where the product was woven "in the gum."

Speed-up Spindles. In the 1870's, machines for weaving the silk were devised. They operated on belt spinners that ran from 11,000 to 12,000 turns a minute. Machines were devised for testing the strength and evenness of the silk thread. Other machines twisted and doubled the filament in one operation. A slack-twisted thread called the "tram," used for the woof on the loom, and a tight-twisted thread termed "organzine," which formed the warp of the future fabric, were produced. As a result of the technological improvements the cost of converting one pound of raw silk into material for the loom was reduced from $4.50 to less than one dollar.

Improved Looms. Improvements were also made in the looms used in the silk industry. Handlooms, which were in general use in Europe, were never popular in America. Here a power loom was devised with a tiny threaded shuttle, which adapted the principle of the sewing machine. Later, this became automatic. In 1900, forty-four thousand power looms were reported in operation, but only two hundred handlooms were listed.

Quality of Machine Products. In silk manufacture quality was not sacrificed, however, for speed. In many respects materials produced in this country were superior, since power-loom production necessitated a better quality of material. American manufacturers were awarded gold medals at the Paris Exposition of 1900 for piece-dyed, printed silks and also for warp-print fabrics. Jacques Huber, a Swiss member from the International Jury on Silks at the Paris Exposition, attributed the superiority of American silk manufacture to, "the American technician who through his inventions of time-saving machinery, is perhaps unequalled. . . . The common sense of the operative makes him especially fit for the manufacturing business." The simplicity of American silk-processing machinery, which became world renowned, enabled the simple weaver to do work which in Europe was done only by the most skilled.

Machinery for Leather Products

Boots and Shoes. The production of boots and shoes was slowly transformed into the factory system during the last quarter of the nineteenth century. The change came gradually, due in part to the great number of hand operations involved in making shoes. The Civil War was a great boon to the shoe industry, because it created a demand that was impossible to supply under the old system of hand-production technology. Lyman R. Blake's sewing machine, developed on the eve of the Civil War, sewed the soles to the uppers, leaving only the heel and toe to be nailed.

McKay Stitcher. In 1862 Colonel Gordon McKay, improving on Blake's machine, introduced the McKay leather stitcher, which greatly speeded up production and led to the invention of other devices. Machines for rolling sole leather, for splitting upper leather, for channeling the soles, and for attaching heels were developed. But the McKay stitcher was applicable only to coarse, heavy shoes which were stiff and unwieldly, and due to heavy inside seams, were also uncomfortable. Finer shoes continued to be custom-made by hand.

The Goodyear Welt. The Goodyear welt machine made possible the production of hand-sewed shoes by machinery. Instead of fastening the upper directly to the sole, as in the McKay process, the Goodyear machine united the upper to an insole, and the stitcher united the welt to the outsole with the welt on the edge. By this process a flexible sole was made with no inside seams. Charles Goodyear, Jr., son of the famous rubber pioneer, was responsible for developing this invention. Years later it was improved upon by Christian Daniel's lockstitch system.

Machines for Lasts. The final step in the mechanization of the shoe industry was taken when Jan Ernest Metzeliger of New York invented a patent laster, known as the "consolidated machine." By this method the upper was pulled over the last by machinery and held there preparatory to applying the sole, either by the McKay or the Goodyear process. Skilled handlasters could attach soles to only sixty pairs of shoes in one day, whereas the machine operators could attach soles to four hundred pairs a day.

Shoes for the Masses. Other mechanical processes included the use of metal fasteners in the McKay machine, rounding, buffing, edge trimming, and heel burnishing—all done by machinery. Some eighty corollary operations required in manufacturing a pair of shoes were, by this time, being performed by machines. The result was production of more shoes and at reduced costs. Good machine-made shoes could be bought for five dollars: in some areas they could be purchased for as little as two dollars, while custom-made shoes continued to cost fifteen to eighteen dollars. Once again the mechanization of an industry had added to the comforts of life, and more people were able to wear shoes. In 1860 the nation's shoe production amounted to $2.90 per capita, whereas in 1900 it averaged $4.90. This meant that for every man, woman, and child almost five dollars worth of shoes had been manufactured. German manufacturers sent their superintendents to this country to study the American system, and shoe machinery was exported to Europe.

Tanning. The increased production of shoes necessitated a change in leather and tanning technology. This was effected in improvements in the chemistry of tanning, in the utilization of new and different hides, and in the development of new leather-working machinery.

Chemistry Aids Tanning. Through the application of chemistry to tanning, special extracts were employed to save the extravagant exploitation of hemlock and oak trees merely to obtain tanbark. The use of these extracts also expedited the tanning process, which often required from twelve to eighteen months when using the ancient bark-tanning method. Among the new processes was the two-bath chrome process invented in 1884 by Augustus Schultz, an American, which made possible an industry that boldly proclaimed it could "shoe the world." Ten years later a simpler, one-bath process was patented, which was reported to have speeded up the manufacture of leather "more than any other invention in the whole history of the leather industry." [3]

Vici-kid. Robert H. Foerderer, after experimenting on a process to utilize goat skin, brought out a new product in 1890 which he called "vici-kid." The manufacture of this leather, which was used in women's shoes and finer leather goods, initiated one of the largest leather establishments in the world for manufacturing imitation French kid.

The use of chemicals made possible the use of sheep skins, horsehides (called cordovan leather), and kangaroo, which was in great demand. In addition to these supplies, more than fourteen million domestic cowhides were used annually, and hides were shipped from Argentina, South Africa, and Mexico.

The leather exhibits at the Chicago World's Fair brought from a German shoe manufacturer the comment that the best machines in the tanning industry were American inventions, and that the art of introducing new and valuable tannages was due exclusively to American tanning technology.

A Well-Clothed Classless Society

With the opening of the twentieth century, the American people were better dressed than they had ever been before. Factory-made clothes were taking the place of home-sewn garments. Domestic dressmakers could no longer hope to compete with the garment makers who, "tied to their machines," could turn out dozens of dresses for every one sewn by hand. Nor could the local cobbler keep pace with the cutting, stitching, and pegging machines in the central shoe factories where output amounted to hundreds of pairs of shoes daily.

Mass production of suits, dresses, shirts, cloaks, and coats, and of boots and shoes, gave rise to that peculiarly American institution—the mail order house. With the coming of the rural mail delivery, and later the parcel post system, the families in millions of American homes were able to "deck themselves out" in the very latest styles and patterns. Foreigners upon coming to our shores frequently expressed surprise when they were unable to distinguish the shop girl from the society matron, or the bookkeeper from the banker. Advances in textile technology and the mechanization of the boot and shoe industry had helped to keep this nation a classless society.[4]

NOTES FOR CHAPTER 27

1. George Sweet Gibb, *The Saco-Lowell Shops; Textile Machinery Building in New England, 1813–1819* (1950), p. 231.

2. This practice in textile technology, similar to that adopted by the steel manufacturers in the 1890's, of scrapping wholesale outdated machinery and substituting the newest, has characterized American industry to this day.

3. *Federal Trade Commission Report on the Leather and Shoe Industry,* I (1909), 156.

4. *United States Industrial Commission Reports,* XVII, XXIII (1900), 669; Frank Allen, "Silk Association of New York."

For developments in the clothing industry see: Zelma Bendure, *America's Fabrics: Origin and History, Manufacture, Characteristics, and Uses* (1946); Frank George Carpenter, *How The World Is Clothed* (1908); Frank George Carpenter and Frances Carpenter, *The Clothes We Wear* (1926); Helena Chalmers, *Clothes On and Off the Stage: A History of Dress from the Earliest Times to the Present Day* (1928); James Franklin Chamberlain, *How We Are Clothed, A Geographical Reader* (1904); *An Encyclopaedia of Textiles from the Earliest Times to the Beginning of the 19th Century,* with an introduction by Ernst Fleming (1927); Ethel Lewis, *The Romance of Textiles: The Story of Design in Weaving* (1937); Robert H. Megrew, *Textiles and the Origin of Their Names* (1906); Perry Walton, *The Story of Textiles: A Bird's-eye View of the History of the Beginning and the Growth of the Industry by Which Mankind Is Clothed* (1936); Owen Osborne, *The Story of the Stocking* (1927).

Chapter 28

BUILDING TECHNOLOGY

Advances in building technology lagged behind those in other fields; and by comparison with food and clothing technology, building technology remained largely in the handcraft, homespun age until the mid-nineteenth century. Building materials were unchanged, and century-old standards of construction continued. The years ahead, however, were to witness many advances in building technology.

AMERICANS DEVELOP NEW ARCHITECTURE

Post-Civil War Demands. An acceleration in the nation's building program began soon after the close of the Civil War. It involved all types of building construction: housing for carrying on the official activities of city, county and state affairs; shelter for the traveling public in hotels and railroad depots; office buildings, stores, factories, warehouses; and homes for both rich and poor. The rapid expansion of the country in other fields necessitated an extensive building program. Larger and costlier public buildings were erected. Instead of the four walls to serve merely utilitarian purposes, there developed a desire for ornamentation and grandeur. Individual wealth found expression in the erection of palaces for the wealthy, and better homes for the poorer classes.

The Utilitarian Principle. In most fields of building development, one feature was distinctly American; namely, that buildings existed for utilitarian purposes, and the form of the building was to be suited to the inherent needs. Principles of engineering and science, rather than art and embellishment, became the basis of sound construction. While architectural forms were borrowed from the classic Greek and from the medieval cathedrals, these were forced to give way to the utilitarian purposes of the structure.[1]

Rise of the Skyscraper. The early skyscraper type of building developed from these basic, utilitarian ideas. This type is distinctly American and is the nation's most distinctive contribution to building architecture. "In the skyscraper," declared Alfred C. Bossom, distinguished British architect, "America has invented and developed a wholly new and revolutionary form and type of building that is absolutely and characteristically her own. . . . These mighty structures proclaim the daring, the inventiveness, the self-confident power of their creators."[2] The skyscraper was not the product of evolution in the art of building, but rather the logical result of America's effort to meet the nation's demands. When needs developed for a new building technology, American architects and builders were prepared to meet them. This architectural-engineering ability represents a formidable constructive force. It epitomizes one significant phase of American civilization.

Nature of the Skyscraper. The skyscraper may be defined as a building erected with a skeleton frame of steel stanchions reaching from foundation to roof, tied together at each floor level by steel girders, around which the walls are hung as a curtain. The function of the metal frame was to carry the weight to the foundation. The walls served only to keep out the weather. Each steel column had its own separate foundation resting on an established footing, usually bedrock. The basement walls prevented water seepage and prevented the ground from caving in.

This type of building grew out of the needs of the time. Concentration of business in congested areas increased land values. Realtors, rather than move out of the congested areas, began erecting taller buildings. But a high building was not a skyscraper. Many European cathedrals were as high as a thirty-story office building, but they were constructed with walls of tremendous thickness to bear the height. This type of building technology was not applicable to office building, for the most valued space, the ground floor, was taken up with masonry construction. The sixteen-story Monadnock Building erected in Chicago in 1882–1883 had walls 15 feet thick at the basement level.

"Elevator Buildings." A forerunner of the skyscraper was the elevator building, erected with solid masonry walls, and rising from ten to fifteen stories in height. These were constructed with cast-iron fronts, iron columns, and wrought iron floors. These new structures imposed a number of pioneer problems in building technology. They provided a training school for the plumbing, heating, elevator, and later, the electrical engineers. The designers of these first tall buildings introduced prefabricated parts, devised specifications for

structural beams, and in general, introduced engineering to architecture. More important than all other contributions, the elevator builders developed a school of architects whose primary interest was not in the dead past, or in exotic forms, but in the inherent nature of the building itself and its relation to society. Among the pioneers were Henry Hobson Richardson, the pathfinder (who did not live to see the fruition of his ideals in the skyscraper), John W. Root, and Louis Sullivan.

First Skyscrapers. The real skyscraper, though unique, was not the product of any one brain. The Home Insurance Building, erected in Chicago in 1884, designed by Major LeBaron Jenny, is believed to be the first in which the basic principles of the skeleton design were utilized. Chicago continued to lead in this type of construction when, two years later, Burnham and Root erected a twelve-story building, using railroad steel rails imbedded in concrete as the footings for each column of steel in the superstructure. Hollabird and Roche the next year put up the Tacoma Building, and used curtain walls without having them rest on the foundation. For the first time in history, bricklayers could be seen beginning the walls midway between the ground and the roof. In 1889 rolled beams were used, and standard bridge shapes were riveted and set up as is done today.

By 1890 the skyscraper had arrived. The Masonic Temple, erected in Chicago in that year, was described as the biggest building in the world, and bore for a time the distinction of "one of the seven wonders of the world." In New York, where the skyscraper reached such renown that many believed it to have originated there, the development was at first hampered by adverse building legislation. The laws were modified in 1892, and in that year Francis Kimball's Manhattan Life Building challenged the supremacy of the Trinity spire, which had vied in height only with the tall masts of the clipper ships. In 1899 New York listed twenty-nine structures, ranging in height from 180 to 386 feet, while Chicago could claim but sixteen, ranging from 189 to 300 feet in height.

Skyscraper Technology. The skyscraper brought about a new development in construction technology that was gradually transferred to other operations and prepared the way for effecting in the twentieth century the mechanization of the entire construction field. Since in skyscraper construction each pier rested on its own foundation, a science was developed for laying foundations. In the past, foundations had been one continuous bed of masonry resting on solid ground, and were required to be proportioned in thickness to the superincumbent load. The new technology was borrowed from the bridge engineers. Roebling's Brooklyn Bridge, with a tower weighing 70,000 tons and rising to a height of 300 feet, rested on a footing made possible by the employment of pneumatic caissons. This principle was employed in sinking the piles for the Manhattan Life Building; and foundation engineering became one of the important features of skyscraper construction.

Expansion of Building Methods. The construction of the skyscraper also brought about a new organization in construction technology. A job superintendent was needed to see to the collection of materials, to organize the work of the several trades, and to supervise the carrying out of exact specifications. Through his supervision it became practical to secure prefabricated parts made to exact specifications. Each beam and girder, and each slab for the curtain wall, was prepared at the foundry, the fabricating mill, or the quarry. Each was marked for the exact use for which it was intended. The construction superintendent followed a maze of blueprints for each new building. From this practice, the standardization of parts was developed, as was the principle of mass production and greater speed in construction. Here was the forerunner of prefabricated mass production.

An American-Style Building. With the advent of the skyscraper a new style in architecture was recognized. It was no longer required that a building be constructed with the classic columns of the Greeks, the flying buttresses of the Gothic style, or the sturdy arch of the Roman. Here was an American style. It found acceptance all over the world wherever civilization had a need comparable to that in America. "No people's innermost self," says Bossom, "was ever more faithfully transmitted to building than was America's to its skyscraper." [3]

Elevators. Skyscrapers would not have been possible without the invention of elevators. Elevators had their origin in hoisting machines. By substituting a platform for the hook on the hoisting machine, this machine was made adaptable to commercial uses. Henry Waterman of New York made this adaptation in 1850, and his hoist was put into operation between two floors in his factory.

Otis Elevator. It was Elisha Graves Otis, however, who developed the forerunner of the modern elevator; and his flare for showmanship popularized the idea. Otis was working in a bed factory in New York when he designed an elevator to move his furniture. He equipped it with an ingenious safety device that clamped the car to the guides in which it ran should the rope break. At the Crystal

Palace Exposition in New York in 1853, Otis dramatized his invention by cutting the ropes, with himself suspended in the cage as a passenger, and to the amazement of the spectators the car was not plunged to destruction. With this success, Otis set up business in the manufacture of elevators and thus established the foundation for this new industry. Otis' elevator operated on a winding drum driven from a line shafting. The first elevators were hand-operated, but steam was soon applied. The cars moved 50 feet a minute. An Otis elevator was installed in the Astor House in 1858.

Baldwin's Elevator. A screw-type elevator was invented by Otis Tufts, and a water-balanced machine was invented by William Hale; but it was not until Cyrus W. Baldwin devised the hydraulic elevator in 1878 that both speed and lifting power, which were the most desired features, were made possible. Baldwin's machine was operated by a system of cables and pulleys controlled by a water-pressure engine. It attained a speed of 600 feet per minute. With the improvements that were effected on this machine, it came to be the most desired type until the turn of the century. An early handicap to elevator development was the absence of a safety feature. Before 1900 statistics listed as many as ten thousand elevator accidents annually.

Electric Elevators. An electric elevator invented by William Baxter, Jr., was installed in an office building in Baltimore in 1884. Five years later electric elevators were installed in office buildings in New York. These were direct-connected, worm-gear, drum-type units with mechanical control. The greatest advance in electric elevators came in 1902 with the gearless traction-elevator machine. A grooved wheel, fitted for a number of cables, each in its own groove, replaced the former spiral drum. This design was adaptable for any height of building, and its speed was limited only by the operator's ability to level the cab at the floors. At first it operated at 700 feet per minute, but as automatic controls were instituted, the speed was greatly increased. Since 1900, elevator development has kept pace with skyscraper construction and has become America's prime vehicle in one type of transportation technology. The beginnings of the modern escalator were seen at the Paris Exposition of 1900 in a display of a link belt, set on a slight incline and with handrails on either side.

Architectural Techniques

As building construction took on more and more mechanical features, engineering technology found its way into architecture. Architecture became a part of the curriculum in American colleges, and by 1900 there were more than a dozen schools of architecture in the country.

Structural Engineering. At first, however, architects disdained structural engineering. Bridges, train sheds, and factory buildings belonged to the field of the engineer—not to the architect. But when it was discovered that the principles of piling used in building bridges were needed in laying foundations for tall buildings, the technique of the engineer was called in. Engineering students studied building construction as a specialty. The science of architecture had become stationary, but engineering was progressive, so that a new term arose, "architectural engineering." The basic union of architecture and engineering was established and the two professions worked together. In describing a meeting of architects, sculptors, painters, and others to design the buildings for the Chicago World's Fair, one of their number reported in 1891 that when the meeting broke up, Augustus Saint-Gaudens leaped to his feet and shouted "Look here! Do you realize that this is the greatest meeting of artists since the fifteenth century?"

Introduction of Steel Beams. The traditional uses of wood and masonry in construction were usurped by other materials. Glass, introduced at the Crystal Palace Exposition and later employed in buildings for other exhibitions, failed to gain a foothold as a basic construction material. On the other hand, the cast-iron front of the pre-Civil War years continued in use because of its fireproof qualities and because of its strength and lightness in construction. Cast iron was employed for structural members, and wrought-iron beams were introduced. At first the iron industry had not fashioned such parts for the building trades, and bridge and railroad parts were adapted as special building beams. In the early period of steel construction, railroad rails were used before the steel mills began rolling structural beams.

The introduction of structural steel brought significant changes in building technology. It marked for all time a turning point in construction science. Andrew Carnegie began manufacturing steel beams for building purposes in the early 1880's at his Homestead Works.

Hollow Tile. Hollow tile, first used in the early 1870's, filled a double purpose. Lighter than other masonry, it was well adapted for use in floor arches. It also provided excellent insulation for steel beams which twisted and warped in intense heat, causing buildings to buckle in case of fire. Balthaser Kreischer, a New York manufacturer of firebrick, was given an early patent for developing hollow tile, and a number of improvements were introduced. The material was first used for interior work only, but when it was produced with a hard-finished surface, it was used for exterior construction and soon became more generally used than stone.

Cement. The need for a material that was fireproof, durable, and adaptable to all kinds of construction was met in the development of cement. Portland cement, first made from rock quarried on the Isle of Portland, England, made possible a complete construction of steel and clay. When reinforced with iron and steel, as devised by Joseph Monier in the 1860's, it marked the end of the old massive, heavy construction.

Portland cement is made by blending limestone and clay at a high temperature to form a new chemical compound. This compound, called "clinker," forms cement when reground. This material was an English product dating back to 1825, when its manufacture was clothed in mystery tinged with alchemy and magic. It was introduced in America in the late 1860's. This first shipment was the forerunner of many heavy shipments of cement that continued until near the close of the century.

An American patent for the manufacture of Portland cement was granted to David O. Saylor in 1871. His product received the highest award at the Centennial in 1876 and was specified by the War Department in the erection of the Eads jetties at the mouth of the Mississippi River. The American-made cement was not widely accepted until the close of the century. Government orders during the Spanish-American War gave impetus to American production.

Soon after the turn of the century, a Portland cement industry was developed in the Lehigh Valley of eastern Pennsylvania, where the limestone, clay, shale, and fuel needed in the manufacture of cement were found in their natural state. Gigantic plants with mechanical equipment were erected. Revolving mills pulverized the materials by action of thousands of steel balls. A rotary kiln, operating at 2800 degrees Fahrenheit, converted these into clinkers; huge grinders pulverized this into cement, ready to be packed in bags for transportation. Early in the new century Portland cement plants sprang up in areas throughout the country where limestone deposits of proper composition were found. The importation of the English product steadily declined.

Reinforced Concrete. In the building industry, cement was used with steel in foundations. It was mixed with sand and gravel, and reinforced with steel for floors and for floor arches. In this form it became a new material—reinforced concrete. This term is applied to a construction in which steel bars or heavy steel mesh is thoroughly embedded in concrete. The steel is first put in position, and then the concrete is poured around and over it, and securely tamped in place so that the steel is completely embedded. When the concrete hardens and sets, there is a resultant material of tremendous strength—the compression strength of concrete, supplemented by the great tensile strength of steel.

It was not until the last decade of the century that this structural concrete came into practical application. Steel construction with reinforced concrete represented the most important building materials invented in centuries, and these materials gave modern architecture its special character. The cheap production of Portland cement, and the economical rolling of the I-beam and other steel shapes, became essential to the skeleton-steel frames. Later developments included factory production of stoneshaped reinforced concrete blocks.

As early as 1888, experiments were carried on in making artificial wood, but it was not until a much later date that a product acceptable to the trade was developed.

Construction Machinery

Improved Tools. Of equal importance to the improved building materials were the improved methods of handling these materials and the new tools that were introduced. As in other phases of technological development, new tools were devised to eliminate the drudgery of arduous hand labor. From cellar to roof, machinery was utilized wherever possible to speed work, reduce costs, and save labor.

Construction Machinery. In constructing building foundations, the steam power shovel, invented soon after the Civil War, was introduced to supplant the horse-drawn slip scraper. These power shovels could fill a wagonbed with earth or gravel in one shovelful. The pile-driving machine, patented in 1883, was used in sinking the piling that supported the skyscraper. These gigantic steam-driven hammers took the place of the archaic system of raising a heavy weight between crude guides by means of horse-drawn cables. A special excavating machine came into use in the mid-1880's. For digging through rock, the steam rock-drill and the compressed-air drill also came into use in the 1880's. The pneumatic drill invented in 1890 could strike a hundred blows to the laborer's one.

On the superstructure, hoists and cranes were adapted to building construction during the 1890's. Operated by steam, and later by electricity, these were developed in many different forms and carried structural members to the desired location. They not only proved to be labor-saving machines, but also eliminated many of the accidents caused by laborers crawling up ladders with loaded hods, or teetering up slippery planks with heavy wheelbarrows. Pneumatic riveters were employed by 1900, when riveting supplanted bolting of steel beams. Later, electric welding was introduced.

Building Tools and Machines. The invention of the concrete mixer in 1900 made possible the use of concrete on a large scale and enabled contractors to use concrete in all seasons. As early as 1870 a machine was used to clean brick, and by 1900 a machine for finishing floors was introduced.

Factory Machines. Not only were improved tools used on the construction site, but factory production of building materials was also facilitated by the introduction of machinery. Quarrying machinery cut the stone at the quarry. A woodboring and mortising machine at the lumber yard lessened the labor of the carpenter. In most carpentry, a change from hand work to mill work was experienced. Scrolls were jigsawed, and windows and doors became standardized and were prepared in the mill. The American house was on the way to becoming a factory product, with the different parts made in the factory and assembled at the site.

Early Prefabricated Houses. Factory production of parts in mass quantities prompted the idea of prefabricated houses. Before 1900 a Massachusetts firm was making prefabricated summer cottages and gardeners' homes, but the conservatism of the public, rather than technical features, delayed the mass production of houses until after World War I.

Painting Spray Guns. Paints and varnishes became factory products. Although prepared and ready-mixed paints were on the market soon after the close of the Civil War, little progress was made before the 1880's. America started growing tung trees to produce tung oil as a pigment carrier for paints and varnishes. Spray guns were invented, which were capable of painting 2,500 square feet per manhour compared with 100 square feet with a hand brush. Since World War I considerable progress has been made in paint development both in pigment and in carrying agents, and a number of varieties have been developed. The advent of plastics added a new phase to the paint industry (see Chapter 39).

Home Construction and Facilities

American Home of 1900. Though the average American home continued to be erected by craftsmen operating much as they had for centuries, the home itself bore little resemblance to the earlier structures. The construction of an American home had by 1900 come to mean much more than the erection of four walls and a roof. Plumbing, heating, and electrical equipment became a part of even the modest homes, and with the advent of these features, science and engineering technology were enlisted in construction technique. The sanitary engineer, the heating engineer, and the electrical engineer became essential to home production.

Heating. Hot air, hot water, and steam for heating had been known from the early days of the nineteenth century, but little progress had been made in their development. Despite the fact that Benjamin Franklin had demonstrated the advantages of a circulating system in his Franklin stove, Americans continued to chill one side of their bodies and to roast the other by the open fireplace. It was not until around the turn of the twentieth century that public opinion in favor of better hygienic conditions and greater comfort brought about a change in the nation's heating system. Advances in furnace heating came slowly, due to the heavy expense of installing central heating and lack of understanding of the principles of thermodynamics. Before American homes could be equipped with adequate heating facilities, science had to find a way to reduce the costs, and had to discover the basic principles in central heating.

Central Heating. The early warm-air furnaces, which marked the first attempt in central heating, consisted of large stoves enclosed in brick. The cold air was admitted through a casing at the bottom, while the warm air emerged from the top. But these "hot air" furnaces embodied none of the principles of science.

The hot-water system was the next method employed. In this system the heat was diffused through loops of pipes. Since it required the installation of a maze of pipes, valves, and stopcocks, this system did not come into general use until late in the nineteenth century.

A steam-heating system was developed from the use of steam for power. At first boilers used in power engines, both high-pressure and low-pressure, were converted into units for heating. In some cases the exhaust steam from the factories was used for heating purposes, first in factories and later in office buildings and assembly halls. In 1870 a steam-heating plant was installed in the Chicago public schools. By 1890 the air-line vacuum had been installed in steam-heating apparatus, eliminating the difficulty of the pipes becoming airbound. Welded steel pipes supplanted wrought iron pipes, and automatic valves supplanted the old-fashioned cocks and valves. The tank for overflow began to be placed in the basement instead of the attic.

The American Radiator. The introduction of the radiator was distinctly an American contribution to heating technology. The radiator began with the pipe-box coil, or the pan radiator made of sheet iron. William Baldwin introduced one of the earliest radiators, the "Bundy" in 1874. It consisted of short lengths of 1-inch pipe screwed into a cast-iron base. It was capped with a cast-iron fretwork. The process of making radiators of cast iron, which was developed in the 1890's, made radiator heat available for the masses. Steel radiators, because of their cost, had been limited mostly to the wealthy, or installed only in large buildings. Mechanical improvements, such as the screw nipple and other control devices for cast-iron radiators, made it possible for homeowners to operate their heating systems without the aid of steam engineers.

American Society of Heating and Ventilating Engineers. The American Society of Heating and Ventilating Engineers, organized in 1895, introduced new principles in heating technology through their research laboratories. This association later became allied with the University of Illinois in carrying out a research program in heating technology. In the decades from 1880 to 1900, American homes were on the way to better heating.

Sanitation and Health Facilities. From the standpoint of health rather than comfort, have come the sanitary improvements in the American home. There was nothing new in the idea of public sanitation, since it had been developed among the ancient nations; but the improvements consisted mainly of new techniques and equipment. These were largely community advancements, dependent on the establishment of public water supplies and sewerage systems, rather than individual accomplishments.

Early in the nineteenth century, individual homes had been equipped with sanitary facilities wherever such were available. In the 1820's there were many homes equipped with baths and running water. A traveler's report in 1836 cited a house in Boston with five bedrooms and two baths. Longfellow's home in Cambridge, in 1840, was equipped with water closets and a kitchen sink. In 1854 Chicago made plans for supplying each city dweller with 9 gallons of water, and by 1855 announced that the city had the first comprehensive sewerage system in the country.

American Association of Public Health. Sanitary measures in the American homes did not reach any degree of general use, however, until the 1880's when an increasing number of communities established water works and sewerage systems. Improvements in sanitary facilities were demanded by public opinion when cities became alarmed over the spread of disease, much of which was attributed to infected water supply and bad sanitation. In 1884 and in 1886, several of the largest cities reported that most of the deaths were caused by communicable diseases. The newly organized American Association of Public Health worked actively to improve sanitary conditions and succeeded in bringing about many reforms in sanitation. By the turn of the century most of the larger cities had enacted sanitary measures and installed extensive sewage systems.

Water Purification. Because the natural water supply was found in many instances to be impure, the system of water treatment by duplicating natural conditions of filtering, first developed in England, was introduced here. In 1884, a distinctly American invention of mechanical filtration through sand treated with a coagulant was found practicable, since it speeded up the process. It was known as "rapid sand filtration." In 1897, a sanitary engineer in Louisville, Kentucky, revolutionized the technique of water purification. He introduced a method of cleaning the filter by backwashing through the sand instead of scraping the sand. Chemical purification, by the addition of chlorine, was introduced after the turn of the century.

Plumbing. Historically, it should be noted that the establishment of sanitary facilities in the home had been the work of the plumber, "the worker in lead." Originally the plumber made by hand nearly everything he used, and the material he worked with was chiefly lead.

Even with the establishment of sewerage systems and public water supplies, the installation of plumbing facilities in homes was slow. Illustrating how reluctant the public was, even as late as 1904, an article in *Everybody's Magazine* set forth the argument on the controversial question under the title "Is Bathing Good For Us."

Meantime, developments in enamelware had progressed to the point where bathtubs, washstands, and sinks were being made from this product. Enamel fixtures came into such high favor around 1900 that American-made products were installed in Buckingham Palace, London. They were also sold to the King of Prussia, to Russia for the hospital at St. Petersburg, to the President of Mexico, to the Federal Government of Venezuela, and in increasing quantities to American householders.

Air Conditioning. Early experiments in air conditioning, which involves heating, cooling, drying, moistening, moving, and filtering the air—a general bettering of the atmospheric conditions—began even before the Civil War. In the textile mills, as early as 1838, an air washer was employed. This involved spray nozzles operated by a rotary pump to humidify the air in order to prevent the threads on the spindles from breaking. A number of contrivances were devised in the period following the close of the Civil War. A complicated cooling system, which involved forcing air over ice racks, was installed in Carnegie Music Hall, New York, in late 1890's, but never proved satisfactory; nor was a similar system that was designed for the New York Stock Exchange in 1900.

Willis Haviland Carrier, soon after the turn of the century, combined the ideas involved in humidifying the air with the heating or cooling features, and introduced modern air-conditioning technique. He is called the "Father of Air Conditioning." Carrier devised a spray-type apparatus designed to humidify, or dehumidify, the air by heating or cooling it with water. He first installed the system in a lithograph plant in 1902, and secured a patent two years later. The term "air conditioning" was first used in the cotton mills. As new refrigerants were devised, and new techniques were added in the compressor, air conditioning systems were improved.

Home Architecture. Building technology as applied to the American homes had not kept pace during the closing years of the nineteenth century with advances in commercial and public buildings. While sanitary and heating improvements found their way into some private dwellings, houses for the masses remained little improved even as late as 1900. City houses were monotonous rows of red pressed brick with oblong openings, sills of white marble, and steps projecting upon the sidewalk. The downpipes discharged the roof water on the footway between the houses, and the cellar was closed with a green trap door. The houses were deep and narrow, and as land values increased and lots became narrower, the house crawled into the back yard, extending four or five rooms deep, admitting little light and air.

Apartment Houses and Flats. About 1880, apartment houses and flats began to appear, the distinction being in the spaciousness and equipment. Apartments, rising eight to ten stories, were equipped with elevators, speaking tubes, automatic door openers and chutes, common laundries, and individual trunk or storage rooms. The flats were crowded over the entire narrow lots on which they were erected. Each floor was occupied by four families. Outside stairways provided entrance in the rear, and narrow shafts leading to the roof provided ventilation. For some rooms, the only light was furnished by light wells.

Farm Houses. In the more prosperous American farm house, the open fireplace was boxed in to provide for the stove and was encased in tiles or surrounded with cabinets and mirrors. Newly added porches were embellished with scroll-sawed pieces. Windows and doors became standardized "mill work," as were many of the fixtures, and the kitchens were smaller and more compact. Halls became more than an entry way as the "reception hall" emerged. Houses were constructed by individual builders, with the lumber sawed on location and the work done by the building tradesmen; little advance in labor-saving was made in either tools or methods.

Such technological improvements as electrical and gas equipment, modern plumbing, and a whole gamut of automatic controls did not come into general use in the American home, office building, and factory until after World War I.

Notes for Chapter 28

1. For an excellent discussion of this development, see Louis Mumford, *The Brown Decade* (1927).

2. Alfred Charles Bossom, *Building to the Skies: The Romance of the Skyscraper* (1934), pp. 9, 14; see also, Thomas E. Tallmadge, *Story of Architecture in America* (1936); W. A. Starrett, *Skyscrapers, and the Men Who Build Them* (1928); Francisco Musica, *History of the Skyscraper* (1930).

3. Alfred C. Bossom, *op. cit.*, p. 14.

The following are helpful in making a study of building technology: Albert F. Bemis and John Burchard, *A History of the Home; The Evolving House* (1933); Edith Louise Allen, *American Housing, as Affected by Social and Economic Conditions* (1930); Frank G. and Frances Carpenter, *The Houses We Live In* (1926); Paul Starrett, *Changing the Skyline, An Autobiography* (1938); Fiske Kimball, *American Architecture* (1918).

Chapter 29

TRANSPORTATION TECHNOLOGY

By the outbreak of the Civil War, most of the basic plans for expanding a nationwide network of railroads and the major technological features to do so had been developed. Double-track, ballasted roadbeds, the "T" rail, wooden ties, swivel-front trucks, and locomotives with sufficient speed and power to pull a freight train of one hundred cars at 55 miles per hour—all these technological features of the American railroad had been developed within thirty years. The thirty-five years following the close of the Civil War saw the railroads reach maturity.

American Railroad History

Railroad locomotives became bigger and more powerful; tracks became smoother, rails of steel replaced straps of iron; coaches became safer and more comfortable; the dispatching and operating systems became more efficient and economical; and greater speed with greater safety was attainable. These technological developments contributed to making the American railroad the nation's greatest transportation system.

Building the Roads and Tracks. While the locomotive had been regarded by power-minded Americans as the most significant part of the railroad, a good track has always been just as important. The technical development of the track was as essential as the evolution of the locomotive. During the Civil War years, because of the heavy demands for iron, there had been few developments in track construction. But following the close of the war, the nation became railroad-minded. Routes for new roads were surveyed across the continent, old roads were repaired, improvements were made in constructing roadbeds, in laying the rails, and in building bridges and trestles. Double-track, and even four-track, systems were introduced, and track construction became standardized.

Improved Roadbuilding Machines and Materials. The first labor-saving devices introduced in track-laying were the steam plow and the steam shovel, employed to make excavations and remove embankments. Next came rock-crushing machines, rock drills, scrapers, and graders. During the 1880's both steam and electricity were used in boring tunnels. Track-laying machines were invented, and derrick booms extending in front of construction trains were used to swing the rails forward and drop them in place.

As seasoned white oak (which provided the best material for ties), became scarce, a poorer quality of wood was used and a system of impregnating the ties, borrowed from a practice of preserving piling in the Mississippi River with creosote, was instituted. This technical method in wood preservation prolonged the life of railroad ties from ten to twelve years.

With the building of larger locomotives and their operation at greater speed, the light flimsy strap rails could not withstand the heavy pounding. Steel rails were the answer, since they lasted about fifteen times longer than iron rails. By 1880 the production of steel rails was more than twice that of iron rails. Five years later almost 1,000,000 tons of steel rails were produced, compared with 13,000 tons of iron rails—just about enough to supply the street railways systems and mines. By 1900 the American rolling mills had captured the markets of the world; and over 6,000,000 tons of steel rails were exported that year, in addition to the amount needed for home consumption.

The establishment of uniform gauge for tracks was a major development in railway transportation technology. With each railroad determining its own gauge (width between the rails), the width varied from 4 feet 8 inches to 5 feet. In 1860 there were eleven different gauges on northern railroads, and a passenger in traveling from Maine to New Orleans had to change trains thirteen times. England had gone through a "war of gauges," and a similar controversy was developing here. With the building of the Union Pacific, 1862 to 1869, the question was brought to the floor of Congress, and was finally settled by President Lincoln who ordered the gauge to be established at 4 feet, 8½ inches. Gauges gradually became uniform for the entire nation.

Railroad Bridges. In bridge building, both for railroads and vehicular traffic, many major engineering feats were accomplished.

While the cheap construction of wooden bridges was continued where roads were extended through sparsely settled territory, large stone and iron structures were erected in the main arteries of travel. The Rock Island Railroad, with the approval of the War Department, built a bridge across the Mississippi in 1855 and a steamboat was wrecked against its piers, the steamboat company sued the railroad for damages. The counsel for the defense was Abraham Lincoln, who by winning the case for the railroad, established the legal right of a company to build bridges across navigable streams.

The Niagara Suspension Bridge. An outstanding achievement in railroad-bridge building of this period was the suspension bridge across Niagara Falls, begun by John Roebling in 1851. It swung from four 10-inch wire cables, which were carried over saddles resting on rollers on top of masonry towers. The cables were anchored to rocks on either side. The bridge had a clear span of 821 feet, stretching 245 feet above the level of the river. When completed in 1855, it was described as "incomparably the greatest bridge in the world." It continued in use for forty years until it was replaced by a steel-arch bridge.

Eads Bridge, St. Louis. Steel-arch construction was used by Captain James B. Eads in building a bridge across the Mississippi River, completed in 1874. Eads employed pneumatic caissons in building foundations, such as the Roeblings used later in building the Brooklyn Bridge. The Eads Bridge, the world's first steel-arch railroad bridge, was hailed by the London *Engineering* magazine as a magnificent achievement, "an example of the most highly developed type of bridge building of the age." This railroad bridge was a vital link in opening up the transcontinental railway system. Captain Eads was later honored by being voted a place in America's Hall of Fame, the first engineer so honored.

Cantilever Bridge. Another type of bridge construction, first modernized in Germany by Heinrich Gerber and often called the "Gerber" bridge, involved the cantilever principle. A bridge of this type was erected in the United States in 1876 across the Kentucky River. Another famous railway cantilever bridge was built for the Michigan Central Railroad over the rapids at Niagara in 1883, and still another one at Cincinnati in 1896.

The Brooklyn Bridge. The best-known bridge in America was the famous Brooklyn Bridge, the crowning achievement of Roebling's career, and called the "Eighth Wonder" of the world. A suspension bridge with a central span, it stretched 1600 feet in unbroken rhythm across the East River; and with the two side spans and approaches it extended to more than a mile (5,989 feet). It represented in size and weight and in endurance a construction never before attempted. It was begun in 1869 and finished fourteen years later. Here were used, as in the Eads' bridge at St. Louis, pneumatic caissons; and here for the first time, steel wire cables were used. These cables stretched over huge masonry towers, 271 feet high, and suspended the floor of the bridge 135 feet above the surface of the river. Each cable was made up of more than five thousand galvanized steel wires, oil-coated and laid parallel. When wrapped, each cable made a compact cylinder 16 inches in diameter. The bridge was erected at a cost of nine million dollars, and its construction involved overcoming many obstacles. It cost the life of the great designer and engineer, John Roebling, and caused the permanent illness of his son, Col. Washington A. Roebling, who from his sick bed superintended its construction until its completion in 1883.[1]

Railroad Tunnels. Another feature of transportation technology developed by railroad construction was the tunnel. Railroads had to pierce mountains as well as go over them, and they had to go under rivers as well as bridge them. The famous Hoosac Tunnel through the Berkshire Mountains, the first long tunnel in the United States, was started in 1848 and was under construction for twenty-five years. The introduction of the pneumatic drill in 1866 facilitated its completion, a task that might well have required another twenty years. As early as 1873 a project was announced for tunneling under the Hudson River, but due to lack of capital and needed machines, construction was delayed until after the turn of the century.

American Railroad Locomotives. The locomotive, more than anything else, is the symbol of the American railroad. Poets and painters alike have attempted to picture this symbol of power and speed. Walt Whitman in his poem, "To A Locomotive In Winter" (1876), described it as "the pulse of the continent."

> Thy black cylindrical body, golden
>   and silvery steel,
> Thy ponderous side bars, parallel
>   and connecting rods, gyrating, shuttling
>   at thy sides.
>
> By day thy warning, ringing bell
>   to sound the notes,
> By night thy silent signal lamps to swing
> Fierce throated beauty.

Law of thyself, complete, thine own
  track firmly holding
(No sweetness debonair of tearful harp or
  glib piano thine)
Thy trill of shrieks by rocks and hills
  returned,
Launched o'er the prairie wild,
  across lakes
To the free skies unpent and glad and strong.

Types of Locomotives. Following the close of the Civil War, locomotives became standardized in speed and power according to type. As greater power was needed for freight locomotives, and more speed required for passenger locomotives, distinction arose between freight and passenger traffic. A new, powerful locomotive called the "Mogul," because of its traction power, had three pairs of driving wheels, and a two-wheeled truck in front, a 2-6-0 type. The "Consolidation" type, a 2-8-0, introduced soon after the close of the Civil War, remained standard until around 1900. The 2-10-0 type, called the "Decapod," had its five pairs of driving wheels connected. By 1890 freight service was handled largely by "Ten wheelers," as well as "Consolidations." [2] The last decade of the century produced two popular types with trailing axles to support wide fireboxes: the "Atlantic," 4-4-2, for passenger service, and the "Pacific," 4-6-2, for freight.

In France, in 1885, an engine was designed to save fuel costs. It involved a compounding principle, in which the steam did double duty. The steam was used first in high-pressure cylinders, and then in low-pressure cylinders. This compounding principle was improved and patented in the United States by S. M. Vauclain of the Baldwin Locomotive Works. Before the end of the century the Vauclain compound locomotives in high wheeled 4-4-2 design were rolling up speed records seldom equaled. The economy in fuel and water, and the efficiency in design in both passenger and freight service, led to the introduction of this type on railroads.

Evolution of Modern Locomotives. Although the essentials of the locomotive had been well determined before 1860, there were many refinements introduced during the next forty years. The adoption of the link valve, a device that provided for cutting off steam in any point in the stroke, was regarded as a dividing line between the early and the modern locomotive. Other changes have been principally in matters of detail. In the gradual perfection of these details, the locomotive evolved into the symmetrical, efficient, and the wonderfully complete mechanism that it is today.

The introduction of steel in the construction of the locomotives marked another advance in railroad technology. Steel boilers made it possible to use higher steam pressure, and therefore achieve more power. Steel fireboxes and American-made steel tires came into general use. Locomotives were built larger and heavier. The invention of the six-wheeled connecting engine during the 1840's made possible heavier locomotives by distributing the weight. By the turn of the century freight locomotives weighing 150 tons and passenger locomotives of 130 tons were in use.

Fuel. As the forests in the eastern states were depleted, and as the railroads extended out into the prairie regions of the West where wood was scarce, the fuel problem engaged serious attention. The change in fuel from wood to anthracite, and bituminous soft coal, did not come until around 1850, and then only gradually after it was demonstrated that a ton of coal was equal to approximately two cords of wood. This was a great saving in carrying capacity, and the use of both anthracite and bituminous coal became practical. By the middle of the 1870's woodburners were almost extinct.

Passenger Accommodations. Improvements in passenger coaches had not kept pace with the technological developments of the locomotives. Horace Greeley, in his *Overland Journey,* complained of the hard benches and the lack of ventilation, "a mephitic atmosphere." He scolded car builders by declaring they had "no right to be ignorant of the laws of life with which they tamper," and he suggested it might be a good thing to bring two or three presentments against car builders and "charge them with manslaughter."

The greatest improvement in the history of luxury travel by rail was the invention of the Pullman car. George Pullman, a young cabinet maker in a small town in western New York, made an overnight trip in a "sleeping car" from Buffalo to Westfield in 1853. He was thinking all night, he said, "of how the car might be improved." The idea remained in his mind, and five years later in 1858, while employed in Chicago, Mr. Pullman with the aid of another cabinet maker had succeeded in converting two day coaches into comfortable sleeping cars. Each car contained ten sections, a linen locker, and two washrooms. These luxurious cars with plush upholstery, and with warm blankets covering soft mattresses, captured the fancy of the traveling public.

The outbreak of the Civil War postponed the use of this traveling luxury, but following the war, Pullman cars began to appear. When the funeral train that bore President Lincoln's body reached Chicago in April, 1865, the body was transferred to Pullman's "Pioneer" for shipment to Springfield for burial. The "Pioneer" became the most celebrated car in America. Within two years, forty-eight Pullman Palace cars were in operation.

Hotel Cars. Pullman created another sensation in 1867 when he introduced the "hotel car." Passengers were accustomed to leave their trains three times a day for quick 20-minute meals served in station restaurants where food was not always appetizing. Pullman designed a "hotel on wheels." Each morning the beds were put away to make room for spacious seats. Small tables were arranged between the seats, covered with tablecloths, and here food that had been prepared in a storeroom kitchen was served. A contemporary newspaper described this as Mr. Pullman's crowning achievement and hailed his success in supplying the car with a cuisine department containing a range, where every variety of meat, vegetables, and pastry, "may be cooked on the car, according to the best style of culinary art." Later the hotel cars were replaced by the Pullman diner, devoted exclusively to serving meals.

In 1870 the first through-train with Pullman cars crossed the continent, linking the Atlantic and the Pacific. Within a few years, Pullman coaches were being shipped abroad to England, France, and Italy. By 1880, sleeping cars and drawingroom cars were in operation on six railroads in England and Scotland.

Vestibule Cars. What proved to be a significant contribution to railroad transportation technology was the vestibule car, which appeared in the late 1880's. "Elastic diaphragms" on steel frames were attached to the ends of cars. When two cars were brought together, the faces of the diaphragms were pressed against each other by strong spiral springs that held them in position. This helped to reduce the oscillation of the cars and prevented telescoping in case of accident. Within a few years, all-vestibule trains appeared on most of the railroads. This feature increased the popularity of the dining cars because passengers could pass from coach to coach with greater safety. The early vestibuled coaches had narrow vestibules, but near the turn of the century the modern vestibule with the trap door over the steps came into general use.

American Mechanics Association. New designs were also introduced in passenger car construction. The formation of the Eastern Car Builders Association in 1867, and the American Mechanics Association in 1872, helped bring about the standardization of parts in the construction of both freight and passenger cars. Cars loaded with freight could be shipped from coast to coast because of standardization. This obviated the necessity of loading and unloading at terminals. Borrowing from the cattlemen, railroad companies branded their cars, and thereafter long strings of freight cars bearing the insignias of roads from all over the country could be seen on many railroads. Repairs of cars could be made in a matter of hours in shops anywhere across the land.

Train Lighting and Heating. Improvements in lighting and heating the coaches went through much the same series of technological developments that occurred in the heating and lighting of homes. The improved oil lamp with reflectors replaced the dripping candles, and in 1883 the Pullman Company introduced the brilliant "Pintsch" light. Gas of high candlepower was manufactured from petroleum and supplied lights that continued to be used well into the twentieth century. The gas was carried in large containers and conveyed through coaches by pipes.

Electric lighting was introduced during the 1880's with the use of storage batteries. In April, 1887, the Boston "Special" of the Boston and Albany Railroad was lighted by electricity and heated by steam, "an arrangement," declared the *Scientific American,* "which added much to the comfort of passengers, and removed the danger of fire, always imminent in the trains lighted and heated in the old way." Each car was equipped with twenty incandescent lamps, of 16 candlepower. In June, 1887, the Pennsylvania Railroad Company installed electric lights on the "Pennsylvania Limited," running between New York and Chicago, and advertised it as, "the first train in railroad history to be fully equipped with electric lights." Steam from the engine was carried to a turbine in the forward compartment of the baggage car, where it drove an electric generator supplying current to the entire train.

Freight Cars. Freight cars followed the same course of improvements as passenger coaches. The early freight cars, which were merely wagon-beds mounted on flanged wheels, were followed by cars designed for special uses. By 1880, boxcars of 10- to 12-ton capacity were being built. Hopper cars were designed to carry coal and ore. Tank cars (see Chapter 23) were built to transport oil. The use of steel to supplant wood in freight car construction was introduced as early as 1892. In 1898, the Pennsylvania Railroad was operating a freight train of all-steel cars. All-steel passenger cars came later.

Safety Devices. During the last quarter of the nineteenth century, a number of safety devices were introduced in railroad operation. The first need in safety equipment was for an adequate device to bring a heavy train to a quick stop. With the old hand brake, it was impossible to operate the brakes simultaneously on a train of several cars, and worst of all, it was dangerous. Various kinds of brakes were tried, including a spiral-spring brake on each car, a chain brake, and a steam brake applicable only to the locomotive. The distinguished scientist Josiah Willard Gibbs, noted more as a theoretical physicist than an inventor, was granted a patent in 1866 for a safety brake [which he designed, the object of which was, "to dispense with the brakeman for railroad cars and to secure prompt action of the brakes."] But none of these inventions was ever universally adopted and the need continued.

Westinghouse Air Brake. The brake problem was solved in 1869 by George Westinghouse, who invented the "Atmosphere Air Brake," commonly called the "Straight Air" brake. Westinghouse gained his ideas for this invention from studying the operation of the compressed-air machine used in drilling the Mount Cenis Tunnel. The principle that he adopted involved a small steam-operated air compressor installed on the locomotive, which provided compressed air to a storage reservoir. This was carried through the train by a continuous pipe, leading to brake cylinders on each car. When the engineer applied the brakes, the plunger was removed from the cylinder, releasing the compressed air which pushed against the brake shoes, holding them against the wheels until released. Flexible hose connections on the train pipes were essential features in enabling this braking system to withstand the shock. Improvements were made later in air compressors, brake valves, brake cylinders, and in lines of pipe, so that the entire brake system came to be automatic. The whole system rested on the original idea in the pneumatic brake.

Block Signal System. Another important safety device in railroad transportation technology was the development of a signal system. With the increase in the number of trains, a block system was worked out whereby consecutive blocks were governed by block signals. The object was to provide a space interval between trains. The first block signals were operated manually. Later, the operation of the block signal was made automatic so that an accident would not be caused by a trainman's neglect in throwing the signal. By 1875 this automatic block system had been designed so that it was interlocking, that is, the movements of the switch and signal appliances were so interconnected that they succeeded each other. Before the turn of the century, mechanical interlocking was replaced by electrical interlocking.

Automatic Car Coupler. Another major safety device was Eli Jenny's invention of an automatic car coupler, introduced in the late 1880's. This coupler, described as "a queer swinging knuckle device which locked hands when pressed firmly together," displaced the need for trainmen to step in between cars and drop in a link pin —one of the most hazardous operations in the history of railroading.

Demand for Speed. Accompanying all these improvements was the constant desire to attain greater speed. Railroad companies catered to a travel-minded public, and each company vied with the other in an effort to establish faster schedules. Faster time, non-stop schedules, deluxe equipment, and dollar dinners were offered to secure greater patronage.

Outstanding in this development was Number 999 of the New York Central, the finest locomotive shown at the Chicago World's Fair in 1894. With a weight of 62 tons, and with driving wheels 84 inches in diameter, it established a record for speed that was unequaled. On the last stretch of its trial trip from Syracuse to Buffalo it attained a speed of 112½ miles per hour, with a few miles being clocked at 38, 41, and 42 seconds a mile.

The New York Central Railroad and the Pennsylvania Railroad set out to capture the New York–Chicago trade. The Pennsylvania promised to make the distance in 30 hours. The New York Central equaled it. The Pennsylvania reduced its time to 27 hours. The New York Central, by installing a track-pan method for scooping up water for the locomotive while on the run, was able to meet that schedule.

Rivalries and speed wars appealed to a travel-conscious public. In her trip around the world to best Jules Verne's vision of eighty days—a trip sponsored by the New York *World* in 1889—Nellie Bly was able to rely completely on the American railroads for the home stretch. She was whisked from San Francisco to Chicago in 69 hours, and although she was unable to make connections with one of the fastest trains from Chicago to New York, the entire run from San Francisco was made in only four days. Jules Verne's dream had been eclipsed by almost eight days.

The Traveling Post Office. The public's desire for a more speedy delivery of news led to another development in railroad transportation technology. Colonel George B. Armstrong, assistant Postmaster General, had introduced a special railway mail service on the Galena line (Chicago and Northwestern) in 1864, and within the next ten years special mail cars designed for use on fast mail trains appeared on a number of lines. Again, the rivalry between the New York Central and the Pennsylvania Railroad led each company to build special railway mail coaches. Commodore Vanderbilt seized upon the idea (not so much for the income it might bring, but for the coveted prestige), regarding it a great honor to be permitted to paint on his train in large letters "UNITED STATES POST OFFICE." In appreciation of this privilege the New York Central built an entire new train known as the "Fast Mail," equipped with hundreds of small mail boxes for letters and newspapers. The initial trip, September 14–15, 1875, covered the distance from New York to Chicago, including thirteen stops and time out for breakfast, dinner, and supper in 26 hours.

Railroad Time Signaling. The spread of railroads throughout the country, with speeding trains delivering passengers on schedule, led to a demand for a standard time system. Samuel Pierpont Langley, noted astronomer and Director of the Allegheny Observatory in Pittsburgh, proposed in 1689 a new system of designating the time changes in the different longitudes. He connected three lines of telegraph with the municipal offices of Pittsburgh and neighboring Allegheny City, and joined them with the lines of the Western Union and with the private lines of the leading railroads. A turret clock in the City Hall of Pittsburgh was equipped with an electrical device so that its time could be regulated from the observatory. Within a year some forty railroad lines established electric connections with the observatory through this system, the first co-ordinated time system in the world. Trainmen were required to carry accurate watches and to check them regularly with the company's clock. For the public, railroad time came to be the source of accurate time information. The United States Naval Observatory in Washington established a similar system a few years later.

Standard Time Zones. The plan of dividing the United States into four standard time zones was brought about by the rapid expansion of railroads. Several zones of time existed in many road systems, with each road operating in the intermediate points on whatever one it chose. They all varied with Washington time. In 1870 there were more than one hundred "sun time" localities. This caused great confusion at overlapping points.

Sandford Fleming, a Scotch-Canadian, in 1878 brought out a feasible plan whereby the globe was divided into twenty-four zones. Four of these fell in the area of the United States: Eastern, Central, Mountain, and Pacific zones. Each community was able to determine its demarcation, and railroads designated the kind of time they were scheduling. In 1883, the General Time Convention of the Railway Managers (now the Association of American Railroads) adopted the system of standard time. At the stroke of twelve noon, November 18, 1883, more than fifty different times were abolished in the United States, and the railway clocks and watches throughout the country were set to one of the four standard times.

Double-track Roads. Another major improvement in railway transportation technology before the end of the century, and the one that marked a railroad as being "grown up," was the system of double-tracking. The first freight trains moved slowly, and before 1900 there seemed to be little necessity for fast freights. As schedules became more crowded, fast deluxe passenger trains were often held up by slow-moving freights. To obviate this, sidings were built on which freight trains waited for the scheduled passenger train to pass. Then came double-tracking, and later a third-track system. The third track was equipped with signals and crossovers with a switch at each end, so that any portion could be used as a relief track, thereby permitting the train to continue at regular speed without backing out. Years later some four-track systems were installed where needed.

The Railroad in United States History. By 1900 the American railroads had achieved first place in the nation's transportation system. Locomotives had developed from an estimated thirty thousand puffing woodburners to about forty thousand powerful coal-consuming giants. In track mileage, the roads had expanded from some 22,000 miles of short stretches, unconnected, and of varying gauges, to almost 200,000 miles of connected tracks—more than one-half the railroad mileage in the world. Records in speed attained during this period have seldom been surpassed to this day. In passenger service the goal of providing greater comfort, efficiency, and ease of travel had been attained. In safety devices and in controlling trains, major technological devices had been installed. The railroads had become a mighty force in American economy. They were the largest users of steel in the country, the greatest consumers of coal, and the greatest single landlord of unoccupied areas.

Builders of a Nation. Not only were the railroads a mighty force in knitting the country together in the nineteenth century; they were also an important factor in opening up vast areas of new territory. Their builders did not wait for regions to be populated and then enter the field to reap the riches; rather, they did the pioneering. James J. Hill, of the Great Northern, after opening up the last section of America to settlers, exclaimed with the pardonable pride of an old railroader: "It is not an exaggeration to say that the railway, next to the Christian religion and the public schools, has been the largest single contributing factor to the welfare and happiness of our people."

Highway Transportation

The story of America's highways does not present such a glamorous picture as the railroads. The efficiency of the American ralroads and the competition that developed in their determination to serve local communities were in part responsible for the country's poor wagon roads. The chief function of wagon roads had been to lead from one farm to another, or from farm to the nearest railway station. If transportation was needed in a section for industrial purposes, it became the practice to run a short railroad spur into that district, rather than build an improved wagon road. Thus, lumber and mining roads were jutting out from main lines in many sections of the country.

The Bicycle Age. The first impetus toward building good roads came from bicycle riders. The period from the early 1880's until the advent of the automobile is known as the "era of the bicycle craze." The two-wheeled, rubber-tired bicycle took America by storm. Everyone who could afford it bought a bicycle—professional men, lawyers, bankers, clerks, clergymen, college professors, and the ladies. For women, the bicycle and the typewriter (see Chapter 28) served to emancipate them from the age-old restricted home life.

At first bicycle riding was confined to the streets of cities and towns. Later, the more venturesome took to the country roads. Champions of long-distance contests and hill-climbing events became national heroes in sporting circles. A "bike-racing" contest held in St. Louis in 1883 attracted riders from a number of states and from Britain. Three years later, the "Century Run" contest (100-mile race) leading out of St. Louis saw sportsmen pedaling along rough, dirty roads and over steep hills, covering the distance in 11 hours and 34 minutes. "On such trips democracy held sway, and class distinctions bowed before the sportsman's ability." [3]

The League of American Wheelmen. In 1887 the League of American Wheelmen was formed, and became a propaganda agency for good roads. The league instituted a countrywide campaign for roadbuilding, the most active pressure group of the kind the nation had ever seen. Its members published articles in newspapers and magazines, of which the "Gospel of Good Roads" was an example. A *Good Roads Magazine* appeared in 1892. The members of the

League of American Wheelmen appealed to city officials, members of their state legislatures, and to their congressmen for better roads. The Chicago World's Fair, 1893, afforded an excellent opportunity to agitate for good roads.

The American Highway Association. Other agencies for good roads became active. The National League for Good Roads was formed in 1893. The same year a federal Office of Public Roads with civil engineer General Roy Stone as director, was established as a bureau under the Department of Agriculture. A few years later the American Highway Association was formed to arouse interest in highway improvement. Road construction soon came to be a state project rather than a local problem. Several eastern states took the lead and formed state highway departments. The New Jersey Legislature enacted a law in 1891 granting state aid to local authorities for road construction. In 1893 the Office of Public Road Inquiries, a division within the Department of Agriculture, was established. Some states enacted laws setting up stone-crushing plants to supply counties that desired crushed stone for surfacing their highways. Convict labor was authorized in many states to maintain roads leading into county-seat towns. As early as 1904 federal aid for road construction was advocated, and a bill for assistance was introduced in Congress. Such aid was not granted, however, until 1916.

Mere agitation, however, did not build roads. Another invention, the automobile, was to appear which made good roads a necessity. Pioneering in building concrete highways began with the laying of a few miles of concrete road in 1908 (see Chapter 33).

The American Navy Modernized

United States Navy in 1880. In the five great naval revolutions of the nineteenth century—the use of steam, shell guns, screw propeller, rifled ordnance, and armor—the United States played a prominent role.[4] But at the close of the Civil War, "silence fell on the United States naval shipyards," a silence unbroken for twenty years. The principle of the ironclad, worked out in the conversion of the "Merrimac" into the "Virginia," and the idea of the revolving turret devised by Ericsson in the "Monitor," were carried on by foreign nations while this country trained her future admirals on wooden sailing frigates. Yet, it was the genius of our people that had contributed the iron and steel armored vessels to the powerful navies of Europe, yet we, ourselves, remained inactive.

Birth of the New Navy. It was not until the administration of President Arthur that steps were taken to improve the situation. When, in 1883, a bill was passed prohibiting the repair of any vessel where the cost exceeded 20 per cent of the cost of the construction of a new ship of the same type, the new United States Navy was born. The new Navy was launched under the direction of Secretary of Navy William M. Hunt, an ardent supporter of a strong navy.

Building the Navy. The United States Navy Yards were ill equipped with tools for working in steel, and the contract for the first vessels was let to a private shipbuilder, John Roach of Chester, Pennsylvania. The Naval Board yielded to the sentiment in favor of imitating the British, and purchased plans in England for the construction of another group of vessels which were authorized in 1887. These were the only vessels of any size that were the product of foreign designers. All others were purely American.

For American shipbuilders, speed and power became the chief objectives. Within a few years warships that were in advance of anything afloat, with power to steam far and fast, and give and take the hard blows of a naval fight, were coming from the American shipyards. At the naval conference at Kiel, 1891, the German Emperor, after critically examining the "New York" battleship, ordered his engineers to board the American vessel and see how perfectly great engines could be made, and how well they could operate. This large armored cruising ship, the "New York," with its ninety-two engines and one hundred and seventy cylinders, won for her builders the prize offered for speed in excess of the required 20 knots. The "Columbia" (1893) held the naval record for continuious sea speed at 22.8 knots per hour for 4 hours, and an average of 18.41 knots per hour while crossing the Atlantic. American excellence was attained when our shipbuilders ceased imitating foreign powers in heavy armor and guns and turned to improving the ship itself. Ship design and equipment were American ideas. Machine shops and blacksmith shops were established on the vessels, and mechanics rather than sailors were sought for seamen.

Engines. Progression in engine development first involved the institution of the direct-acting type in screw propellers. By 1870 compound engines were used in which the working fluid in the high-pressure cylinders was again used in the larger, low-pressure cylinders. In the early 1880's, triple expansion engines were instituted in which the steam was utilized three times. By the turn of the century, Parson's turbine had been made adaptable to marine engines. In all this the aim was increased power with low fuel consumption, a feature much more essential on sea than on land.

Melville, Marine Engineer. America's new Navy utilized the triple-expansion type of engine with a number of innovations and improvements. In 1887, George Melville, Chief Naval Engineer, introduced dozens of innovations designed to promote efficiency. He designed machinery which made our ships famous, and which placed his name foremost among the world's marine engineers. Boilers were improved to permit greater steam pressure. The old flat-sided, box-shaped boiler gave way to the cylindrical type. Pressure was increased from 60 to 100 pounds per square inch. With the triple expansion engine, a cylindrical boiler with three internal corrugated furnaces became general. Then came the coil boilers, with pressure up to 200 pounds.

Steel Construction. With the discarding of wood as a material for construction, a controversy ensued as to whether iron or steel should be used. The advocates of steel held that wrought iron was too expensive, and rolled iron was not sufficiently uniform in structure. The new "mild steel," which they advocated, was questioned because of its newness. Finally, the advocates of steel construction won out.

Armor. It was in the armor sheathing that the greatest technological development occurred. With the advent of armored vessels, the progression had been from compound armor (wrought iron on which was plated a hard steel face of one-third the thickness of the iron), to solid steel, to nickel steel, to Harveyized armor. A race now began between the development of the power of projectiles and the strength of armor. As guns were improved, armor was strengthened to resist them. When the new Navy was undertaken there were no armor-plate mills in the country, and the armor for the turrets on the first vessels was obtained from abroad. The demand for armor plate led to the establishment at the Bethlehem Steel Works of one of the world's largest and best-equipped steel forging plants.

The first plan was to make armor plate so thick that it could withstand the penetration of all shells. Plates from 18 to 24 inches thick, of tremendous weight were hammered out. Because of this weight disadvantage, an invention of Hayward Augustus Harvey, patented 1888, was adopted. This involved heating the plate in charcoal, thereby increasing its impenetrability. With Harveyized steel, armor plate was reduced to 12-inch thickness. A further improvement by William Corey in 1895 made American armor plate equal to the products of the famous Krupp plants in Germany.[5]

Naval Guns. While developments were being made in armor plate, the power of the gun was steadily increasing, and the effectiveness of the projectile and the powder was keeping pace. Guns were made with twice the velocity of those of 1882. They could fire with greater rapidity, and had penetration and range not possible with weapons of the old type. Steel projectiles were fitted with steel caps so that they could penetrate Krupp armor when given sufficient velocity. Philip Hichborn devised a balanced turret by shifting the weight of the gun mounts and recoil apparatus so that the guns could be turned in any direction. Bradley A. Fiske devised precision instruments for range finding, gun sighting, and gun pointing, all of which were described as major advances in naval technology.

Modern Training and Equipment. Naval training kept pace with naval equipment. A college for training naval officers was established at Newport, Rhode Island in 1884, and a plan was introduced that provided for sending high-honor graduates from the naval academy abroad for study. A naval training school in electrical engineering was established at Boston, 1898.

The new Navy kept abreast with other technological advances, such as the use of telephone, wireless, and electrical power, so that by 1900 the United States possessed a small but compact and homogeneous fleet, containing some of the most efficient ships in the world. By the time of the Spanish-American War, when the new Navy was called into action, the American shipbuilders had acquired the technological know-how for constructing efficient fighting vessels. Although the American Navy did not yet rank with the great navies of the world, it possessed individual ships that were comparable to any others. The technological advancement had been rapid, and naval architects of other nations took warning.

Notes for Chapter 29

1. David B. Steinman, *The Builders of the Bridge, The Story of John Roebling and His Son* (1945); and David B. Steinman and Sara Ruth Watson, *Bridges and Their Builders* (1941).
2. For an excellent history of the major technological developments in railroading, see Robert S. Henry, *This Fascinating Railroad Business* (1948).
3. *Missouri Historical Review,* XLVI (July, 1952), No. 4, 355-356.
4. James Phinney Baxter, III, *Introduction of the Ironclad Warship* (1933); and Frank M. Bennett, *Steam Navy of the United States* (1897).
5. "Naval Supplement," *Scientific American* (April 30, 1898).

See also J. E. Watkins, "Development of American Rail and Track," *Smithsonian Institute Report* XL (1888–89) Part 2; S. K. Farrington, *Railroading from the Head End* (1943); Richardson and Blount, *Along the Iron Trail* (1938); D. M. Rice, *Romance of the Railroad* (1941); Frank P. Morse, *Cavalcade of Rails* (1940); J. B. Husband, *Story of the Pullman Car* (1917); Edward Hungerford, *Locomotives on Parade* (1940); *A Yearbook of Railroad Information,* Eastern Railroads President's Conference (1945).

For naval technology, consult James P. Baxter, III, *Introduction of the Ironclad Warship* (1933); "Resumé 1883–1924," *Navy Year Book,* Senate Naval Affairs Committee, 65th Congress, 3rd Session and 66th Congress, 2nd Session: Doc. 418, VII, 428; XIV, 302; II; John D. Long, *New American Navy* (1903); Frank C. Bowen, *A Century of Atlantic Travel* (1930); Henry Hall, *Report on Iron Shipbuilding Census* (1880); Donald Mitchell, *History of the Modern Navy* (1946).

Chapter 30

COMMUNICATION TECHNOLOGY: FROM OCEAN CABLES TO WIRELESS

The new instruments of communication technology that were developed during the period from 1870 to 1900 included ocean cables, the telephone, the typewriter, photography, and improvements in newspapers and printing.

The idea of laying a cable across the Atlantic had received strong support during the years immediately following the invention of the telegraph. Professor S. F. B. Morse had made frequent requests during the 1850's for federal aid to experiment with cable communication. He believed that telegraph wires, properly insulated, laid across the ocean floor, and equipped with powerful sending and receiving instruments would work as successfully as the land telegraph. This idea was promoted by Cyrus W. Field.

Transoceanic Communication

The question of a submarine cable came up during every session of Congress for the next few years, and finally Congress passed an Act in 1857 authorizing the President to negotiate contracts with England for jointly laying a cable between Newfoundland and the coast of Ireland, and provided for equal use of the cable by both British and American subjects.

Laying the First Cable. Two warships, the "Niagara" of the United States and the "Agamemnon" of Great Britain, manned by seasoned sailors, were dispatched to lay the cable. The "Agamemnon" departed from Valentia on the coast of Ireland, while the "Niagara" set out from Trinity Bay, Newfoundland, laying a cable made from wires insulated with gutta percha and encased in a protective covering of steel wires. After repeated failures, due to the cable breaking, the two ships met in midocean and set out in opposite directions, laying cable as they sailed toward their respective shores.

On August 16, 1858, Queen Victoria cabled greetings to President Buchanan, expressing the fervent hope that, "the electric cable which now connects Great Britain with the United States will prove an additional link of friendship between the nations. . . ." President Buchanan replied, "this triumph is more glorious than was ever won by conqueror on the field of battle," and he expressed the hope that, "this electric telegraph [cable] will prove to be a bond of perpetual peace and friendship, an instrument destined by Providence to diffuse religion, civilization, liberty, and law throughout the world."

Public Interest. Public interest in this new instrument of communication technology was widespread. The New York *Times* (Sept. 1, 1858) stated that the cable was fast entwining itself around the affections of our people. "Every little hamlet and all the pretentious towns in the country are going mad on the strength of this wonderful experiment." Newspapers of the inland cities, as well as those on the eastern seaboard, devoted long articles to the newest achievement in telegraph communication. "King Cyrus of old conquered the world for himself," declared one headline, "our King Cyrus [referring to Cyrus Field, often called the 'Father of the first Atlantic cable'] conquered the ocean for the world." Clergy from their pulpits proclaimed the cable to be an instrument that would help bring peace to mankind. *Living Age* magazine declared it would be "the starting point of the civilization of the latter half of the nineteenth century."

Failure of First Cable. Despite the glowing prophecies, however, the cable engineers had difficulty in keeping it in operation. Late in October, 1858, the cable engineers feared that the fading signals signified its approaching death. During the Civil War years neither Great Britain nor the United States was willing to undertake the expense of repairing the cable, and further cable communication was delayed until after the close of the war.

Laying New Cable. In 1866, and in 1868, two new cables were laid across the Atlantic. In 1869 France laid a cable connecting her capital by telegraph and cable via Nova Scotia with New York City. The first wholly American-owned cable was laid across the Atlantic by Jay Gould in 1881. Three other American-owned cables were laid within the next few years. During the next half-century, ocean cables became the chief agency in overseas communication technology.[1]

Telephone and Telegraph Systems

Intracity Telegraph Systems. Meantime, the telegraph system was becoming an indispensable agent of intracity communication. Fire and police departments in some of the larger cities had established telegraphic signal systems in the 1850's. After the close of the war, financial houses in a few of the larger cities installed telegraphic machines to transmit vital information immediately to local subscribers. By 1870 printing telegraph companies were established in New York to provide access to a central office, from which messages were relayed by "long distance" telegraph lines.

The Telephone. Following the rapid expansion of the telegraph came growing demands for an instant, simple, direct system of communication—one that would enable one party to communicate directly with another. The demands increased with the growth of cities and the expansions of business firms. They presented a challenge to a number of inventors, chief of whom were Thomas A. Edison, Elisha Gray, and Alexander Graham Bell.

Elisha Gray. Elisha Gray of Chicago, later a Professor in Oberlin College, invented in 1873 "an automatic self-adjusting telegraph relay, a telegraph switch, and an annunciator," adapted for use in hotels and large business offices. He observed that sound produced in the presence of a magnet would cause another magnet of similar adjustment to respond to the tone, and that by the same wire, another sound might be transmitted by the same magnet; this, in turn, could be reproduced by a receiving magnet adjusted in harmony with the transmitting magnet. This could be repeated indefinitely over the same wire by a constant current used to augment the current produced by the vibrating membrane. Gray hoped by this method to be able to transmit tones of the human voice, and continued to carry on a number of experiments.

Alexander Graham Bell. While Gray was working on his invention, Alexander Graham Bell, a Scotsman who had arrived in Boston in 1873, was also experimenting on a device aimed at transmitting the actual sound of the human voice by means of an electrical current. Bell had spent a few years in London where he had learned of the experiments of the German physicist, Helmholtz, who had demonstrated how tuning forks would vibrate indefinitely by the use of electromagnets. The blending of the tines of several tuning forks had produced a complex sound with qualities not unlike the human voice. Bell reasoned that if a magnet or an electrified wire could cause a tuning fork to produce a musical sound, a musical telegraph could be constructed which would send all the musical notes produced on a piano keyboard over a single wire at the same time. From that idea, Bell ultimately developed the telephone.

Soon after arriving in Boston, Bell was appointed professor of speech and "vocal physiology" in Boston University's School of Oratory. He also opened a private School for Vocal Physiology, taught lip reading by a method he had learned from his father, who had pioneered in that work, and offered lessons in "visible speech for the deaf."

Bell's Interest in Multi-telegraph. While developing a new technique for teaching mutes, Bell also became interested in a project that was engaging the attention of a number of Boston bankers and business men, namely, the possibility of developing a multi-telegraph. There was need for a telegraph system that would carry a number of messages over the same wire at the same time. With the financial aid of Thomas Sanders (who had a deaf-mute son), and Gardiner G. Hubbard (whose deaf-mute daughter later became Bell's wife), Bell began a series of experiments that led to the invention of the telephone.

Others had attempted the electrical transmission of speech, but none had had Bell's experience or knowledge of the science of sound and speech. Bell believed that he had to discover a mechanism that would produce a continuous current, and also would "vary in intensity, as the air varies in density when sound is passing through it." After months of experiment with what he called the "harmonic" telegraph, by which he hoped to transmit more than one Morse code message simultaneously over the same wire, Bell had a mental picture of an apparatus that might possibly produce the wavelike current required to transmit speech.

Experiment of June 2, 1875. The possibility of transmitting speech, the human voice, and having it heard distinctly claimed Bell's attention throughout the years of 1874 and 1875. With the aid of a young mechanical assistant, Thomas A. Watson, Bell carried on a number of experiments with pairs of telegraph instruments. One day, June 2, 1875, an experiment was performed which made history. Bell was sitting in one room with a receiving instrument pressed to his ear, while Watson was in another room operating the transmitters. Noting that the contact point of one of the transmitters was screwed too tightly to the reed to operate, Watson tapped it with his finger to start it vibrating. Bell heard the faint sound, a "twang," as he called it. He even recognized the quality, the pitch of the twang of the vibrating spring. For the first time in history, "a human ear heard the tones and overtones of a sound transmitted by electricity."

First Telephone. Bell asked Watson to design an instrument, which they called the "harmonic" telegraph. It consisted of a reed mounted over a magnet, with one end of the reed fastened to a membrane drumhead. When the membrane was set in motion by the voice, the reed vibrated over the pole of the magnet, "and set up in the circuit an undulatory current that was a copy of the waves in the air caused by the human voice." That was Bell's (and Watson's) first telephone. While this instrument transmitted the tones of the human voice, it was not until March 10, 1876, that Bell, using a different form of transmitter "sent over an electric wire a complete sentence of speech." Mr. Watson, while working in a basement room, heard the historic words through the telephone receiver, "Watson, come here, I want you."

Bell's Patent. For this invention, Bell was granted the first patent on a telephone (No. 174,465), which read in part, "The method of, and an apparatus for, transmitting vocal or other sounds telegraphically, as herein described, by causing electrical undulation, similar in form to the vibrations of the air accompanying the said vocal or other sounds, substantially as set forth." This patent, granted in 1876, has been described as the most valuable single patent ever issued. It was the beginning of a new age in electrical communication technology.

First Public Demonstration. The next development in Bell's telephone was characterized by one of history's unusual dramas. Upon the insistence of his fiancée, Mabel Hubbard, and her father Gardner Hubbard, Bell's telephone was exhibited at the Philadelphia Centennial in 1876. Located in an obscure corner of the Education Building away from most of the other exhibits, weeks passed before the new invention attracted attention. Then came the day of the judges' inspection. They examined the queer looking exhibit, made a few remarks, and started to leave. At that precise moment, Don Pedro, young Emperor of Brazil, who years before had been in Boston and had watched Bell teach a class of deaf mutes, was then visiting the Exhibition. Anxious to know about Bell's new invention, he placed the receiver to his ear while Bell went to the other end of the line and spoke to him. In amazement, the Emperor exclaimed, "My God—it talks."

Grouped about the Emperor were a number of spectators, most prominent of whom were two distinguished scientists, the venerable

Joseph Henry, America's dean of electrical engineers, and Lord Kelvin of England. After each in turn listened to the voices coming over the wire, they announced to the judges, "Here is the greatest marvel ever achieved in electrical science." As one observer noted, "Of all the gifts which young America received on its one-hundredth birthday, the telephone proved to be the most valuable."

Public Ridicule. In the eyes of many doubters, however, Bell was a humbug, an imposter, a windy ventriloquist. He was subjected to ridicule similar to that heaped upon Westinghouse when he proposed "to stop a railroad train with wind." But the public demonstration that Bell and his friend Watson gave before the numerous audiences in Boston, Salem, Charlestown, and elsewhere soon became front page news. The *Boston Advertiser* (October 19, 1876) printed an extended telephone conversation carried on between Bell in Boston and Watson in nearby Cambridge. A band playing the "Star Spangled Banner" in Boston was heard by an audience of two thousand people in Providence via the telephone receivers attached to telegraph wires. The evangelist Dwight L. Moody and singer Ira Sankey, currently engaged in a Boston revival, used this new instrument of communication to exhort a group of unseen worshippers a few blocks away. The practical use of the telephone had been demonstrated.

First Telephone Exchange. New developments came rapidly. In June, 1877, five Boston banks installed telephones, each connected with one central switch wire. These were used as burglar alarms and constituted the first crude telephone exchange. Other cities were anxious to have telephones also; within three years over fifty thousand were in service, and telephone conversations were carried on between cities 40 and 50 miles apart.

First Telephone Company Organized. In 1877, the Bell Telephone Association was organized. A group of able young men—including Bell, the inventor; Watson, the builder of the instrument; Sanders, the financier; Hubbard, the salesman; Theodore N. Vail and W. H. Parks—laid plans that led to the building of the vast telephone network in America. William H. Forbes, son of a wealthy East India merchant and son-in-law of Ralph Waldo Emerson, was elected president of the National Bell Telephone Company. This group, with a few able assistants, pioneered the telephone system into one of the nation's major industries.

The Transmitter and Switchboard. The first major improvement upon Bell's telephone was the invention of the transmitter by a young German immigrant, Emile Berliner. Bell had designed the receiver, which was the most sensitive part of the entire telephone mechanism. It, together with the transmitter, represents "as complete an artificial tongue and ear as human ingenuity can make."

Next came the switchboard, an indispensable part of the telephone system and one of the most intricate pieces of mechanism known to science. Although Charles Scribner of Toledo and Chicago is usually given credit for this important invention, it actually resulted from a number of inventions and from thousands of individual parts. Without the switchboard there could be a telephone, but no telephone system.

Telephone Exchange. From the switchboard, with its numerous signals and batteries, there developed the telephone exchange. It was the home of the switchboard. Two or more phones joined by a wire cord established a "connection." The first exchanges were operated by boys, but it soon developed that girls, "possessing docile dispositions, soft voices, unlimited patience, and nimble fingers," proved vastly more efficient. The efficiency and quality of telephone service, before the installation of the dial system, depended upon the girl at the end of the wire who dispatched the "talk trains," and ruled the wire highways.

Expansion of Telephone Industry. Once the technical problems of building a telephone had been achieved, the next major step was one of expansion and integration. Few industries have evolved so rapidly from so few inventions as did the telephone system. As individual systems were organized, the need for connecting them into one unit became obvious. Long-distance lines were constructed, switchboards were expanded, and telephone conversations between persons separated hundreds of miles became common. By 1900 1,000,000 miles of wire had been strung, and one-half of the people of the United States were within talking distance of one another.

First Presidential Use of Telephone. President McKinley was the first chief executive to use the telephone extensively. During the campaign of 1896, he and his party managers frequently discussed campaign plans over the "phone." McKinley regarded the telephone as an agency that "brought us closer together." President Theodore Roosevelt used the telephone very infrequently; whereas President Taft conducted a considerable amount of official business by telephone; he had the first telephone exchange installed in the White House.

News Reports by Phone. Business firms, corporations, and newspapers soon came to regard the telephone as a necessary part of their organization. For quick newsgathering the telephone was indispensable. The *Boston Globe* claimed to be the first newspaper to have received news dispatches by telephone. The *Central Press* established a co-operative service for telephoning news items to a number of newspapers at the same time, and did it with greater speed than could be achieved by the telegraph. By the turn of the century the one remaining significant technical advance to be made in the central office was the dial system, an accomplishment that was not universally introduced until fifty years later.

Telephone, Instrument of the People. The telephone, unlike the telegraph and the cable, was the instrument of, and for, the people. It served individuals of limited means as efficiently as the man of wealth. It aided in assimilating the various dialects of different nationality groups. It was a stimulus to both local and national business. Small and large organizations alike proudly advertised the fact that they could be reached by telephone.

Typewriters Spur the Written Word

The typewriter, like most inventions, was the product of evolution, and many inventors contributed an important share in creating a machine that has become an indispensable piece of equipment in every business and professional office in the land. A number of inventors during the half-century before 1873 had produced experimental machines for typing letters and numerals. A patent was granted to William Austin Burt of Detroit as early as 1829 for his "typographer," but the time was not yet ripe for such an invention.

Teletype System. A quarter of a century before any inventor had developed an accurate typewriter, machine writing was carried on over long distances by electromagnetic communication. What was virtually a teletype system preceded the typewriter by almost thirty years. Samuel Morse and Alfred Vail, in their work on the telegraph, developed a keyboard before 1845 that looked much like the typewriter keyboard. In 1846, Royal Earle House patented a machine that recorded messages, not in dots and dashes but in Roman type. This was improved by David Edward Hughes in the next decade. They were only one step away from electric typewriters.

Typewriter Pioneers. Credit for the invention of the modern typewriter goes to C. Latham Sholes of Milwaukee, printer, editor, and inventor; and to those associated with him, Carlos Glidden, attorney, and Samuel W. Soule, machinist. Sholes was engaged in making a numbering machine to number tickets consecutively when Glidden suggested he should make a machine that would print letters as well as figures. Sholes began work on the idea. He named his machine the "typewriter," and the first words written were, C. LATHAM SHOLES, SEPTEMBER 1867. The Sholes machine had an up-strike type bar, a circular type basket, finger keys arranged in rows in the back, and a cylindrical platen roller. There was little that was really original in Sholes invention, but he continued his experiments until he had evolved a keyboard system and linkages to the type bars, coupled with the *universal* arrangement of the keys, which enabled the device to type accurately and rapidly.

James Densmore's Remington Machine. Another pioneer was James Densmore, an oil man from Meadville, Pennsylvania, who spent his fortune in producing model after model before he had a machine perfect enough for the manufacturer, E. Remington and Sons of Ilion, New York. This firm manufactured firearms, sewing machines, and agricultural implements; but in 1873 they bought the Densmore patent rights to make the typewriter. The first typewriters were made in the sewing-machine department. The early machines were ornamented with a gay and flowery design, and were fitted to a sewing-machine stand with a treadle operating the carriage return. The public regarded this new invention as an interesting novelty, but few were willing to pay one hundred dollars for a typewriter when they could purchase a pen for a penny.

Mark Twain's Typewriter. The first sales efforts for typewriters were not directed toward the commercial world, but to court reporters, editors, writers, and clergymen. The typewriter was regarded as too impersonal for business use. Mark Twain was one of the earliest users of the typewriter, and his *Adventures of Tom Sawyer* was claimed to be the first typewritten manuscript to go to the printers. Twain called his machine a "curiosity breeding little joker," but liked it, "because it piled an awful stack of words on one page," and because he could "lean back in his chair and work it." The typewriter attracted little attention at the Centennial in 1876 where it was exhibited, although a special inducement was made of typing messages to send home on payment of twenty-five cents.

Sholes Keyboard. Sholes and Glidden, who had built the first machines, were compositors and were little interested in the alphabetical arrangement of type. They based their arrangement of the keyboard on the frequency of the juxtaposition of the letters in writing, placing them where there would be the least likelihood of type-bar collisions. The keyboard of today is almost identical with Sholes design of 1873. The first typewriter was limited to upper case composition (capitals), but this deficiency was overcome in the second model brought out in 1878. It employed both large and small letters on the same type bar, which was operated by a shift key.

Improvements in Machines. By 1884 a number of inventors had entered the field and many wars in ideas and designs resulted. There was the battle over a double or single keyboard with shift key; and an argument over ribbon inking or direct inking. Then came the innovation of visible writing, made successful in a Fitch typewriter in 1886. The Oliver machine, the invention of Reverend Thomas Oliver, introduced a successful down-stroke-from-the-side type of operation. The Smith Premier, the invention of A. T. Brown in 1888, put on the market by Lyman C. Smith and Company of Syracuse, became the champion of the double-keyboard style. The sale of the Smith Premier spread to all parts of the world. The Underwood machine, developed by Franz X. Wagner, was a pioneer in the front-strike method and permitted fully visible writing. This machine later included the tabulator as an integral part. The Remington machine adopted visible writing in 1907. From these improvements a machine that met all the needs for mechanical writing was developed.

Social Effects of Typewriters. The typewriter came to be an indispensable machine in the business and commercial world. It affected the social life of the country as well. It opened up a new field of employment for women, as thousands of feminine typists invaded the offices of business and professional men and took up secretarial positions. The typewriter has been cited as being more powerful in the emancipation of women than were the crusades of Susan B. Anthony and her militant army of feminists.[2]

Photography

Another major development in communication technology occurred with the improvements in photography. Pictures, images, and sketches of people and objects have a universal appeal. John Richard Green, distinguished English historian, once wrote that he considered the six-penny photograph to be the greatest boon ever conferred on the poorer classes. "Anyone who has seen the array of little portraits of the boy who has gone to Canada," he continued, "the girl who is out at service, the little one with the golden hair who sleeps under the daisies, or the old grandfather in the country, will feel with me that the six-penny photograph outrivals such philanthropic claims as schools, museums, and shoe-black brigades."

Mathew Brady, 1823–1896. As early as 1850 daguerreotype photography (see Chapter 17), which had recently been introduced from France, was engaged in by a number of traveling artists. With their cumbersome and costly equipment, often requiring a two-horse wagon to haul it about, these artists traveled across the country making pictures of persons who could afford the price. Mathew Brady had given the nation a priceless collection of pictures of the Civil War, but the heavy expense involved in procuring cameras and the cumbersome photographic equipment that was needed to develop pictures reduced him to poverty, even though the federal government had paid him twenty-five thousand dollars for a set of two thousand selected photographs.

John W. Draper. John W. Draper, distinguished chemist, historian, and pioneer photographer, had tried for years to bring the art within reach of a greater number of people. To ease the strain of sitting immobile for several minutes while facing the camera, Draper introduced the headrest to clamp the sitter's head gently in a vise. He experimented with silver bromide and discovered that the use of this chemical reduced by one-half the time formerly required to develop his pictures.

Wet-Plate Photography. Following the close of the Civil War a number of discoveries came into general use. Among these was the wet-plate process involving the use of collodion. This was poured on a glass plate, used while still sensitive and produced negatives from which any number of photographs could be made. At the Philadelphia Centennial in 1876 a special building, Photographic Hall, was required to exhibit the daguerreotype prints, portraits, and the latest photomechanical equipment. These large and varied displays marked the ultimate achievement of the American wet-plate photographer. Included in the exhibits was the new dry-plate process, which was to supplant the collodion negative "as the wet plate had in turn supplanted the daguerreotype." [3]

The Dry Plate. The dry-plate gelatin-bromide process was first developed in England in the late 1870's, but American photographers were slow in introducing it here. When it was introduced, says Taft, "it brought with it a new name, that of a man who eventually proved to be a commanding figure in American photographic history." [4] The man was George Eastman, a frugal bank clerk in Rochester, New York, whose early hobby was photography.

George Eastman, 1854–1932. Attracted by a formula that appeared in an English almanac which suggested the possibility of reducing the size and weight of outdoor equipment, Eastman envisioned the commercial opportunities offered by the dry-plate process. In his search for a transparent flexible film, he prepared and patented a paperback film in 1884. This, the first commercial film, was cut into narrow strips and was wound on a roller device. It was made on coated paper instead of glass—a technological achievement that wrought a revolution in photography. Special roll holders were designed to handle the new film.

Kodak. A few years later Eastman manufactured the Kodak, the first hand camera, for all other cameras had required a tripod. The Kodak was advertised as an essentially portable camera, intended for making instantaneous exposures. Its simplicity and lightness were its chief features. The name Kodak was Eastman's own, "a word which he chose out of the clear sky, a strong, distinctive trade name, not easy to confuse with others."

This new small camera used a roll of the newly patented film sufficient for one hundred exposures. The film was inserted at the factory and came with the camera. When the pictures were taken and all the film was used, the camera was returned to the manufacturer to have the pictures developed and to be reloaded. "You press the button—we do the rest," became a household slogan, and was introduced even in England by Gilbert and Sullivan in their new opera, "Utopia."

Transparent Film: Celluloid. Two other advances in the technology of photography were the development of a bromide paper, which Eastman helped to develop and put on the market, and the introduction of the celluloid film. This latter development, pioneered by Hannibal Goodwin of New York and one of Eastman's chemists, Henry M. Richenback, resulted in producing a transparent flexible film. It had a nitrocellulose base.[5]

Forerunner of the Motion Picture. At the very time the transparent film was being developed, Thomas Edison, who was experimenting with motion pictures, found his greatest difficulty was that of obtaining a satisfactory film. When he heard of the Eastman transparent film he ordered a roll, and upon seeing it, is reported to have told his laboratory assistants, "That's it: now work like hell." The moving picture industry was close at hand.

Newspapers and Printing

Improvements in the rapid dissemination of ideas through the newspapers resulted from a number of technological developments that occurred during the last quarter of the nineteenth century.

Technical improvements were introduced in printing presses, in typesetting, in typecasting, in stereotyping, in the manufacture of paper and ink, in the methods of newsgathering, and in the circulation departments.

Bullock's Web Press. When Hoe's Lightning press had appeared in the late 1840's (see Chapter 15), it was hailed as the ultimate triumph in mechanical genius. It printed only on one side at a time, however, and the machine had to be fed one sheet at a time. Each sheet passed through the press twice. A continuous feed was needed and a perfecting operation that would print both sides of the paper. This need was supplied by William Bullock who in 1863 introduced what he called his web-perfecting press, first used by the Philadelphia *Ledger* in 1865. In this press the paper was fed from a continuous roll by means of a feed roller and was carried between impression cylinders and type cylinders. The paper was automatically cut into sheets and carried by grippers to the endless belt that conveyed it to the flyboard. Bullock's press involved all the principles of an automatic press. It was described as "the simplest and most enduring machine that ever printed paper, thoroughly American and most original."

Double-Perfecting Web Presses. In 1871 Hoe and Tucker continued the web-perfecting press and installed their first one for the New York *Tribune.* The entire operation was mechanical, save folding. This task was for many years performed by the newsboys who were the carriers. In 1881, a triangular "former," or folder, was devised. When larger papers were needed, wider cylinders, printing four pages at a time instead of two, were instituted. These were called double-perfecting web presses. In 1887 came the quadruple press, or four four-page cylinders; then the sextuple, with six four-page cylinders capable of producing forty-eight thousand twelve-page papers an hour. The era of press monstrosity had arrived.

The New York *Herald* said of its new sextuple press in 1891, "it is nothing less than a miracle of mechanism." The press was fed from three rolls of paper, each more than 5 feet wide, and used 26 miles of paper in one hour. The machine weighed 58 tons and was described as having "the strength of a thousand giants, rapid as lightning, and yet so delicate in its operation that it performed on fragile paper without tearing it." [6]

Electric Presses. There followed the octuple press with ninety-six thousand twelve-page papers an hour, four rolls, and four pages wide. Then in 1904 came the double-sextuple press, using twelve plate cylinders, twelve impression cylinders, employing three webs of paper, four pages wide. This machine also fed, cut, folded, and counted the papers. By this time electricity was the motive power.

Stereotyping. Press improvement could never have been possible had the stereotyping process not been adapted to newspaper work. Stereotyping was used during the nineteenth century in making books, but the matrices, made of plaster of Paris, required days to dry. In the early 1860's Charles Craske made these matrices of papier-mâché, from which metal curved plates of page size were cast to fit the revolving cylinders. This supplanted Hoe's original wedged-column system that held the type in shape on the type cylinder. Stereotyping effected great savings in typesetting and in typecasting. As many plates as the press could employ were made from a single form, and wear and tear on type was saved. Stereotyping revolutionized advertising technique, since it made possible full page displays without the necessity of column rules. A major change in stereotyping occurred in 1901 when H. A. Wood developed the autoplate for making the stereotype. This was a casting machine that could be operated at five times the speed of the hand-casting box.

The Halftone Engraving. The invention of the halftone photo-engraving process opened up a big field in newspaper illustrating. Stephen H. Horgan introduced a halftone plate which appeared in the New York *Daily Graphic,* March 4, 1880, but it was not until the late 1890's that Whitelaw Reid of the New York *Tribune* introduced Horgan's halftone on a stereotyping web-perfecting press. The application of the halftone to the large newspaper presses came in time for the Spanish-American war, when news photography became a feature. Hearst sent some camera men to Cuba, while other publishers sent photographers with their reporters to the Philippines. Camera reporting was carried out more extensively in the Russo-Japanese War of 1904.

Mergenthaler's "Lin-O-type." As speedier presses developed and the stereotyping process was made applicable to newspaper work, the old method of setting type by hand, letter by letter and stick by stick, was the bottleneck in the operation. Printers dreamed of a machine that would do this work for them. Ottmar Mergenthaler, an employee in the United States Patent Office, devised a machine which came to be called a "Lin-O-type." The machine was operated from a keyboard, similar to that of the newly invented typewriter. Each key, instead of reproducing a letter on paper, brought into line an individual matrix which, when assembled with others, produced a mold for casting a slug or line of type from a pot of hot metal. This machine met with quick success. Whitelaw Reid of the New York *Tribune* used it in an issue of July 3, 1886.

Other Typesetting Machines. Other typesetting machines followed, such as Lanston's Mon-O-type, Wilbert S. Scudder's Intertype, and another by Washington I. Ludlow. Machines proved a great boon to typesetting. They permitted greater speed and there was no type to distribute after the printing was done. Wear and tear on type was eliminated, for the slugs of type were used but once and the metal remelted for the next casting.

Paper Supply. Today's newsprint could not roll from the giant presses without new sources of paper supply. The foundations for modern papermaking were laid by the work of two Germans who, according to tradition, observed from watching the work of a wasp that paper could be made from wood. Their process, known as the "mechanical pulp" process, was introduced into this country by Rudolph Pagenstecher, who brought from Germany a pulp machine and set up a small factory at Interlaken, Massachusetts, to grind up wood fiber. This wood pulp, mixed in small quantities with pulp made from rags and other materials, slowly found acceptance with the paper manufacturers in this country.

Soda Pulp. A number of other methods based upon chemical formulas were introduced in an effort to speed up wood-pulp production. The most important were the soda, the sulfite, and the sulfate, a later improvement on the soda. The soda process, introduced into the United States by Hugh Burgess of England, involved the grinding of logs into a pulp and then boiling the pulp in a soda solution. The product was then passed through a filter of coal ashes. Paper made in this manner had a tendency to rattle, but when short-fibered hard wood was used a more satisfactory paper resulted.

Sulfite Pulp. The sulfite process is said to have been discovered accidentally by Benjamin Tilligman. Tilligman dropped a burned match in a sulfuric acid solution; the next day he observed that the match had become a pulp from which paper could be manufactured. This process became commercially successful in the late 1880's. Paper mills were built to grind wood and reduce it to pulp in lead-lined towers, where it was heated and rotated in a solution of bisulfite of magnesia. With the introduction of this process the pulp-paper industry began to boom.

Technology Speeds the News

Advances in communication technology brought about many changes in the newspaper world by 1900. The telegraph and the telephone hastened the dispatching of news, whether it happened in the local community or in some distant point. The ocean cable carried news events from across the seas. As the coded dispatches arrived they were transcribed by the telegraph operator. In the composing room, the copy was cast on a typecasting machine into metal type, from which the papier mâché stereotyped page mat was made. Huge presses rolled out thousands of copies per hour, folded, and ready for the mail chutes and the corner newsboy. For the out-of-town subscribers, special editions carried by speeding mail trains were delivered a few hours after the city edition had appeared. Advances in communication technology made it possible to disseminate news to the American public within a matter of hours after the events had occurred. American technology had contributed much to the theory that an informed citizenship is democracy's best hope.

Notes for Chapter 30

1. See Philip Bayard McDonald's *A Saga of the Sea: The Story of Cyrus N. Field and The Laying of the First Atlantic Cable* (1937); George A. Schreiner, *Cables and Wireless, Their Role in the Foreign Relations of the United States* (1924); For contemporary interest in the laying of the early cables, see files of *Harper's New Monthly Magazine;* the *Nation;* and contemporary newspapers.

2. The following are helpful in making a study of the typewriter: *The Story of the Typewriter,* Herkimer County Historical Society, N. Y. (1923); George C. More, *History of the Typewriter* (1909); A. C. Reiley, "Fiftieth Anniversary of the Typewriter," *Current History,* No. 5 (1923); Bruce Bliven, Jr., *The Wonderful Writing Machine* (1954); Richard Current, *The Typewriter* (1954).

3. Robert Taft, *Photography and the American Scene* (1936), p. 363. (Used by permission of Mr. Taft.)

4. *Ibid.,* p. 378.

5. Taft says that the lawsuits that followed between Goodwin and Eastman, in which each accused the other of infringements, comprise the most important legal controversy in the whole history of photography.

6. New York *Herald* (May 10, 1891).

For some good histories dealing with newspapers and printing, see Frank Luther Mott, *History of American Magazines,* 3 vol. (1921) (1928); Alfred Lee, *Daily Newspaper* (1937). For history of typewriter, see Henry W. Roby, *The Invention of the Typewriter* (1925); Charles Edward Weller, *The Early History of the Typewriter* (1918). Some books on history of the telephone are: Arthur Pound, *The Telephone Idea; Fifty Years After* (1926); Catherine Dunlop MacKenzie, *Alexander Graham Bell* (1928); Herbert Newton Casson, *The History of the Telephone* (1910); Frederick Leland Rhodes, *Beginnings of Telephony* (1929); *Annual Reports,* American Telephone and Telegraph Company. For history of Atlantic cables, see Charles Bright, *The Story of the Atlantic Cable* (1903); Henry Martyn Field, *The Story of the Atlantic Telegraph* (1892).

In addition, for newspapers, see: Dard Hunter, *Paper Making* (1943); Kenneth Olson, *Engine and Mechanics of Newspaper* (1930); Oswald Garrison Villard, "Some Newspapers and Newspaper Men," (1923). On photography, in addition to Taft's book, above mentioned, see also, Joseph Maria Eder, *History of Photography,* translated by Edward Epstein (1945).

PART IV

WORLD LEADERSHIP: from 1900 to the present

During the first half of the twentieth century the United States rose to world leadership, and her scientific and technological achievements were of such magnitude as to fit her for this responsibility.

Food production was mechanized from the farm to the family table. America's food surpluses began to flow to hungry peoples throughout the globe. Houses were erected by mass-production methods and equipped with conveniences that a century earlier not even kings could enjoy. Scientists developed methods and materials that made it possible for man to clothe himself with fabrics other than woolen, linen, cotton, or silk.

In transportation technology the horse and buggy were supplanted by the automobile, and the airplane reduced distances from a matter of days to hours. In communication technology, instruments were developed that carried man's voice around the world.

In the field of power production, men captured the potential of the atom, and released a source of energy greater than that of all human slaves in all history.

With the development of a system of automation and control engineering, man began to relieve himself of the major tasks of physical drudgery and to move into an age when his spiritual and creative powers foretold the coming of a new world.

Chapter 31

TECHNOLOGY AT THE TURN OF THE CENTURY

At the opening of the twentieth century, numerous efforts were made to appraise the achievements that had come about in American science and technology during the preceding century, and to forecast trends for the future. Newspapers and magazines conducted surveys that summarized the accomplishments in various fields of science and industry, and then predicted that the next half-century would bring about technological achievement hitherto undreamed of. Educators, scientists, and philosophers were likewise aware of these new forces. Journals of scientific societies published lengthy articles reviewing the progress of the last half-century and the last quarter-century. American ingenuity had already passed a milepost, it was pointed out, by abundantly providing man's three basic needs —food, shelter, and clothing—and was now pressing forward to new goals.

A Fifty-Year Appraisal of America

Illustrating these views was the publication of the prize winning essay in a contest conducted by the *Scientific American*, in 1896 (the year of its semicentennial), on the subject, "Progress of Invention During the Past Fifty Years." The winner of the award, Edward M. Byrn, characterized the period as, "an epoch unique in the world's history . . . , when nine-tenths of all America's riches and physical comforts had come into existence." He likened the late developments to "a gigantic tidal wave of human ingenuity so fruitful in its wealth and so beneficial in its results" that it was impossible for his generation to properly comprehend.

American Inventions. In referring to the period fifty years earlier, Byrn noted there were no telephones, no cables, no electric lights, photoengraving, or photolithographing. There were no cameras, no gas engines, no web-perfecting printing presses, no linotype machines, no passenger elevators, no asphalt pavements. Neither were there triple-expansion engines, nor barbed-wire fencing, nor self-binders; there were no oil or gas wells, no ice machines, no phonographs, cash registers, typewriters, aluminum ware, Bessemer steel, sanitary plumbing, coal tars, or electric machines. All these had come into use within the past half-century.

National Bureau of Standards, 1901. These developments during the closing years of the nineteenth century were officially recognized by a number of Acts passed by Congress, one of which was the establishment of The National Bureau of Standards in 1901. The National Academy of Sciences took the lead in urging the creation of such a bureau, pointing out that there were no facilities at the disposal of government or the scientific men of the country, "for the standardization of apparatus used in scientific research. . . ." Scientists and inventors were finding it necessary, in most instances, to send much of their apparatus abroad for comparison.

Bureau's Aims. The Act creating the bureau of standards, signed by President McKinley, March 2, 1901, declared, that the functions of the bureau would be that of establishing standards for "the comparison of the standards used in scientific investigations, engineering, manufacturing, commerce, and educational institutions with standards adopted or recognized by the Government." The bureau was also authorized to construct, when necessary, standards to test and calibrate standard measuring apparatus, to determine properties of materials "when such data are of great importance to scientific or manufacturing interests. . . ." The inventive minds of the country were now ready to begin mass production. The one basic handicap, that of standards of precise measurement for tooling to mass production, had been removed.

New Concepts of Technology

The scientific and technological developments around the turn of the century had not been confined exclusively to material and physical advances. The progress in science and invention had brought new concepts regarding this nation's position in the new world. The half-century that had closed had been an age in which man's dreams had been transformed into practical inventions aimed at taking the drudgery from man's back and loading it on machines, which were producing in mass quantities the things man needed. In this new world of invention, man had tapped the hidden forces of nature and put them to his use. He had become a participant in a new and expanding creation.

Mission of Science in Education. The awareness of this rapid scientific and technological development went beyond the industrial world. John M. Coulter, distinguished scientist, in a commencement address at the University of Michigan, June 21, 1900, spoke on the subject, "The Mission of Science in Education." Science was introducing new values in education. It was developing a spirit of inquiry which was helping to break down dogmatism. It compelled teachers and students to stick to facts, and it aided in cultivating the habit and power of observation. Science was revolutionizing educational methods by introducing the laboratory techniques, which, in turn, led to the seminar method. The laboratory method in science could be applied to other disciplines as well.

Chemistry, Engineering. In speaking before the Society of Chemical Industry in London, July 18, 1900, Charles F. Chandler, noted American chemist, surprised his British friends when he told them that schools of chemistry were so numerous in the United States that it was impossible to state accurately their number, and that almost every large college, university, and technical school had a chemistry laboratory. He told them there were more than ten thousand students enrolled in engineering schools. Research in applied chemistry was developing rapidly, especially in the fields of metallurgy, petroleum, gases, paper manufacture, celluloids, coal tar, rubber, and cotton seed.

"Made in America." Britain's awareness of the growing might of America was evidenced when Lord Roseberry told a group of British manufacturers in 1901 that it was not military conflicts for which the British should prepare, but rather an economic struggle. The United States was feared, "because of her unmatched ability to manufacture and her unlimited power to produce." Pointing to the invasion of American goods in Britain, Lord Roseberry noted that the typewriting machine used in preparing his address, as well as thousands, "nay hundreds of thousands of other machines throughout the world" were made in America. The same applied to the table on which the machine stood, his office furniture, bookcases, and chairs. "All these," he added, "can not be made anywhere in Europe so practical, so convenient, or at a similar cost as in the United States." [1]

Historians' Awareness of Science, Technology. It was not only the scientists themselves who sensed the significance of these changes at the turn of the century, but the historian, the economists, and the sociologists were likewise cognizant of them also. Charles Francis Adams, in his presidential address at the American Historical Association, December 27, 1901, pointed out that it was the technologist who determined public affairs, not the political leaders. Commenting on the unimportant influence of congressional and presidential campaigns (observing that he had participated in twelve presidential campaigns himself), Mr. Adams said that, "Congress has all along been but a clumsy recording machine of conclusions worked out in the laboratory and machine-shop. . . ." In anything affecting more than a division of the spoils, he said that it was "the student, the man of affairs, and the scientist who, today in last resort, closes debate and shapes public policy."

As an illustration, Adams noted that technology had obviated the whole basis of the free silver controversy which had been the crux of the 1896 presidential campaign. While orators waxed eloquent over the relative merits of the double standard versus single standard in what he called the "chatter" of debates raging around the issue of bi-metallism, gold was discovered in Alaska. Adams noted, "the discoverer and the scientist, moving hand in hand, had been getting in their work." It was under "the magic touch of their silent influence that the world's gold production, between 1896 and 1899, had been increased from less than ten million ounces to nearly fifteen millions." In money value alone, he pointed out, "it exceeded the combined value of gold and silver production of the earlier period." [2]

Fred Morrow Fling. Other historians began to note the impact of science upon the public mind. Fred Morrow Fling, in a stimulating article, "Historical Synthesis," in the *American Historical Review* (October, 1903), urged his fellow historians to recognize the logic of the new historical concept, "a concept that must be enriched by the idea of evolution." Observing that this new concept went beyond the Darwinian theory and spread to the natural sciences, Mr. Fling believed it would next affect the social sciences. These facts had already been recognized by the sociologists and economists, and henceforth, said Fling, they must claim the attention of fellow historians as well.

George Baxter Adams. Five years later George Baxter Adams, in his presidential address entitled, "History and the Philosopher of History" (Richmond meeting in 1908), posed the question, "What should the historian do in view of the threatened invasion of his domain by ideals and methods not his own?" [Referring to the inroads of science.] Noting that the political scientists, geographers, economic historians, and sociologists had become aware of the influence of scientific developments, Adams wondered what the historian would do about it. "It is useless to pooh-pooh this movement," he declared, "or to underestimate it." The younger historians would have no choice. The new concept of the role of science would compel their early consideration.

Ambassador Dr. Joaquin Nabusco. That America's achievement in science and technology would rate as a major contribution to western civilization had become manifest to statesmen and philosophers in American republics other than the United States. The Brazilian Ambassador to the United States, Joaquin Nabusco, in a Baccalaureate address at the University of Wisconsin, June 20, 1909, on the subject, "The Share of America in Civilization," referred to the five major forces which President Charles W. Eliot had enumerated, as being the major forces contributed by the United States. These were arbitration of all international disputes, religious toleration, manhood suffrage, proof that all races could live together peaceably, and diffusion of man's material products among the world's population. The Ambassador declared that America's inventions had made possible a standardization and a unification that could be called American, and that America's trains, fast boats, aeroplanes, cables, wireless, "all seem to announce that the new tendency of mankind in every direction is to merge." Although he was uncertain as to whether or not all these developments were for permanent good, of one thing he was sure—"The Age of Franklin will not end as the Age of Midas. You have changed the rhythm of life, you write it in quick tempo, and the world is catching from you the spirit of rapid transportation, and is writing it also in the American *prestissimo* instead of the old adagio." [3]

Thorstein B. Veblen, 1857–1929. The emergence of these scientific and technological forces was also commanding the attention of the economists and sociologists. Thorstein B. Veblen, known for his savage attacks upon big business, the profit system, and vested interests, nevertheless, viewed realistically the impact that technology was having upon the entire economy of the western world. In his article, "The Place of Science in Modern Civilization" (1906), Veblen noted that science and technology combined had come to be the dominant force in modern life. People in other ages had excelled in other achievements, he said, "but these had dwindled with the passing of time, while the achievements of science are held higher as time passes." He observed that scientific theories had already diverged from the formulation of pragmatism to the matter-of-fact method known as technology. In "our modern culture," Veblen added, "industrial processes and products have progressively gained upon humanity until these creations of man's ingenuity have latterly come to take dominant place in the cultural scheme. They have become the chief force in shaping man's habits and thoughts."

Result of New Technologies. Among the significant forces that were shaping man's thought habits during the first decade of the twentieth century were the coming of wireless telegraphy, the automobile, the airplane, the motion picture, and developments in power technology. Accompanying these new and hitherto unknown technological achievements were the rapid improvements that were occurring in agricultural technology, in food production, and in the output of factories, mines, and mills.

The Science of Conservation

Many of the recent advances in American technology had made such exorbitant demands upon the nation's natural resources that conservation-minded people became alarmed.

Economists, geologists, foresters, and many industrialists were disturbed over the waste and the useless exploitation of all natural resources across the virgin continent. Dr. Charles R. Van Hise, President of the University of Wisconsin, distinguished geologist and a pioneer in the conservation movement, took the lead in warning the American people of the impending danger when he declared that "the conservation of the nation's natural resources is more important than all the political or social questions upon the solution of which we are now engaged." [4] He ventured the prophecy that a thousand years hence when the history of this period is written (the opening years of the twentieth century), that the amount of space to be given to current political and social questions, "which are so puzzling now, will seem relatively small." On the other hand, he made the prediction that the space that would be given to the conservation movement will be far greater than that accorded all other undertakings. "If this movement succeeds," he continued, "the names of the men who have led it to success will loom large."

Pioneer Conservationists. Van Hise had been impressed by the works of Major John Wesley Powell who, in his *Lands of the Arid Region,* had emphasized the need for large irrigation projects and reservoir sites in the south central areas of the United States. Gifford Pinchot, Chief of the Forest Bureau of the Department of Agriculture; Dr. William J. McGee, distinguished geologist and hydrologist; F. H. Newell, Charles D. Walcott, and Senator Francis G. Newland, joined forces with Dr. Van Hise and others to establish a conservation movement on a national scale.

Depletion of Natural Resources. They undertook a quick survey of the status of the nation's resources and predicted that the country's coal supply, iron ore deposits, natural gas, and petroleum supply would all be exhausted before the middle of the twentieth century. Some predicted that a few of the reserves would not last even that long. At the rate in which the forests were being slashed, it was predicted that they would be depleted within forty years. The loss of the virgin forests also meant the loss of millions of tons of top soil every year. As to the natural gas deposits, Israel Charles White, noted geologist of West Virginia and one time chief geologist for the Brazilian Coal Commission, estimated that in his state, alone, the daily loss of gas from burning wells and leaking pipes amounted to more than 200,000,000 cubic feet. "What an appalling loss to transmit to posterity," he wrote.

First National Conservation Congress, 1908. Aware of the growing interest in the conservation movement, which he himself had helped to create, President Roosevelt issued a call for a national conservation conference to meet in May, 1908, in the East Room of the White House. The governor of each state and territory, members of Roosevelt's Cabinet, justices of the Supreme Court, senators and congressmen, presidents of various national societies interested in reclamation and conservation, and a number of special guests, including William Jennings Bryan, Andrew Carnegie, James J. Hill, and John Mitchell, were invited. "It is doubtful," wrote Roosevelt in his *Autobiography,* "whether, except in time of war, any new idea of like importance has ever been presented to a nation and accepted by it with such effectiveness and rapidity as was the case with this conservation movement. . . ." And Dr. Van Hise declared that never before in our history had the scientific men of the country met upon equal footing with those engaged in politics.[5]

State Action. The results of this national conference were far reaching. As President Roosevelt expressed it, not only did the conference draw up the first inventory of our natural resources, but also the reports were unique in the amount and variety of information brought together. They laid squarely before the American people the essential facts regarding our natural resources, "when facts were greatly needed as the basis for constructive action." Following the national meeting, the conservation issues were taken up by the states. By 1910 thirty-eight of the forty-six states had created agencies (they were not all called Conservation Commissions) to work with the national organization in projects that affected their local areas.

Inventory of United States Resources. Problems like soil conservation, irrigation, and forest preservation were tackled by both state and government scientists. The National Conservation Commission, appointed by President Roosevelt following the close of the conference, prepared an inventory of the nation's natural resources, which in turn was submitted to the President. Mr. Roosevelt described it as "One of the most fundamental documents ever laid before the American people." For the first time in history the American public was appraised of the extent of their natural resources, and were likewise warned that the raw materials upon which the nation had built a prosperous economy were fast approaching depletion.

In an effort to check these wastes, the committee of scientists recommended that government agencies and private corporations be directed to institute a rigid program of conserving the remaining natural resources, and to seek new and untapped sources of supply. The forces of government, science, and industry united their efforts in a nationwide movement. Geologists, petroleum technologists, and physiochemists, equipped with scientific instruments, were able within a few years to assure the American people that most of the nation's natural resources, particularly coal, petroleum, and the supply of natural gas, were virtually unlimited.

Industry Becomes Science-Minded

Meantime, a number of industrial and manufacturing enterprises set their scientists to work to discover new uses for the many by-products from their plants, and to find ways for reclaiming discarded waste materials.

Scientists Seek Salvage Techniques. Scientists in the paper industry discovered that hemlock trees, which had been slashed merely to obtain the bark, affored excellent raw material for pulp paper. Distilling interests discovered that the residue from tons of waste products could be turned into valuable food for cattle. A meat-packing industry reported that by employing scientific methods in reclaiming heretofore waste products, it had been able to salvage every part of the animal carcass "except the squeal." The cement industry reported shortly after the turn of the century that its scientists had discovered a method of making cement from blast-furnace slag.

Development of By-Products. Products derived from petroleum that were once discarded as waste began to rival illuminating oil in importance, while scores of substances for medicinal uses were also derived from petroleum. The electrical industry found uses for paraffin wax as insulating material. Products were procured from corn that far surpassed the value of the original uses of the grain, and corn oil was produced which provided a substitute for linseed oil. Oil from corn was also used in the manufacture of vulcanized rubber. The application of science and new technologies pointed the way for conserving waste materials, and offered formulas for creating new industries. The new age in which science and technology were destined to play a significant role had arrived.

Notes for Chapter 31

1. Charles R. Flint, "Business Situation in the United States and the Prospect of the Future," *North American Review* (March, 1901), p. 381 *ff*.
2. *Annual Report of American Historical Association* (1901) (1902); p. 63.
3. *American Historical Review,* XV (1909), No. 1, p. 3.
4. Charles Richard Van Hise, *The Conservation of Natural Resources* (1911), p. 375.
5. See Chapter 11 for discussion of "Columbian Institute"; an earlier attempt to unite the forces of science and government.

See also Chauncey Mitchell Depew, ed., *One Hundred Years of American Commerce; A History of American Commerce, by One Hundred Americans; A Chronological Table of the Important Events of American Commerce and Invention within the Past One Hundred Years* (1895); Wright Byrn, *The Progress of Invention in the Nineteenth Century* (1900); William Henry Doolittle, *Inventions in the Century* (1902); George Sutherland, *Twentieth Century Inventions, A Forecast* (1901).

Chapter 32

SCIENCE AND TECHNOLOGY HELP MAKE AMERICA A WORLD POWER: 1900 TO WORLD WAR I

It was during the opening years of the twentieth century that the United States emerged from a position of isolation to that of a world power. For the first time in its history, the United States, following its victories in the Spanish-American War, extended its powers over lands and people beyond our continental borders (other than Alaska). The time had come, we were told, when America should "Take up the White Man's Burden," and carry the products of our industrial age to the people of backward nations.

The Road to World Leadership

Behind this rise of imperialism lay the forces of science and technology. The nation's emergence into a world power was due primarily to the recent advances in industrial, engineering, manufacturing, and agricultural technology. As a result of these advances the United States passed its nearest competitor, Great Britain, in iron and steel in 1898. It took the lead in coal production the following year. In the invention and production of machine tools, the United States, had by 1900 passed Germany, its closest rival, and in 1901 we passed England in the value of export trade.

The developments in science and technology following the turn of the century were important, not merely because of the improvements they brought, but also because of their entirely new, and in some instances revolutionary, characteristics. Instead of merely creating machines to perform mechanically the same operations formerly done by hand, new machines were devised for automatic, mechanical performance. In its technological achievements, as in diplomacy, the nation was on its way toward world leadership. This was demonstrated in its ability to provide food, shelter, and clothing in abundance, and to develop surplus production in most of its major industries.

Agricultural Superiority

Agriculture at the turn of the century continued to be the nation's leading industry, with one-third of the nation's population still engaged in farming. And the nation's farmers continued to contribute more to the national income than all other industries combined. During the first years of the new century, the farmers were able to report a decided increase, both in production and in income. In 1905, the Secretary of Agriculture, James Wilson, reported that the value of the total production in agriculture amounted to more than $6,400,000,000, which was 36 per cent above that of 1899. This increase had been due primarily to the introduction of new and improved machinery, "and the application of scientific methods in all branches of farming."

Increased Production, and Income. This increased production, with a corresponding increase in income, led the Secretary to declare that, should the increase continue for another three years, "the farmers will have produced an amount of wealth within a ten-year period (1898–1908) equal to one-half of the entire wealth produced by the toil and savings of three centuries." Farmers were growing richer. Their bank deposits had increased 129.2 per cent within eight years, and in some states their deposits showed an increase of more than 350 per cent. "The man with the hoe," declared the Secretary, "has become the man with the harvester, and he is the depositor and shareholder of the bank."

Science in Agriculture. The reasons given for these mounting production records and profits were the introduction of improved methods of farming, intense crop cultivation, improved machinery, importation of seeds and plants which could be grown successfully in the United States, and the scientific services rendered by the Department of Agriculture. Of the five thousand persons employed by the department in 1905, some one thousand five hundred were classified as scientists and scientific assistants, whose entire services were devoted to helping the farmer find new and improved methods of farming and to discovering means for improving the breed of livestock.

The recently established graduate school of agriculture at Ohio State University was training teachers of agriculture and agricultural experts needed in all sections of the country. Enrollments in land-grant colleges had increased rapidly, and even secondary schools had introduced courses in agriculture, with Wisconsin, Minnesota, and Nebraska taking the lead. A dairy school established in Wisconsin had sent out eight hundred trained butter- and cheesemakers. Louisiana had a school for training sugar experts.

President Roosevelt's Report, 1906. President Theodore Roosevelt, in his annual message to Congress, December 6, 1906, reporting on the scientific experiments carried on by the Department of Agriculture, referred to the Department as an educational institution with "a faculty of 2,000 specialists making research into all the sciences of production." He reported that of the more than five thousand students in attendance at state agricultural colleges, many were preparing for positions as teachers of agriculture. The increase in enrollment in agricultural schools continued, and by 1910 thirty-nine state agricultural colleges and forty-two privately endowed colleges were giving instruction in various branches of agricultural science to more than eighteen thousand students.

Bureau of Chemistry. The scientific work carried on by the Bureau of Chemistry was increasing the products of the farm and forests. Its scientists examined all imported plants and foods, tested seeds, soils, and moisture contents for every section of the country. They also worked with the forestry division in the study of tannin, gums, rubber, and wood pulp, and tested road materials. They analyzed melons, sugar beets, and other sugar-producing plants. They had found a powder and spray product to rid the potato patch of the potato bug, and had announced that a method had been discovered for eradicating the pear blight.

Entomologists Scour the Globe. Entomologists returned that year (1906) from interior Asia, where they had discovered a parasite near the Great Wall of China, which they brought back to fight the ravages of the San Jose scale. They also had discovered a species of ladybird in the same region which preyed upon a plague similar to San Jose scale. Secretary Wilson expressed the hope that we would receive from these ladybirds "the same good results which were accomplished in China and Japan." A ladybird of another species had been imported from eastern Europe to feed upon plant lice that were destructive to our nation's pears, melons, and other fruits. A parasitic fly from South Africa had been brought in to fight the black scale, the worst pest of the orange and lemon industry in California.

Figs in California, Rice from India. Fig fertilizing insects imported from Turkey in 1909 helped establish the fig-growing industry in California. Fifty tons of figs were produced in one year in an orchard near Fresno, California. Scientists had also succeeded in introducing into the rice fields of Louisiana a special specie of rice seed from India, which doubled the production per acre. Almonds from Spain, dates from Algeria, and mangoes from India were added to America's food crops during the early years of this century. "Our government," declared the Secretary in 1909, "is doing more to help the farmer than all other nations combined."

Cotton Culture. In 1911, a staff of scientists who had been engaged in cotton research announced that some twenty types of white American cotton had been prepared for storage in vacuum to determine the permanency of the type for future years, "a method believed to be of the most fundamental importance in the cotton industry." Further research bearing on the length and strength of cotton fibers had helped the cotton growers in the selection of their seeds.

New Paper Materials. A valuable study entitled "A Technological Work on Crop Plants which May Be Used for Making Paper" (1912) revealed the possibility of utilizing such by-products as cornstalks, rice straw, flax and hemp waste, jute, wild plants, and certain grasses, rushes, and canes. Broomcorn stalks, with yields as high as 42 per cent of available fiber, produced a good quality of book paper when combined with an equal quantity of wood pulp.

Farmers Well Informed. The farmers of America, then as now, were better informed on subjects relating to their calling than were other farmers in the world. The research staffs in the Department of Agriculture prepared bulletins on hundreds of subjects and distributed ten million copies in 1912. Many of the studies were prepared with the view of having them used as textbooks in schools of all grades. The Secretary, in commenting on these scientific and technical publications, asserted that in no other country was so much information made available to so many people on so many subjects.

Food Technology

The scientific preparation and preservation of food, as important as the raising of food, was marked by a number of developments during the first two decades of this century. Chief attention was given to guarding the people's health by ensuring a pure food supply.

Pure Food and Drug Act, 1906. For years food technologists had been urging food producers to be more careful in preparing their products. The movement was led by Dr. Harvey W. Wiley, who headed the Bureau of Chemistry in the Department of Agriculture. Members of Congress also became interested, and an Act, referred to as the "Pure Food and Drug Act," was passed in 1906. It called for a rigid inspection of all foods, drugs, and medicines, required that labels state the ingredients, and imposed penalties for false advertising.

National Canning Association, 1913. Canners, too, had been equally active in campaigning for pure foods and began to establish research fellowships in university laboratories for the purpose of studying food poisoning and methods for preventing food spoilage. Bronson Barlow made a study, while holding such a fellowship in the University of Illinois, on the relation of the losses in the food industry to bacteria. His findings (that the action of thermophilic bacteria resulted from filled cans being attacked while still hot) marked a great step forward in the science of canning. (Bronson later became president of the National Canning Association.)

Pasteurization of Milk. Modern pasteurizing plants introduced from 1907 to 1910 changed the entire trend of handling milk products. Robert Koch's discovery, that heat would stop the fermentation and putrefaction caused by germs, was of basic importance to the milk industry. By 1914 New York City's sanitary code made pasteurization mandatory for all milk unless certified or tuberculin tested.

The pasteurization process involved heating the milk to 145 degrees, and then rapidly cooling it so that objectionable micro-organisms were destroyed without otherwise altering the properties of the milk. Various types of machines were designed for the pasteurizing process. The distribution of milk was made more sanitary. Paper milk bottles appeared on the market. Bottling equipment was installed in dairies. Rotary fillers came into use around 1910, and other machines applied the caps. The bottling process became completely automatic. A milk powder came into use before 1910. Dried milk found a new use in World War I, when a dehydrating process was reversed and milk was liquefied by adding butter and water. Sailors enjoyed milkshakes and ice cream even when aboard ships in the South Seas.

Mechanical Household Refrigerator. Modern mechanical refrigerators appeared around 1912 to 1914. A. H. Goss, after years of experimentation, was given credit for developing the principle on which the Kelvinator was based. Its introduction was slow until a widely circulated treatise, "The Invisible World," by Dr. W. L. Mallinson of Michigan State College, appeared. The effect of this treatise, which was one of the early presentations in popular language of bacteria presence in food, reached multitudes of people and made them microbe conscious.

New Foods in Quantity Production. A number of new foods began to appear on the canner's list during this period. Science developed a way to raise asparagus that resisted the rust epidemic, and "the queen of the vegetables" came on the market. The Dole Company, long interested in the pineapple industry, installed equipment that completely modernized the process of peeling, slicing, and processing the pineapple. Quantity production resulted, and the pineapple became a part of America's daily food.

By 1917, when the United States entered World War I, the farmers of America and the food processors, canners, and distributors were able to feed not only our own service men and women, but also to send vast quantities of surplus food and agricultural machinery to our allies overseas. Recent advances in science and technology had, in large part, made this possible.

Twentieth-Century Building Technology

In the early years of the twentieth century, building became an outward expression of success. The whole nation throbbed with an active building program. Skyscrapers became taller, not so much because of economic reasons to secure increased rentals, but because they stood as huge signboards to advertise the success of the owners. The Woolworth Building towered upward 792 feet to tell the world of the "five and ten" stores. Structures became bigger. The Pennsylvania Hotel, with twenty-two hundred rooms, boasted of being the largest hotel in the world. The Lincoln Memorial had the largest content of marble ever used in one building. Construction work was done faster. Contractors assured the owners that cold weather, blistering heat, or storms would not stop their work.

American Construction Abroad. American building technology spread to other lands. American contractors were engaged for construction work in Canada, in Cuba, in European cities, and in far off Japan. In Canada, despite the subzero weather, work proceeded, and to keep the concrete from freezing while being poured, heated sand and gravel were mixed with boiling water, and the entire building was enclosed in a huge tarpaulin. In Japan, American construction technique was faced with the paradox of using Japanese hand labor. American machinery was stored while piles were driven with wood framing. A rope over a pulley at the top, tied to a heavy hammer at one end, and lines of twenty or more Japanese women pulling at the other end to raise the hammer and let it down, drove the pilings into place.

Despite the frenzy of building, in the quiet of the laboratory engineers were studying such matters as stress and strain, and were scientifically tackling other engineering problems to determine how to make metals stronger and construction safe and sound. Daring, inventive, boastful of bigness, eager for novelty, and spurred on by a conquering vitality, the nation's builders began to express the spirit of twentieth century America.

Significance of Baltimore Fire, 1904. Justification for steel construction was tragically demonstrated in the Baltimore fire of 1904 in which the steel and concrete structures, standing across the line of the fire, stopped the conflagration. Although the woodwork of these buildings was gutted, as were the sleepers imbedded in the concrete to which the flooring was nailed, and the melted lamp bulbs hung like icicles, the steel frames remained straight, unwarped and untwisted. Fireproof buildings of steel construction had been put to the test, and they measured up to the engineers' expectations.

Building Machinery. As the skyscraper craze continued, New York, with her Flat Iron Building of sharp, thin, clean construction design (since razed) surpassed Chicago. The *Times,* symbol of the spectacular city life, moved uptown but was challenged by the Metropolitan and the Woolworth Buildings. In other cities, the Munsey in Washington, the Frick Building in Pittsburgh, and similar structures in Philadelphia, St. Louis, and in San Francisco after the earthquake, all presented a challenge for the construction of more and greater building machinery. The hauling of heavy steel beams necessitated an advance survey of city streets in order to avoid a cave-in, made possible because of the subterranean maze of pipes, wires, and sewers. An increasing amount of detail work had to be completed at the fabricating plants due to the narrow streets; and building construction became more and more an assembling operation, since the cities afforded no storage space.

Mechanization was increased. The advent of the motor truck proved a great boon. Bigger steam pipes and bigger water pipes were designed. In new construction, equipment was devised to lift loads greater than were ever carried before, and "relay hoisting" was resorted to by lifting the materials half-way and then transferring them to other machinery. Industrial cars, on narrow-gauge tracks ran around the perimeter of each floor to distribute materials from the hoists. Cranes, operating from a small monorail on each floor, lifted the load from the truck cars and swung it into position. Hoists carried the men, as well as materials, to their positions.

Concrete Construction. Concrete was used in increasing applications. In 1911 a concrete block machine was invented that cast blocks of standard size. They were more durable than the terra cotta of previous years and were applicable to many more uses. Homes, warehouses, and even factory buildings were constructed of concrete before World War I.

Elevators. The passenger elevators were a never ceasing wonder to visitors from abroad. The hydraulic elevator, operated by a system of cables and pulleys and controlled by a water-pressure engine, continued in use for several years. The greatest advance was the improvement in the electric elevator by the introduction of a gearless traction machine. A grooved wheel, or sheave, fitted for cables, each in its own groove, replaced the former worm gear and spiral drum. This design was adopted for buildings of any height. Speed was limited only by the operator's ability to level the cab at the floors. American elevators were installed in London before 1912, replacing the customary plunger type in use there.

Sanitary Features. Though the introduction of sanitary features did not become general until after World War I, the people of no other country have done so much in perfecting the accessories of the bath as have the Americans. In 1913 gaseous chlorinization of water as a means of purification was introduced, and at the same time steps were taken toward the purification of our streams by removing the sources of contamination. Septic tanks were in use, as were improved cesspools by 1910. Copper, bronze, and galvanized iron were introduced to replace lead pipe. Laundry trays, kitchen sinks, washstands, and other similar appurtenances were all a part of the sanitary equipment long before World War I.

Heating. Although central heating systems had been established before the turn of the century, there had been little advance made in installation technique. As a result, in many instances the systems had fallen into disrepute. In 1914 the National Heating and Ventilating Association was formed to promote the proper installation of central heating units and to aid in educating the public in every branch of the heating industry. This Association, by a co-operative agreement with the University of Illinois, was able to use the laboratory of that institution to perform scientific experiments on heating requirements and to give special training to heating engineers.

Air Conditioning. Improvements in the science of air conditioning continued in a small way until the outbreak of World War I. Progress was then halted until the 1920's, when motion picture theaters began to install air-conditioning units and widely advertised the comforts they offered to their patrons.

American Textiles Lead the World Market

Developments in the textile technology in the early years of the century represented more accurately America's advancement to world leadership than did any other technological achievement. Prior to this period, leadership in textiles belonged to England and to the Continent. While their output could not compare in quantity production with that of the manufacturers of this country, their cheap labor offset this disadvantage, so that the clothing trade looked to imported goods as superior. America was thus forced to be content with production of cheaper grades of goods in quantities, and thereby held that market.

Europe Buys American Textile Machines. It had been only in the closing decade of the nineteenth century that the superiority of American textile machinery had been recognized. The appearance of the Northrup loom had posed a challenge to European textile superiority, but the so-called fancy weaves could not yet be produced. The Crompton and Knowles loom, which could produce the pattern-chain weaves, was not completely automatic and therefore could not compete with foreign labor. The invention of a device for the pattern loom to detect the time for shuttle changing made the production of the fancy loom automatic. Then the problem of fancy weaves was solved. This invention came in 1907; from then on American mills could produce the finest materials in the fanciest patterns and in the greatest quantities. After that date, European textile manufacturers came to American manufacturers to purchase their machinery, just as Europeans had been buying our typewriters, sewing machines, and reapers for years.

Woolens. In the woolen industry the Noble comb, with improvements that produced the finest woolen yarns, made it possible for us to compete with European markets. The object of combing was to straighten out the fibers, and to comb from the slivers all the fibers less than a certain length. The Noble comb was wholly a mechanical device, continuous in operation. All the movements of this machine were circular, so that it was described as the "almost perfect machine." Added to the efficiency of the American machines was their simplicity. Their operation was so simple that American manufacturers were able to employ untrained workmen for the same type of work that required the most skilled labor in England.

On Eve of World War I. By 1917, when the United States entered World War I, advances in textile technology, the boot and shoe industry, and in rubber goods enabled the manufacturers of these products to meet what at first appeared to be impossible demands. The ability to gather up millions of pounds of raw cotton, flax, wool, and leather, and to process the raw materials into finished uniforms, boots, shoes, belts, and tents, and to keep supplies in transit to the battle areas was proof of the technical skills and methods of mass production that had come to be part of the nation's economy.

Steel Technology Develops By-Products

The first two decades of the twentieth century witnessed a number of developments in the steel industry. None of these was as revolutionary as the changes that were made in the 1930's. However, the character of the improvements seemed to forecast America's destiny for steel production in two World Wars.

Consolidation of Steel Industry. First, from the economic standpoint, there was the consolidation of the various enterprises. The giant United States Steel Corporation was formed in 1901. Lesser companies consolidated into independent corporations. Also, there was the rearranged geographic distribution of the industry. The sand dunes of Gary, Indiana, on Lake Michigan, became a steel metropolis in 1908. Increased production followed in the wake of this extension. In 1900 the production of all kinds of steel in the United States was 10,000,000 tons, but in 1918 it was 45,000,000 tons, far in excess of Germany's 14,000,000, England's 9,000,000, and France's 2,000,000 tons.

Metallurgical Research. More important, however, were the technological developments. Before the end of the period there had grown up an appreciation for, and an application of, the technical, scientific knowledge required by metallurgical research. While veteran steelmen at the turn of the century scoffed at the handful of metallurgists then connected with the industry and feared they "would ruin the business," by 1917 the metallurgist was indispensable. His services were needed in making chemical analysis of steel, in determining its physical structure, and in measuring the reactions to heat treatment. Although the blacksmith had for centuries quenched the hot metal in water to "temper" it, he never understood what was accomplished by this operation. This the metallurgist discovered in heat treatment. The empirical methods of the past, though often producing a fine product, were giving way to scientific practice.

Alloys. It had been discovered in the laboratory many years previously that alloys could be added to steel to alter its properties, but the discovery had had only limited usage. With increased use of the open-hearth process, the utilization of alloys became more practicable. The alloys employed were confined principally to those found in the various ores: the vanadium ores from Peru; nickel in the ores from Canada; and titanium, copper, and adamite (a supercarbon alloy of chromium and nickel). These alloy products were first squandered in gadgets instead of being put to the use for which they were so admirably fitted. But during this period the industry became aware of the uses of alloys, and alloy steel began to make its appearance.

Electric Furnace. The increasing use of open-hearth furnaces as compared with the Bessemer furnace had gained such ascendancy that by 1910 the tide had turned, and 66 per cent of production was from open hearths. Looming on the horizon was a new process in steel technology, borrowed from the French, which by 1916 had made phenomenal progress. This was the electric furnace. When the United States entered World War I, there were three hundred and twenty-three installations of electric furnaces in this country and only eight hundred and seventy-five in the entire world. The largest electric steel plant in the world was in Chicago.

The original electric furnace, the Héroult furnace, in which the principle of operation was the same as the arc lamp, was introduced in 1909. In construction it included a steel shell containing a basin for the molten steel, similar to the open hearth. Electrodes lowered into this basin, through which the electric current was passed, created intense heat as the electric current jumped from the positive to the negative electrode through the iron. This heat was sufficient for the steelmaking operation. Electric furnaces were expensive to operate because of the high cost of electric current. The products were much the same as had been made by the crucible process. By 1915 the crucible process was almost discarded. Until after the war, electric furnaces did not compete with the open hearth.

By-product Coke Ovens. The early years of the twentieth century saw the introduction of by-product coke ovens as a part of the steel plant. Experimental ovens had been instituted previous to this period, but not as a part of the steelmaking industry. When they became a part of the industry, the spare gas from the ovens was used in the open-hearth furnaces, and the by-products (benzol and toluol) were diverted to war purposes. It was found that ordinary "steam" coals could be used in the by-product ovens by employing a blending and washing process and removing the sulfur. By this process the shortage of metallurgical coal was overcome. Larger ovens were developed than were used in the old beehive ovens, and the coking time was shortened. This method effected economies in addition to those realized through the utilization of the by-products.

Coke Methods and Products. The process of making by-product coke was one of distillation. The apparatus was a type of retort with a coking chamber, from which the gases were drawn through uptake pipes into water-sealed collecting troughs. Thence, they were drawn through a series of condensers and scrubbers from which the various products were deposited. The principal discharges from the distillate were tar, ammoniacal liquor (ammonia water), benzine, and surplus gas. The ovens were fired from heating flues in the walls. The old arduous task of drawing the coke, experienced in the beehive ovens, was eliminated by a ram operation. The ovens were constructed with openings at each end, which were closed during the coking operation. When the process was completed, a ram operated from a car at the rear forced the coke out on a cooling platform. It was then ready to be loaded in cars and shipped to steel plants.

Puddling. Other changes were effected in the technological process of steelmaking prior to World War I. These included greater mechanization, enlargement of equipment, and economies in operation. One of the more important was mechanical puddling. Puddling was one of the most arduous of steelmaking operations. Many mechanical puddling furnaces had been devised, but none had been able to replace the skill of the experienced puddler in determining the timing of the operations. The new device with electric controls was short lived in its effectiveness, however, since the open-hearth process in making steel eliminated much of the need for puddled iron.

Steel Consumption. While a number of new uses had been found for steel, many more were to be discovered. Up until 1920, many parts of the automobile continued to be made of wood, and all-steel passenger cars were still in the future. Per capita consumption of steel had increased, however, from about 500 pounds in 1900 to about 1,200 pounds in 1920. The United States had become the greatest steel-consuming nation in the world.

Power Machinery Predominates

The production of power, always of prime importance in the highly mechanized operations of the mills and factories, became more urgent during the opening decades of the twentieth century. The use of electric controls mounted rapidly. In the new steel plant at Gary, Indiana, all machinery was driven by electric power, and wherever possible, hydro-electric power was used. By 1920, 40 per cent of all electric power was hydro-electric.

Steam Turbine. As a generator of steam power, the huge reciprocating engine was doomed to fall before the new steam turbine which made its debut around 1900. The small compact steam turbine took up one-tenth the space and weighed one-eighth as much as the engine it replaced, and it produced more power at less operating cost. The principle of the turbine was based on rotating motion, rather than reciprocating motion. This motion is effected by forcing a jet of steam against a series of curved blades attached to a shaft, which causes them to rotate at great speed and thereby rotate the shaft from which the power is harnessed. The turbine was utilized in operating electric generators, and thus electricity was used instead of belts and drives as a means of transmission of power. Before World War I some thirty thousand turbo-alternators in 10,000-watt size had been built. These large construction units proved so economical that it was no longer advantageous for a manufacturer to produce electricity with his own power plant.

Diesel Power. Other prime movers established in this period were the internal combustion engine (see Chapter 33) and the diesel engine. The diesel, the invention of Dr. Rudolph Diesel, used low-grade petroleum for fuel. This fuel was injected by means of a spray nozzle into the cylinder, in which ordinary air had been compressed to one-sixteenth its volume. Attempts were made to utilize the diesel as a stationary engine in the early years of the century, but no extensive use of the principle was made until after the all-American improvements were introduced, when it was applied to marine engines and later to railroad locomotives.

Glass and Rubber Industries Expand

The glass and rubber industries had marked improvements during the early years of the twentieth century, although neither experienced the radical changes that were effected in the later decades. Glass making, while one of the earliest industries introduced in America, was one of the latest to reach the mechanized production procedure characteristic of American technology.

Machinery for Glassmaking. The revolution in glassmaking did not occur until after World War I, but in the early years of the century certain advances were made that promoted this revolution. Until the twentieth century, glass was a product of manual labor exclusively. It was a craft cunningly guarded, in which American workmen were supplemented by skilled craftsmen, largely from Belgium. High labor costs and increased demand for the several glass products forced the plants to introduce machines for handling, and to adopt other quasi-mechanical methods. The introduction of the tank for the melting operation, replacing the old handmade pots, so increased the quantity of glass available at one time that shorter methods of production were found necessary.

In glass manufacture the molten glass must be made up into the finished products immediately, or the "batch" is lost. To meet this demand semi-automatic machinery was introduced, or machines that simulated the old handcraft operations. Even with this introduction, however, glassmaking failed to reach the mass production status, for a scientific knowledge of glass itself was lacking. Chemists in the industry were few, and glass manufacture was carried on by secret formulas and on unionized craft bases until after World War I.

Lubbers Blowing Machine. Among the semi-automatic machines that were developed was the Lubbers blowing machine for window glass. This was a huge machine that substituted compressed air for lung power. It blew better glass and larger cylinders than the glassblower could produce. But the machine was expensive and its introduction was slow.

Owens Machine. The most unique glass machinery that was introduced in the first decades of the century was the Owens bottle machine. Owens' name ranks in the glass industry with Whitney's in cotton and Goodyear's in rubber. The first invention was a semi-automatic bottle-blowing machine. This machine was in fact a collection of inventions, and it introduced machine operations for the several steps in bottlemaking. One machine made a small cavity in a gob of glass, another dropped this into a mold, and a third injected compressed air which blew the glass into the pattern of the mold. Carrying and "take-out" devices were introduced, and electric controls played a part, so that finally it became a "one-man, no-boy machine" doing the work of seven men. There remained the gathering problem, that of taking the molten glass from the furnace and starting it on the blowing operation.

Bottlemaking Technology. The Owens machine first appeared in 1904. Within five years, due to a number of improvements, the machine was made so completely automatic that it even gathered the gob of molten glass from the furnace by suction. A mechanical arm dipped into the boot at the end of the melting tank and swung the gob into position, where the mold to form the neck of the bottle seized it. When this operation was completed, the blow-mold encompassed the gob of suspended glass, and the lip and neck mold relinquished its hold, ready for another gob. Compressed air distended the suspended gob to the pattern of the mold and then the completed bottle was transferred by machinery to the conveyor belt for its journey through the annealing oven, thence to the packing house. This machine produced two hundred forty bottles a minute.

Light Bulbs. Until around 1900 electric light bulbs were made by hand. But after this date, when the Corning Works "blew a bubble for Edison," a skilled blower and his helper were able to produce twelve hundred light bulbs in an 8-hour shift. Owens introduced a machine that performed the blowers operation mechanically, and although the process was not completely automatic, it eliminated the work of skilled artisans and increased production to more than three times that of the hand shop. The real revolution in electric light bulbmaking was to come in a later decade.

Plate Glass. Plate glass, unlike other glassmaking processes, was never conducted in the laborious manner in this country that prevailed in Europe. From the time of the introduction of plate manufacture in this country, American factories were equipped with elaborate operating machinery. While the early plateworks had their special cranes, transfer tables, and automatic belts used to handle the heavy plates, technically the process was the same as handcraft procedure. Many minor improvements were introduced in the several phases of glassmaking, but it remained for the glass shortage experienced during World War I to make this country aware of the unprogressive state of its glass manufacturing establishments.

Early Use of Rubber Limited. Rubber, which had been taken out of the bauble stage by the discovery of the process of vulcanization by Goodyear, remained one of the minor industries, too insignificant to list save in the miscellaneous classification in the Census report of 1900 (and then it was fifty-fourth in a list of eighty industries). It was the advent of the pneumatic tire for the automobile that brought rubber into prominence. Until then, crude rubber was used mostly for boots and shoes, rubber blankets, hose, and in the 1890's for solid rubber tires for bicycles and tricycles.

Rubber Processing. Rubber technology, after the introduction of the Dunlop pneumatic tire in 1898, was marked by advances in skill in operation, in the use of machinery in manufacture, and in the development of science in the understanding of its properties. The processing of the coagulated latex, the product derived from the rubber tree, involved a series of operations, such as washing the raw rubber, plasticizing it or softening it into a gummy stage, compounding with the desired ingredients, calendering, sheeting or tubing, fabricating it into the desired article of commerce, and finally vulcanizing to render it durable. The first rubber to come from the wild rubber regions was received as "biscuits." This supply, due to the wastefulness of the exploiters and the treachery of the natives, was wholly inadequate. The first supply of plantation rubber, grown in groves on the Malay Peninsula, arrived in 1907. By 1914 these rubber plantations were producing one-half of the world's supply.

Mechanized Rubbermaking. American genius and skill was enlisted in developing labor-saving machinery to perform the simple but arduous operations in processing the raw rubber. Streamlining methods were introduced to effect mass production. Power operation was employed for the mixers, electric controls were instituted for the machinery, mechanical belts and conveyors were introduced for handling operations, and machines were devised to fabricate the various products.

Contributions of Science. Science, too, played a part in the early years of rubber manufacturing. Vulcanizing is a scientific operation, and though Goodyear discovered the process, he was not aware of the chemical changes effected. The problems confronting the rubber manufacturer were to offset the deteriorating effects of heat and oxygen and to prevent aging deterioration. As heat was essential in the vulcanizing process, accelerators were sought from the scientists to hasten the process. Manufacturers soon realized they needed the chemists.

The first contribution of science to the industry was the discovery that certain substances speeded up vulcanization. Though these substances were introduced as accelerators, it was discovered that they improved the product as well. This opened the way for the addition of other ingredients, which meant that more chemists were needed. Organic accelerators were discovered that not only shortened the time required for vulcanization, but which also made the process possible at a lower temperature. To offset the deterioration from aging, various resins, camphor, coal tar, creosote, and other ingredients were introduced. All this had a stimulating effect in bringing the scientist to the study of rubber, a study that expanded rapidly in the next quarter-century.

Industry Creates a World Power

America's rise to world power by 1917 was recognized not only in terms of protocols, trade agreements, and international treaties, but even more dramatically because of her industrial might. Advances in agricultural technology, in food production, in steel production, and in textile output resulted in building up vast surpluses in each of these industries. With the outbreak of war in Europe in 1914, England, France, and their allied powers turned in desperation to America for material aid. As the war dragged on, Germany, in turn, realizing that it was America's industrial power which was preventing her from knocking out her enemy, turned her U-boats loose in the Atlantic, hoping to destroy the flow of American supplies and thus starve her enemies into submission. It was American industrial power that had brought the United States into world prominence and ultimately to world leadership.

Notes for Chapter 32

Scores of books and hundreds of magazines, United States government reports, and proceedings of engineering and scientific societies describing America's progress in every field of activity, appeared in 1900 and the year following. It would be useless to attempt to list them here, since it would assume card catalogue proportions.

For references relating to the advances in agriculture, the main sources are, of course, the annual reports of the United States Department of Agriculture.

For references on building technology, consult notes for Chapter 28.

For references on clothing technology, consult notes for Chapter 27.

For references on steel technology, consult notes for Chapter 22.

For references on glass technology, consult United States Census Report, VIII (1900); Helen and George L. McKearin, *American Glass* (1941) and (1948); Warren Scoville, *Revolution in Glass* (1948); Alexander Silverman, *Bulletin American Ceramic Society,* V (1938).

Chapter 33

TRANSPORTATION AND COMMUNICATION TECHNOLOGY: 1900 TO WORLD WAR I

The developments in transportation and communication technology that occurred during the opening two decades of the twentieth century were destined to change the future of American history —indeed, the history of the entire world. In the field of transportation technology, it was the automobile and the airplane that brought about the greatest changes. Improvements in railroad transportation, while important, were not revolutionary. In the field of communication technology, this period witnessed the beginning of the wireless age, the radio, and the motion picture.

The Beginning of the Automobile Era

While the automobile was not born in the United States, nevertheless, this is the country where it has had the greatest development. In 1769 Nicholas Cugnot, a Frenchman, built a three-wheeled carriage, installed a steam-power plant, and propelled the carriage for a few miles. In England steam coaches operated as public conveyances and made daily trips between villages as early as the 1830's. Electric propulsion was used in the early 1870's, when a Londoner operated a vehicle powered by an electric battery. In France the internal combustion engine first came into the picture in 1863, with Jean Joseph Lenoir's one-cylinder engine mounted on a carriage type of vehicle.[1] In Germany, N. P. Otto's four-cycle internal combustion engine, fueled by carbureted gasoline (that is, gasoline mixed with air) became a power plant for automobiles. Two other Germans made important contributions: Karl Benz developed a four-stroke engine for a car, introduced the differential, and a special electric spark system; while Gotlieb Damlier developed a high-speed combustion engine. Prior to 1890 automobiles were being manufactured in France, Austria, Germany, and Italy.

America Adopts the Automobile. The United States became automobile-conscious in the early 1890's. Prior to this, steam vehicles had been attempted and a few electrics had been built, but the internal combustion engine existed in this country only on paper—in George Selden's patent rights. There was no one inventor of the automobile, either in Europe or America. Cars resembling buggies were assembled from a variety of parts drawn from dozens of different sources. A gasoline engine was attached to a buggy, and then by such devices as chain drive, gears, and sprockets the buggy was made to run. Outstanding among the pioneers in the automobile industry were the Duryea brothers, Hiram Percy Maxim, Elwood Haynes, Ransom E. Olds, Roy M. Chapin, Alexander Winton, Henry M. Leland, the Dodge brothers, James W. Packard, John W. Willys, Howard E. Coffin, and best-known of all, Henry Ford.

Pioneer Automobile Builders. In 1892 Charles and J. Frank Duryea built and operated a gasoline-propelled car at Springfield, Massachusetts, which they claimed to be the first successfully operated automobile in this country. It had a one-cylinder engine, mounted on a four-wheel buggy. In 1892 Henry Ford built a gasoline "buggy," but "it did not run to my satisfaction," declared Mr. Ford, "until the spring of the following year." Hiram P. Maxim, famous for his invention of the Maxim silencer, began his experiments by trying to motorize an old-fashioned tricycle, and believed he was alone in this field. In 1894 Elwood Haynes of Kokomo, Indiana, a metallurgist and petroleum engineer, built and operated a car. R. E. Olds, who had first built a steam carriage, turned his attention in 1896 to a gasoline-propelled vehicle, a two-passenger Oldsmobile runabout. Within five years he was producing cars and selling them for six hundred and fifty dollars each.

Alexander Winton, known as the builder of the first "big" car, had a bicycle repair shop in Cleveland, where in 1897 he built two phaeton-type "horseless carriages." His position was unique, in that he was one of the first to build cars with the idea of selling them. Winton is also credited for later introducing spare parts, and for organizing service stations to keep his cars in condition after they left the factory.

The early builders of America's automobiles worked alone. They were mechanics, bicycle builders, and repair men. H. H. Franklin, the pioneer of the air-cooled system, had been a die caster. Henry Ford was a machinist in an electric power house in Detroit before he thought of building automobiles. George N. Pierce had started his career building bird cages, bicycles, and refrigerators. Elwood Haynes had been an employee of a natural gas company. The Studebaker brothers were buggy and wagon manufacturers. And, as Mark Sullivan once observed, "The Peerless car came from a clothes wringer factory, and the Stanley steamer from a photographic dry-plate factory."

Types of Power Propulsion. During the experimental years, opinion was divided over the advantages of different kinds of power for automobiles—whether steam, electric, or gasoline. The simplicity of the steamer was its chief feature; it had no transmission or gears, no ignition, no vibration, and was easy to lubricate. Those who favored the electric car pointed to its easy maneuverability, which enabled women to drive it, and its freedom from noise and vibration. Also, the "electric" required little attention other than recharging the battery—but therein lay its chief difficulty. The advantages of the gasoline engine, on the other hand, far outweighed the disadvantages. The recent developments in petroleum technology assured an abundance of fuel supply. Most country stores kept a supply of gasoline on hand, but most important of all, the inventors and technicians saw in the internal combustion engine a potential power unit with unlimited possibilities.

Automobile Public Events. Great impetus was given to building automobiles when on Thanksgiving Day, 1895, the Chicago *Times-Herald* sponsored an early automobile race, a 55-mile run from Chicago to Evanston, and return. It took place during a blizzard, over roads covered with snow and ice. J. Frank Duryea won the race with a car that he had designed and which was propelled with a two-cylinder engine. He was awarded a cash prize of two thousand dollars. *The Horseless Age,* a magazine that had just been launched, ventured the prophecy that a new industry was about to come into being.

The rivalry between the early builders of automobiles had a stimulating effect on the industry and caused builders to compete with one another by placing their cars on display. The first automobile show was held in 1900 in Madison Square Garden, New York, and the same year the *Automobile Magazine* was established. Alexander Winton, one of the exhibitors, set out from Cleveland over rough roads that hardly deserved the name, and when he finally arrived in New York in his mud-spattered, weather-beaten car, he had difficulty in convincing doubters that the car had made the entire trip under its own power.

The first automobile show was patterned largely after the bicycle shows. The first horseless carriages resembled the carriages and buggies of that day, so much so that the vehicle often retained the old-fashioned dashboard with a whipholder socket. The early car builders also borrowed ideas from the bicycle manufacturers. The first cars had wire wheels like the bicycle wheel, and single-tube pneumatic tires. Many of them also had a chain drive.

Racing Publicity. In an effort to further attract public attention, the automobile builders began racing their cars with rival competitors. Long-distance tours of 200 and 300 miles were scheduled. Hill-climbing contests were announced. In 1901 a Locomobile car ascended Mt. Washington in New Hampshire and Pikes Peak in Colorado. In 1903 the Glidden tour across the continent was arranged, and a Winton car crossed the United States in sixty-three days, nineteen of which were spent in idleness, due to car and tire trouble, running out of gas and oil, and waiting for new supplies. Barney Oldfield, whose name was to become synonymous with auto racing, drove Henry Ford's famous "999" to victory over Alexander Winton's famous "Winton Bullet." The next year Oldfield set another world record when he drove the revamped "Bullet" against Tom Cooper at the speed of a mile in one minute. The following year, William K. Vanderbilt drove his Mercedes (an imported car) a mile in 41 seconds.

A New Industry Emerges. By 1900 the United States had developed a technological background upon which the automobile industry could be built. Petroleum technology had made possible an abundant supply of kerosene and gasoline. The system of interchangeable parts had reached a stage in numerous enterprises where special machines had been designed to manufacture spare parts in quantity production. The machine-tool industry had developed to the stage where lathes, planes, screw-cutting machines, dies, steam hammers, and presses were being manufactured for many different industries. These varied technologies needed only the trained engineer and businessman (the engineer-entrepreneur) to launch a new industry on its way.[2]

This need was met in part when trained engineers like Hiram P. Maxim, graduate of M.I.T., Howard C. Coffin, engineering graduate from University of Michigan, James W. Packard, graduate in mechanical engineering from Lehigh University, and such highly skilled mechanics (although not university-trained engineers) as R. E. Olds, Henry M. Leland, the Dodge brothers, Alexander Winton, John N. Willys, and Henry Ford entered the automobile industry.

Evolution of Designs. Each of these men contributed new ideas and designs which went into the making of the automobile. Around

1900 two-cylinder cars were supplanted by four-cylinder models. The engine was moved from under the seat to the front of the car. Front radiators were installed to add to the appearance of the car, shock absorbers were introduced to add to the comfort of the rider, and tool boxes, a necessity before service stations appeared, were conveniently placed on the running board. Other advances included the lengthening of the wheel base, longer bodies, and doors installed in the side of the car instead of placing them in the rear. The tiller gave way to the steering wheel; the limousine, the closed car, made its appearance, although it did not come into general use until years later. The air-cooled engine came next, bevel gears replaced chain drives, and the disc clutch, which permitted three speeds forward and one in reverse, was introduced.

Automobile Mass Production

Henry Ford's Model T. Henry Ford, whose name was to become synonymous with mass production, began around 1900 to concentrate on building a simple, useful, standardized car at a cost low enough for people with an average income to buy. In 1903 he organized the successful Ford Motor Company, naming himself as vice-president, designer, master mechanic, superintendent, and general manager. The growth of the Ford Motor Company is one of the epics of the industrial world. During the first years of operation the Ford company sold a car with a two-cylinder 8-horsepower engine for eight hundred and fifty dollars, considerably below the average of the time. By 1908 Ford had introduced his famous Model-T. His whole life had gone into the design of this car and he continued its manufacture unchanged for twenty years. Built for farmers, shopkeepers, and professional men, with mechanical parts so simple that almost any one could make repairs, the high, awkward Model-T chugged through mire and swamp, and bumped over ruts, rocks, and stumps. Its low horsepower engine, its planetary transmission, and its magneto ignition system were the outstanding features.

Standardized Parts. Standardization and mass production became Ford's basic policy, and to this end he introduced improved machinery for manufacturing interchangeable parts in quantity production. Later, Ford discovered that he could purchase parts from small manufacturing companies as cheap, in some cases cheaper, than he could manufacture them. Ford was a stickler for standardization and was quoted as telling his customers they could have any color of car they desired, "so long as it is black."

Conveyor Assembly-line System. To speed up production, Ford next utilized the conveyor assembly-line system. In the early method of production a group of four or five men huddled around a frame, adding parts which were brought on trucks from other sections of the factory—a method wasteful of time and human energy. When the conveyor or belt system was introduced, workers stood on each side of the belt and each worker performed a special task. Parts—such as carburetors, complete and ready to be installed, batteries, motor blocks, wheels, gasoline tanks, screws, bolts, nuts—all arrived at the appropriate moment and were assembled in proper order into a complete car. Ford described this technique as one that, "lifted the hard work off the backs of men and laid it on steel and motors." When the car reached the end of the line it was driven off on its own power. Mass production had arrived.

With these technological improvements, the time required for producing a car was greatly reduced. Lower selling prices followed. Cars thus produced sold for as little as four hundred and fifty dollars. Ford's aim, expressed in his own words, was to "build a standardized, practical low cost car for every adult person in America." Later, he expanded his aim to include the whole world.

Automotive Engineering Becomes Professional. Around 1910 the automobile industry entered upon a period of rapid expansion. Americans, with rare exception, stopped importing European cars. The gasoline car, due to its mechanical simplicity and power, had largely displaced the steam and electric cars. Automotive engineering became a profession, signifying a new group of technical experts.

Electric Self-starter, 1911. Of the several advances in automobile engineering, the introduction of the electric self-starter stands out. Pioneering in this field were Vincent Bendix and Charles F. Kettering. Bendix's "mechanical hand" twirled the motor of the car to start it and was automatically released as the engine picked up speed. Kettering's starter, introduced in 1911 and based on electrically driven gears, turned the engine over just long enough to set it firing under its own power. This device marked a major technological development in automobile history. The ordeal of starting the engine by hand-cranking had been a deterrent in attempts to popularize the automobile. Women now began driving cars. Within a few years the numbers of car sales more than doubled, due in part to this new self-starting device.

Automobile Parts. The manufacturing of auto parts became a major industry. During the early years of the automobile industry, the whim of each car designer dictated the style of his product—a practice that was not only costly, but also prevented standardization. Finally, when the Society of Automotive Engineers was organized, attempts were made to introduce a system of standardization. This marked another technological milepost in automobile history. With the introduction of the standardization system and the universal adoption of the assembly-line technique, cars were produced more speedily. Quantity production had been achieved, and cars became cheaper. Prices were reduced so that the workmen who assembled the cars, along with tens of thousands of other wage earners, were able to buy them.

Major Improvements. A number of advances in motor transportation technology occurred in the period immediately preceding World War I. The introduction of all-steel bodies, twin-six engines, and V-8 engines became the craze. Multiple cylinder cars were the highlight of the nationwide automobile shows of 1916. Cord tires were introduced, as were steel-disc wheels. The good road movement was under way, which more than anything else furthered the use of closed cars. The closed car became standard equipment during the 1920's.

Trucking Industry

Early Trucks. The automobile, like the first railroad, was built primarily to carry passengers. But automobile manufacturers, again like the railroad companies, discovered a demand for speedy shipment of freight. To meet this demand, specially designed chassis and bodies were built. A few electric-powered trucks and steamer trucks were manufactured before 1900, but these were slow in coming into use. It was generally believed that while the buggylike motor vehicles might eventually replace a few carriage horses, they could never compete with the sturdy dray horses.

"Delivery Wagons." In the face of these predictions, however, a few manufacturers began advertising "delivery wagons," powered by one-cylinder motors. They offered many advantages for "trucking" products. The editor of *Horseless Age* wrote in one of its early issues that trucks offered the most alluring feature of the motor vehicle industry to the far-sighted capitalist or manufacturer, "provided they know the difference between a Paris fashion and a labor-saving machine." The pleasure car was not to be ignored, he observed, but it would be a minor part of the automobile industry. Although his prediction proved incorrect, nevertheless, it represented a widespread belief of that day.

Motors versus Horses. It was not until around 1910 that trucks began to compete with horse-drawn vans and wagons. Motor-truck manufacturers pointed to the many advantages of the truck as compared with the horse. They popularized Edison's remarks that "a horse was the poorest and most expensive motor ever built: he eats ten pounds of feed for every hour he works. Motor trucks, on the other hand, need feed (fuel) supply only when they are working." Horse-drawn wagons operated in a limited area only, whereas trucks could operate between distant cities; that is, if the cities were connected by roads and highways. Statistics were published showing that truck transportation was cheaper than horse-drawn transportation, since one truck could displace ten horses.

First Use of Trucks. The larger cities were the first to welcome the use of trucks. Many of the cities passed ordinances prohibiting the stabling of horses within city limits. Bakers introduced motor driven "pie wagons," dairymen substituted gasoline milk wagons, department stores promised quick delivery with electric trucks, and in some of the larger cities, the United States Post Office Department authorized the use of trucks for collecting mail. The American Railway Express early introduced motor trucks for delivering express parcels.

"Ship by Truck." Before World War I trucks had made little inroad upon the railroads. But with the outbreak of the European war in 1914, when American railroads were overtaxed, manufacturers turned to the trucking industry for aid, especially for short hauls. "Ship by truck" became a nationwide slogan. Automobile manufacturers concentrated upon truck production and in 1917, the year the United States entered the war, some three hundred thousand trucks were manufactured.

In addition to the demands for trucks in America, our allies placed large orders for American-built trucks, and the United States Army began replacing four-mule teams in a number of army units, notably the Commissary Department, with motor trucks. In 1916 a United States Ambulance Corps went to France equipped with a motorized truck and with wheels mounted on pneumatic tires (see Chapter 40 for use of trucks in World War I).

Highways Span the Nation

The introduction of the automobile was accompanied by a good-roads movement; indeed the two developments came simultaneously. Roadbuilding had always been a state and local project. At the beginning of this century none of the states had adopted a basic policy relating to roads or highways. Roads spread out from cities and towns into the rural communities, with the sole purpose of enabling farmers to "go to town." Plans for connecting distant cities had not yet developed.

National Good Roads Association. With the advent of the automobile, however, came a desire to travel greater distances and at greater speed. Car owners joined forces with public-spirited persons in urging organized action. A National Good Roads Association was organized and attained prominence soon after the turn of the century. At the meeting in St. Louis in 1903, President Theodore Roosevelt, William J. Bryan, and General Nelson A. Miles, addressed the session, emphasizing the value of a national network of good roads. A number of leading railroads joined the movement. They outfitted Good Roads Construction Trains equipped with the latest roadbuilding machinery, and traveled about the country giving actual demonstrations in road improvement.

American Road Congress, 1912. As agitation for good roads continued, Good Roads Associations sprang up in many states. Woodrow Wilson, Governor of New Jersey, addressing an American Road Congress in Atlantic City in 1912, stressed the advantages that were derived from carefully planned roads built by competent engineers. In the same year, a Good Roads Convention held in Washington called attention to the need for federal aid in building new roads and in improving those already in existence. The federal Office of Public Roads urged the need for improved roads for rural mail carriers, now that they had to carry heavy parcel post packages as well as letters.

National Highways. In 1913 an ambitious program was proposed for building a hard-surfaced all-weather highway across the continent. Led by Carl G. Fisher of Indianapolis, and backed by a number of automobile companies, a group set out to map the route from ocean to ocean, giving it the magic name of The Lincoln Highway. Lacking road maps, they seized upon railroad time tables for mileage and traced a red-line route which was to become the nation's first transcontinental highway. Next came a north and south route, extending from Michigan to Miami, to be known as the Dixie Highway. Then came the Meridian Highway, sketched on the map from the Rio Grande to Winnipeg, Canada; and the Jefferson Highway from New Orleans to Canada. For many years these roads were little more than lines drawn on maps.

Federal Highway Financing. The mounting interest in good roads was reflected in Congress by the appointment of a Committee on Roads in 1913, which after three years of study brought about the Federal Aid Act of 1916. Its enactment into law was endangered, however, by President Wilson's unwillingness to sign it on the grounds that there was no constitutional authority for giving federal aid to build highways. The appearance of a German submarine in the harbor in Baltimore on the day Wilson planned to issue his veto caused him to change his mind. Deciding that the bill could be incorporated in a national defense act, he signed it into law. America's entrance into the war the following year, with the accompanying demands upon all roadbuilding material, retarded the good-roads program for a period. During World War I the need for improved roads became more apparent when trucks were dispatched from munition centers to training camps.

Coming of the Airplane

Early Experiments. The next revolutionary development in transportation technology to occur around the turn of the century was the invention of the airplane. Man's desire to fly dates back many centuries. As early as the fourth century B.C. the scientist Aristotle advanced some vague theories regarding the possibilities of human flight. In the thirteenth century the noted English scientist, Roger Bacon, foresaw the possibility of man inventing a mechanism that would lift himself from the earth and propel him through the air. Leonardo da Vinci, in the fifteenth century, worked out a number of devices for transporting man by air, and had he been able to invent a satisfactory steam, gas, or jet power unit to propel his planes, da Vinci might well have ushered in a new age of flight.

Balloon Flights. During the latter part of the eighteenth century, French scientists, using balloons inflated with heated (rarefied) air, made a number of flights, on occasions reaching a height of more than 5,000 feet. Balloon ascents were also made in England, Scotland, Germany, Italy, and in the United States before 1800. During the American Civil War, balloons were used for reconnaissance purposes (see Chapter 20). While the introduction of the balloon enabled man for the first time in history to ascend thousands of feet into the air, he still found himself at the mercy of the winds and unable to travel in any desired direction.

Early Gliders. Meantime, numerous persons were experimenting with other possible methods of flight. Sir George Gayley, William S. Hanson, John Stringfellow, Francis H. Wenham, all of England, and

Alphonse Penaud of France built glider planes of various designs during the last half of the nineteenth century. The German engineeer Otto Lilienthal and the French-born Octave Chanute (who is best known, however, for his work in the United States) succeeded in piloting gliders for several hundred feet. Lilienthal controlled the flight of his glider by shifting his body from one side of the plane to the other. Chanute, by installing a flexible tail on his plane, was able to pilot his craft without the necessity of moving his body.

Americans Become Interested. This was the stage which gliders had reached when a few American men of science, among them Samuel Pierpont Langley, became interested. Langley, who had been an assistant on the staff of Harvard Observatory and had taught in the United States Naval Academy, was serving as Director of the Allegheny Observatory and professor of physics at the University of Pittsburgh when he took up his studies of designing model planes. He designed an efficient supporting and guiding surface, mastered the problem of equilibrium, and constructed an adequate launching device.

Samuel Pierpont Langley, 1834–1906. Scarcely had he started to work on his first models in Pittsburgh when his appointment as Secretary of the Smithsonian Institution took him to Washington in 1887. There he worked intermittently for the next few years and tried out more than one hundred models. The model frames were built of light wood, with two propellers attached, driven by twisted rubberbands. From these experiments, Langley learned that he needed a stronger source of power for propulsion. This caused him to construct and install a small steam engine on his model planes.

On May 6, 1896, at Quantico, Virginia, Langley made two launchings of a model plane, with no one aboard. During each launching the plane remained aloft for about 90 seconds. Six months later he launched another model plane which traveled almost three-quarters of a mile. When the fuel was exhausted, the plane dropped gently into the Potomac River. Having succeeded in flying a heavier-than-air model plane propelled by motor power, Langley felt that his first mission in the science of aviation had been accomplished.

Charles M. Manly. Although he had passed the three-score year milepost, Langley still had hopes of building a man-carrying flying machine. He was encouraged in this undertaking when President McKinley and the War Department's Board of Ordnance allocated fifty thousand dollars to be used in constructing a successful flying machine. After years of experimenting, aided by Charles M. Manly, a graduate engineer from Cornell, and after a trip abroad to study the latest development in steam engines, Langley and Manly developed a model 52-horsepower internal combustion engine to be used in propelling a man-carrying flying machine.

Langley's Failure. October 7, 1903, was the date set for the first trial flight at a point a few miles down the Potomac River from Washington. The plane was mounted on top of a houseboat, ready to be hurled into space by a coiled spring (Langley called it a catapult car). Upon releasing the spring, the plane in which Manly was sitting raced down a two-rail runway and plunged into the water. Manly escaped, fortunately, with little more than a ducking.

Jeers greeted Langley's and Manly's failure. Newspaper headlines—such as "Buzzard A Wreck," "Airship Fails to Fly," "Professor Langley's Airship Goes to River Bottom," "Costly Airplane Wreck Travels Only As Far As Catapult Sends It," "Langley's Aerodrome Swoops Into Potomac and Ducks Professor Manly"—appeared in the press for days following the failure. As Mark Sullivan said: "The stories actually seemed to gloat in affirmation and repeated reaffirmation of the unqualified completeness of failure." [3]

Two months later, on December 8, 1903, Langley made another attempt, and again the launching apparatus failed. As before, when the plane reached the end of the runway, the rear of the machine in some manner was caught by a part of the launching car. The rear portion of the plane and wings were broken and the plane plunged into the river. Again, the skeptics jeered and newspapers screamed at such crazy, costly experiments! The failure of the second trial ended Langley's experimental work in aeronautics. As Manly later wrote in describing this last effort—

> Success seemed only a step away, and his age was such that any further delay in achieving success increased the probabilities of his not living to see it, but he maintained positively and absolutely that if neither the War Department nor others felt sufficient interest in the work to provide the small amount of funds necessary to continue the experiment he would rather deny himself the hope of living to see the machine achieve success.[4]

Langley's Contribution. Langley died February 27, 1906. His contributions included the formulation of principles relating to wind resistance and air pressure; the building of a whirling table equipped with wooden arms driven by engines capable of rotating the arms; the development of an instrument for determining the various centers of pressure on planes moving horizontally at different angles of inclination; and the designing of instruments for registering wind fluctuations.

The Wright Brothers Triumph. Nine days after Langley's second failure, Wilbur Wright flew a heavier-than-air plane which he and his brother Orville had designed, a distance of 852 feet, kept it in the air for 59 seconds and brought it down safely. This marked the first time in history that, "a machine carrying a man and driven by a motor had lifted itself from the ground in free flight."

Wilbur Wright, born April 16, 1867, and his brother Orville, four years younger, had for years operated a bicycle shop in Dayton, Ohio. During the 1890's they became interested in aviation and began a serious study of the mechanics of flight. Like other pioneers, they too, began experimenting with gliders. In May, 1899, Wilbur wrote to the Smithsonian Institution, declaring that while he was enthusiastic about aviation, "but not a crank . . . ," he wished to obtain any papers which the Smithsonian had published on the subject, and a list of any other available material relating to mechanical or human flights.[5]

Early Attempts. After reading articles relating to flying by Otto Lilienthal (who had been accidently killed in 1896), Samuel P. Langley's *Experiments in Aerodynamics,* and Octave Chanute's *Progress in Flying Machines,* and recent copies of the *Aeronautical Annuals,* the Wright brothers began building gliders of their own. Not satisfied with the idea of maintaining equilibrium while in flight by merely shifting one's body from side to side, as most experimenters had done, the Wrights set to work on an entirely different principle. The result of their experiments marked a milepost in aviation history.

Using a small pasteboard box, with opposite sides removed, Wilbur Wright observed that by holding the top forward corner and the rear lower corner of one end of the box, he could warp it so that simultaneously the surface on one side (forward) inclined downward, while on the other side it inclined upward. In this manner, "the upper and lower surfaces of the box were given a helicoidal twist, presenting the top and bottom surfaces of the box at different angles on the right and left sides."

Applying these principles to the surface of gliders, warping them in the same manner, the Wrights discovered a method for "lateral balancing based on the application of varying vertical wind pressures on either wing." Keeping a glider plane on even keel was the chief problem that had to be solved before man could safely fly.

Prototype Gliders. After building a model plane (glider) which they were able to fly, Wilbur Wright wrote to the United States Weather Bureau, inquiring about prevailing wind conditions in different sections of the country. He was informed that the region around Kitty Hawk, North Carolina, was one of the windiest places in the eastern part of the United States—the wind averaged about 16 miles an hour. Also, he was told that there was a long, sandy beach located there, about one-half mile wide, clear of grass and high hills, which extended a distance of some 60 miles.

Experiments at Kitty Hawk. The Wrights made yearly trips of a few weeks each to Kitty Hawk, first in 1900, then in 1901, 1902, and in 1903. After a few weeks of experimenting with a glider at Kitty Hawk (they built and took a new one with them each year), they would return to Dayton and concentrate on a number of improvements. What had at first been a hobby, gradually became a matter for serious scientific study. Among the studies that they made were a number relating to wind resistance, the law of lift and drift, the proper arrangement of spars, braces, balancing, curved surfaces against plane surfaces, testing Lilienthal's tables, experimenting with wind tunnels, and testing all types of wing surface. The knowledge they gained in the latter experiments marked a turning point in the history of aviation, since it is the design of the wings that makes flying possible, not the engine or the propellers.

After developing a proper wing design that enabled a plane to fly, the Wrights next sought a satisfactory motor for the plane. Unable to secure one from any automobile manufacturer, they designed one of their own. Next, they designed propellers, which they declared were more efficient than any they had read about.

The First Motor-Powered Airborne Flight. In the fall of 1903, the Wright brothers returned to Kitty Hawk with their latest plane. After some three weeks of work in uncrating and reassembling the plane, they installed the motor, and on that historic day, December 17, 1903, Orville "got on the machine at 10:35 for the first trial." The machine lifted from the truck, rose suddenly to about 10 feet in the air, and flew approximately 100 feet. Man and machine had been in the air about 12 seconds. Next, it was Wilbur's turn, and he flew the plane about 175 feet. Then Orville made his second flight, which was about the same distance. Then Wilbur made his second flight; he remained in the air 59 seconds and flew a distance of 852 feet. Strangely enough, however, this history-making event attracted little public attention.

After dismantling their machine and shipping it back to Dayton, the Wright brothers continued their practice of flying in or near their home town. Their aim was to develop a plane that could fly anywhere and in any kind of weather. Since they had developed a motor-propelled plane, they no longer needed the strong prevailing winds and steep sand dunes around Kitty Hawk. Their first "flying field" was an open meadow, belonging to a farmer living a few miles out of Dayton, where they continued their experiments.

Early Appraisal of Wright Invention. Here, during the next few years they made a number of improvements. They learned how to fly the plane in circles, stay in the air for a half-hour or longer, cover distances up to 25 miles, and land the plane at will. All this time the public did not appear to be interested. Early in 1905 the two brothers wrote to their local congressman, informing him of their success in inventing a flying machine, and suggesting it might have some practical use. One of the uses, they said, might be "that of scouting and carrying messages in time of war." Secretary of War William Howard Taft and his Chief of Ordnance informed the Wrights that since flying had not been brought to a practical stage, it would be unwise to allot any funds for experimental devices for mechanical flights. The Board of Ordnance added, however, that should the Wrights succeed in making their machine practical, it would be pleased to hear from them.

Federal Aid—Foreign Interest. Two years passed, and in May, 1907, the United States Board of Ordnance and Fortification informed the Wright brothers that the Board was undertaking studies relating to the construction and testing of flying machines. The Wright brothers informed the War Department they had some planes in course of construction and, "would be pleased to sell one or more of them to the War Department if any agreement as to terms can be reached." [6]

Meantime, aviation enthusiasts in England and on the Continent were hearing of the successful flights of the Wrights. Both France and England sent agents to Dayton to secure the exclusive rights for their planes. A syndicate in Vienna hoped to purchase a plane and present it to Emperor Franz Joseph on the occasion of his sixtieth jubilee in 1908.

The Wrights went to Europe in 1907 and took one of their planes with them. After visiting Paris, Berlin, and London, and talking with a number of capitalists about organizing companies for producing their plane abroad, they organized a company in Paris. In the spring and summer of 1908, Wilbur made a number of flights at Le Mans which stirred the French people as nothing had for years. Russian and Italian aviation enthusiasts also came to view the flights. French newspapers gave columns to an account of the flights, and a song "Il Vole" (he flies) was dedicated to Wilbur. European capitalists attempted to secure exclusive control of the Wrights' patents.

First War Department Contract. Meantime, Orville Wright had returned to Dayton where, in February, 1908, the United States awarded a contract to the Wrights for twenty five thousand dollars to build an airplane that would fly. The contract called for delivering a plane capable of carrying two men and sufficient gasoline for a flight of 125 miles, capable of attaining a speed of 40 miles per hour and of maintaining sustained flight for one hour, and having the ability to land undamaged.

Barnstorming Draws Public Interest. Other air-minded persons entered the field. Popular interest in aviation was first aroused by the barnstorming tours of a veteran showman, Thomas Scott Baldwin, who gave the American people their first opportunity to see a dirigible. With the experience carried over from balloon ascensions, Baldwin equipped a "rubber cow" with a motor and thrilled the populace at the St. Louis Exposition in 1904. The next year he won a prize at the Lewis and Clark Exposition in Portland, Oregon, and in 1908 he furnished the War Department with its first dirigible.

Glenn Curtiss. Baldwin brought Glenn Curtiss, a manufacturer of motorcycles, a stunt rider and speed king in motorcycle races, into the field of aviation. Curtiss built motors for Baldwin's airship and other dirigibles. In 1907 he furnished a motor for Alexander Graham Bell's man-flying kite, and worked with Bell in the Aeronautic Experiment Association. Curtiss, a self-taught aviator, built his own planes and engines. In 1908 in his "June Bug" he captured the trophy offered by the *Scientific American* in a demonstration at Hammondsport, New York, witnessed by more than a thousand thrilled spectators, including a number of scientists and aerial enthusiasts. The next year in his "Gold Bug" he gave the New Yorkers their first view of an airplane in flight. In the late summer of that year, Curtiss broke the world's speed record, flying at the rate of 47.65 miles an hour, and captured the James Gordon Bennett trophy. He then took his plane abroad and won a forty thousand dollar prize at the first international aviation meet held at Rheims, France.

Flying Becomes a Profession. Following the exhibition at Rheims, Americans began to take a serious interest in the progress of aviation. The fact that a quarter of a million Frenchmen had paid to witness the event at Rheims caused some American promoters to offer lucrative prizes with the result that a group of enthusiasts took up flying as a profession, devoting their entire time to it. Prior to this, aviation had been dominated by a handful of inventors and designers like Curtiss and the Wright brothers, but now aviation meets became the order of the day. In the absence of organized exhibitions and valuable prizes, aviation teams were formed and made their living as stunt fliers at circuses and fairs.

Aviation Events Bring Recognition. In the United States the first aviation meet was held at Los Angeles in January, 1910. Aero clubs were formed. Aeronautic magazines appeared, and aviation schools were established to train fliers. Spectacular flights were inaugurated. In 1909 it was hailed as a feat when Wilbur Wright flew from Governors Island to Grant's tomb and back, and when in 1910 Curtiss flew from Albany to New York, a milepost in aviation was achieved. The following year an even more spectacular flight was performed when Galbraith Perry Rogers flew from New York to California in forty-nine days and sixty-eight hops, with slightly over 82 hours in the air. He averaged 51.59 miles per hour. In 1910 the first degree in aeronautical engineering was given at Columbia University. The American public had accepted aviation as another triumph in transportation technology.

Pioneer Pilots. A few daring women entered the field. Harriet Quimby, a serious minded editor of *Leslie's Magazine,* in 1911 was the first woman licensed pilot. A few weeks later Matilda Moisant, whose brother was a flier, received her commission and teamed up with Miss Quimby in an exhibition team. Ruth Law in 1912 made a nonstop flight from Chicago to New York.

Men aviators continued to startle crowds by their stunt flying. None was more daring as an exhibition pilot than Lincoln Beachley who earned the title, "the flying fool." He flew over the mists at Niagara Falls and was the first to imitate the French flyers in "looping the loop." By 1911 there were some three hundred airplanes in the United States, and by 1913 planes were flying at a speed of 100 miles an hour.

Lessons Learned by Barnstorming. Through experience gained in barnstorming stunts, considerable skill in handling planes was acquired. Methods for overcoming wind variations were studied, and the need for sturdier planes was appreciated. The first planes were built of spruce wood, steel wire, and fabrics, and powered by small gasoline engines. Materials had limitations as to strength and stresses, and none had been yet developed to withstand the pressure caused by flying at high speed. Planes often crashed. Engines stalled and went into deadly nose dives. Exhibition flights failed in stormy weather and fatalities were great. It required an emergency like World War I to cause designers and builders to seek the aid of scientists in solving their technical problems.

Early Commercial Flying. Even before World War I, however, attempts were made to introduce the airplane in commercial operation. The Wrights in 1910 established an aerial express between Dayton and Columbus. The Post Office Department swore in a pilot to drop mail in flight at Mineola, Long Island. In 1914 a scheduled passenger service operated for three months across Tampa Bay in Florida. For the most part, however, commercial use of airplanes, as well as major aeronautical technological improvements, were postponed until after World War I. (For the part played by the airplane in World War I, see Chapter 34.)

The first two decades of the twentieth century had witnessed the birth of two inventions, the automobile and the airplane—inventions that were destined to bring about the most far reaching revolution in the whole history of American transportation technology.

Railroads Modernize

There were no marked technological developments in the railroads in the opening decades of the century. "Bigger" was the slogan, but coupled with this was "better" also. The limit of track extension was reached in 1916. After that mileage began to recede due to consolidation and elimination of spur roads.

Railroads Begin to Electrify Equipment. Electrification was introduced on some of the main roads for passage through cities, at terminals, and in tunnels to eliminate the smoke nuisance. The Norfolk and Western Railroad, in introducing electric engines for freight through a mountain tunnel in which there was a considerable grade, found the electric motor an advantage in power. Twelve electric locomotives did the work of twenty Mallet-type steam engines, and did so in less time. Because of the advantage of electric power on grades, the Chicago Milwaukee and St. Paul electrified its road from Chicago to Avery, Idaho, in 1917. Construction costs were heavy, but a saving was effected on operating costs. In electrification, both third-rail and overhead-wire systems were employed.

Improved Locomotives. Improvements in the steam locomotive were made to bring about increased power, fuel economy, and increased efficiency. Fast freight was needed in the twentieth century to bring citrus fruits from Florida and California, and to carry the products of the great packing houses from Chicago to market. Lengthy trains of refrigerator cars could no longer be relegated to sidings to let the fast passenger trains go by. Up to World War I locomotives underwent a gradual evolution from the older types, in that they developed a tractive force almost four times that created in the earlier period, and they almost doubled in weight.

Communication Reaches a New High

Newspapers. In the development of communications through the printed word, newspapers quickly took advantage of the new technological developments in newsgathering by utilizing the wireless telegraph and the radio, just as they had done with the telegraph and telephone decades earlier. Wireless dispatches were received as early as 1901. (See page 497.) The first radio-transmitted news story to be printed came in 1909 from Captain Robert Edwin Peary, when he hoisted the American flag at the North Pole.

Unit Press. In printing technology, streamlining and efficiency rather than new techniques marked the major development. The monstrous quadruple, sextuple, and octuple presses, with their mountains of plate cylinders and impression cylinders and ever-extending decks, were being streamlined into compact units. Each unit printed sixteen pages. These could be established in batteries to suit the needs of the publication. Greater precision in operation was effected and efficiency was increased. The autoplate casting machine for the stereotype mentioned earlier was invented and was heated by electricity. This was a great labor saver and removed an arduous and hot job from the printing industry.

Offset Printing. The greatest development in printing in the early decades of the century was in the institution of offset printing, in the introduction of the rotogravure press, and in the advances in illustrations. Offset printing belongs to lithography, in which transfer patterns were made from a lithograph drawing on stone, aluminum, or zinc and from which impressions were made. This was applied to textiles as well as paper. In the early 1900's a lithographer accidentally discovered that an impression could be taken on a sheet of paper from the blanket that was used in the lithographing operation. This procedure was cheaper and was a great timesaver. It was first applied to the newspaper field in comic and color supplements, but was improved and applied to finer work.

Rotogravure. Rotogravure was borrowed from the textile industry and was first adapted to the newspaper field in 1912. The New York *Times* has been credited with having the first rotogravure illustration in its Christmas issue of that year. The principle of rotogravure is opposite from the stereotype method in printing, in which the impression is made from a raised surface. In rotogravure the design is sunken, or intaglio, so that the surface is the nonprinting area. In producing the printing plates, the design was etched or produced by photo-engraving, and the depressions thoroughly inked, while a "doctor blade" scraped the surface clean. At first, rotogravure printing was done on sheet-fed presses, but rotary presses have been designed for operating at high speed for tabloid or newspaper products.

Photographic Reproduction. Many demands were made for newspaper illustrating, and the halftone for reproducing photography that was introduced in the late 90's became highly developed. News photography from the Spanish-American War, from the Philippines, and from the Russo-Japanese War greatly increased the use of illustrations. The technology of newspaper illustrating became characterized by the same features of speed and quantity production that belonged to the other departments. Little of the old handcraft was left in newspaper illustrating. Many of the advances in printing were curtailed by World War I.

Telegraphy and Radio Embrace the World

Wireless Telegraphy. Wireless telegraphy as a means of communication became a reality in the first years of the twentieth century. Communication between points not connected by wire had been the toy of the laboratory both in the United States and in Europe. Joseph Henry, in this country, had discovered the electrification of needles by the discharge of an electrical machine 30 or more feet distant. It was James Clark Maxwell, British theoretical physicist, often called the "Father of Wireless," who advanced the theory of electromagnetic waves. And it was Heinrich Hertz of Germany who unlocked the secret of these elusive electrical waves. Hertz demonstrated that electromagnetic waves could be sent through space at the speed of light, and that they could be intercepted at a position remote from the source by an intercepting wire system. Electromagnetic waves became known as "Hertzian" waves; the interceptor of the waves was called the "Hertzian" loop.

Branly's Coherer. Another step was taken when Edouard Branly, a Frenchman, devised a superior interceptor which Sir Oliver Lodge called the "coherer." This was a thin glass tube like a thermometer and was filled with metal filings which would cohere when subjected to electromagnetic waves. It thus became a conductor, but would decohere with a slight tapping, and thereby break the circuit. By the addition of an automatic tapper attached to the tube, the tube became an effective means for the transmission of successive impulses, which could be dispatched in the dots and dashes of Morse's telegraphic code.

Marconi. Combining these scientific discoveries into a workable system of communication, Guglielmo Marconi, in 1896 on his father's farm in Italy, put them to practical use. Marconi patented his device in England in 1897, and organized the Wireless Telegraph Company. He added the Morse telegraph key to make and break the circuit. He grounded one-half his aerial, and projected the other half into the air, greatly increasing the distance for the transmission of signals. Because his improved aerial resembled the feelers of a monster insect, it was called the "antenna." Marconi improved Branly's coherer by adding nickel and silver and by making the decoherer automatic. The first device constructed transmitted the signals a distance of only 2 miles, but by 1897 Marconi sent the signals 18 miles. Two years later messages were sent across the English Channel. In 1901, the first trans-Atlantic signal was heard, and by 1903 newspapers were being printed at sea with dispatches supplied by wireless. The Marconi Wireless Telegraph Company of America was organized in 1899.

Lee DeForest. Less than two months after Marconi had taken out his American patent, Lee DeForest, a young Yale graduate, had assembled a complete setup in wireless telegraphy and had invented an electrolytic anticoherer. Eager to try out his ideas in a big way, he secured a reporting contract with the Publishers Press Association to cover the International yacht races in 1901, but the conflicting electrical waves sent out by rival interests in their endeavor to report the event caused so much interference that no one was able to report the race successfully. President Theodore Roosevelt, an ardent believer in the future of wireless, gave DeForest an opportunity to demonstrate his equipment to the United States Navy, and he installed several sets of wireless equipment on a number of naval vessels. He also equipped the private yacht of Sir Thomas Lipton, and secured a contract with *The Times,* London, and the New York *Times* to cover the Russo-Japanese War.

Public Wireless Communication. By 1904 a system of wireless communication was in operation. Stations had been established at Block Island, Point Judith, and Coney Island. The wireless stations were among the chief centers of interest at the St. Louis Exposition in 1904. By 1907 wireless had proved it was more than a dream of visionary scientists. In February, 1908, trans-Atlantic wireless stations were opened to the public for the transmission of "Marconigrams."

Wireless on Ships. As wireless seemed to work better over water than over land, and since there was no other rapid communication system operating over the ocean, it was from vessel to vessel and from land to ships at sea that wireless had its early installations. The sinking of the "Republic" in 1910, and of the "Titanic" in 1912, brought dramatic attention to the importance of wireless for the safety of life at sea. In 1910 Congress passed a radio Act making it unlawful for any passenger ship with more than fifty persons aboard to leave one port for another more than 200 miles away unless the ship was equipped with wireless apparatus and a skilled operator.

Wireless Improvements. Many improvements were made in wireless telegraphy in the next five years. A high-frequency alternator was devised, which transformed the current into a continuous flow rather than a succession of pulses. Also, improvements were made in the detectors which rendered them more sensitive. The outbreak of World War I found the attention of all belligerents focused on the importance of wireless communications. Unprecedented demands were made to rush developments. Wireless proved of inestimable value in the war at sea, and was used effectively by President Wilson in making his war aims known to the world (see Chapter 37).

Radio Communication. Meantime, American inventors had demonstrated that sound could be transmitted without wires as well as with wire connections, and had succeeded in developing a workable system of radiotelephony. Pioneering in this field was Reginald Aubrey Fessenden. He and Lee DeForest, Edwin Howard Armstrong, and E. F. W. Alexanderson were often referred to as the "Big Four in Radio."

Wireless Telephony versus Wireless Telegraphy. Successful transmission of speech and music required a more complicated technique than that required for transmitting dots and dashes. Sounds entering the transmitter were converted into electric impulses, carried through the air, and reconverted into sound waves at the receiving end. A technique based on the comprehension of wavelengths was developed. The pulsating waves of the transmitter were converted into continuous waves for the reception. From five to fifteen times more power was required for telephony than for telegraphy.

Fessenden's Alternator. Fessenden was a distinguished scientist and technologist, acclaimed by Elihu Thomson as "the greatest wireless inventor, greater than Marconi." He was concerned not merely with scientific discoveries, but also, like Benjamin Franklin, with their practical application. He was born and educated in Canada, then came to the United States, and in his first years here he acquired practical experience with Edison. In 1892, he became professor of electrical engineering at the University of Pittsburgh, where he conducted his earliest experiments. His first success was in December, 1900, when, using a spark transmitter, he transmitted speech through the air that was picked up by wireless telegraph stations, "poor in quality but intelligible." The first improvement Fessenden planned was a high-frequency dynamo which he called an "alternator." He made application for a patent in May, 1901, asking that a patent, "for improvements in apparatus in wireless transmission of electro-magnetic waves and improvements relating to the transmission and reproduction of words or other audible signals" be granted. The patent was approved in September of the same year, and was the first ever made for voice transmission by electromagnetic waves.

Liquid Barretter. Fessenden met with delay in the construction of his alternator, but in 1902 he devised an improved receiving instrument, or interceptor of the electromagnetic waves, which he called a liquid "barretter." It was an electrolytic detector. It resembled a miniature light bulb containing a platinum wire, one end of which protruded into a weak nitric acid solution. This device formed a reliable rectifier, changing the pulsating waves into a continuous wave and making possible the use of headphones.

By 1903 Fessenden's first high-frequency alternator was built by Charles Steinmetz at the General Electric laboratory. This was a generator of alternating current, the idea for which had been conceived by Nicola Tesla. Fessenden took up where Tesla left off. Before this, wireless transmission had been made through a spark apparatus which, because of the din, rendered reception of sound waves of the voice impossible. With this instrument the electric impulses were sent out in a serpentine curve (sinusoidal), traveling out in all directions, "forming the ethereal bridge which provided the passage for the electrically converted sounds from the studio to the receiver." Fessenden's first alternator operated on 10,000 cycles. In 1906, Alexanderson completed a 75,000-cycle one for Fessenden, and in 1909 Alexanderson built his own with improvements of 2 kilowatts and 100,000 cycles.

Heterodyne Reception. Another invention of Fessenden's that was basic in the development of radio was heterodyne reception. This is regarded as the second greatest invention in radio art. It arose from the quest for a more effective type of receiver than the coherer. It acted by the joint operation of two alternating currents, one from the transmitter, the other from the receiver. By the superimposing of the one frequency on the other, a third frequency was formed that was superior both in sensitivity and in static control. Fessenden produced the heterodyning frequency with an arc generator. The heterodyne receiving system was years ahead of the industry. It was improved later by the Armstrong superheterodyne.

From a laboratory at Brant Rock, Massachusetts, December 24, 1906, Fessenden made his first all-radio broadcast, a program of music and speech which was heard by surprised listeners as far down the Atlantic Coast as Virginia.

DeForest's Audion Tube, 1907. What has often been described as the truest little giant in all history, for it set unbelievable currents in motion, was DeForest's three-electrode vacuum tube, which he called the "audion." This was used as a detector and amplifier. Since the electrical waves that fall on the aerial wire at the receiving station are identical with those sent out from the transmitting aerial wire, they are a feeble copy on a reduced scale. DeForest's tube performed the double function of detecting these waves and releasing a much larger amount of electric power from a source at the receiver, which amplified or magnified the signalmaking power. The device, patented in 1907, was an evacuated (vacuum) tube, in which were a sealed filament, a square of platinum for a plate, and between these a nickel grid. It acted on the principle of a thermal detector. This tube was effective in both wireless telegraphy and telephony, and came to be used as generator and detector, also. Without it, long distance radio would have been impossible.

DeForest made a successful demonstration of his invention on the yacht of Peter Cooper, which led the U.S. Navy to make a trial installation on two battleships. This resulted in the installation of radiotelephony on the twenty naval vessels that made the memorable cruise around the world in 1908–1909. DeForest attracted international attention when he broadcast from the Eiffel Tower, using phonograph records with such success that his broadcast was heard as far as Marseilles. In 1910 he persuaded Caruso to broadcast an opera over the air. In 1916 DeForest opened up broadcasting from the Columbia Phonograph Laboratories, and broadcast the Presidential election results of that year.

Armstrong's "Feed-back" System. The true miracle of the vacuum tube was not realized, however, until suitable circuits had been designed for its employment. Edwin Howard Armstrong, believed by many to be the greatest American inventor since Edison, while an undergraduate student in electrical engineering in Columbia University, developed the famous "feed-back" or regenerative circuit, patented in 1914. This device gave undreamed of amplification to weak signals, and went far toward eliminating the squealing noises due to continuous oscillation.

Armstrong's Superheterodyne. During World War I, as an officer in the Signal Corps, Armstrong modified Fessenden's heterodyne system to produce the second great radio circuit, the superheterodyne system, which was to become standard equipment. It revolutionized the production of radio sets. The antennas were shortened and receivers achieved a sensitiveness and selectivity previously unknown. A fourth and an equally outstanding contribution of Armstrong's was developed in the 1930's, that of frequency modulation, known as FM. (See Chapter 39.)

Alexanderson's Magnetic Amplifier. Another pioneer in radio development, as noted earlier, was the Swedish-born Ernst F. W. Alexanderson, who came to this country to work with the electrical wizard, Charles Steinmetz, in the General Electric laboratories at Schenectady. There, in 1909, he improved on Fessenden's alternator with one of his own. More powerful than Fessenden's, it was capable of producing continuous waves. In 1913 Alexanderson invented a multistage radio-frequency magnetic amplifier, a "harnessing link" between antenna and microphone. His multiple-tuned antenna more than doubled the efficiency of the antenna at Schenectady. It was only 900 feet long as contrasted with the mile-long antenna in use. His installation at the Marconi station at New Brunswick increased the power there tenfold. Alexanderson later became the chief engineer for the Radio Corporation of America.

The growing interest in radio was spurred on not only by highly trained scientists, but also by the growing demands of amateurs. Schoolboys began to experiment with wireless, and when music and the spoken word could be picked up, along with the weather signals, and occasional calls for help from ships at sea, an army of enthusiastic listeners developed.

The Crystal Set. The advent of the crystal detector was in part responsible for this increased interest. General H. C. Dunwoody of the United States Army is credited with this invention. Carborundum crystals served as a valve in suppressing high-frequency waves coming from the antenna, thereby changing these impulses to those that could come within the audible sound range of the ear, to which headphones would respond.

Greenleaf Pickard. Greenleaf Pickard similarly discovered the rectifying properties of silicon. This proved to be a simple, inexpensive device that boys could fashion for a few cents, and it played a remarkable role in the forward march of radio. Manufacturers were not prepared for radio-reception and had no receiver sets ready for the trade; consequently homemade receivers proved a great boon. With them the amateur experimenter entered the field and boosted the industry at a period when it was important that the general public be educated to the possibilities of radio.

Radio Captures a New Generation. The years from 1910 to 1915 saw rapid development in the scientific approach of the general public towards radio. Boys who started out as "ham" operators had an opportunity to study radio engineering seriously as courses were introduced in colleges and universities. Dr. Michael I. Pupin, professor of electrical engineering at Columbia, was noted not only for inspiring his students, but also for his own contributions to radio science.

A Radio International Conference was held in London in 1912, designed to promote uniformity in practice of wireless operators. In 1914 the Institute of Radio Engineers was organized. The John Wanamaker Company, appreciating the advertising value of wireless, installed a wireless system between its New York and Philadelphia stores, with a brilliant young engineer, David Sarnoff (later to be president of Radio Corporation of America), in charge. By 1915 American wireless engineers, by connecting three hundred thermionic valves, had succeeded in transmitting speech across the Atlantic to Paris. The cost was several hundred dollars a word.

This newest instrument in communication technology was to be put to a practical test during World War I.

Notes for Chapter 33

1. An internal combustion engine is defined as an engine in which the flywheel draws forward a piston, which in turn draws a charge of gas and air into the cylinder. When the stroke is about half accomplished, the valves close, an explosion is caused by an electric spark, the piston returns and the waste gases are expelled. This was the

original invention by a Frenchman around 1860. There have been hundreds of improvements since.

2. I am indebted to John B. Rae of Massachusetts Institute of Technology for ideas on this particular subject. He kindly presented me with a copy of a scholarly paper entitled "The Engineer—Entrepreneur in the Automobile Industry," which he read at the annual meeting of the Mississippi Valley Historical Association, St. Louis, April 26, 1955.

3. Mark Sullivan, *Our Times,* II (1927), 562-563.

4. Charles M. Manly, *Langley Memoirs on Mechanical Flight* (1912), 251.

5. Fred C. Kelly, editor, *Miracle at Kitty Hawk: The Letters of Wilbur and Orville Wright* (1951), p. 16; (by permission of Farrar, Straus and Young, publishers).

6. For a discussion of the airplane's role in World War I, see Chapter 35.

For a further discussion of aviation technology, see Jerome C. Hunsaker, *Aeronautics at Mid-Century* (1952); Oliver G. Sutton, *The Science of Flight* (1949); Henry Ladd Smith, *The History of Commercial Aviation in the United States* (1942); Frank Wead, *Wings for Men* (1931); Alexander Magoun and Eric Hodgins, *A History of Aircraft* (1931); Marvin W. McFarland, editor, *The Papers of Wilbur and Orville Wright,* 2 vol., (1954).

For a further discussion of automobile technology, see David L. Cohn, *Combustion on Wheels, An Informal History of the Automobile Age* (1944); E. D. Kennedy, *The Automobile Industry—The Coming of Age of Capitalism's Favorite Child* (1941); R. M. Cleveland and S. T. Williamson. *The Road Is Yours, The Story of the Automobile and the Men Behind It* (1951); Arthur Pound, *The Turning Wheel* (1934); Eugene W. Lewis, *Motor Memories, A Saga of Whirling Wheels* (1947); Bellamy Partridge, *Fill 'er Up, The Story of Fifty Years of Motoring* (1952); Clinton Edgar Woods, *The Electric Automobile, Its Construction, Care and Operation* (1904); Floyd Clymer, *Those Wonderful Old Automobiles* (1953); Philip Van Doren Stern, *A Pictorial History of the Automobile* (1953); Allan Nevins and Frank E. Hall, *Ford, The Times, The Man, The Company* (1953); Jean Labatut and Wheaton J. Lane, editors, *Highways in Our International Life—A Symposium,* sponsored by the Bureau of Urban Research, Princeton University (1950); Ralph Cecil Epstein, *The Automobile Industry; Its Economic and Commercial Development* (1928); Hiram Percy Maxim, *Horseless Carriage Days* (1937).

Also: *The Scientific American* has brought out special issues devoted to the automobile; *Moody's Manual of Investments* (1947), (1948), (1949), (1950); *Greyhound: The Motor Bus and Modern Highway Transportation,* The Greyhound Corporation, Chicago (no date). Articles of special interest include one in *The Saturday Evening Post* (April 20, 1946); two in *Fortune Magazine* (August, 1934) and (September, 1944); and one in the *National Geographic* (July, 1950). See also an article by Frederic L. Paxson, "The Highway Movement, 1916–1935," *The American Historical Review,* LI (January, 1946), No. 2, 236-253; Thomas C. Murchison, *The Romance of the Development of the Automobile* (1933).

The following are only a few of the many excellent books suggested for a study of wireless communication, including early developments in radio. Lee DeForest, *Autobiography* (1950); Orrin E. Dunlap, *Almanac* (1951), and by the same author, *One Hundred Men of Science* (1944); W. Rupert MacLaurin, *Invention and Innovation* (1949); Donald McNicol, *Radio Conquest of Space* (1946); R. C. A., *Thirty Years of Pioneering* (1949); Gleason L. Archer, *History of Radio* (1938); Herbert L. Marx, *Radio and American Life* (1953).

Chapter 34

MOTION PICTURE TECHNOLOGY

The motion picture, universally valued for entertainment and for educational uses, has proved one of the world's most effective means of communication, utilizing oral as well as visual media in later years. In its role of dissemination of ideas, it has had no bounds. Nationality, language, education, or economic status have presented no insurmountable barriers. Born in the early years of this century, in 1910 it was still a novelty, but by 1920 it was the most popular form of commercialized amusement in America. At that time there were an estimated forty thousand motion picture theaters in the world, of which some seventeen thousand were in the United States. The weekly attendance in this country was reckoned at ten million. By 1950, the weekly attendance was estimated at seventy-eight million. Since 1915, the United States has been the world's greatest motion picture producing country.

"Persistence of Vision" Theory. Truly a product of science and technology, this modern technological Hermes, the motion picture, was created from the scientist's laboratory and the engineer's study. Emanating from the science of optics, it has embodied a philosophy developed from the days of Aristotle, based on the theory of "persistence of vision," i.e., that an image does not cease when the object is removed from sight. In modern times this idea was toyed with by Michael Faraday and Sir John Herschel. Joseph Antoine Plateau of Ghent in the 1830's constructed a twirling disc which he named the "Fantascope." W. G. Horner of England had a revolving drum lined with a sequence of pictures, viewed through a slot in the surface, which he called a "Zoetrope." A number of other scopes and graphs were laboratory novelties.

Projection. Equally important was the development of the art of projecting the image beyond its fixed position. A Jesuit priest, Athanasius Kircher, who produced a monumental work in 1645, *The*

Great Art of Light and Shadow, devised a magic lantern. This was improved by a number of later scientists, evolving into Sir Charles Wheatstone's "stereoscope" in 1838. The idea of projecting the moving image was developed by an Austrian military officer, Franz Uchatius, who in 1845 devised a machine to show drawings for military instruction, combining Kirschner's magic lantern and Plateau's twirling discs.

Camera. In the realm of applied science the development of a high-speed camera played a major part. Such a camera was put to the test in the famous sporting event staged by Leland Stanford in 1872, where a wager was laid on the gait of a running horse. With the itinerant photographer Edward Muybridge manning a battery of cameras for which John D. Isaacs, engineer, had devised a series of electric controls, not only was the wager settled, but the possibility of instantaneous photography of objects in motion was proved. Janssen's "photographic gun," 1875, eliminated the necessity of a battery of cameras by substituting a rotating, sensitized, disc for the photographic plates. It was controlled by a gear mechanism.[1]

Eastman's Film and Edison's Kinetograph. When Eastman brought his roller film, made of paper coated with an emulsion, for his newly invented Kodak to take the place of cumbersome plates, the genius of Edison was drawn into the field. He seized the idea as a means of improving his phonograph, i.e., to give it eyes as well as ears. Edison saw the possibility in Eastman's film and made true records of motion on the 50-foot strips of film, sixteen images to the foot. This was equivalent to what could have been less perfectly done in eight hundred poses on a still camera, or with a battery of eight hundred cameras. Edison called his motion picture camera the "Kinetograph," the first that could photograph a few seconds of action at a time.

Edison's young assistant, William Kennedy Laurie Dickson, was commissioned to construct a small photographic laboratory for the production of the first Edison films. In 1893, at a cost of a few hundred dollars, he built "Black Maria." It was merely a shed painted black inside and out, and rested on a revolving base which could be turned to follow the sun. In this manner he kept the actors brightly lighted against a black background. Fred Ott's "The Sneeze" was one of the first films. For this, the only prop necessary was a can of red pepper.

The Kinetoscope; Peepshow. To "put eyes in the phonograph," the kinetoscope, or peephole machine, was devised. It was a simple instrument, operated by a battery-impelled motor, by which the 50-foot film was run between an electric light and a rapidly revolving shutter which exposed the picture by flashes to the viewing lens into which the spectator peered. The film was wound in an endless loop over a series of small rollers known as the "spool bank." It ran continuously and was designed for a single observer.[2] On April 14, 1894, the Edison-Dickson kinetoscope was used to open a peepshow in New York City, and a device was added whereby the observer dropped a coin in a slot to start the picture in motion.[3]

Projected Pictures. The technique was left to others, however, for projecting pictures on a screen. The Lumière brothers, Louis and Aguste, in France, produced what they called the "Cinematograph" in 1895. In this country Woodville Latham, using Edison's kinetoscope, claimed he was the first to show a projected picture commercially when he presented the Jim Corbett and Pete Courtney fight.

Another of the early attempts was that of C. Francis Jenkins of Washington, D. C., said by many to have priority in developing the motion picture projector, for "reproducing animated pictures on the screen." Jenkins in 1894–1895 produced a camera with a revolving lens, which he called the "Phantoscope," and a projecting machine, the "Vitascope," with which in 1896 a motion picture was shown as part of a vaudeville show in New York City. Thus, it became customary to show pictures at the close of vaudeville shows. They became known as "chasers."

The Public Supports Commercialized Amusement

Film Exchanges. Soon after the turn of the century motion pictures were being shown in several cities throughout the country. The technological developments that enabled pictures to be projected upon large screens and viewed simultaneously by hundreds of patrons made it necessary to change the program frequently. Exhibitors accumulated a number of films which, after being shown to their local patrons, could be exchanged with exhibitors elsewhere. This practice led to the film exchange, a system that became a major part of the early motion picture industry.

The Nickelodeon, described as "the world's first all-moving-picture theater," was opened June 19, 1905, in Pittsburgh, Pennsylvania. The owner, J. P. Harris, fitted up an old storeroom, installed a white linen sheet, some opera chairs, a crude phonograph, and "added a lot of stucco, burlap and paint," and a myriad of incandescent lights. The show was continuous from nine o'clock in the morning until midnight, and the entire program consisted of one or two reels of film, running for a period of 15 to 20 minutes. Day and night attendance often reached five thousand. The uniform price of admission was a nickel, and there were no reserved seats. A double-feature bill, "Poor But Honest," and a serio-comic, entitled "The Baffled Burglar," featured the opening week's program.[4] Within three years there were an estimated eight thousand nickelodeons in the nation.

"Making a Movie." For the most part movies were made in the streets, since only a few studios had been built. Skylight lofts and tops of tall buildings were the filming centers. The camera man was director, photographer, laboratory expert, and sometimes the leading actor. The only artificial light employed was the mercury vapor lamp. The camera was bulky, its machinery often got out of order, and carrying it through the streets was a laborious task. A great improvement in the early 1900's was the flexible tripod head which made panoramic views possible. Lenses were slow and the chemical operation in printing and developing was equally primitive. The first pictures to reach the screen showed little more than simple movements.

Types of Film Productions. Great stimulus was brought to American film producers by the famous director George Melies, who produced the first "feature film." His filming of "Cinderella" was a landmark in motion picture production. Inspired by Melies, Edwin S. Porter produced "The Great Train Robbery," which proved to be a classic and opened up an avenue of subject matter that appealed to American audiences.

To offset the influence of such thriller-type pictures, producers turned to history for their subject matter, and sought the services of actors from the legitimate stage. Sarah Bernhardt in 1912, induced by the plea that her art should be preserved for posterity, appeared in the title role of "Queen Elizabeth." Other feature films which followed were "The Prisoner of Zenda," "Tess of the D'Urbervilles," and "The American Citizen." The feature pictures were presented with an extravagance and lavishness that could not be matched on the legitimate stage. "Quo Vadis," presented in 1913, lasted for 2½ hours.

David W. Griffith made film history when, in 1915, he produced the elaborate motion picture, "Birth of A Nation." Griffith introduced a number of new techniques in production. He included not only stars, but also mobs of people. Herds of livestock were filmed, stampeding from one location to another. Homes and barns were burned, farms were ravaged, and the country laid waste. To catch the action of all these happenings, his cameramen moved into the midst of the exciting occurrences and registered effective close-ups.

Griffith's Technique. Griffith established film technique as evolving around the unit of the shot. He differentiated the screen from the stage. He evolved the full shot, the medium shot, the close-ups, the spot, the fade, and the fireside glow. He determined that the subject matter of a film should be conceived in terms of the camera's eye and film cutting. The manipulation of the shot built the scene, said Griffith, and the continuity of the scene built the sequence. When the final film was presented, it was accompanied by an orchestra with special musical selections, aimed to further excite the audiences.

Hollywood, Movie Capital. During the early years of the motion picture industry, most of the films were produced in New York City, but California's sunshine and even climate lured the producers. Gradually they converged on Hollywood and developed there an entire community devoted to motion pictures—the American "movies" and Hollywood became synonymous.

Modern Motion Picture Technology

Effect of Movies in World War I. The influence of the motion picture as a propaganda agency was recognized early after the United States entered World War I. One of the first acts of George Creel, Chairman of the Committee on Public Information, was to organize a Division of Films which was used to tell the American people why this nation had entered the war, and to unite the people in an all-out effort for speedy preparedness. Pictures of patriotic rallies showing the boys leaving for training camps and the naval stations were shown in theaters all over the land. The war bond drives were accompanied by pictures of enemy atrocities (not always documented), aimed at arousing public support in financing the war.

Improved Technique. A number of technical improvements occurred during the war. Cameras were refined and elaborated under the stimulus of war needs. Other developments included more accurate field finders, more flexible tripods, and longer focal length lenses. Telephoto lens, called the "long tow," made it possible to photograph a distant scene so that it would appear within close range on the screen as an image seen through the telescope. A gyroscope camera was developed which was operated by compressed air.

After the war, army signal corps cameramen became movie directors. The flexibility of the camera turned the cameraman's attention to composition. Light was keyed to the mood of the scene, gauzing the lens was introduced to soften the features, and art directors became necessary members of production staffs. Film cutting became a specialized craft. Yet, the motion picture was not too far removed from the early nickelodeon days. In 1918 the entire equipment for a well-equipped projection booth could be obtained at a cost of a few hundred dollars.[5] But a new era in motion picture technology was about to arrive.

Sound Accompanies Motion

Sound came to the silent screen when the Warner Brothers produced "Don Juan" in New York City, August 6, 1926. Although other producers claimed they were making "talking pictures" before this date, the showing of "Don Juan" is generally regarded as marking the arrival of the "talkie" as a commercial success.

Early Talkies. The feature, "Don Juan," was itself a silent film, but New York's Philharmonic Orchestra was heard, although it was not in the pit. Will Hays, first czar of the movies, talked to the audience at the opening of the program from the screen; and a series of vitaphone shorts presented such stars as Anna Case, Mischa Elman, and Marion Talley. The next year when the voice of Al Jolson was heard from the screen in "The Jazz Singer" the future of the talking pictures was assured, and all Hollywood turned to the production of sound pictures.

Edison and the "Talkies." Edison had always been interested in the phonograph rather than moving pictures, and in 1906 announced his "camerophone" for synchronizing pictures with the phonograph. In 1908 his "cinema-phone" appeared for adjusting the film speed to that of the phonograph, and a few years later Edison's "kinetophone" was projecting talking pictures on the screen with the phonograph. The horn was placed behind the screen and a synchronization system of ropes and pulleys was strung from the projecting machine in the rear to the screen in the front. Edison produced a number of "talkies," with subjects ranging from "Brutus and Cassius" to "College Days." Edison's real contribution to talking pictures, however, was in the phonograph and the incandescent light.

Science Produced the "Talkie." Preceding the talking moving picture, years of research had been given to the study of sound waves, notably in the phonograph, the telephone, and wireless. Electroplating the selenium cell had been introduced, and the photoelectric cell had been developed. From these came the condenser, the microphone, the vacuum tube, the amplifier, electrical recording and electrical reproduction, radio broadcasting and transmitting pictures by wire. It was from these developments that the talking picture evolved, rather than from the early attempts to give sound effects to the projected silent picture. Among those who brought the sound picture to America were Alexander Graham Bell, Thomas A. Edison, Thomas Armat, Lee DeForest, Theodore Case, George Eastman, Sam Warner, and many others who worked in diversified fields.

Film Recording. The technique of recording sound on film required long years of research. By 1926 the General Electric laboratories had developed a sound-on-film device that was practical. A great improvement had occurred when DeForest introduced his phonofilm. By this method, the original sounds were converted into electrical currents which photographically varied a stream of light on a film. When the sound was reproduced, a photoelectric cell was used to convert the variations in the light beam, as it passed into the sound track of the moving film, back into electrical currents. These currents in turn were changed into sound at the loudspeaker. Vacuum-tube amplifiers were used in both processes to provide the needed energy, since the energy of the voice and the output of the electric cell were too minute.

Technical Features of the "Talkie." The "talkie" met with popular success, and the search continued for improved recording and reproduction techniques. The microphone was hung on a boom which carried it to the actor, and thus avoided the defects of the fixed microphone. Portable recording equipment restored outdoor shooting and gave the sound movie the necessary scope and flexibility. Soundproof camera coverings permitted as much mobility as the microphone boom. The mixer panel, which enabled technicians to blend dialogue and modulate volume, allowed the use of more than one microphone. The use of fine grain developer reduced surface noises of the film. All these were among the techniques developed in Hollywood laboratories.

Educational and Industrial Films. The talking motion picture was hailed by educational leaders. Films made especially for classroom teaching, and adapted to every grade, were accompanied by the spoken voice. A new type of pedagogy evolved. Dr. George F. Zook, president of the American Council of Education, declared that, "the sound motion picture is potentially the most revolutionary instrument introduced in education in this generation." [6]

American industry had early realized the possibilities of the film. As early as 1915, the H. J. Heinz Company began advertising its "57" products by means of the motion picture. Within a few years other industries were proclaiming their products from thousands of movie houses throughout the United States, and in many foreign countries. More recently, virtually every major industry and corporation has developed films of their own for use in training their employees, and for educational exhibits in schools, churches, clubs, and for other interested groups.

Color and Three-Dimension Photography

Introduction of Technicolor. The next major development in motion picture technology was the introduction of the colored motion picture, i.e., technicolor. During the 1890's, P. E. Ives of Philadelphia, devised a plan to use a still camera for producing images in colors approximating those of the original object. He was able to show that a transparency displaying all colors in nature could be produced on the same principles which made three-color printing possible. With a device which he called "photo-chromoscope," a camera was constructed to make three negatives for trichromatic photography by using three-color filters, one each for red, blue, and green. From these he prepared three positives, illuminating each by a red, green, or blue light. When visually combined, these showed the colors of the object. Ives' principle was made a commercial success in projecting lantern slides.

Kinemacolor. A two-color process, red and green, the Kinemacolor, was developed in England and was used to film the Coronation of George V in 1911. The picture was taken through two-color filters and exposed alternate spaces on the sensitized film to red and green rays, respectively. The camera took sixteen pictures a second through each filter, or thirty-two in all. Each pair of frames represented all the color value that could be derived from two primary colors, and resulted in a black and white negative. In projecting the image, a revolving filter corresponding to that in the camera was used, which produced alternating green and red pictures. This succession of pictures blending on the screen through "persistence of vision" gave a reproduction of the scene photographed in color. To the human eye it appeared the same as those in the original. Kinemacolor met with great popularity even though it was not entirely satisfactory. Its essential features had an important bearing on later experiments.

W. J. Peck. In 1921, W. J. Peck of New York produced a two-color method in which the color was combined in the film and could be exhibited through an ordinary projector. Peck photographed the pictures on two frames, developed them, and then printed the green negative film on one side of the positive film and the red negative on the other. When successful, the results approximated the requirements of ideal color.

Pioneers in Technicolor. Three outstanding contributors to the development of color photography were Dr. Leonard Troland, who developed a camera that photographed three pictures simultaneously; George Eastman, who developed the Kodachrome film with five coatings; and Herbert T. Kalmus of Massachusetts Institute of Technology. Kalmus, known as the "Doctor" and "Mr. Technicolor" in the film colony, profited by both the Troland and Eastman inventions, and devised a dye-transfer method of printing which brought success to colored picture photography. Walt Disney's "Three Little Pigs," an early cartoon presented by this new method was widely acclaimed. This picture was soon followed by "Snowwhite and the Seven Dwarfs," which became a classic. Then came "Becky Sharp," "The Viking," and "Gone With The Wind," all in full length photoplay in technicolor. A new era in motion picture technology had arrived.[7]

Three-dimensional Technique. The latest technical achievement in filmdom was the development of films with three dimensions, or the illusion of three dimensions. Three dimensions, sometimes termed "natural vision," or stereo-vision (seeing in the round as opposed to flat surface), goes back to the stereoptican technique of Wheatstone in the early nineteenth century. It employed the principles of binocular vision. As this idea was applied to the screen, two images were projected simultaneously from separate projectors upon a single screen. Instead of appearing side by side as in the stereoptican, one was projected on top of the other. To view this film, it was necessary for the viewer to wear polaroid glasses. To shoot a "3-D" film, a special camera was used with two lenses about 3 inches apart, (like a person's eyes) and two rolls of film (like the retina of the eye). A series of three-reelers called "audioscopics" was produced in the mid-1930's. "Man-in-the-Dark," featuring new monocolor in sepia, appeared in the early 1950's in 3-D film that produced realistic effects. Due to the expense in properly equipping motion picture houses, this latest technique came slowly.

Cinerama. Two other forms of projection were developed that were not three dimensional, but gave the illusion of depth. These were the Cinerama and the Cinemascope. Fred Warner developed what he termed the Cinerama, evolved from the great panorama shown at the New York World's Fair in 1939. The picture was projected on a large curved field. As originally developed it was filmed with a camera that had eleven lenses, and was cast on the screen by eleven precisely synchronized projectors. The Cinerama camera was improved and simplified by the United States Air Force. More recently the film was taken with three wide-angle lenses, still requiring three projectors. The large curved screen extended 160 degrees in width and 60 degrees in height, so that the viewer had the sensation of being in the midst of the action. The initial film produced was, "This is Cinerama."

Cinemascope. The Cinemascope, when introduced, also gave the illusion of depth, but not as sharply as 3-D. It was classified in the category of "squeezed movies," since it was based on the principle of the anamorphic lens, which compresses panoramic scenes onto a standardized movie film. When projected, the scenes spread over a great concave surface, ranging up to 50 feet wide to 20 feet high. The spreading out was accomplished by a special projector with a cylindrical wide-angle lens. The picture presented the illusion of depth through side scenery or peripheral vision, and further realism was created by "stereophonic" sound.

The idea for the Cinemascope came from an optical device invented in 1925 by Professor Chretian of the Paris Optical Institute, who devised a camera lens that compressed a picture almost 50 per cent horizontally; a compensating lens on the projector stretched it out again on the screen. The Cinemascope required only one lens to take the picture, and only one projector properly equipped, to show it. An early production of the Cinemascope was "The Robe," in 1953.

Growth of the Industry. The motion picture soon became the nation's most extensive medium of entertainment, and the most popular agency for disseminating ideas yet invented (for radio and television, see Chapter 37). People in Manchester, England, became more familiar with New York, Chicago, and Los Angeles than they were with London. Britishers and west Europeans were as familiar with America's Grand Canyon and her Rocky Mountains as they were with the Alps or the Apennines. To the slogan, "The sun never sets on the British Empire," was added the phrase, "or on an American film." In the 1920's 70 per cent of the films shown in England were American-made. The same was true in France.

Trade Follows the Film. During the 1920's the American movies were doing for this country what the then Prince of Wales (later, Edward VIII, today the Duke of Windsor) was doing for England on his many good-will tours—advertising American products throughout the world. The Prince of Wales in addressing the British National Film League, had openly declared that, "trade follows the film."

If the United States abolished its diplomatic and consular services, kept its ships in harbor, and its tourists at home, and retired from the world's markets, its citizens, its towns, its countryside, its roads, motor cars, counting houses and saloons would still be familiar in the uttermost corners of the world. . . . The film is to America what the flag was once to Britain. By its means Uncle Sam may hope some day, if he be not checked in time, to Americanize the world.[8]

Foreign Competitive Inroads. Lord Newton informed his colleagues in the House of Lords (1925) that British cinema producers were losing out to the American producers. He told how the Americans had seized upon the cinema as, "a heaven sent instrument" for advertising their country, their products, wares, ideas, "and even their dialect." "America," he continued "is using this invention to persuade the whole world that it (America) was the only country that counted." He observed that the peoples of the Near East, after seeing American films which showed American styles in clothes, caused the manufacturers in the Midlands and Yorkshire areas to change their plants so that they could produce clothing, boots, and shoes similar to American-made products.

The American film was also invading South American countries during the 1920's, and such far away lands as Java, Sumatra, and India. In Brazil, the showing of American-made racing cars brought an increase in orders from an average of four or five a month to four or five a day. The same was true in the matter of sewing machines after the latest models were shown on the screen in the movie houses of southeastern Asia, and the island countries of the Pacific. The films advertised American habits, particularly American industrial customs, better than any other agency.[9]

American bungalows began to be introduced in Brazil following the showing of a movie featuring their development in the United States. American business firms and corporations started producing special films showing the advantages of their special products, and the trade index of foreign exports revealed rapid increases in sales, ranging all the way from farm machinery and roadbuilding equipment to typewriters and adding machines.

The Public's "Open Sesame"

The growth of the motion picture industry during the first half-century of its development had been phenomenal. Beginning with the penny arcades and nickelodeons of fifty years ago, it developed into a multibillion-dollar industry. Since the close of World War II, statistics reveal that approximately seventy-eight million Americans attend movie theaters weekly, and pay an estimated billion and a half dollars a year in admissions.

While it is impossible to appraise the total effects of motion pictures upon American life, or determine whether it mirrors the current social forces, as an agency in comunication technology it has become a major factor in disseminating information and ideas. Before the development of the motion picture, no person, no product, had ever been known to so many people in such an intimate way. With the aid of such technical devices as high-speed cameras, projectors, and a silver screen, millions of Americans, young and old, were able to see people and places of which they had never heard.

While the world has recognized the far-reaching effect of the motion picture, and while the screen stars are more universally known than statesmen or benefactors, yet the greatness of the American film must be credited to the scientist, the technician, and the craftsman behind the scenes. Albert Warner paid tribute to these men, without whom the motion picture could not exist, when he was quoted as saying, "You men perfected the machine. We supplied the showmanship. The importance of the engineer in the field of motion picture development has never been fully appreciated. Few persons know how much we owe you men."

Notes for Chapter 34

1. For histories on the beginnings of the motion picture industry, see such works as Martin Quigley, *Magic Shadows* (1908); Terry Ramsaye, *A Million and One Nights: A History of the Motion Picture,* 2 vol. (1926); Jean Benoit-Levy (Theodore Jaeckel's Translation), *The Art of the Motion Picture* (1946); J. P. Kennedy, editor, *The Story of the Films, As Told by Leaders of the Industry to Students of the Graduate School of Business Administration,* Harvard University (1927); Frederick A. Talbot, *Moving Pictures* (1923); Margaret F. Thorp, *America at the Movies* (1939); William M. Seabury, *The Public and the Motion Picture Industry* (1926); Lewis Jacob, *Rise of the American Film* (1939); George Ralph Doyle, *Twenty-five Years of Films: Reminiscences and Reflections* (1936); Will H. Hays, *See and Hear, A brief history of motion pictures and the development of sound* (1929); K. M. MacIlvain, *Sound Picture Recording and Projection* (1931); Gilbert Vivian Seldes, *An Hour with the Movies and the Talkies* (1929).

2. See Terry Ramsey, *A Million And One Nights, A History Of Motion Pictures,* 2 vol. (1926).

3. Carl W. Ackerman, *George Eastman* (1930), gives an excellent account of Eastman's many developments in the field of photography. See also, Lester Cowan, *Recording Sound for Motion Pictures* (1931).

4. Eugene LeMoyne Connelly, "The First Motion Picture Theater," *The Western Pennsylvania Historical Magazine,* XXIII (March, 1940), No. 1, 2. Mr. Connelly, the author of this article, relates this incident: "To me fell the task of coining a new name. My suggestion was that the price of admission and the old Greek word for theater—odeon—be combined to form a name, using the slang term 'nickel' for five cents, and 'odeon' for theater. The combination 'Nickelodeon' was so euphonious that in some quarters it has stuck to this day." *Ibid.,* p. 3.

5. James Ross Cameron, "Introduction," *Motion Picture Projection* (1942).

6. Quoted in *School and Society* (July 19, 1941).

7. Benjamin Hampton, *A History of the Movies,* Covici–Friede (1938), pp. 407-408. Hampton gives credit to William Francis Jerome for being the first to develop color photography. Hampton summarizes the progress in color photography under three developments. The first technique involved separating the three primary colors—red, blue, and yellow—in camera work and then joining them together again in projection. The analysis part of the technique, consisting of segregating the three fundamental colors, was developed in a process known as the "Lumiere," "Autochrome," or the "Kinemacolor," and resembles the system used in making color halftones for printing; the second method was accomplished by supplying colors to films by stencils, or by other processes which may be compared with lithography; the third method called for the application to the celluloid strip of two coatings of film, one in front, the other in the back—the back coating carrying the color separations.

8. Quoted from London *Morning Post* by Edward G. Lowry in an article written for *Saturday Evening Post* (Nov. 7, 1925) entitled, "Trade Follows the Film."

9. Article by N. D. Golden, Assistant Chief of Motion Pictures, Bureau of Foreign and Domestic Commerce, in *Congressional Digest* (November, 1928), pp. 296-299. There is still another impression made by American motion pictures which Mr. Golden did not mention, namely, that of picturing America as a land of cowboys and Indians, of gangsters shooting policemen, and of pranksters slapping one another in the face with custard pies.

Chapter 35

TECHNOLOGY IN WORLD WAR I

What appeared to be impossible demands upon America's technology were made in World War I—a war that was fought on land, in the air, and on and in the sea. While not a "total" war to the extent that World War II was, yet it was to involve, as President Wilson pointed out in his first war message to Congress, April 2, 1917, the combined mobilization of "all the natural and material resources of the country." The European allied nations had been bearing the brunt of battle for almost three years, and their situation was desperate. Daily cablegrams were warning America that unless men and materials in unlimited quantities were forthcoming at once, the end was near.

State of the Union: World War I

Council of National Defense. Following America's entrance into the war in early April, 1917, plans were drafted for speedy mobilization of the nation's total resources. A Council of National Defense, authorized by Act of congress, August 29, 1916, marked a new venture. The act provided for the organization of the nation's entire social and industrial strength, aimed at prosecuting the war at greatest possible speed. The Council in turn created an Advisory Commission of specialists in the various fields of industry. Walter S. Gifford, an engineer, was selected as director of this commission, and each member was made chairman of a committee according to his specialty. Daniel Williard, railroad president, was national chairman. A Naval Consulting Board was formed with Thomas A. Edison as chairman. This Board surveyed the industries of the country for the needs of the Navy, listed all machinery and machine tools suitable for naval use, card-indexed thousands of firms that could manufacture naval material, and acted as a Board of Inventions for the country. Edison, himself, worked on the problem of the detection of torpedoes, and the quick-turning of cargo ships.

Demands for War Material. The arrival of General John A. Pershing with a small contingent of troops in France in early June, 1917, bolstered the ebbing morale of the Allies, but it was the urgent need for equipment and supplies of war material, both for our forces and the allies, that posed the greatest demand. The United States was called on to supply immediately great quantities of propellants (both smokeless and black powder) and high explosives, and beginning in July, 1918, to supply all the medium and heavy field artillery. The United States met these demands.

As Benedict Crowell, Assistant Secretary of War and Director of Munitions, later wrote, "the United States soon developed the mightiest industrial organization in the world." He described the Americans as a people traditionally versatile in applying the forces of machinery to the needs of mankind. "Possessing unusual engineering skill, inventive genius, and a capacity for organization, the people of the nation," declared the Assistant Secretary, "speedily mobilized both its man power and production power so as to insure ultimate victory." [1]

Munition and Ordnance Production

Powder. By the time the United States entered the war, the existing powder and high-explosive plants were already working to capacity on production for the Allies, but within a few months more than fifty new plants were built. At Nashville, Tennessee, Old Hickory, the largest powder factory in the world, extended over 5,000 acres. Here, raw crude cotton was fed into a plant that produced the necessary acids and solvents and then converted the ingredients into finished smokeless powder at the rate of 100,000 pounds a day. At Nitro, West Virginia, another plant almost equalled this output.

High Explosives—TNT. These achievements represented more than mere mass production. New methods and processes were developed. Ammonium picrate, known as explosive "D," which was used before the war, continued to be used as the bursting charge for shells for guns of 10-inch size or larger, while a new explosive, Amatol, made of ammonium nitrate and trinitrotoluene (TNT), a British development, was adopted for other uses. To offset the shortage of TNT, a nitro-starch explosive was devised for use in hand grenades, and a product called lyconite was authorized for loading drop bombs. Toluene was procured from by-product coke ovens, and new methods were developed from natural gas and from petroleum by a cracking process.

Supply of Nitrogen. The supply of nitrogen which was curtailed by the U-boat campaign was augmented by using the modified Haber process, employed at Sheffield, Alabama, and by a cyanamide process employed at Muscle Shoals in the oxidation of ammonia to make nitric acid. To expedite manufacture, a new powder-drying process was developed by installing a system of centrifugal force. Extruding machines were designed for loading shells.

A new streamlined shell, which could be aimed more accurately, was designed by Major F. R. Moulton, a former professor of astronomy at the University of Chicago. He applied the principle of the projectile force of comets in developing his design for 6-inch shells for coast guns. New bullets developed during the war—tracer bullets, incendiary bullets, and armor piercing bullets—were quickly adapted to our manufacturing program.

War Production Program. Once the program for war mobilization got under way it resulted in many technological advances in the manufacture of artillery weapons, small guns, mobile field guns, railway artillery, and motorized guns. Tanks and tractors marked an innovation in war technology, and trench warfare made new demands. Aircraft came into use for the first time. Quartermaster's supplies, signal corps materials, engineering equipment, all represented technological advances.

Guns, Recuperators. The French and the British put at our disposal the plans and specifications of their latest designs in artillery. With the American system of mass production with interchangeable parts, and with prompt tooling, the French recuperators (the mechanization for the counter recoil of the howitzers) were speedily produced. Special recuperator plants such as operated by the Dodge brothers, the Otis Elevator Company, and the Singer Manufacturing Company, equipped with machine tools and dies and gauges were assembled, and quantity production achieved. The first recuperator was finished in September, 1918, and within sixty days almost five hundred had been produced. The production of this French recuperator in quantities was described as a technological triumph in the development of American ordnance, and ranks as the top American achievement.[2]

Mobile Field Artillery. The manufacture of gun carriages presented many engineering and production problems. Caterpillar tractors or trucks were used for drawing guns up to 240 mm, while larger guns were mounted on railway cars. High-grade steel was used in carriage castings, and wheels of various types and sizes were made of hickory or oak, fitted with rubber tires—an American innovation. Carrying out the idea of mounting heavy artillery on railway cars, an innovation resulting from the American Civil War, the United States shipped heavy coastal defense guns abroad, where they were immediately mounted on railway cars.

Motorized Transport. Motorized transport was a new technological development. In former wars the mobilization of armed forces had been limited by the pulling power of the horse. The longing of generals for an animal "with the speed of the deer and the power of the elephant" was satisfied in the development of the motor truck, the caterpillar tractor, and that monster of the battlefield—the tank. The truck and the tractor performed much the same function as in the industrial world.

Tanks. The tank was an armored caterpillar motor car mounted with machine guns. It was evolved as an engine of destruction by the French and the British, and spread havoc among enemy troops. In tank construction, American manufacturers supplied the motors and machine guns, the British constructed the hulls, and the tanks were assembled in France.

Machine Guns. World War I saw the first extensive use of the machine gun, which had been invented by Richard Gatling during the American Civil War, and later improved by Hiram S. Maxim and John M. Browning. Both the Allies and the Germans had realized its value. But the excessive weight and the intense heat produced by rapid firing had retarded its popularity. To overcome these imperfections, Colonel I. N. Lewis invented a light air-cooled type that was hailed as one of the greatest inventions of the war, and the English Vickers gun proved an improvement on the Maxim gun. Although the United States had to build completely new equipment for the production of machine guns, such progress was made that in a little more than a year our output was greater than that of any other country.

Trench Warfare. Trench warfare called for new weapons. Borrowing from ancient times, hand grenades were used, and more important, the fragmentation grenade. These were built of stout metal that would scatter into pieces when the interior charge was exploded. This was used only from the protection of the trenches. America early began mass production of these, and in addition devised an offensive grenade that troops could use in the clear as they advanced to rout the enemy from their positions. This weapon, made of laminated paper spirally wound and waterproofed with paraffin, produced a deadly effect of flame and concussion over a small area. Other types of grenades were those filled with phosphorus, an incendiary type for starting fires, and the gas varieties, both poison and tear gas. By the end of the war the United States was manufacturing grenades in greater quantities than our Allies combined.

Trench Mortars. American scientists employed electric arc and acetylene welding in the manufacture of gas drums used in the Livens projector, a British instrument for hurling bombs from the trenches. Trench mortars, a new weapon to the United States, quickly reached quantity production after the difficulties in adapting foreign designs had been overcome and plants were equipped for their production. In making shells for these mortars the United States manufacturers introduced a system for electrically welding pressed-out plates. Shells for the 6-inch mortars were made by American stove manufacturers, using cast iron instead of steel for the body.

Pistols, Revolvers. Pistols or revolvers, looked upon by other nations as primarily a military ornament and carried only by officers, were everyday weapons for Americans. The automatic Colt 45, described as "the straightest shooting pistol ever produced," was already in quantity production in the United States, and by December, 1918, more than three thousand revolvers were being manufactured daily. Improvements in the automatic revolver enabled experienced marksmen to fire twenty-two shots in 12 seconds.

Rifles. The rifle production of the United States was regarded as another major accomplishment of the war. Although we sent to France upward of two million soldiers, each infantryman as he stepped ashore carried his own gun of American manufacture. Often times it was the peculiarly American Springfield rifle, but more generally it was a modified Enfield, since the American factories had been geared to quantity production of the Enfield for the British before the United States entered the war. This rifle, with slight modifications, was widely adopted by American troops.

Helmets and Armor. The United States' fighting forces in World War I were better fed and better equipped than were those in any previous war, and this despite the fact that never before had supplies been transported so great a distance to so many men. Scientific and technological research played as important a role in the personal equipment of the fighting man as it did in the instruments of warfare. This war saw the revival of the old custom of wearing personal armor. Scientists went scurrying to museums throughout the country to study the designs of armor used in the Middle Ages, with the view of adapting it to modern lightweight, alloy steel. Though light front-body plates, weighing but 9½ pounds and lined with sponge rubber, and breast plates with thigh guards of heavier construction were made, these went little farther than the test stage. The American soldier never liked them and they hampered his mobility. The helmet, however, was adopted and became standard equipment. The British helmet design was altered by American manufacturers, who designed their own original type of lining, made of woven cotton twine, so constructed as to fit the head tightly and distribute its 2-pound weight evenly. Mass production of helmets was reached long before the end of the war.

Feeding and Clothing the Armies

Food Rations. That the average fighting man at the close of the war weighed from 10 to 12 pounds more than when he entered the service proved the adequacy of food supplies, and gave further proof of the advances in food technology. Most of the food consumed by the American armed forces, with the exception of some fresh vegetables purchased in France, was produced in this country and shipped abroad.

Dehydration of Foods. Shortage of ships and limitations in space led to new developments in food technology, particularly in dehydrated vegetables, boneless beef, dried eggs, and instant coffee. "The road to Berlin was paved with tin cans," observed a war correspondent. Dehydrated vegetables, prepared by a process whereby the moisture was removed with no loss in nutritive values or flavor, were shipped in hermetically sealed cans. Beef was shipped in refrigerator cars from the packing plants in Chicago to the company kitchens at the front, and kept in cold-storage warehouses. For economy in space the beef was boned, and sinews and inedible portions removed. The instant-coffee industry, which was in its infancy in this country at the outbreak of the war, made rapid advances. An improvement was made in the handling of green coffee by shipping it unroasted, and then setting up coffee roasting machinery abroad to ensure its freshness.

Emergency Rations. Special army rations—the emergency ration which every soldier carried with him, and the reserve ration for use in the trenches—were scientifically prepared. These provided a maximum of nourishment, compressed in a minimum space. Packing plants were specially equipped for packaging these rations on mass production lines. The food items were carried along on a conveyor belt, packed into containers, and hermetically sealed, all as an integral operation. The idea introduced in one of these plants of substituting tobacco and cigarettes instead of excelsior in packing, led to the adoption of the custom of including them in the ration. The age-old chewing tobacco, once used by almost every man and boy who wore the country's uniform, gave way to candy and chewing gum which came to be regarded as virtual necessities.

Clothing. Second only to the subsistence division was the clothing and equipage division in enlisting the services of scientists, designers, and technologists. Working together they produced serviceable equipment in the shortest time possible with the greatest saving in scarce materials. Shortages in wool, leather, and metals were reckoned with. The demands made by the mud of the trenches, and the contrasts in climate from the Russian winters to the scorching Mediterranean summers had to be met. The Pershing overcoat was shortened using less wool. Trousers replaced the old breeches. The Pershing shoe, and later the Victory shoe, was made with three heavy soles screwed together and reinforced with steel on heel and toe. Muskrat, marmot, and wolf fur were secured in quantities for garments for the soldiers in Russia, and for airplane pilots.

Substitutes. Substitutions were made where shortages existed. Vegetable ivory from the tagua nut was used for buttons instead of brass. Spirally wound leggings of canvas replaced the leather puttees. Cotton thread was used instead of linen. Synthetic dyes were introduced. A waterproofing grease for shoes and other materials that would not turn rancid was devised. New technological processes for waterproofing fabrics to conserve rubber products were developed. The Quartermaster's Department became a mecca for inventors for real or fancied improvements in every line.

Army Engineering Techniques

In this mechanized war, the role of the engineering corps was greatly expanded. The combatant engineer took his place with the infantry and comprised approximately 8 per cent of the combatant troops. He advanced with the tanks, prepared the way for the infantry, cut barbed wire entanglements, improvised temporary foot bridges, destroyed mines and traps, constructed miles of corduroy roads, filled crater holes, and often times fought alongside the infantryman.

The Noncombat Engineer. The technical noncombatant engineer was assigned to supplies and administration, and was often placed in charge of production and construction. He set up electrical and repair shops near the front, constructed barracks, cantonments, and hospitals. He built and operated the light narrow-gauge railroads for hauling troops and supplies from the main depots to the front lines. He built roads and bridges—pontoon bridges of axle strength to support the artillery, and sectional bridges that could be thrown up in a matter of hours. The engineer that remained at home often supervised construction of roads and means of transportation. An innovation in transportation technology was introduced when engines were shipped "on their wheels" across the Atlantic, ready to haul freight cars within an hour after landing. From base port to the front line of assault, the engineers were constantly in action, with their motto of "Essayons," constantly before them.[3]

Naval Warfare Technology

Action on the sea in World War I was primarily warfare against submarines. The German submarines sought to win the war early by sinking all Allied merchant shipping, and thus force a hasty surrender. Nothing like this new form of warfare had ever before been known, and the United States was not prepared for this kind of naval fighting, because, said Admiral Sims, "we were a more civilized people." The planned battleship construction was temporarily suspended, and orders were given for the construction of destroyers. In meeting this program, American shipbuilding technology astonished the world. At the Mare Island Shipyard, the destroyer "Ward" was launched in seventeen days after the laying of her keel. In prewar days it had required from twenty months to two years to build such a destroyer. The new shipyards at Squantum, Massachusetts, beat this record with the first ship, the "Reid," which was finished and made ready for her trials in forty-five days. The plant at Squantum was only one of several where destroyers were built. Here, keels for five ships could be laid in one day. The yards covered 18 acres under one roof, where boats were built by assembly-line construction. Raw materials were assembled at the places where they were needed, and a finished destroyer came out at the end of the line. These latest American destroyers were of light draft, easy to manipulate, and speedy. They carried 4- and 5-inch guns and twelve torpedo tubes, each of which launched torpedoes weighing more than a ton.

American destroyers differed from the British in that they were long and slender, sleek, and as the Navy put it, "feminine rather than masculine in appearance." They were built for wide range, and could keep to the sea longer than could the British.

United States Submarines. The American submarine, the L-boat, was feared by the Germans more than any other naval weapon. The early submarines, David Bushnell's "Turtle," 1777, and Fulton's "Nautilus," 1803, had the fundamental principles of buoyancy, stability, and control of depth which were found in the modern submarines designed by John P. Holland. Modern boats differed in that they attacked enemy ships by firing torpedoes as well as by ramming. They were powered by two engines, oil burners for surface cruising, and electric motors for submerged travel. They could remain submerged for longer periods than the German boats, because there was no need to economize on storage batteries. German boats were forced to the surface after 50 to 60 miles to recharge the batteries by means of oil-burning engines.

Subchasers. The most satisfactory answer to the enemy submarine was the submarine chaser, a staunch little wooden craft, no larger than a motor launch, called the "Cinderella of the Fleet." These were equipped with listening devices, and could search, pursue, and attack enemy vessels. Described as the eyes and ears of the anti-submarine forces, they were rightly feared by the U-boats. They crossed the Atlantic under their own power, although they had to be refueled en route. They were equipped with wireless, guns, and a depth charge.

Listening Devices. The listening devices on these submarine chasers were radio tubes developed from an American invention, the submarine bell and the "vibrator." The "C" tube, useful at short distances, and developed by Dr. William D. Coolidge, resembled a telephone receiver. The "K" tube could detect the sounds of larger vessels at a distance of 20 miles. It was an adaptation of the rotary compensator devised by Professor Max Mason of the University of Chicago.

Eagle Boats. The Eagle boats, built by Henry Ford on the principle of mass production, were employed as submarine chasers. They were armed with 4-inch, 50 caliber guns, discharge projectors, and antiaircraft and machine guns. Though only a few dozen saw action before the armistice was signed, the quantity production principle was effective. In late 1919 Ford completed twenty-three boats in one month. Trawlers or mine sweepers were used to clear the channel of German mines. American yachts were stripped of their luxurious fittings, armed and converted into men of war for special patrol duty.

Converting Enemy Ships. American technology made possible a special supply of ships by speedily repairing the dismantled German liners that had scurried into American harbors before the United States entered the war. Among these was the "Vaterland," the largest liner afloat, which was named the "Leviathan." More than a half-million American troops were taken across the Atlantic in converted German ships. The efforts of the Germans to put their vessels completely out of commission were thwarted because of new devices in ship repairing. With the aid of electric welding and oxyacetylene welding, boilers and pipes were repaired in position, obviating the installing of new equipment. Wrecked vessels were speedily made seaworthy.

Naval Convoys. The Navy's carefully executed convoy system made it possible to transport troops and supplies to the European battlefronts in overwhelming numbers. The system was likened to America's method of handling freight cars on transcontinental railways. A complicated but perfectly organized system that reached the remotest corners of the earth was effected. All shipping followed mapped routes carefully, meeting at specified "gateways" for dispatch on schedules as fixed as those of a great railroad. A convoy included all types of craft from the slow freighters to the fastest on the seas. Its speed was geared to the slowest vessel. As it proceeded it seemed to sprawl over the sea in ungainly fashion, twenty to thirty ships, spaced 900 to 1,000 yards apart, taking up 10 to 12 square miles of ocean surface. As the ships entered the danger zones, infested by submarines, they were met by destroyer escorts, (which proved to be the protecting arm of the convoy), submarine chasers, and submarines, which conducted them by signal code, zigzagging, moving during the night's black-out in close formation to within range of the harbor guns.

London Admiralty Operations. Numerous technical devices were developed to make the convoy system successful. The complete operation was regulated from the Admiralty office in London with every ship movement chartered on a huge map of the world where paper boats were moved on the charted courses as the ships proceeded on the high seas. Radio communication made this possible, and radio information received there of submarine danger was dispatched to ships at sea directing the convoy's route to safe channels. All ships were equipped with wireless, with special clocks set at exact time signals. Ship captains, even of the smallest freighters, were instructed in special seamanship, sailing information, zigzagging, and station-keeping.

Special Purpose Arms. Specifically adapted naval weapons included the depth charge that carried a charge of explosives which detonated on reaching a given depth. It was dubbed by the sailors as "the ash can." American improvements included a Y-gun which hurled the huge charges with great accuracy and at less risk. A depth-charge barrage could be fired in a pattern around the vicinity of a submerged submarine. Other inventions included a nonricochet shell, a flat-nosed projectile, which was effective in firing at a submerged boat. A twin gun was produced for destroyers, two barrels on a single mount, which could be fired alternately.

Mines and Mine Laying. Mine laying had been practiced in warfare for many years, but mines had never been laid in ocean depths such as that leading from the North Sea. American naval technology introduced a new type of mine. Instead of the huge steel globe filled with explosives and set off by contact, the "Mark VI," the invention of Ralph C. Browne, was a spherical steel case fitted with a copper wire, which served as an antenna supported by a metal buoy. Any metallic substance that came in contact with the antenna set up an electric current that caused an explosion. This made it impossible for enemy U-boats to avoid the mines by passing under or over them, and made necessary only one-fourth the number of mines required in the old type. A system of mine laying was evolved whereby four planters were able to plant approximately one thousand mines in 30 minutes. "Nails in the Kaiser's coffin" as the sailors termed them, were laid as a barrage in and around the North Sea, covering a distance of several hundred miles.

U.S. Naval Contributions. Although England continued to be recognized as the world's first sea power, the contributions of the American Navy toward winning the war should be remembered. The United States naval forces at Queenstown escorted through the danger zone almost one-half of all cargoes that left American ports. In the movement of troops, American destroyers did the largest share of the work, although there were fewer than one hundred destroyers in the service as compared with Britain's four hundred. The sinking of more than two hundred German submarines was attributed to American forces, and the auxiliary functions performed by submarine chasers, mine sweepers, mine layers, aircraft, patrol boats, and even of our capital ships as they merged with the Grand

Fleet at Scapa Flow, all played a part in the great Allied scheme that brought the enemy to terms of surrender.[4]

Air Warfare Becomes Reality

World War I brought aviation to sudden maturity, condensing into two years the normal progress of fifteen years. Although mechanical flight was born in the United States, yet when we entered the war, no American-built plane had ever carried a machine gun or anything other than the simplest of necessary implements. The task of directing airplane development was placed in the hands of the Signal Corps, for the function of the airplane was regarded as principally reconnaissance. The plane was looked upon as the eyes of the army and navy, and was expected to operate in an auxiliary capacity. In the years prior to our entry into the war, some one thousand planes had been built in this country, largely as training planes for European countries.

Early Construction. For our own force, the first construction was the training plane, but it was not until the builders turned to the production of service planes that they came to realize the magnitude of the engineering and manufacturing problems involved. When the armistice ended the fighting, more than eleven thousand planes had been produced, of which eight thousand were training planes. The industry was equipped and geared to production at a rate of twenty thousand planes a year. Forty-eight flying fields had been constructed; more than twenty thousand air service officers, and one hundred and seventy five thousand enlisted men were enrolled in the service.

In building the pioneer planes, lumbermen were called on to supply light, seasoned, yet strong wood. Strength was the main objective, and the procurement and treatment of spruce wood for airplanes became a major problem. Only a small portion of the lumber logged was satisfactory. A method of splicing short lengths to make wing beams was devised. Had the war continued another few months, laminated beams would have become universal. Log sizes were standardized, and the Forest Products Laboratory in Madison, Wisconsin, worked out scientific methods for seasoning.

Covering for Wings. Another problem was that of providing sufficient linen for airplane wings. With each plane requiring from 250 to 500 yards of cloth, the supply of Irish linen was soon exhausted. The Bureau of Standards came up with a new "dope" to treat cotton cloth. It was made from acetate of lime and had the advantage over the nitrate compounds formerly used, in that it was noninflammable. Cotton with this treatment proved superior to the linen cloth. Sea Island cotton from Egypt was stockpiled to be ready for any emergency. To assure sufficient quantity of castor oil, the lubricant then used for airplane motors, the government secured enough beans from Asia to plant 100,000 acres.

The Liberty Engine. The Liberty engine was America's distinctive contribution to war aviation. The American automobile industry had developed an engine-building capacity that formed the basis for this program. The aim was to produce an engine that would meet the army's specifications, and also be capable of quantity production, that would give a maximum of power with a minimum of weight, and one that would consume oil and gas economically. When the federal government agreed to pay royalties for use of patents, the aviation engineers reached out and took what they deemed best regardless of patents, with the result that the Liberty engine was a composite type, embracing the best features of all the engines ever built.

An Assembled Engine. The first eight-cylinder Liberty engine was delivered in Washington, July 4, 1917, less than six weeks after the first plans were drawn. Its parts had been made by factories all over the United States and hurried to the Packard Motor Company, the assembling center. To produce every part of the Liberty engine, some three thousand small jigs, tools, and fixtures had been employed. The horsepower, at first 300, was stepped up to 400, and a twelve-cylinder rather than the eight-cylinder model was constructed. By June, 1918, quantity production had been reached when more than one thousand engines a month were being produced, and by November they were being turned out at the rate of one hundred fifty a day. By the time the armistice was signed more than fifteen thousand Liberty engines had been built, of which some five thousand were used in planes built in the United States. Another five thousand were installed in planes built overseas for our forces; some four thousand were used in the navy planes, and a thousand were sold to our Allies. Both France and England had placed large orders for our "Liberties" by the close of the war.

Airplane Weapons. The installation of the Lewis machine gun made the airplane an efficient fighter, as well as an observer. The gun and airplane propeller operated on synchronized gears to facilitate firing without hitting the propeller blades. Tracer bullets, incendiary bullets, and armor-piercing bullets were devised. Bombing came late in the war. Bombsights were developed, adjusted to height, speed, and strength of wind. These had a quick release mechanism operated by a lever.

Photography. Photography was developed to enable the reconnaissance plane to secure pictures of the entire enemy terrain, an innovation in military photography. As antiaircraft guns forced the planes to go higher, special equipment for cameras was devised for long distance. Longer focus lenses, special dry plates, and special color filters were invented. Pictures were often taken at an altitude of 3 and 4 miles. At first these cameras were equipped only with plates, but the Eastman Company devised a film that would overcome the static electrical charges which affected films in cold dry regions. Cameras were equipped with films permitting from 80 to 400 feet of exposure.

Aviator's Clothing. Flights at 20,000 feet made it necessary to equip aviators with warm clothing, and oxygen masks were a requirement for flights at high altitude. A one-piece, waterproof fur-lined garment was devised, with an interlining of flexible electrical heat units connected with the engine. The fur of the Chinese dog was found to be light and warm, and was used for this lining. Gloves and moccasins were electrically heated.

Airplane Radio. When in May, 1917, American scientists were called upon to produce a radio for airplanes they were asked to perform a task which the allied nations had been unable to accomplish during the first three years of the war. Although radio communication was being used in ground fighting, it had not been adapted to the air, where the need was infinitely greater. Radio telephony had been well established in this country, but its installation, to be effective in the din of battle and in rapid flight, presented a much greater problem (see Chapter 41). The vacuum-tube device that was already in use in both transmitter and receiver was adopted. A headset was developed to exclude all undesirable noises, and a filter was invented for the transmitter that was insensitive to engine and wind noises, yet was responsive to the human voice. The equipment was light in weight, but strong enough to protect the delicate mechanism from the vibrations of the engines. A dynamo was invented to be operated by a wind propeller, driven by wind rushing at the speed of the plane, yet regulated to provide a constant current. This equipment was ready for installation in the planes as they were being produced in late 1918.[5]

Balloons. Balloon production ranked as one of the most successful of all of World War I projects. April, 1917, production of military observation balloons was at the rate of two or three a month, but by the close of the war some seven hundred had been manufactured. The average life of a balloon in service was fifteen days. Balloon cloth of rubberized cotton was made in a special weave. New rubber compounds were developed. Gas and electric windlasses for propulsion were evolved. Improved wire cables were constructed, and portable hydrogen gas plants were built. To ensure a supply of gas "nurse balloons," storage balloons from which the active balloon could be fed, were designed.

Major Achievements. While in many respects America's aviation strength was still potential rather than actual, the achievements during the eighteen months of the war compared favorably with that of our allies. When hostilities ceased the United States had almost one thousand airplanes at the front, compared with France's three thousand, Great Britain's two thousand one hundred, and Italy's six hundred. Brigadier General William (Billy) Mitchell described the new American models as equal to any of the European planes and superior in maneuverability. It was one of these new American planes, the NC-4, which six months later made the first flight from America to Europe, via Nova Scotia, the Azores, Portugal, and on to Plymouth, England.

Chemical Warfare Technology

While World War I was often called a chemist's war, and World War II a physicist's war (see Chapter 41), both sciences, physics and chemistry, were called upon in both wars. Realizing the many contributions which the chemists could make, President Wilson created a special Chemical Warfare Service in the spring of 1918, and appointed Major General William L. Sibert as director. In calling upon the seventeen thousand members of the American Chemical Society for aid, General Sibert said, "This is a chemical war: therefore, the War Department must have immediately available all possible information regarding chemical materials and chemical man power."

Scope of Chemical Research. The contributions of the chemists affected almost every phase of the war. They worked on such projects as high explosives, sanitation, foods, soldiers' rations, optical glass, nitrates, and numerous metallurgical problems. They discovered methods for preventing meat spoilage, studied the catalytic action relating to yeast in breads, and the properties of various proteins. In an effort to discover the causes for such diseases as scurvy, pellagra, and beri-beri, the chemists made thorough studies of the effects of desiccated vegetables. Hundreds of chemists were assigned to the Quartermaster Corps to study such materials as canned food, the storage of food, containers, leather, and paper products.

Chemical Industry Developed. Prior to the war, foreign competition had prevented the organic chemical industry from becoming established in the United States, and this country was entirely dependent on Germany for her source of supplies. When the war cut off this supply, America was forced to develop her own industry, which proved to be a prodigious technological achievement. This was demonstrated in accomplishments during the war years, and later to a greater extent, in their application to peace time projects.

Dyestuffs. Germany's monopoly of the dyestuffs industry gave her an advantage in time of war because high explosives were prepared from the same crudes and by the same reactions, and with the same apparatus as were dyestuffs. Dyestuff plants were easily converted to produce smokeless powder, TNT, and picric acid. The necessity of manufacturing these war essentials forced us into experience with the coal-tar crudes, and from these the infant dye industry was born. In 1914, America produced less than 3 per cent of the national consumption of dyes. At the close of the war there were five dyestuff companies, which amalgamated with the companies making coke, alkali, acids, and various products from coal tar. The American Chemical and Dye Corporation came into being. Within this brief five-year span, the United States could lay claim to a greater chemical industry than that of any other country, having surpassed both her German and British competitors.

New Processes. A number of new chemical processes were developed. In 1917 Gibbs produced in the color laboratory at Washington the hydrocarbon anthracene, the parent of the madder dyestuffs. A series of pigments known as "vat dyestuffs" was developed and economic processes for their manufacture devised that were equal, and in many cases superior, to the German manufactured colors, and at greatly reduced costs.

Even the oldest chemical industries—the alkalies, sulphuric acid, nitric acid, muriatic acid, bleaching powder, soap—all underwent a great change when the usual raw materials were not available and were produced by other means, usually by the action of a catalyst. Glycerol was made from glucose by fermentation; nitric acid and ammonia from coal tar; furfural used in making resins from oat hulls; and lacquer from the waste product from corn. In medicine, the antipyrites (the fever reducers) and the antiseptics were derived by synthesis from coal tar.

Optical Glass. Another technological achievement was the development of optical glass. Here again, Germany had the monopoly, but it was necessary that this country have lenses for telescopes, binoculars, cameras, and for gun sights and periscopes. Not only was the formula evolved for the composition of optical glass, and production plants and equipment devised, but technological processes were improved over those of the Germans. Hand methods of stirring were displaced by machines. The time of furnace operation was shortened from 60 hours to 24. The annealing period was reduced from four weeks to three days. Rolling into sheets instead of sawing was introduced. The Geophysical Laboratory assumed the leadership. Scientists sought out the few optical engineers in the country. In a brief nineteen months, 95 per cent of the optical glass used in the war was produced in this country.

Gas Warfare. When in April, 1915, Germany introduced the use of toxic gas, contrary to her pledge signed at the Hague Conference, she started a game both sides could play. The first gas used at Ypres was chlorine, but there was not a sufficient amount used for Germany to blast her way to Paris. Later the so-called mustard gas was the principal one used by the Germans. When the United States entered the war, we were prepared with the manufacture of poison gases and with gas masks for protecting our soldiers from the effects of enemy gas. The American gas mask afforded complete protection against chemical warfare. No American soldier was ever gassed due to the failure of his gas mask. Such casualties as did occur were from the soldier's failure to adjust the mask quickly enough, or to his ignorance of the presence of the gases.

Tear Gas. American chemists quickly prepared new gases to combat the Germans. Chloropicrin was effective as a tear gas rather than as a poison. It was inexpensive to produce, and it compelled the enemy to wear gas masks continually with debilitating results. Phosgene was a deadly gas that did not linger. It could be used in assault, permitting our troops to advance immediately. Other new and powerful poisons, some thirty in all, were introduced which outdistanced Germany's production, both in quantity and in the effectiveness of their uses. Chemists also aided in developing the incendiary bombs, smoke funnels, smoke screens, smoke grenades, colored rockets, flame throwers, and gases for balloons. They worked on many raw materials, especially in the production of toluene by cracking petroleum.

Chemistry Unlimited. The chemists purified vast water supplies both for the training camps in the United States and for the front-line forces in Europe. They developed new substitutes for rare medicines, and the biochemists assisted the medical corps in developing new methods for helping wounds to heal and fractured bones to knit. Professor Charles L. Parsons, in writing on the contributions of the chemists, said late in 1918, "Without chemistry the continuation of the war would be impossible."

Role of the Physicist

Sound Devices. The physicists also played a major role in World War I. As camouflage developed, a need was felt for a means of detecting the presence of the enemy. A group of physicists and inventors met this need with the development of a long-distance eardrum known as the "geophone." With this instrument, the location of the enemy could be detected, subterranean mining operations could be determined, and the site of the enemy cannon and the position of his airplanes could be located. Other sound-ranging implements (a combination of microphones and recording machines), so delicate they had never before been used outside of the laboratory, were developed. By introducing special filtering devices, all useless sounds were eliminated, even amid the din and concussion of heavy bombardment, and only the sought-for vibrations were recorded. This was a scientific instrument which the Germans never succeeded in producing, nor were they able to conceal their guns from its range. Sound ranging for the detection of airplanes at night was effected by an instrument consisting of a sound-ranging device and a listening mechanism.

Flash Ranging. Gun positions were also determined by flash ranging, whereby observation telescopes mounted on tripods watched for the flashes of enemy guns. A group of American physicists improved the searchlight equipment by the invention of metal mirrors to take the place of glass, and by improving the lamp by reducing the size of the carbon. Trench periscopes were popular with the fighting men, for they enabled them to look above and beyond the top of the trench without exposing themselves to enemy fire.

Gunsights. Other scientific devices took the guesswork out of gunfire. These were precision instruments that enabled the gunner to direct his cannon accurately at an unseen enemy many miles away. They included the aiming circle, rangefinders, telescope, elevation quadrants, deflection boards, wind indicators, and of utmost importance, panoramic sighting instruments.

Signal Corps. The time-honored communication devices of using runners, flares, wigwagging, and carrier pigeons were completely outmoded in World War I by the combination of the telephone, radio, and wireless. Areas stretching a hundred or more miles could be operated as a unit, with every part in immediate touch with every other part through a system of scientific signaling. A number of new devices were introduced to adapt these instruments to war purposes. Among these was the outpost wire, a two-way wire that could be flung to the front lines and did not need to be grounded. Grounding made eavesdropping possible, while the two-way wire was operated in code and could not be detected by the enemy. These wires were of sturdy construction; they lay on the ground and were subject to troop movements across them. Another adaptation was the buzzer phone, a telephone-telegraph system with a buzzer attachment by which the messages were transmitted in code. The mobile switchboard made connections possible in the outposts and in the trenches.

Radio in World War I. Radio connection was most effective with ships at sea and with airplanes. The wide application and improvement of the vacuum tube, in both receiving and sending apparatus, was an innovation of the war that greatly enhanced the development of radio (see Chapter 36). For all these new forms of communication, electric batteries were employed, bringing about many improvements in the power and durability of the storage battery. The developments made in the Signal Corps communications system were carried over to peace time uses after the armistice was signed.

Summary of American Contribution

In summarizing the role of American technology in World War I, the following developments are a few of many that should be remembered: the size and efficiency of the American Expeditionary Forces; the rapid production of equipment, food, munitions and material; the speed with which troops and supplies were dispatched to Europe, a distance of 3,000 miles, and the increasing amount of supplies that continued to flow until the end of the war. Great as all these developments were, however, they were only "pilots of the quantities to come." As Benedict Crowell, Director of Munitions said, "Those years [1917–1918] will forever stand as a monument to American genius of workshop and factory, which insured victory by insuring the timely arrival of the overwhelming force of America's resources in the form of America's munitions."

Notes for Chapter 35

1. See Benedict Crowell, *America's Munitions; 1917–1918* (1919), Chapter 1.
2. A brief definition of a recuperator was given by describing it as a device that absorbed the gun recoil, and then properly restored it to position.
3. Crowell, *op. cit.*, p. 366.
4. See William Snowden Sims, *Victory at Sea* (1920); and Josephus Daniels, *Our Navy at War* (1922).
5. For a detailed discussion of this novel device, see Crowell, *op. cit.*, p. 329.

Of the hundreds of books written on World War I, the following are only a few that are helpful in making a study of America's scientific, industrial, and technological developments: *The Third Annual Report of the National Research Council;* Frederic L. Paxson, *Economic Mobilization for the War of 1917* (1929); also by the same author, *America at War, 1917–1918* (1939); also *Postwar Years Normalcy 1918–1923* (1948); Robert M. Yerkes, *The New World of Science; Its Development during the War* (1920); Benedict Crowell and R. F. Wilson, *How America Went to War,* 6 vol. (1921); Bernard M. Baruch, *American Industry in War: A Report of the War Industries Board* (1921); Arthur Bullard, *Mobilizing America* (1917); Frederick Palmer, *Newton D. Baker—America at War,* 2 vol. (1931); On the United States Navy's part, see W. S. Sims and Burton J. Hendrick, *The Victory at Sea* (1920); Elting E. Morison, *Admiral Sims and the Modern American Navy* (1942).

Chapter 36

COMMUNICATION TECHNOLOGY SINCE 1920

The improvements that have been made in communication media have been among the outstanding examples of the application of science and technology to the everyday needs of life. A comparison of the isolation of the communities scattered along the Atlantic seaboard in Colonial times—dependent on the arrival of a ship from overseas, or a letter by an infrequent post, or the sparse news in the biweekly newspaper—with conditions today reveals the extent of the improvements in communication.

Significance of Modern Communications

In the early days of the United States, Napoleon's defeat went unhailed for months because the news had not reached our shores, and Andrew Jackson unknowingly fought the Battle of New Orleans after the peace treaty had been signed; but today, in the remotest rural areas as well as in the metropolitan districts, news is instantly available. This development in communications has been a mighty unifying force in our history and has tended to standardize our civilization. Now that a revolution has taken place in communication technology, taking in the entire world, the question arises, "Will it have a similar unifying global effect?"

Effect on General Public. To the post office and the newspaper, which the people of the United States have enjoyed from the earliest period, have been added the telegraph, the telephone, wireless, radio, television, and radar. It has been only in the last thirty years that wireless, radio, and television have come into general use. Today, practically every man, woman, and child in the country has felt their influence.

Postal Efficiency. Least significant of the technological advances in the last thirty years have been in the postal innovations. However, since the early 1920's when a few daring fliers traveled the uncharted sky road across the continent, carrying the United States mail, air mail has grown into an accepted and commonplace service universally enjoyed. The technological improvements in the handling of mail in the great centers—with the intricacy of devices for distributing, stamp canceling, and in the performance of other functions—have been made with each advance as it developed. The giant United States Post Office system has maintained to this day its record of efficiency, a record instituted by Benjamin Franklin, first Postmaster General.

Printing

Printing Presses. Printing during the last three decades has adapted the new developments found in other fields. The huge presses, made of superior metal and adapted to improved electric controls, perform automatically with a battery of units. An example is the press installed in *The New York Times,* capable of printing two and a half million papers of thirty pages each, in one night, with one press rolling out thirty-five thousand copies an hour. This is the equivalent to the work of two hundred thousand men working with the equipment of Gutenberg's pioneer press.

Advances have been made in color printing with the carefully mechanized color presses, which can turn out the specialized work for which they are designed at a rate almost equal to that of the regular presses. Four-color rotogravure was made possible after World War I, and the new dyes developed since 1920 have made possible a vast range in color design. For finer color work than newspaper illustration, presses have been designed that make reproductions of three-dimensional qualities as beautiful as engravings.

Papermaking. In papermaking, machines have been developed to satisfy the presses, so ravenous that they have ground up the nation's forests at an alarming rate. The Great Northern, a new grinder of 2,400 horsepower introduced in 1934, had an output of 20 tons of pulp a day and required the services of only four men to operate it. To meet the challenge of depletion of timber supply, an experiment of Dr. Charles Holmes Herty of Georgia's State Department of Forestry resulted in the discovery that southern pine could be used in paper manufacture. This resulted in a threefold benefit. It increased the use of the sulfate pulp process, which had not been applicable to other timber, but which was cheaper and simpler in production. It provided a new crop for southern waste land, and it saved the monarchs of our forests. The use of reclaimed paper by a de-inking process has been made possible and has been applied wherever economically advantageous.

Telephoto. The most radical changes, however, in printing technology have been effected by the adaptation of electronics to the publishing business. To meet the growing demand for illustrations, together with the necessity for speed in publication, the transmission of pictures by wire, and later by telephoto, was introduced. In 1924 a picture from the Republican National Convention in Cleveland was successfully transmitted by wireless. Two years later a halftone and manuscript service between New York, Chicago, and San Francisco was inaugurated by the American Telephone and Telegraph Company. The transmitting of spot-news color pictures from a distance occurred on the occasion of President Franklin D. Roosevelt's welcome to George VI and Queen Mary in Washington, June 7, 1939.

The Teletype. The new technological devices forced the passing of a unique position in the editorial rooms, that of the telegraph operator. In the place where he once sat the entire working period (day or night) with earphones clamped over his head while he tediously transcribed the code that came in over the wires, a little machine was installed that tirelessly turned out reams of copy which the copyreader picked up at his convenience. This device was an electrically controlled typewriter that was connected with a type-operated sending machine, the teletypewriter, or teletype. In 1935 the Associated Press eliminated its last telegrapher.

The same kind of machine next invaded the composing room, where the teletypesetter, first used in the late 1920's, automatically operated the linotype machine. This machine was capable of turning out twelve lines of type a minute.

Photoelectric Typesetting. Another machine was introduced which could profitably by-pass the linotype machine. This new device sets type photoelectrically on a film, which is developed automatically, and then converted into a photoengraving ready for the press. One such process is operated by a special typewriter and engraves on a magnesium plate which weighs less than 1 pound as compared with the conventional stereotype plate weighing nearly 50 pounds. Shift keys on the typewriter provide a variety of sizes and fonts of type. Other innovations have involved casting type from plastic, which weighs only one-tenth as much as metal. Typesetting has ceased being a bottleneck in printing.

Radio Broadcasting

By 1920 radio broadcasting as a means of communication had been established on a thoroughly practical basis. The Radio Corporation of America was formed in 1919, and took over the Marconi Wireless Company, uniting radio interests with the General Electric Company. Following the close of World War I, an army, not of the radio tinkerers that existed earlier but of radio technicians, returned to civilian life. Radio became the hobby of thousands of young Americans. But there were no commercial broadcasting stations. Radio "fans" listened for hours, day and night, to pick up some chance messages.

Birth of Broadcasting. One of these operators was Dr. Frank Conrad of Pittsburgh, Pennsylvania, who had devised programs which he put on the air at regular periods. Realizing the possibilities in broadcasting, the Westinghouse Company employed the services of Dr. Conrad and erected a powerful transmitter in East Pittsburgh. This inaugurated regular broadcasting over station KDKA, hailed as the pioneer broadcasting station of the world. Other broadcasting stations were erected. Station WAS was built in Detroit in the same year. Broadcasting of the returns of the Harding-Cox election in November, 1920, is cited as a milepost which revealed some of the possibilities of radio.

Stations Increase. Broadcasting had a spectacular growth. At the end of the year 1922, there were thirty licensed stations. In 1924 there were five hundred. A spectacular feature of that year was a special program broadcasting the beat of the human heart. During the same year, the ceremonies at the opening of the United States Congress were sent over the air. Commercial sponsors proved helpful in financing "commercials," and the American public accepted them. A multiple station network was set up and linked together by wires. In 1926 the National Broadcasting Company was organized, and before the close of the year the Blue Network was established. By 1930 round-the-world broadcasting was effected.

Receiving Sets. The equipment available for receiving sets at the beginning of broadcasting was the homemade crystal set and the one-tube regenerative set. The power was supplied by a battery, and listening was done through headphones or the sound was amplified by a horn borrowed from the phonograph equipment. Since the crystal sets had no amplification power, they were discarded. The vacuum tube produced continuous oscillation in the circuit and this resulted in squealing noises that presented a problem.

Alexanderson's superheterodyne receiver could have eliminated this obstacle, but his receiver was not suitable for the public because of the intricacies in tuning. Refinements were rapidly introduced. In 1923 a neutrodyne receiver came out which had the advantage of receiving the same station at a fixed position on the dial. Conrad's a-c tube made it possible to operate from the household electric light current. The headphones and tin horn gave way to the electrodynamic speaker. A screening grid and an improved high-vacuum tube helped eliminate the static noises. When a means was found for tuning the fixed intermediate frequency of the superheterodyne receiver, with its multiple transformers, the fine sensitive receiver that is still in use today was evolved.

Electronic Amplifier. Alexanderson struggled to overcome static that could not be filtered out in the receiver sets, since static was identical with the radio waves used for broadcasting. He tried to correct the trouble at the transmitter and devised an electronic amplifier to take the place of the old magnetic amplifier. That stepped up the power in terms of kilowatts, not merely watts, and could be controlled by an ordinary microphone. It became the basic principle of the modern broadcasting transmitter in the a-c stations.

FM Broadcasting. Alexanderson continued his experiments and concluded that the solution lay in finding a new kind of wave. In 1933 he perfected a system of frequency modulation which he substituted for amplitude modulation, and which overcame natural static and many forms of man-made static. This was the major technological innovation in the 1930's. Alexanderson worked single handed to introduce the FM wave, which was based on pitch rather than strength, but the major broadcasting companies were not interested.[1] Other developments of the 1930's were minor or gadgety improvements, rather than basic engineering innovations.

Alexanderson finally obtained backing in some small independent stations in New England where static was especially annoying, and by 1940 some one hundred and fifty applications had been filed for FM broadcasting licenses, and a number of concerns had been licensed to manufacture FM receivers. FM is rated as superior in tone, as well as in freedom from static, and permits higher fidelity transmission and less interference. More channels are available with FM than with AM, for the FM spectrum has room for three to six thousand stations as compared with AM's one thousand.

Short Waves. The United States later realized the advantages of short-wave broadcasting (waves of 100 meters or under). Experiments with the short wave were undertaken, and in 1924 the system was used by KDKA in broadcasting to the British Isles. In 1929 and in 1930 the National Broadcasting Company and the Columbia Broadcasting System experimented with short wave, but its use was discarded. It was Frank Conrad's discovery of the characteristics of short wave for long distance transmission that made radio communication worldwide. For this, Conrad was awarded the Edison medal, the highest award of the American Institute of Electrical Engineers. Short wave permits a speed of signaling many times that in longer waves. It is much freer from static and is effective in daylight transmission. During World War II the government took over the operation of short-wave stations, and by 1945 there were thirty-six government transmitters operating on round-the-clock broadcasting.[2] Long wave AM commercial stations use the short wave for point to point communications and in national hook-ups.

Radio Today. Radio today has come a long way from the crystal sets and the school boy tinkerers of pre-World War I days. In 1955 there were more than one hundred million radio sets in the United States, more than twenty-three hundred AM broadcasting stations, and more than six hundred FM broadcasting stations. Science and technology had given to man, in the brief span of fifty years, his greatest systems of mass communication.

Telephone Communication

While progress was being made in radio, scientists in the laboratories of the Bell Telephone Company were engaged in research in both pure science for fundamental principles and in applied science for practical application.

Application of Radio Principles. Principles of radio were adapted to communication by wire, and the vacuum tubes and amplifiers utilized in radio were found to be applicable to the telephone system. Moreover, an interchange of wire and wireless telephony was in many uses effected.

Transcontinental Service. During the 1920's transcontinental service was extended to the Pacific Coast. After World War I international service, combining radio and wire transmission, was established. In effecting the long-distance transmission, relay or repeater stations were installed at intervals. Here the three-electrode valve was effective in amplifying the electric impulses and in transferring speech currents to the wire circuits, and vice versa.

A vast number of improvements were effected by the utilization of both mechanical equipment and the new scientific discoveries. The dial system was established, supplementing mechanical operation for the manual. Improved equipment was provided in telephone sets by combining transmitter and receiver into one piece. Numerous improvements were made in the exchange centers.

Coaxial Cable. Outstanding among the improvements was the coaxial cable. This resulted from the efforts of the American Telephone and Telegraph Company engineers to install a cable which would permit several hundred telephone conversations simultaneously. By 1936 an installation between New York and Philadelphia had been completed. The transmission was made on the beam system from a high sending tower, and was effected by carrying the electric impulses created by the voice on a high-frequency band which can be broken down. Each cable was made to carry a number of different frequencies without interference from each other. By 1947 the coaxial cable was found applicable to television as well.

Trans-Atlantic Cable. The latest engineering advance in telephone technology was the laying of the first deep-sea trans-Atlantic telephone cable from Newfoundland to Scotland, a distance of almost 2,000 nautical miles. The project was begun in June, 1955, and was undertaken as the joint operation of the American Telephone and Telegraph Company and the British and Canadian Communications Agencies. Service over the cable is scheduled for late 1956. The cable was unreeled from the large tanks in which it was stored on the stern of the "Monarch" as the vessel sailed across the Atlantic. Amplifiers, called repeaters, especially constructed to withstand the enormous water pressure (up to 3 tons per square inch), amplify the voice currents every 40 miles.

Television

Television, like radio, had its origin in the laboratory. Scientists had made the basic discovery that light could be translated into electrical impulses, and thus transmitted either over a wire or broadcast through the air. The research engineer went further and devised means for translating these concepts into practical operations. To marry sight with sound and to transport both across the invisible waves of ether had been man's dream for ages. The realization of this dream was accomplished by the combined efforts of many minds in many countries over a period of years.

Early Scientific Discoveries. The early laboratory discoveries that later proved fundamental in television included Sir William Crookes' discovery of cathode rays (1876), now known as the electron beam, and his invention of the Crookes' tube, first looked upon as a scientific toy; the photoelectric cell (1888), a device which changed light intensity of color variations to corresponding electrical energy; the mechanical scanner invented by Paul Nipkow of Germany in 1884, a perforated, spiral distributing disc based on the principle of persistence of vision; the oscilloscope of Karl Ferdinand Braun of the University at Strasbourg, which showed the variations of the alternating current in a vacuum tube; Einstein's dictum that the camera could turn a picture into electricity; the use of cathode-ray tubes for electrical transmission evolved by A. A. Campbell-Swinton; and the application of the DeForest three-electrode tube to amplify electrical impulses. From here the industrial laboratories took over the major developments.

Early Technological Developments. Meantime, tinkerers, inventors, and technicians were making their discoveries. While it was known that phototelegraphy was not the same as television, it was discovered that many of the techniques were sufficiently similar to enable ingenious telegraphers to contribute to the art of television. As early as 1842, Alexander Bain of Scotland achieved a system of facsimile that was a forerunner of modern facsimile methods of sending photographs. For this discovery Bain has been called the "Father of Facsimile." Many telegraphers found out that selenium had increased power in transmitting electrical impulses when exposed to light, and they toyed with the idea of utilizing these light-sensitive properties of selenium in transmitting images.

From the telegraphers came the first recognized system of television, or distant seeing. As early as 1875 they had a "Telectroscope," with selenium as the operating device. With the invention of the telephone, inventive minds turned to wired-image transmission as well. John Kerr and Alexander Graham Bell took out patents relating to television. Later numerous schemes of "seeing by telegraph" appeared, but before 1909 interest waned.[3]

Mechanical Scanning. The first demonstration of practical use of electrical transmission came with the employment of a system of mechanical scanning in the early 1920's, when C. Francis Jenkins of Washington, D. C., sent a picture of President Harding by telegraph from Washington to Philadelphia in 1923. In his scanning system, Jenkins used (rather than Nipkow's disc) a system of prisms for deflecting the light beams.

John Baird of Scotland in 1928 with his own system of mechanical scanning, transmitted a trans-Atlantic telecast.

Herbert E. Ives. Herbert Eugene Ives of the Bell Telephone Laboratories, who had worked on wire-photo service, successfully showed President Hoover on a large neon tube screen as he was broadcasting. He was viewed at a distance of several hundred miles.

Ives also presented a two-way television demonstration in which the speakers at either end of the telephone saw each other as they conversed.

Alexanderson. Alexanderson, using a perforated scanning disc and high-frequency neon lamps, forecast commercial television when he demonstrated in Schenectady a television system of transmission to his home. In 1930 he presented a theater demonstration given on a 7-foot screen which showed an orchestra performing in the laboratory miles away.

Improvements in Scanning. Television was still far from being practical, for the images remained crude and indistinct. The fault was attributed to imperfect mechanical timing. Mechanical scanning was not sufficiently rapid to break down the image into the number of lines required. The two scientists most responsible for the revolution in scanning were Vladimer K. Zworkin and Philo T. Farnsworth.

Zworkin's Electronic Scanning. Zworkin, born in Russia, came to the United States after the close of World War I and became associated with the Westinghouse laboratories, and later with Radio Corporation of America. Within a few years he produced the iconoscope (the "eye") of the television camera. With this device he performed the double function of scanning the image and converting it to electrical values by means of electronics. This was done with the cathode-ray tube, in which was mounted a mosaic plate that performed the functions of the photoelectric cell, and an electronic gun which took the place of the mechanical scanner.

Farnsworth's Image Dissector. Farnsworth, likewise, developed an electronic scanner, which he called an "image dissector." It operated on principles different from the iconoscope, but served the same purpose. It served as a replacement for the mechanical scanner, and it produced several stages of amplification as well. With this dissector the electronic image was produced with greater rapidity, resulting in several times as many lines as the mechanical scanning device. His inventions became widely adopted during the middle 1930's.

TV Research. It was during the 1930's that television began to emerge from the laboratories. Radio Corporation of America in 1935 built a transmitting station on top of the Empire State Building and announced its new television research program. David Sarnoff and Allen B. DuMont were leaders in introducing television into the theaters and the homes. DuMont made improvements in radio tubes and became a manufacturer of commercial television sets. He also developed "magic eye" tuning and improved the clarity of the pictures by what was termed "high definition," and "electrostatic deflection." Following a number of improvements, on April 30, 1939, at the New York World's Fair, with President Franklin D. Roosevelt participating, Radio Corporation of America and National Broadcasting Company introduced television as a service to the public.

War Interruptions. Two years later, however, when the United States entered World War II, the National Defense Communications Board stopped all construction of receiving sets, and television again retired to the laboratory. Several studies were made meantime of the use of high frequencies and color production, and a system of intercity relays was developed. By the time the war ended, the American Telephone and Telegraph Company began constructing a coaxial cable for television use. In the late 1940's the public was clamoring for television sets.

The growth of television communication was phenomenal. By the mid-1950's more than five hundred television stations were in operation, and an estimated twenty-seven million homes were equipped with television sets. In some cities of the nation there were more radio and television stations than newspapers. Science had wedded sight and sound and transported it across space.

Radar

Radar, often called the "electronic watch dog," is based on the principle of echo, and grew out of directional short-wave radio. Its source is the electronic magnetic spectrum. The science behind radar involves the same principles as radio, and its adaptation by American scientists resulted from discoveries made in the early 1920's. While experimenting with directional radio, scientists observed that moving objects between the transmitter and receiver disturbed the reception. Because of its use in national defense, the early developments in science were kept from the public, and it was not until it was a full-fledged achievement that the public became aware of its existence. The result was that its accomplishments seemed miraculous.

Radar Technology. Directional radio as contrasted with broadcasting was designed to keep radio messages secret. It was carried on with the waves of ultra-high frequency, utilizing a narrow beam. In its use the observations were made that a "Heaviside layer" existed in the atmosphere from which a directional beam was reflected, and that a directional beam was reflected from a distant object. It was then determined to place the high-frequency sending and receiving apparatus side by side, and broadcast the radio wave so that the receiving set caught the rebound or the echo. Radio equipment consisted of a transmitter and a receiver. The transmitter was capable of hurling into space the pulses that traveled in a narrow beam of high-frequency energy which was several times as powerful as the output of the largest radio station. The receiver was capable of detecting the return echoes.

With radio waves traveling at the speed of light, the time of travel of radio waves could be determined. This made it possible to measure distances by noting the time elapsing between the broadcast wave and the reflected reception. Many uses of radar have already been evolved. In meteorology it has been applied to detect storm centers. It has also been employed to plot undersea oil wells. With its use, a topographic mapping of the moon has even been attempted. It has been used in surveying by observing the time required to bounce an echo back. Medicine, marine navigation, and aviation all have discovered essential peace time uses for radar. Radar differs from optical range in that it operates in any kind of weather and during the night as well as during the day. It has become a main contribution to communication technology.

Science, during the first half of the twentieth century, had given to man the miraculous means to instantly communicate ideas, words, and images from any one spot on the globe to any other.

Notes for Chapter 36

1. W. Rupert MacLaurin, in his book on *Invention and Innovation* (1949), p. 187, described this system in these words: "Frequency modulation differs from amplitude modulation in the following respects: In standard broadcasting an electrical pump at the base of the antenna pumps electricity into the antenna and sucks it back again a great many times a second. The current is modulated by superimposing a voice current which changes the signal strength in accordance with the fluctuations of the voice. Throughout this process the electrical pump continues to operate at a constant speed. In frequency modulation, the strength of the antenna current does not vary, but the speed of the electrical pump is changed in response to the fluctuations of the voice. . . . When the FM waves enter the receiver they pass through a part of the set not found in AM receivers, the "limiter," which is a combination of tubes that chops off the tops and bottoms of the waves, so that they are all of the same amplitude or height. The FM receiver is designed to be responsive to changes in the frequency of the incoming signals rather than changes in amplitude; and since most static is predominantly amplitude-modulated, the static signals are eliminated."

2. This system was used by the Voice of America in broadcasting overseas programs.

3. See Stanley Kempner, *Television Encyclopedia* (1946); Philip Kirby, *The Victory of Television* (1939); R.C.A., *Electronics Research and Development* (August, 1937); Herbert L. Marx, *Television, Radio and American Life* (1953); Charles A. Siepmann, *Radio Television and Society* (1950).

Chapter 37

A NATION BETTER-FED, BETTER-HOUSED, BETTER-CLOTHED

During the thirty-year period from 1925 to 1955, American technology was put to a greater test than at any other period of the nation's history. It was during this period that a severe economic depression occurred, and a second World War. But, despite these two tragic developments, the American people continued to prosper. They achieved the highest standard of living in all history. Due to new technologies that supplied the people with their three basic needs of food, shelter, and clothing, they were not only able to provide for themselves an even higher standard of living, but also to produce huge surpluses for other people of the world. Advances in food technology, building technology, and clothing (textile) technology made these conditions possible.

The Tools of Agriculture

Farm Mechanization. The most significant advance in agricultural technology during the 1920's was the introduction of the tractor. When the Farm-all tractor appeared in the early 1920's, a new era in farm mechanization had arrived. The Farm-all tractor did more than pull. It pushed, carried, and hoisted both farm implements and farm products of all kinds. It was so constructed that attachments which sawed, planted, cultivated, harvested, and threshed could be added.

The Secretary of Agriculture reported that there were more than a half-million tractors of all kinds in use in 1925. Motor trucks and automobiles, along with tractors, had increased the mechanical horsepower on the farm five times beyond that of 1920, according to the Secretary's report. As a result, hired labor had decreased during the preceding five-year period by more than 12 per cent; but despite all this, the nation's farm production had shown a total increase of 17 per cent during this period.

The Combine. Farm mechanization occurred in other branches of farming. The Secretary of Agriculture, in 1927, devoted an entire section of his annual report to "The Eastward March of the Combine." Although the combine had found its greatest use in the wheat-growing areas of the Midwest—Kansas (for example) before the close of the 1920's was harvesting more than one-half its wheat with combines—this new labor-saving machine began its eastward march into Ohio, Pennsylvania, Delaware, New Jersey, and neighboring states. The combine, powered by a tractor, lowered the cost of harvesting and threshing, reduced the amount of labor required, shortened the harvesting and threshing season, and reduced waste of grain. It practically eliminated the need for transient harvest hands. Between 1920 and 1927, the nation's farm population had decreased more than three million "due primarily," said the Secretary of Agriculture, "to the introduction of scientific methods, and to the displacement of hand labor by modern machinery."

Rubber Tires for Tractors. A major improvement in the history of the farm tractor occurred around 1930 with the introduction of low-pressure rubber tires. This new feature rendered the machine 25 to 27 per cent more powerful. Equipped with rubber tires, tractors could travel faster, plow more ground on less fuel, and perform numerous operations more speedily than when fitted with steel tires and lugs. As rubber-tired tractors began to win all plowing contests, sales increased, and by 1940, 90 per cent of all tractors were equipped with rubber tires. The tractor led the march of machines in the entire farm mechanization program.

Agriculture during World War II. In 1955 there were more than four million, five hundred thousand tractors on American farms. The mechanization of agriculture during the pre-World War II period found the American farmers ready to meet unheard of demands when the United States entered the conflict. Mindful of the early years of World War I, when this nation was called on to feed our allies as well as our armed forces, the American farmers expected similar demands to be made with the outbreak of World War II in Europe. Accordingly, they sought frantically to mechanize every phase of agricultural production. The most modern tractors, plows, seeders, harrows, cultivators, mowers, and combines were purchased, and crop acreages were increased at an amazing rate.

While few technical improvements in agricultural production were introduced during the war years that could be called revolutionary, the farmers were able, by working longer hours and making use of the latest mechanical improvements, not only to produce sufficient supply for our own needs, but also to ship vast surpluses to our allies. The demands of the war years had brought forth the maximum of production which the latest methods and machinery could supply. Further scientific and technological advances had to wait until after the close of the war.

Mechanization in Postwar Years. No sooner had the nation returned to a peacetime economy, however, than new advances began to show up in agricultural technology. Farm engineers and manufacturers of farm equipment turned their efforts toward plans for further mechanization of the farm. In his annual report for 1948, the Secretary of Agriculture described the advances made since the close of the recent war as having "ushered in the greatest era of farm mechanization in history." Referring to the new equipment as, "the kind that excites men's imagination," he mentioned new cultivators that used a flame to kill weeds; a potato harvester that dug, graded, sacked, weighed, and delivered potatoes to a waiting truck; a planter developed by Department engineers that planted corn and peanuts, opened the bed, planted the seeds, and placed the mixed fertilizer, all in one operation; sugar cane harvesting machines that cut and loaded the cane at an estimated saving of one-half to two-thirds of hand-harvest labor.

Cotton Technology. The design of new and improved tractors and their further use continued to challenge the skill of the manufacturers. The crawling tractor, an adaptation from war service, became a rubber-tread truck. Hay harvesting became mechanized. Mechanization in the growing and processing of cotton, long a dream of southern planters, had become a reality by 1950. Such improvements as modern cotton pickers, new devices for cleaning, extracting and removing foreign matter, methods for storing the cotton, drying it and restoring the moisture needed for dry fibers during the ginning were introduced to ensure a continuous cotton production.

Fewer Farmers, Greater Production. The amazing productivity of American agriculture continued through the first half of the decade of the 1950's. Farm efficiency rose faster during the post-World War II decade (1945–1955) than the efficiency of the rest of the nation's economy. Due to a vigorous, progressive program in agricultural technology, the average American farmer in 1955 was 110 per cent more productive than he was in 1935. As a result, some 36 per cent fewer farmers, working fewer hours, were producing 43 per cent more than farmers produced twenty years earlier. Less than 14 per cent of the nation's population lived on farms in 1955 as compared with 25 per cent in 1930. In 1955 only one out of every nine employed Americans was working on a farm. Farm mechanization, more than any other factor, explains the reason for this rapid rise in production with reduced man power.

The increase in the use of tractors (more than four and a half million in 1955) was more than equaled by the introduction of other machines. Close to one million combines, seven hundred and fifty thousand milking machines, more than five hundred thousand corn pickers and two million five hundred thousand trucks were reported in use in 1955. Approximately 92 per cent of farm homes and buildings were wired for electricity. Electricity was used not only for lighting, but also to operate small machines such as grinders, milking machines, saws, and drills used by the farmer, while the farmer's wife used electricity to operate the household and kitchen equipment.

New Role for Farmers. Technology had brought about many fundamental changes. The era of old-fashioned dawn-to-dusk hours of drudgery had passed. The farmer had more leisure time—time to think, to plan, and to conceive of newer, more efficient machines. He passed his ideas on to inventors, and occasionally helped in building new machines. A new generation of farm boys was receiving more intensive and practical training in operating machinery than had been given to any earlier generation.

New Strains of Grain. Production of grain crops, due to fertilization and intensive farming, increased from 25 to 30 per cent during the second quarter of the twentieth century. Hybrid corn, introduced around 1930, increased the per-acre yield from 30 to 33 per cent. In 1955 more than 85 per cent of all corn was of hybrid origin, and man-hours needed for producing crops had declined. In 1930 it required approximately 120 hours to produce 100 bushels of corn. In 1955 it required only 34 hours. It required 70 man-hours to produce 100 bushels of wheat in 1930; in 1955, only 26. For potatoes, the number of hours required to produce 100 bushels in 1930 was 63; in 1955, 27.

The Farmer's Income. The number of farms declined from six million and three hundred thousand in 1930 to five million and four hundred thousand in 1952. Farm population declined from approximately thirty million and five hundred thousand to twenty-one million and nine hundred thousand in the past two decades. Because fewer farms and farmers produced more, the farmer's average net income rose from $631.00 in 1930 to $2,268.00 in 1954, a 259 per cent increase, while the income for the average manufacturing employee rose only 146 per cent.

Nutrition Values Increased

The Food Technologist. Food technology during the thirty-year period following World War I was marked by the increased application of science and by advances in food technology. Before the war there were fewer than a dozen food technologists in the country to offer their services to the processors, but within ten years there were more than three hundred. In 1955 no processing plant could exist without a well-staffed scientific laboratory. Improvements in methods of food processing had been devised, and more foods have been added to the list of processed products. Rivalry existed among food packing companies in bringing new foods to the market, and in all of this, scientists led the way. The big three of the early days of the canning industry (corn, tomatoes, and peas) have many competitors on the grocer's shelves today.

Effect of National Habits. Science, too, has played a part in the food demands of the consumer. The percentage of young men rejected by the army in World War I awakened the nation to the need for improving its health habits, especially for increasing food values. Statistical reports, showing the large number of school children suffering from malnutrition, aroused the citizenry to an investigation of food needs. Paul de Kruif in his *Hunger Fighters,* 1928, revealed the inadequacy of diet in a large class of our population. It was not enough to grow vegetables, a directive to the Victory Gardeners of World War II set forth, but equally important was the production of vegetables that would build the greatest degree of health.

Nutrition Foundation. The Nutrition Foundation, established in 1942 by the food industry of the United States and Canada, set up independent research and educational projects in the science of nutrition. Nutrition became the newest science. Dieticians prepared food charts showing the tissue-building factors and the energy-building factors in foods. Newspapers published articles on food values. Public utility companies distributed "balanced" menus to the citizenry on all levels. The American public, which once considered a meat and potato diet with a piece of pie for dessert as adequate, came to demand fruit juices, a variety of vegetables, whole grain cereals, and balanced meals. The public learned about vitamins and enzymes and demanded food that would include them. The scientist pointed the way and the technologist prepared the food to meet these requirements. Nutritive values came to be an objective of canners and food processors, as well as flavor, color, and attractiveness of the product. Vitamin-enriched foods appeared. Low-calorie and high-protein foods were emphasized. Balanced meals were packaged in one container. To meet the need for speedy preparation of meals created by women entering industry during the war, ready-mixed foods were put on the market. Packaging with the many new foils and plastic wrappings was carried into all food lines. By 1950, America was nutrition-conscious.

George Washington Carver, 1864–1943. The nation's vast crops and food production, ranging all the way from the grain cereals to the lowly peanut, has recently moved into a chemical age. Typical of the role of the chemist in agriculture was that played by the late scientist, George Washington Carver. A one-time slave, he rose to world renown, honored at home and abroad for his many scientific discoveries. To find a crop other than cotton for the South, Carver developed new uses for peanuts, for sweet potatoes, and pecans, and brought these crops to great commercial importance. He devised more than three hundred ways of using the peanut, ranging from an oil used for medicinal purposes to dyes produced in more than a dozen shades. He also derived from the peanut such by-products as milk, ink, flour, breakfast foods, and face creams. Because of these developments, peanuts became one of the chief crops in the United States, with annual production yields of more than 2,000,000 pounds. From sweet potatoes and yams more than a hundred products were obtained, varying from flour to shoe blacking. Carver also discovered some sixty uses for the pecan nut.

Frozen Foods. As the demands for nutritive values were being felt, a new process of freezing foods was developed that kept foods indefinitely, preserving all the pristine freshness and original color and aroma of the fresh product, and all the vitamin and calorie content. This was the frosted foods or "Quick Freeze" process developed by Charles Birdseye. Birdseye conceived the idea while on a trip in Labrador with Sir Wilfred Grenfell in 1912, when he noted the quality of meat and fish quickly frozen in the Arctic air. It was not until the mid-1920's, however, that the foods prepared by a process devised by Birdseye appeared on the market to compete with the fresh foods and the canned products as the most acceptable type of preservation.

Quick-Freeze Products. The essential feature of this process was that it be done quickly. Less than 4 hours elapsed from the time the product left the field until it was prepared, frozen, boxed, ready to keep indefinitely. While a distinct process was designed for each food, the general procedure was to subject the food products to a temperature of 10 to 15 degrees below zero centigrade. Usually the food was subjected to agitation so that it was not frozen solid, but each particle remained whole. Some methods included freezing in air or direct immersion, as in commercial icemaking; indirect contact either single or multiple by contact with cold plates; or spray or fog freezing.

Distribution of Frozen Products. The distribution system for getting frozen foods to the consumer presented as great a technological problem as processing the foods. A continuous "pipe line" of zero temperature from processing plant to the home refrigerator was necessary. This included refrigerated transport from the packers warehouse to a cold-storage plant in the wholesale warehouse, and then by refrigerated trucks to frozen-food cabinets in the retail food stores. It was the designing, the installation, and the financing of these frosted food cabinets that presented the greatest challenge. It was accomplished on a nationwide scale by the services of refrigerating engineers, and after an investment of millions of dollars.

In the first twenty-five years of its history, frozen-foods processing had a steady development. Several brands were placed on the market besides the original Birdseye products, and the contents went beyond the original limited number of fruits and vegetables to the fruits more difficult to process. They were expanded to include fruit-juice concentrates, fish, poultry, and precooked specialties. Among the latter have been frozen fish sticks, frozen chicken pies, and frozen meat pies to meet the housewives' demand for speed and ease in preparing foods.

Frozen Food Lockers. Along with the development of frozen foods came the frozen food locker. It originated as a community refrigeration storage plant to which individuals, especially farmers, had access for storing meat rather than putting it down in brine. Community lockers became a factor in food preservation and were popular in the rural areas. The farmer took his corn-fed hogs and cattle to the locker plant where he could even secure butchering service, and have the meat chilled, frozen, and stored; and have chops and steaks cut to his demand. This was far easier than to salt, smoke, and process his product in the old-fashioned laborious way.

Deep Freeze. Freezing cabinets for household use, which appeared in the 1940's, have had a wide development. These were miniature quick-freezing plants in which foods were given the same quick-freeze treatment they received in the large frozen-foods industries, and this enabled the housewife to preserve food in this manner rather than by canning. These also served as storage plants, where commercially frozen foods were stored for household use. Deep-freeze cabinets have been made in many sizes and are operated on the household electric current.

Packaging. Packaging frozen foods has become a separate industry. First developed for the commercial packer, a fiber and metal can was put out by the manufacturer of cans. Paper products manufacturers developed a package for home use. The success of frozen-food products has been due in part to the careful packaging and handling in the distribution.

Birth of a New Industry. By whatever process food has been preserved—canning, dehydration, or freezing—food preparation and preservation have gone out of the home into industry, and have entered the assembly line of mass production. It has passed from the arts into the sciences with the engineer, the food technologist, the bacteriologist, and the food chemist working together to assure the nation its supply of fresh foods. The industrialization of food technology reached its highest peak when the wife of the farmer who "reaps and sows" began buying her daily bread from the traveling grocer who delivered it to her door already wrapped and sliced. Technical advances in food production have been marked by more than the attainment of producing more with less labor. For a public that has always been skeptical about a change in food habits, quality has ever been a guarantee of the packer. Even in the old days before the exactness of science could guarantee that the product would be properly preserved, the processor stood ready to make good the losses. Food technology has not only taken the drudgery out of food production, it has made it possible for the populace to be nourished better on a scientific diet than was ever done before.

Housing the Nation's Millions

Better Housing. Developments in building technology in the past thirty years, involving architecture, building materials, and construction methods, have presented a maze of innovations. These vary from the introduction of modernized cabinets in the streamlined kitchen to such glass and aluminum structures as the United Nations building. Behind all these innovations, however, lies the basic idea that the functional purpose of building should be the criterion in construction. Building technology within the past three decades came to grips with realism, and the use to which a structure, a material, or a method was to be put has been the determining factor in its employment.

Dependence on Science. The daring adventures in many innovations in the use of new materials, new designs, and new fabricating methods have sprung from an increased faith in, and reliance upon, science. The engineer and the architect have worked together, and the number of architectural engineers has greatly increased. Research laboratories have taken over the testing of materials, and the determination of the stress and strain that could be exerted. Structural members are cast to meet this tension. Savings have been effected also in not making structural members excessively strong. Building, like many other arts, has entered the realm of science.

Structural Beauty. Office buildings that stood stark and bold in their simplicity and lack of ornamentation revealed an honest beauty in their structural strength and utilitarian purpose. This idea of structural strength, inaugurated in the buildings of the Chicago World's Fair, 1893, was concealed there under an outward shell of classic design with its rhythmic columns. Today, these outward shells have been discarded and the iron trusses and steel and concrete construction, with their functional sheathing of metal, glass, steel, or stone are recognized as having a beauty of their own in their strength, simplicity, and devotion to utilitarian purpose.

Office Buildings. Construction methods that developed with the skyscraper have evolved into more efficient and more highly mechanized procedures. The ultimate of height was reached with the Empire State Building, erected in 1931, which towered 1,472 feet and contained one hundred and two stories. Since then, emphasis has been put on new materials and simplification of construction. Aluminum, glass, and steel have been introduced for the skinlike covering. Units of one whole story in extent have been precast, fitted with windows and glass, delivered on the site, and swung into place by huge cranes without the use of scaffolding. Units for each floor have been suspended independently, and any unit could be removed without damage to the others. Cellular steel panels, welded to the structural steel frame, served as conduits for the utilities. A large office building with aluminum facing was erected in New York in 1950 with the speed of an automobile assembly line. The job was completed in six and a half working days after the steel framework was erected. The eighteen hundred panels, each two stories high, were swung into place without the use of exterior scaffolding.

Prefabricated Houses. Following World War II, technical innovations were extended to include the construction of home dwellings. Government housing projects and large-scale private enterprise housing developments stimulated the utilization of prefabricated parts, and in some cases entire prefabricated construction. Many of these constructions, though intended to be temporary, were found to be durable and provided many families with greater home comforts than they had ever known before. For economic reasons they were architect-engineer planned, and construction was standardized. After the war, similar type housing was continued under government authorities, and private enterprise continued on a greater scale. The dweller had come to see the advantage of large-scale developments. Many of these were located in suburban districts. Skilled workmen found vent for their construction abilities and relief from the monotony of machine operation in the factories by building their own homes on the "do-it-yourself" plan. Houses sprang up all over the countryside. A low ranch-type evolved that could be built from a stepladder by the owner working after hours. Despite the standardization of parts, individualism became a feature. Site planning, or the integration of the placement of the building with the architecture and landscaping, developed individualistic features.

New Materials. New materials were utilized; plywood, stronger than wood, was used for exterior walls by some of the largest construction companies. Steel panels, aluminum siding, sawed stone, and wood so impregnated as to be flame resistant, termite proof, and even preshrunk, was made available to the builders. The new lacquers made it possible for the owner to do the painting. The lacquers, a product of World War I, of a nitrocellulose base combined with resin, pigment, solvents, and a plasticizer, were applied with a spray gun, or by a brush if a delaying ingredient was added to slow up the drying process. Steel, wood, and plaster could be painted with the new lacquer which was one of the great boons of the postwar era.

Home Furnishings, Equipment. New technologies have also revolutionized the furnishing and equipment of the American home. Automatic equipment has transformed the heavy drudgery of housework into relatively easy and speedy tasks. Developments in plastics, aluminum, magnesium, and stainless steel have brought about major changes. The automatic washers and dryers, dishwashers and disposals, electric mixers and food choppers, pressure cookers, and multipurpose gadgets have streamlined the work in the American household. Refrigeration technology (noted earlier) has provided food lockers, and deep-freeze units have enabled families to store up quantities of food sufficient to last for months at a time.

Heating Technology. Technology has also added to the comforts of modern homes and buildings by improvements in heating, lighting, and insulation. Although the housing data included in the Census of 1950 revealed that only half of the nation's homes were equipped with central heating plants, that ratio represented a 50 per cent increase over the preceding ten years (some parts of the country showed a much higher percentage). While coal was still the predominant fuel used in heating, yet its use had dropped in the last ten years from 55 to 34 per cent. Gas for fuel, on the other hand, had tripled during the last decade. Oil and electricity were also employed, with infrared radiant heating being hailed as a technological wonder. At Hanford, Washington, an atomic reactor was utilized for space heating, an infant step on the way to the realization of deriving more use from atomic fuel.

A recent development in out-of-door heating technology involved the installation of underpavement pipes, heated by steam or electricity, to clear sidewalks of snow and ice. This marked another development in heating technology.

Lighting. The development of fluorescent lighting has marked a departure from either flame or filament light source. Although the first mercury vapor lamp was introduced by Peter Cooper Hewitt around the turn of the century, the adaptation of fluorescent substances to general lighting utilities has come largely within the last two decades.

During the early 1950's, teams of scientists took up the fight for pure air, particularly in the metropolitan areas where air pollution had threatened not only human health, but also had become a threat to industry's very existence.

Man-Made Fibers Clothe the People

Clothing Technology. Resulting from the concentrated interest in chemistry, which was stimulated by World War I, the first of the man-made fibers was made marketable in the postwar period. These became rivals to the four basic fibers of wool, linen, cotton, and silk that had enjoyed the monopoly since civilization began. These first products comprised the rayon group, and though not synthetic in the technical sense of the word, they were a chemically produced substitute for a natural product. They represented many years of research since Wohler in 1828, in Germany, began the synthesis of organic compounds.

Rayon. The laboratory background for rayon originated in Germany, France, and England. The first rayon, the nitrocellulose type, was the "dope" used on the wings of the airplanes during World War I. It was denitrated to render it less flammable. Viscose, made from an English formula, from cotton linters, a nitro-cellulose product that is similar to the food of the silkworm, comprised the greatest quantity of the early product. Cellulose acetate, likewise, was produced in the 1920's. It more nearly resembled silk. It could be woven into finer threads and produced a thin but durable fabric. It was marketed as "celanese." The name "rayon" was applied to these products in 1924.

Nylon. These were followed in the 1930's by a growing pattern of truly synthetic products made from coal tar, phenol, water, air, petroleum by-products, or other hydro-carbon substances, by a process known chemically as polymerization. DuPont's "nylon" received the greatest acclaim when it appeared during the shortage of materials just before World War II and was fabricated into women's hosiery. It was still a laboratory product in the mid-1930's. The difficult process of spinning had to be mastered, for neither the viscose wet spinning or acetate dry spinning was applicable to the new substance. A melt spinning was devised. Then a hosiery machine that could handle the tough but fine yarn was devised, since existing knitting machines clogged under the impact. Dyes presented a third problem. Nylon showed the way technically, chemically, and economically to the process of developing other synthetic fibers.

Wool Substitutes. While the early fibers took the place of silk, substitutes were sought for cotton and wool, and the chemist's laboratory was the field for synthesizing these textiles. Orlon came from the laboratory in 1941 with qualities resembling wool, but it was not until ten years later that all the difficulties in manufacture had been overcome, so that the customer could purchase Orlon garments. The creation of new man-made fibers in quantity production signaled another triumph in textile technology.

Unprecedented Prosperity

President Franklin D. Roosevelt, in a moment of despair during the early years of the depression of the 1930's, described one-third of the people of the nation as ill fed, ill housed, and ill clothed. But he little realized that despite the tragedy and destruction of a second world war, our people would within the next twenty-five years, become the best fed, the best housed, and the best clothed people in the world. Applied science and technology had made these conditions possible for the American people, and by producing surpluses for the peoples of other lands.

Notes for Chapter 37

For advances in agricultural technology the *Year Book* of the United States Department of Agriculture continues to be a gold mine on all developments relating to any phase of agriculture.

Also, the files of *Agricultural History* contain numerous articles covering the major advances in agricultural technology.

On building technology, the notes for Chapter 28 are helpful for this period also.

On textile technology see: Zelma Bendure, *America's Fabrics; Origin and History, Manufacture, Characteristics and Uses* (1946); Frank George Carpenter, *How the World Is Clothed* (1908); Frank George Carpenter and Frances Carpenter, *The Clothes We Wear* (1926); Helena Chalmers, *Clothes, On and Off the Stage; A History of Dress from the Earliest Times to the Present Day* (1928); Ethel Lewis, *The Romance of Textiles; The Story of Design in Weaving* (1937); Perry Walton, *The Story of Textiles; A Bird's-eye View of the History of the Beginning and the Growth of the Industry by Which Mankind Is Clothed* (1936); *An Encyclopaedia of Textiles from the Earliest Times to the Beginning of the 19th Century, with an introduction by Ernst Flemming* (1927).

Chapter 38

TRANSPORTATION TECHNOLOGY SINCE WORLD WAR I

It was during the period following World War I that Americans began to travel on a scale never before dreamed of. The war, itself, was primarily responsible for this. Millions of service men had for the first time traveled beyond the community in which they had been born and reared. Civilians, too, moved from farms to cities and sometimes to factories located several hundred miles distant to take up peacetime jobs in the industrial centers that sprang up after the war throughout the country.

The American Riding Public

Many of the men in the armed services, upon returning home, were not content to settle down to the routine of their former jobs. The popular song, "How 'Ya Gonna Keep 'Em Down On The Farm," carried a meaning more significant than at first realized. Tens of thousands of service men who had never owned an automobile had learned how to drive cars and trucks, while many others had learned how to fly airplanes. They and their "buddies" would soon be "on the go."

Americans "Hit the Road." Not only ex-service men, but civilians, too, got the wanderlust. People were of a mind to travel. The one thing needed was a dependable, sturdy car, capable of traveling at good speed, and one that would sell at a reasonable price. Accompanying this demand was also the need for good roads. Both of these came rapidly during the next three decades.

Development of Modern Passenger Cars

Balloon Tires, Four-wheel Brakes. Automobile engineers began making a number of improvements on cars. Low-pressure balloon tires, aimed to provide more comfortable riding, were introduced during the early 1920's. The growing traffic congestion, caused by the increasing number of cars, made quick stops necessary, and this led to the introduction of the four-wheel brake. Safety glass was installed in doors and windshields. A new high-compression antiknock gasoline, ethyl gas (tetraethyl lead) was introduced, which developed greater engine power. The automobile engines (power plants) of the 1930's were a great improvement over the sputtering engines of the preceding years. The chief mechanical advances were in connection with the four auxiliary functions of the engine: carburetion, ignition, lubrication, and cooling.

Improved Comfort in Automobiles. The automobile held a commanding position in the industrial world by the close of the 1920's. Thousands of miles of paved roads, filling stations, repair shops, and the perfection of workmanship, enabled car owners and their families to travel without fear of breakdowns or of being stuck in the mud. Closed cars came into quantity production, and with the installation of heaters, winter travel offered little less comfort than summer travel.

Fluid Drive, Hydromatic Drive. While the manufacture and sale of cars declined somewhat during the lean depression years of the 1930's, improvements in construction continued. Production had reached a peak in 1928 when more than four million cars were made, a figure not again attained until after the close of World War II. The expense of retooling eliminated the early possibility of radical changes, but many refinements were effected. The trend to automatic drive was definitely established. By 1931 free wheeling was available on some makes of cars, and by the close of the decade, fluid drive and hydromatic drives had appeared. The gearshift lever was taken from the floor and put first on the dashboard and then on the steering column. The all-steel turret top appeared, the running board was eliminated, and cars were made wider and lower to hug the road. Ford had so completely divorced himself from the Model-T in the early 1930's that he came out with a V-8. In 1941 the twenty-ninth million Ford came off the assembly line. The decade of the 1930's saw the diesel-powered truck introduced.

Automobiles After World War II. Soon after the United States entered World War II, the design of cars was "frozen" and a strict program for rationing the number and sale of cars was established. Few major technological innovations were introduced until the mid-1940's. When production was resumed, there were still no immediate major technological changes, although rumors persisted of great achievements in experimental cars. Fiber glass and plastic supplanted steel for body structure in a few show cars, but manufacturers feared to risk mass production. Cars up to 190 horsepower were put on the road. New streamlined bodies, molded in one piece of plastic, were tried in experimental cars, but these did not appear in production until the mid-1950's. Diesel, turbine, and improved steamers passed the stage of engineering ideas and evolved into actual working models. Prospects at the beginning of the last half of the century were that manufacturers were ready to go ahead as soon as they felt assured the public was ready to buy.

The "Firebird" Auto. Specially designed automobiles, powered by gas turbines, resembling rocket ships, and operating somewhat like a jet airplane, made their appearance in the mid-1950's. The pioneer "Firebird" was used to explore the commercial possibilities of the gas turbines. Operating somewhat like a jet plane with the exhaust from its rear, an "engine" was used to trap and drive a turbine, which in turn propelled the wheels. The Firebird's power-package engine marked a new development. The engine consisted of two mechanically independent parts, one called the gasifier section and the other, the power section. The gasifier section provided the source of compressed hot gas; energy from this gas was delivered by the power section to the car's rear wheels. The design of the car marked a revolutionary advance. Its needle nose, delta wings sweeping back along the rear half of the body, a vertical fin, and the body made of plastic and reinforced with fiber glass, all introduced a new era in automobile engineering.

By the mid-1950's a number of new technological gadgets had been added, aimed to improve the riding comfort of passengers. Power steering and power brakes were introduced. Push-button controls were installed for raising and lowering the windows, cleaning the windshield, and for raising and adjusting the tops of convertible cars. Air conditioning systems were installed—for a price—in some models. Tubeless tires to ensure greater safety had become standard equipment.

Motor vehicle registrations passed the fifty-eight million mark in 1954 of which over forty-eight million were passenger cars, with more than nine million trucks, and close to two hundred and fifty thousand buses. An anticipated sixty million registered motor vehicles was predicted by 1956.

Nationwide Trucking

All-purpose Trucks. Following World War I, business and commercial firms used trucks in greater numbers. Large department stores catered to shoppers outside the city. Contractors and trucking companies promised their customers round-the-clock trucking service. Trucks equipped with hoists and lifts, with "dump" bodies, and unloading devices dumped thousands of brick where needed, untouched by hand, and undamaged. Trucks began to replace narrow-gauge railroads in hauling logs and lumber.

Farmers shipped their livestock by truck and were able to rush their animals to market without loss of weight. Truck gardeners delivered the produce of their fields and gardens while still fresh, and thus commanded higher prices. Dairymen used trucks to haul their milk to the local stops on "milk-train" routes. All these uses led to a rapid increase in truck production, and by 1930 the United States was manufacturing more trucks than the rest of the world combined.

Trucks Compete with Railroads. This newcomer to transportation technology began to make inroads on the nation's railroad business. Manufacturers started specializing in truck design with models ranging in size from small half-ton pick-up trucks to the huge 20-ton boxcars. In 1940, before the outbreak of World War II, there were more than four million trucks registered in this country. In 1950 a new monster appeared, the world's first turbine truck, a Boeing 175-horsepower gas turbine, less than one-tenth the weight of the conventional diesel engine which it replaced. In design the gas turbine was similar to the jet airplane engine, with the power-producing section being in fact a small jet engine.

By 1953, when the fiftieth anniversary of the beginning of truck transportation was celebrated, statistics were released showing the phenomenal growth of the industry. Seventy-five per cent of all freight in the nation was moved on trucks at some stage of transportation, and more than twenty-five thousand communities depended wholly upon trucks for every article of produce that was moved in or out of those communities, a form of transportation technology that began only a half century earlier.

Buses Carry the Public

While the huge motor trucks with boxcar carrying capacity were bringing changes in freight transportation, the passenger transportation was also being affected by the introduction of motor buses. The motor bus industry, like many other enterprises, had an humble beginning. A few car owners, notably persons who had once been engaged in operating livery stable service, saw the advantage of replacing their horse-drawn vehicles with automobiles. This practice sprang up in a number of cities and towns before World War I. The more enterprising auto livery companies began to design special jitneys and small coaches to accommodate more passengers. For a period motor buses were used only for local transportation, and utilized chiefly by railway companies. As improvements were made in body designs and riding comfort, however, and as roads were developed into smooth highways, bus service between cities grew rapidly. Within the ten-year period between 1920 and 1930, the beginning of the motor-bus era, the registered number of commercial buses increased from approximately ten thousand to almost a hundred thousand. By 1950 many cities were entirely dependent upon motor-bus service to nearby railroads.

Passing of the "Little Red School House." Motor buses brought about major changes in the rural schools, when in the 1930's they were used to transport children from their homes to school, and return. School districts with small populations merged with adjacent districts to organize township consolidated schools. This resulted in building better schools, with better equipment, and led to the employment of better qualified teachers. The "Little Red School House," symbol of America's rural educational system for over a century, was on the way out.

Change in body designs and the introduction of improved motor power resulted in a rapid expansion of the motor-bus industry. The capacity of the buses was increased by moving the motor over the front axle. Diesel motors were introduced. This meant that more miles per gallon of fuel could be obtained.

Motor Carrier Act. In the larger cities, large double-deck sightseeing buses were introduced. Bus routes were next extended to outlying towns and by 1930, it was possible to cross the continent by motor bus. By 1935 the motor-bus business had so developed that the Interstate Commerce Commission assumed jurisdiction of this vast network of transportation system under the Motor Carrier Act.

The motor-bus business expanded rapidly. The Greyhound system, with its affiliates in Canada, extended its network into the provinces of Manitoba, Saskatchewan, Alberta, and British Columbia, and then into the eastern provinces of the Dominion. In the 1940's it extended its lines into Mexico and established a major international transportation system. Other lines consolidated into a nation-wide network. Improvement in schedules, refinements in equipment, and improvements in service had placed the motor bus in a major role in the country's transportation system by the mid-century.

Highways Span the Continent

Following the close of World War I, the army in the summer of 1919 added its support to the good-roads movement by staging a transcontinental run from Washington to San Francisco. An army supply train consisting of seventy-five cars, trucks, ambulances, and repair cars, carrying two hundred officers and enlisted men, left Washington on July 7, 1919, and after sixty-two days of travel through mud, dust, rutted roads and sand, arrived at its destination. When a similar feat was repeated the following summer, pressure groups descended upon Congress asking for aid in a good-roads program.

Bureau of Public Roads, 1921. Late in 1921 a Federal Highway Act was passed, establishing a Bureau of Public Roads. A map of arterial highways was prepared, designating roads leading to every city of fifty thousand population or over. A ten-year roadbuilding program was adopted. The individual states united with the Joint Board of Interstate Highways in selecting routes that would best serve nationwide interests.

Roadbuilding Technology. The inauguration of such a vast network of highways brought about a new development in roadbuilding technology. Highway building and automobile manufacturing became twin giants in a new American industry. Colleges and schools of engineering introduced courses for training highway engineers. Roadbuilding became a science. Road testing and research laboratories were organized to aid highway engineers in selecting the best grades of stone, sand, gravel, bituminous materials, cement, concrete and brick. Studies were made showing the effects of temperature changes on road surfaces; subsoils were analyzed; and experiments were undertaken relating to the effects of the flow of water through drainage structures and of the run-off water on adjacent lands.

New Materials Sought. New road-surfacing materials were sought, suitable for automobile and truck. Pneumatic tires, rolling at high speed, stripped the waterbound macadam roads of the fine dust essential to binding the road together. New materials, such as asphalt, heated oils, and refined coal-tar products, made possible hundreds of miles of new construction. In 1914 of the several thousand miles of better-than-clay roads in the country, three-fourths had been constructed with better-type materials than macadam.

In the meantime, roads of concrete construction had been undertaken as an experiment. A concrete rural highway, one mile in length, was laid in 1908 in Wayne County, Michigan. The next year, 4 miles were built, and in 1910 20 miles were added. By the close of 1914, there were more than 2,000 miles of concrete construction on the rural highways of the nation. In addition to the roads built of concrete, there were some 1,500 miles of road built of brick by 1917.

Roads of Concrete. But the greatest development in roadbuilding technology was made possible by the introduction of concrete. After years of unsuccessful experimenting with asphalt, coal tars and oils, in an attempt to bind together the stone and gravel of macadamized roads, and to prevent clouds of choking dust, road builders turned almost exclusively to concrete. The quality of concrete construction was steadily improved. Improved methods for mixing and laying the concrete were developed. Mechanical mixers appeared during the early 1920's, capable of mixing a cubic yard of concrete. They operated automatically. Machines were designed to pour the concrete, and to smooth the surface. The ancient method of building roads with pick and shovel, plow and horse-drawn scraper was rapidly passing. The United States Bureau of Public Roads laid out test roads for experimental purposes. The longitudinal center joint was introduced, as were cross sections of expansion and contraction joints.

New Engineering Principles. The automobile brought about changes in engineering principles, as well as in construction methods. Tile drainage was installed to take care of underground water. Ditches were enlarged. Concrete box culverts, end walls, and drop inlet culverts made new demands upon highway engineers. The arched surface, or camber, was reduced. Instead of sloping 1 to 24, as in the macadam roads, it was reduced 1 in 72. Safety became an important feature. Sharp curves were widened and banked, or eliminated altogether. Highway engineering, formerly frowned upon by civil engineers, became highly technical and a highly respectable profession. A system of "stage" construction (so named because funds permitted the construction of a high type of road only in stages) was adopted. Surveys, grades, drainage, rights-of-way, and alignments were designed for future rather than immediate needs.

Roadbuilding Machinery. Equally significant was the introduction of specialized roadbuilding machinery. Mechanized equipment came with the advent of the high-powered gasoline engine. Excavating machines, rollers, graders, and scrapers, powered by gasoline motors, drove the four-horse teams off the highways. The day of the brawny laborer stirring cement with a heavy hoe, shoveling it into a wheelbarrow, and dumping it on the highway had passed. His work was measured by yards, whereas the work of the cement mixers was measured by miles. Construction companies turned to tool and machine manufacturers for new kinds of equipment, designed especially for excavating, grading, hauling, and dumping dirt and gravel.

Varied Equipment. Modern pneumatic drills, operated by portable air compressors, opened up holes several times faster than the old hand-held, hand-hammered drills. More powerful explosives blasted out tons of rock, ready for the huge tractor trucks to haul away for needed fills. With such technical devices as tractors, power shovels, trench cutters, graders, asphalt spreaders, rollers, dump trucks, and concrete mixers, highways of better grades and wider construction were laid out in hitherto inaccessible areas.

Pennsylvania Turnpike. The demand for road improvement led to the adoption of the old system of turnpikes and toll operation. An early successful example of this was the Pennsylvania Turnpike, constructed as an all-weather, high-speed road, and hailed as "The World's Greatest Highway." In its construction, all the known principles of highway engineering and skill, including the use of all modern machines and roadbuilding materials were employed. Authorized in May, 1937, by the Pennsylvania Legislature, the first major link of the highway, extending from a point near Harrisburg to a point near Pittsburgh, was completed and opened for traffic in October, 1940. Further building was halted during World War II, but soon after the close of the war, construction was resumed. The eastern extension from Harrisburg to Philadelphia, and the westward extension from a point near Pittsburgh to the Ohio State line, were completed and opened to traffic in 1951.

Novel Features in Road Planning. The major features of this all-weather Pennsylvania Turnpike were: a dual four-lane highway, 24 feet wide with a 10-foot divider. It was projected through rolling farm lands, and tunneled through seven mountain ranges, with no grade on the entire highway of more than 3 degrees, and no curve of more than 6 degrees. The road was built without permitting a single intersection or grade crossing along its entire length. The 200-foot wide right of way was fenced the entire distance to prevent hazards from pedestrians and stray animals. Included among the major construction features, besides the tunnels, were bridges and viaducts, the longest of which was a 600-foot span, 60 feet above the valley floor beneath.

Construction Technique. The construction equipment included huge trucks, scrapers, loaders, light railways for moving material and supplies, with diesel engines as the motive power. Giant stationary compressors were installed to furnish air for the rock-drilling machines, pumps, hoists, and other equipment. A drill carriage, traveling on rails 22 feet apart, permitted the loading shovels and trucks to pass through. While excavating the tunnels (some of which had originally been built for a railroad, later abandoned) ventilating systems were installed to protect the workmen.

The basic purpose of the Pennsylvania Turnpike was to shorten the driving distance between the eastern and western borders of the state, to enable cars to travel at high speed with the utmost safety. The Pennsylvania Turnpike became a model for a larger system of superhighways throughout the country. "It marked a mile-post in our forward march," declared Congressman Cartwright of Oklahoma, member of the House Committee on Roads, "toward overcoming natural obstacles and conquering the many complexities and uncertainties that prevail in this man-made world in which we live."

By the mid-twentieth century the nation had become turnpike-conscious, and toll roads, which took a heavy traffic from the state highways and city streets, were favorably received by travel-minded citizens. In the mid-1950's approximately five billion dollars annually were being spent on highway construction, and President Eisenhower, in an appeal to the governors of the forty-eight states, recommended a nationwide highway program extending over the next ten years, to cost more than one hundred billion dollars. It was needed, declared the President, to keep pace with the nation's expanding economy.

Railroads Meet Competition

Railroads, Diesels. Numerous advances in railroad transportation technology occurred during the second quarter of the twentieth century. The introduction of diesel locomotives brought many improvements. Their greater efficiency resulted in lower fuel costs. The diesel goes to the repair shop less frequently than the old coal-burning monsters, and it proved easier on tracks and roadbeds. In 1940 fewer than eight hundred diesels were in operation, and those were used mostly for switching. In 1955 diesels supplied the motive power for approximately 90 per cent of all railroads in the United States.

In the mid-1950's an improvement over the diesels was introduced, the gas turbine-electric locomotive. This new power unit, called a "second cousin" to the jet plane, operates on lower maintenance costs, cheaper fuel, and possesses far greater pulling power.

Track maintenance had likewise been largely mechanized. The former drudgery tasks performed by men swinging picks, shovels, and hammers are performed by numerous machines, some of which seemed almost human. Power hammers on wheels drove spikes in railroad ties. Machines for cleaning ballast rode the rails under their own power, screened the ballast, and dropped it back in place. Mechanical tampers, with short steel fingers packed the ballast evenly. This provides smoother roadbeds.

Safety Devices. Technical devices to ensure greater safety were introduced. With radio telephones, locomotive engineers could talk to engineers and conductors on other trains. Along with radio came a system of microwave communication, used where telegraph lines were difficult to install and maintain because of terrain and weather. Automatic block signals informed the engineer of track conditions miles ahead, and on some lines automatic controls stopped or slowed the train when approaching danger zones. Where the modern system of centralized traffic control had been installed, one operator in a central office maintained control of the movements of trains over hundreds of miles of track.

"Piggy-back" Freight Cars. A recent advance in railroad technology that has affected freight shipments was the introduction of the "piggy-back" service. By this method, highway trailers were moved on to flatcars, hauled to any desired destination, dismounted, and sent on their way to nearby areas that do not have railroad lines.

Passenger Comfort. Air conditioning to provide more passenger comfort had become almost universal by the 1950's. Another feature aimed at passenger appeal was the introduction of skylighted dome cars, which enabled passengers to "look up as well as out" while speeding across the countryside. For overnight passengers "roomettes" provided all the comforts and privacy of a hotel room. The railroads of America instituted an intensive research program in the early 1950's, aimed at improving transportation technology on what was still the nation's greatest freight carrying system.

A New Era of Aviation

At the close of World War I, there was a great surplus of planes and aircraft equipment, and thousands of young American boys who knew how to fly, but there was no place in the economy of the country to absorb a new aviation industry. Planes were sold at prices lower than automobiles, and many pilots, loathe to give up their wings, attempted to keep alive their interest in aviation. Airports and flying fields sprang up with little financial backing, and short air lines were established. These efforts, while not lucrative for the aviators, kept aviation in the public eye. But public interest in airplanes for commercial use had not reached the point where the industry could be supported.

Army and Navy Flights. With the traditional desire of the American people to return to peacetime conditions, interest in military aviation also dwindled. The Army and Navy continued to train flyers, however, and established a series of trans-Atlantic and transcontinental flights. In 1919 the Atlantic was spanned in a navy plane in three hops, with actual flying time of 26 hours and 45 minutes. The transcontinental race that was undertaken the same year was even more significant. It involved the establishment of a transcontinental airway, 3,000 miles long, across deserts, mountains and forests, with airports, landing fields, and radio and telegraph connection. Although these establishments were not continued, they laid a foundation for commercial aviation in the United States.

United States Air Mail. Following the close of World War I, it was the United States Post Office Department that took the initiative in commercial aviation. Air mail service between New York and Washington was begun in 1918 and transcontinental service in 1924. Although Postmaster General Burleson was a champion of air service when he advertised, December, 1918, for pilots having more than a thousand hours flying experience, there were only forty such in the country. The first venture resulted in piling up a huge deficit. In September, 1919, a route from New York to Chicago over the terrifying "hell stretch of the Alleghenies" was undertaken.

Frank Knight, Night Flying. In 1921 air mail service was extended to Omaha, and the first transcontinental mail flight was completed, February 22, 1921. This was made possible by the heroic venture of Frank Knight, a lone pilot, who carried the mail from San Francisco to Chicago, flying by night over an uncharted route with no radio communications, radio beams, instruments for blind flying, emergency landing fields, or lighted airways. Spurred by the historic slogan, "the mail must go through," Knight delivered his mail to his relays in Chicago, who carried it on to New York, completing the trip in 33 hours, 20 minutes. These flights led to the erection of lighted airways. Equipment for night flying was developed, and emergency landing fields were established. In 1924 night flying of the mails across the continent was recognized as a milepost in aviation history.

The Kelly Act, 1925. Postal revenues for mail service were far below cost, and it was not until the Kelly Act of 1925 that the foundation for commercial air transportation was laid. By this Act, transport of the mails was turned over to private enterprise under government contract. The following year the government had fourteen commercial groups flying the mails, and the same year Western Air Express opened regularly scheduled flights to cross the country in 48 hours.

Generals Billy Mitchell and Doolittle. Meantime, General William Mitchell, the distinguished flying general in World War I, and the commander of the first mass air attack in history, campaigned vigorously during the postwar years for an expanded air force. His demonstrations of the need for strong air defenses attracted wide publicity, and his bombing tests on four seized enemy war vessels, which had been turned over to America for experimentation, proved that aircraft could sink a dreadnought. In 1922 "Jimmy" Doolittle, who had taken a degree in aeronautical engineering at M.I.T., spanned the continent in less than 24 hours. In the next year the first nonstop transcontinental flight was accomplished. The year 1924 found the navy fliers undertaking the odyssey of a round-the-world flight.

Lindberg's Trans-Atlantic Flight, 1927. Great impetus was given to private flying by offering prizes for specific achievements. When in 1927, the Lone Eagle, Charles A. Lindberg, spurred by a twenty-five thousand dollar prize, spanned the Atlantic in the first nonstop flight, the American public hailed the young aviator as one of its immortals. Overnight, millions of Americans became aviation-minded. For aviators the flight of the Lone Eagle stimulated a rush for new records, and for youth it set aviation as a goal. For business, it opened up new channels. Thirty-one crossings were attempted in 1928 alone, only ten of which were successful. Amelia Earhart, Wiley Post, Frank Hawks, Howard Hughes, Jacqueline Cochran, all set new standards in flying and accomplished feats believed to be impossible.

Civil Aviation Act, 1929. The Air Commerce Act of 1929 became the legislative cornerstone of commercial aviation legislation. It was the nation's first federal civil aviation law. A Bureau of Aeronautics under the Department of Commerce was created, all planes and pilots were required to have a license; inspection and registration was authorized, and an aeronautical research division was established. At the same time the Army and Navy were given authority to initiate costly research beyond the capacity of private firms to operate, resulting in a number of improvements in plane design.

Improvements in Planes. With the growing interest in aviation came improvements in planes, both in the engines and in construction. The "Wasp," an air-cooled radial engine generating 400 horsepower, was described as the first American engine to rank with European competitors. Among other technological advances were improved designs for increased wing loading, multimotor equipment, new methods of cowling (covering) the engines, and the introduction of new antiknock high-octane fuel. Other technical advances included the introduction of a retractable landing gear (tuck-away wheels), wind flaps designed to permit fast ships to take off speedily and land safely, the addition of the controllable-pitch propeller, and in response to passenger demand, a system for reducing the deafening noises.

Metal for Planes. Planes constructed of metal began to appear in the late 1920's. An all-steel plane, save wing and tail covering, was built of stainless steel in 1931, but by the mid-1950's aluminum was largely replacing steel. Air-cooled radial motors supplanted the water-cooled motor. Improved landing instruments, the aid of radio, the altimeter, and blind flying instruments added to air-travel safety. Pilots began to rely on their instruments and flew above the mountains, not between them.

New and better instruments increased control and safety; the supercharger was improved, allowing airplanes to fly at higher altitude; the variable-pitch propeller gave the pilot greater control over the speed of his plane. Planes became lighter when aluminum construction was introduced. The designs were simplified and the engines cowled to reduce drag, thereby increasing the economy of operation, which in many instances spelled the difference between profit and loss. Self-starting engines were installed; planes were operated by radio beam; and air conditioning made high-altitude flying physically comfortable. With the introduction of all these technical improvements, an increasing number of travel-minded people took to the airways.

The DC-3, 1936. In 1936 the modern airliner made its debut when the DC-3 appeared. Carrying a payload of 9,000 pounds and twenty-one passengers, it cruised at 180 miles an hour. The plane measured 64 feet in length, had a wing span of 95 feet, and weighed 24,000 pounds. It became the predominant type in commercial aviation. The DC planes freed the airlines from dependence on government subsidized mail cargo. They could make money by hauling passengers. Next, when a four-engine giant appeared, the "Stratoliner," with accommodations for thirty-three passengers and with a cruising speed of 250 miles per hour, the age of transport liners had arrived.

Civil Aeronautical Act, 1938. Airplane regulation was codified by the Civil Aeronautics Act of 1938, whereby a Civil Aeronautics Board, a policy-making authority, was created and a Civil Aeronautics Administration was empowered to certify routes, to license pilots, and to carry on administrative work.

Global Service. In the meantime intercontinental and overseas air transportation was established. The Pan American set out to secure large mail contracts, and expanded from an airway of 250 miles to a trunk line encircling all Latin America. In 1931 it launched its first "American Clipper," and Buenos Aires was reached by air for the first time. In 1935 the company extended its service to the Philippines, and the next year to Hongkong. By 1937 trans-Atlantic flying had been established by both European and American airlines, with planes passing over midocean. Overseas flying stimulated radio navigation and called for a standardization of flying technique. The pilot became a captain, and his plane was equipped with a myriad of costly electronic gadgets.

Airplane Put to Other Uses. It was during the decade of the 1930's that the many uses of the airplane were discovered. In the winter of 1932–1933, planes dropped food to the snowbound inhabitants in Colorado, New Mexico, and Arizona. Fodder was dropped for sheep, and hay for cattle. Aerial service was instituted for the forest fire patrol. Mapping was done by plane, and aerial pictures were taken. In agriculture, spraying from planes was instituted to kill pests and to distribute fertilizer. Planes became cargo ships, and anything small enough to fit into the hold of the airplane that was wanted in a hurry became available by means of air transportation.

Research in Aviation Technology. The amazing technological advances in aviation in the 1930's, despite the nationwide depression, arose from the co-operation between the centers of basic research of university laboratories, Wright Field, and the National Advisory Committee for Aeronautics. The aviation industry profited greatly by scientific research. The small funds allotted to research during the early part of the decade mounted into the hundreds of millions of dollars by 1940. At the close of World War I, the United States was at the bottom of the list as far as air power was concerned, but before the outbreak of World War II, this country had the largest air force in the world.

The airplane had ceased to be a one-man product, or a one-industry product. Metallurgists developed light alloys; chemists improved the gasoline; civil engineers designed new strong, light structural shapes; mechanical engineers developed the efficiency of the engine; the electrical engineers devised radar, the radio, and countless electrical gadgets to aid the pilot. The National Advisory Committee for Aeronautics, organized in 1915, with laboratories in Virginia, Ohio, and California, had succeeded in making the American transport planes standard equipment for most of the world's airlines.

Aerodynamics, Thermodynamics. The fields of research included aerodynamics, a study of the effect of mechanical forces upon air in motion; thermodynamics, the improvement of the engine and the engine parts; and the study of the structure of the plane. Any part of the plane that set up resistance to the airstream, and thus reduced its aerodynamic efficiency, was studied with the view of improving it. The engine was cowled; and a system of ducts, inlets, and scoops for inducting the air won the Collier trophy for the National Advisory Committee for Aeronautics in 1929. Multiple engines were introduced, and engine design was streamlined and simplified. Speed records were increased from 200 miles per hour to close to 500, and cruising speeds increased from 120 miles per hour to 250 miles per hour.

Structure. More spectacular than any other improvement was that of design. Wood and steel gave way to aluminum. Boxlike wings gave way to the internally braced thick wings with metal skin. A commercial plane with a single-shell construction, the Lockheed "Voga," was built in 1927. Most of the features utilized in the 1930's had been discovered years earlier, but it was in that decade that the best from each former type was adopted and assembled into the efficient air transport of that period.

The Turbo-Prop Engine. Evolving from World War II came a new type of engine which brought about a change in aerodynamics. The reciprocating engine used to drive a propeller had been the principle of propulsion since first utilized by the Wrights, but further improvements on the reciprocating engine had reached the stage of diminishing returns. Propellers became inefficient at a speed over 400 miles per hour. The turbine, used to drive a propeller in the turbo-compound engine (piston engine), greatly increased the engine's power. This type of engine was readily adapted to planes in the regular passenger liners. The turbo-propeller engine was developed. This dispensed with the piston engine entirely. The propeller was operated directly from the turbine. This engine proved to be much faster than the piston engine and developed amazing horsepower.

Jet Engines. A radical change came with the advent of the jet engine which eliminated the propeller and propelled the plane by thrust. The jet engine is so called because it concentrates a jet of gas under great pressure on the blades of a turbine. The exhaust of hot gasses pushes the plane along like the recoil of a gun. The jet principle was applied in the turbo-jet, the ram-jet, and the rocket. The development of jet engines necessitated a change in the design of planes. Wings were made thinner, short and stubby, and swept back sharply from the fuselage. Jets, operating at high speed, performed best at high altitudes where the air is thin. A jet engine is not measured in horsepower but in pounds of thrust, because it has no power except when it is moving.

Supersonic Jets. The speed of the jet has made it possible to break the sound barrier. In October, 1947, Captain Charles Yeager's rocket ship flew faster than 1,000 miles an hour, far exceeding the 760 miles per hour of sound. In 1951 William Bridgeman pushed a Douglas "Skyrocket" to 1,258 miles per hour, and in 1953 Scott Crossfield almost doubled the speed of sound when he flew 1,327 miles per hour. Jet-propelled planes have been adapted to sustained flights as well. A National Guard aviator started the year 1954 with a new record when he made the flight from New York to Los Angeles in 4 hours, 8 minutes, and 5 seconds—an average of 615 miles an hour.

Helicopters. Of growing importance is the helicopter, called the "ugly duckling" of the aeronautical family. Igor Sikorsky is regarded as the pioneer builder of the helicopter in the United States. The helicopter differs from the rigid-winged plane in that its wings are the rotors, or airfoils, which whirl above the fuselage. The wings are narrower and thinner than those on a conventional plane. It is powered by an engine mounted on the fuselage. Its unique device is the rotor head on the shaft that runs from the engine to the whirling rotors. This device enables the pilot to tilt the rotors in the direction in which he wishes to fly.

Uses of the Helicopter. The helicopter is the most adaptable of all aircraft—it can do anything a bird can do. It is capable of going up and down and sidewise; and it can hover. It operates at slow speed, or it can travel 100 miles per hour. It can fly under other planes, and yet rise above surface obstructions. It does not require large landing fields with runways. A helicopter can take off from a tennis court or from the roof of a properly constructed building. It has been called a versatile "windmill."

The helicopter has been put to many uses. In the Korean War it was used to evacuate wounded troops from inaccessible places and to transport a company of marines fully equipped to a rugged mountain top, ready for combat. For peacetime uses, it has been adapted to commercial transportation service, relieving the congestion on crowded city streets. It has been used in New York City for police-patrol work, and to perform missions to otherwise inaccessible locations. In August, 1953, a helicopter was tested on a hop to Europe, and it successfully completed the trip, with the longest flight from Iceland to Scotland, a distance of 940 miles.

Science Creates a Giant Industry. In the brief fifty years since flight was inaugurated, air transportation has become one of the nation's giant industries. In 1951 it passed the billion dollar mark, with revenues totaling well over one billion dollars. Airports have become busy terminals where planes arrive and depart every few minutes. By the mid-1950's, airlines were carrying some forty million passengers over many billions of passenger miles, and the aviation industry was employing close to one million workers. An era believed by many to be the final development in transportation technology has arrived.

Notes for Chapter 38

For the history of the automobile, see the following: G. B. Glasscock, *The Gasoline Age—The Story of the Men Who Made It* (1937); David L. Cohn, *Combustion on Wheels, An Informal History of the Automobile Age* (1944); E. D. Kennedy, *The Automobile Industry—The Coming of Age of Capitalism's Favorite Child* (1941); R. M. Cleveland and S. T. Williamson, *The Road Is Yours, The Story of the Automobile and the Men Behind It* (1951); Arthur Pound, *The Turning Wheel* (1934); St. John C. Nixon, *The Invention of the Automobile* (London, 1936); Joseph P. Black, "How To Please 210 Million People," *Travel* (November, 1949); Eugene W. Lewis, *Motor Memories, A Saga of Whirling Wheels* (1947); Stephen Longstreet, *A Century on Wheels, 1852–1952* (1952); Bellamy Partridge, *Fill 'er Up, The Story of Fifty Years of Motoring* (1952); Victor Longheed, *The Gasoline Automobile, A Practical Discussion of the Development and Present Status of the Automobile* (1912); Howard W. Troy, "The Four Wheel Drive Story," *Moody's Manual of Investments* (1947), (1948), (1949), (1950); *Greyhound, The Motor Bus and Modern Highway Transportation,* The Greyhound Corporation, Chicago (no date). Articles of special interest include one in *The Saturday Evening Post* (April 20, 1946); two in *Fortune Magazine* (August, 1934, and September, 1944); and one in the *National Geographic* (July, 1950). Literally hundreds of articles on the automobile, referring to its invention, introduction of parts are found in the proceedings and transactions of scientific societies. Most helpful are the *Transactions of the Society of Automobile Engineers.*

The following are helpful on certain phases of technological developments in railroad history: J. E. Watkins, *Development of American Rail and Track,* Smithsonian Institution Report, V (1888–1889), pt. 2, 44; S. K. Farrington, *Railroading from the Head End* (1943); Richardson and Blount, *Along the Iron Trail* (1938); D. M. Rice, *Romance of the Railroad* (1941); Frank F. Morse, *Cavalcade of Rails,* (1940); J. B. Husband, *Story of the Pullman Car* (1917); Edward Hungerford, *Locomotives on Parade* (1940); *A Yearbook of Railroad Information,* Eastern Railroads President's Conference, 1945 edition.

The following are helpful for tracing the advances in aviation technology: Archibald Black, *The Story of Flying* (1940); Roger Burlingame, *Engines of Democracy* (1940); Chelsea Fraser, *The Story of Aircraft* (1939); G. W. Gray, *Frontiers of Flight* (1948); Charles Grey, *The History of Combat Planes,* (1942); Alden Hatch, *Glenn Curtiss, Pioneer of Naval Aviation,* (1942); Douglas J. Ingells, *They Tamed the Sky* (1947); Bernard Jaffe, *Men of Science in America* (1944); Samuel Paul Johnston, *Flying Fleets* (1939); G. B. Manly, *Aviation from the Ground Up* (1940); Henry L. Williams, *Fundamentals of Aviation,* (1946); Shafto Dene, *Trail Blazing in the Skies,* (1943).

Chapter 39

TECHNOLOGICAL ADVANCES IN METALLURGY SINCE 1920

Metals provide the very sinew and strength of our modern civilization. The vast increase in their use has been a major force in developing the strength of America today. The abundant use of metals has long been recognized as a measure of the standard of living of a people. Agricola's monumental work of the sixteenth century (translated in 1912 by Herbert C. and Ann Hoover), which was the standard for two centuries, advanced the thesis that metals are the greatest necessity to man, and that art is not of much importance if it lacks metals.

Metallurgy Key to American Prosperity

In America the contribution of metallurgy has been distributed over all fields of activity—agriculture, communications, transportation, manufacturing, power generation—and these metals are being used not only in larger amounts, but in a greatly diversified number of ways. While the quantity phase of the use of metals was inaugurated with the Bessemer converter, the quality phase did not come until about the third decade of the twentieth century, spearheaded by metallurgical science.

Science in the Iron Industry. Steel, which represents 90 per cent of the nation's metal industry, has undergone a revolution in the last thirty years under the guidance of science. The United States struggled through World War I with a shortage of surgical instruments, specialized cutlery, and poor razor blades because the finer steels had come from Germany. But when large research laboratories were established, metallurgical changes were introduced, and a new age of steel dawned. In the new age, steel was not the heavy industry it had once been, in which massive products were turned out in great quantities, but it had become an industry where exactness and precision were demanded. Scarcely one-fourth of the production was fabricated in the same fashion in 1937 as it had been in 1927. A steel expert of the 1920's would have been lost in a steel plant of the 1930's, the change was so radical.

Improvements in Steel Technology. While the basic processes of making steel continued, these were so improved, from mining the ores to fabricating the finished product, that they bore little resemblance to the methods of earlier days. Anticipating the day when the supply of iron ore would be exhausted, tests of low-grade ore were made which showed some of these to be superior to the high-grade ores. Blast furnaces were operated by push-button control and became gargantuan structures with a capacity of 1,200 tons, as compared with Carnegie's huge (for that day) Isabella furnace of 100 tons. The Bessemer converter gave way to the open-hearth and electric furnaces. It, however, was revived in the war emergency, and the electric eye was introduced to perform the critical operation of determining the time of the blast, taking the place of the former skilled operators. Induction-type electric furnaces were put in operation as well as the arc furnaces for certain types of steelmaking. Plant production was integrated, that is, all operations were performed in one plant, from the ore to the finished product. Plant capacity was increased 55 per cent over 1917, and eight times that of 1901. A steel plant in the 1930's with an output of a million tons of ingots represented a capital investment of one hundred and sixty million dollars, with the rolling mill alone costing sixty million dollars.

Improvements in Steelmaking

Continuous Rolling. One of the most radical improvements was the introduction of the continuous rolling mill. This mill produced sheet steel in coils hundreds of feet long and wider than was possible in old mills. From this an automobile top could be pressed out in one operation, or a front fender drawn out in one piece. Such steel was not only larger in size, but also was made with a tensile strength to stand the strain of terrific stretching; it was fine grained, dense and flawless, to take a polished surface. In 1947 a continuous mill was turning out steel strip at the rate of a mile a minute. The continuous mill marked the first improvement in sheet rolling in a hundred and fifty years, and eliminated one of the most arduous tasks in the whole steel industry, that of rolling steel. The new rolling process made possible tin cans of finer grain, so that some fruits which hitherto could not be processed were successfully canned in the new type steel.

Heat Treatment. The metallurgist introduced an exactness and precision in heat treatment so that such properties as toughness, hardness, and resiliency could be accurately predicted and controlled. This was accomplished by fitting the treatment to the steel and by instituting controls in both the heating and cooling processes. Such precision had been unknown in the early 1920's.

Alloy Steel. Alloy steels, or those in which some elements other than carbon are employed in sufficient amounts to modify their properties, had been used in the first decades of the century for paring knives and cutlery; in the 1920's they were used to make parts. By 1930 there were some two hundred trade names for specific steels. In an automobile there were thirty-two distinct types, each specifically made to order. Chrome was added for toughness, nickel for surface hardening, chrome molybdenum for strength, chrome vanadium for resiliency, tungsten for efficiency in machine tooling, and copper for resistance to corrosion. These several alloys found uses in engine parts, gears, machinery parts, locomotive frames and parts, and tool points and edges. Some 300,000 tons of alloy steel were turned out from the electric furnaces in 1930. Metallurgists envisioned even greater uses for alloy steels. Ralph G. Budd conceived of an alloy-steel airplane twelve years before it became a reality as a cargo plane in the World War II. Budd's idea for an alloy-steel train met with more immediate success when the Burlington "Zephyr" won fame in the 1930's in breaking records between Chicago and the Pacific coast.

Electric Welding. In the 1930's the uses and applications of alloy steel were enlarged. A process for welding alloy steel by electric current was devised, making possible the construction of larger members with this metal. Turbo-generators were made to withstand a pressure ten times as great as those used during the 1920's. Bulkheads and seawalls were made with sheet piling using corrosion-resistant steel alloys. Seamless tubes made it possible to drill oil wells to a depth of 3 miles, as compared with 8,000 feet in 1924. While a Liberty engine weighed 1.7 pounds per horsepower, the new alloy steel reduced this to 1 pound per horsepower. The automobile took 20 per cent of the steel output, while other huge users were railroads, container industries, shipbuilding, and office and household equipment. Ten million dollars a year were spent on research, and by 1940 in the steel laboratories of the country there were some twenty-five hundred metallurgists, chemists, physicists, and engineers, all devising new properties to be used in tailor-made steels. Physical properties were tested with the microscope and X-ray, and chemical qualities were tested by analysis.

Steel Goes to War. General Brehon Somerville was quoted as saying that when Adolph Hitler put his war on wheels, "he ran right down our alley." No people could have been better prepared in the development of the steel industry to fight a war than was the United States. World War II has been called "a war of alloy steels," and the great advances made by this country in the 1930's paid off. The ingenuity of the steelmakers was taxed to the utmost, however, when the supply of alloying metals was cut off. When the Japanese gained possession of the islands in the Pacific, tungsten, chromium, nickel, and manganese (for which the United States depended on foreign importation) were cut off. With the pressure of war needs, stock piles diminished rapidly. The situation was critical. The scientists worked feverishly in the laboratory to devise a way to stretch out the reserves of the scarce alloy metals. Formula after formula was tried. Again the years of research paid off when science discovered that by using smaller percentages of two, three, or even four of the more abundant alloys, rather than larger percentages singly of the critical ones, an alloy steel could be produced to serve the purpose. The wartime steel was called "national emergency steel," but to the steelmen it was known as "dirty steel." By 1943 it constituted one-third of the total alloy steel production and it is difficult to see how America could have built up her tremendous fighting power without it.

Postwar Steel. Since the war, the expected slump in the steel industry was not of material consequence because the industry was ready to fashion products with especial adaptations for peacetime needs. The great building boom created new demands in steel siding and walls. Transportation needs, such as automobiles and railroad cars, which had been neglected to meet the wartime emergency, required vast quantities of the special steels designed for specific purposes. Factories making peacetime products had to retool. The great steel laboratories entered into a new field of expansion to meet these new demands and to cope with the challenge presented by the nonferrous metals and the newly developed plastics and other materials.

ALUMINUM

One of the challenges to steel in some of its uses was presented by aluminum, which came out of the cooking utensil stage to take its place in industry in World War I. A slump that occurred in the industry immediately after the war, due to difficulties in machining the product, was offset by extensive research begun about 1920.

Science came to the rescue and new technologies spurred the aluminum industry to enter new fields. Great advances were made in developing strong and relatively hard alloys, and in developing methods and applications of heat-treating them to improve their qualities for specific uses.

New Uses. Aluminum alloys were used in both wrought and cast form. The first uses for aluminum were found in the automobile where forged aluminum was used in connecting rods and reciprocating parts, and finally in pistons. With the advent of the trailer in the early 1930's, the use of aluminum was increased. Railroads used aluminum first in seats and nonfunctional parts. All-aluminum car bodies, superstructures, and underframes were built in 1931. In aviation, aluminum was first used for structural members and wing cover. With the development of better alloys, the Douglas airliner was made possible and the 1-ton Boeing clipper by the mid-1930's. Building construction found uses for aluminum, first in roof trusses and framing for domes; then in windows; and by mid-century, in siding and walls. Aluminum was used for bulkheads for dams and for paving in bridge construction.

Uses in World War II. From the attack on Pearl Harbor to V-J Day more than 11,000,000,000 pounds of aluminum were produced for war purposes. Seventy-five per cent of the weight of the average plane was made of aluminum, and the extra speed that was made possible by the reduction in weight spelled the difference between success and failure in many missions. Aluminum was used in the superstructure on ships, for pontoon bridges, for parts for jeeps and tanks, and for aircraft landing mats.

Postwar Development. Out of the war came a know-how in workmanship that was transferred to peacetime uses. Refinements in processing were developed. New alloys were discovered. Aluminum foil, a great boon in packing and a noncombustible wrapping, was taken over by a number of industries, especially the food-packing and processing firms. Aluminum has continued to serve in nonspectacular ways corresponding to the little combs and pin trays with which it started out, but it has also filled some of the most critical and dramatic needs in the metal industry of the nation.

Charles Hall. When Charles Hall in 1886 gave the world the first piece of shining metal produced in a process that was applicable commercially, he envisioned great uses for his discovery. Time was required, however, for other advances to be made to make his vision possible. Hall's process for making aluminum was a dual method in which, first, a powder called "alumina" was produced from bauxite ore, by a chemical process in a digester. The second process was transforming this powder, the oxide of aluminum, to the pure metal by a smelting process in a flux of cryolite, activated by means of an electric current in "pot lines" of steel cells. An enormous quantity of electric current was required to effect this second operation, and the development of the process was dependent on the development of greater turbines to generate the current. Supplies of bauxite, originally from France, were discovered in this country, chiefly in Arkansas, but improved refining processes were worked out before they could be utilized. Fully sixty such processes have been developed.

Fabrication. Fabrication of aluminum proved a stumbling block from the beginning. The Pittsburgh Reduction Company, which first took up the development, could find no industries to handle the product, and therefore the pin tray and cooking utensil business arose. As new methods of fabricating metals developed, however, many of these were applied to aluminum. One of the best suited to aluminum was die casting. Cast products of various alloys were extruded through dies. Aluminum was rolled into sheets, some of gossamer thinness, only five-thousandths of an inch thick. As a paste or powder it was mixed with a solvent and was used as a leafing agent. The demands of the radio increased the output in die casting. Forging aluminum, a difficult operation because perfect homogeniety was required in the billet to be forged, was effected by the 1930's. Heat treatment greatly enhanced the quality of aluminum alloys. Wire-drawing technique made aluminum applicable for fences and screens, and brazing surpassed welding in certain uses.

Production of aluminum has made great strides in its brief history of seventy years. It has attained a technological standard equal to the oldest metals and has pioneered in developing new processes that have been transferred to other fields.

Modern Application of Various Metals

Other nonferrous metals have come into increased use as metallurgy made rapid progress. Some of these are the age-old metals which have been put to new uses. Others, hitherto little known outside the laboratory, have come to take their place in industry, still submerged, because their role has been that of an alloy in steel. Still others, brand new metals, finding their place in the periodic table of the laboratory for the first time, have immediately exerted a great influence. Development of these metals has been called the "miracle of metals."

Role of the Chemist. In these developments the chemist has played the chief role. He engages not only in producing the materials in the greatest quantity at the least cost, but also attempts to discover the best utilization of the nation's mineral resources. Efficient utilization has been the primary objective. By substituting the abundant metals for the scarcer ones, by obtaining the metal product from the less convenient sources, and by salvaging precious metals from dumps and waste products, the metallurgist has placed conservation on an equal basis with supply. Secondary sources have been developed, and economies in smelting and processing have been effected.

Copper. The age-old metals, copper and lead, for which no new uses had been found for hundreds of years, came into new uses with the electrical age, and attained greater eminence than they had ever achieved in their long history. Copper wire was used for transmission of electrical current, and in many other electrical functions. Since the supply of copper is limited, and it is feared it will become exhausted, conservation methods were adopted. The great smelting plant at Bingham Canyon, Utah, stands as a model of modern engineering and metallurgy. Here, a whole mountainside was blasted away so that ore could be scooped up in a triumph of mass production. A by-product plant was erected with a half-mile of flues, and a 585-foot smokestack was built from which a variety of products are obtained in fragmentary bits. Most ores contain small percentages of many metals. In the smelting plant, a reverberatory furnace and a Bessemer converter, borrowed from the steel industry, have supplanted the old wasteful blast furnace. Here, in a dual operation, a coarse metal, the "matte," is produced first. In the converter, a chemical change occurs, resulting in 99 per cent pure copper. Efforts to salvage copper have been so intense that it has even been collected from the drainage of the watermains.

Lead. Lead, while more abundant, exists in a limited quantity in the earth's crust. By a process called "cupelling," lead oxide has been extracted in separating silver from the lead ores. This oxide was used in paint. Lead smelting has been one of the simpler operations carried on in a reverberatory furnace. The role of lead in the storage battery, as insulation in the use of X-ray, and as sheathing for the great Atlantic cable, is of utmost importance.

Flotation. A twentieth century innovation, the outgrowth of American invention and initiative, has been the flotation process used in securing metals from low-grade ores and from the waste piles of former processes. Metals of great value have been recovered by this method, which has been especially applied to recovering copper. In this process, finely crushed ore is churned to a froth with water and oil. By what is termed a "surface tension," the particles of metal cling to the oil as it rises to the surface and can be removed, while the rock and debris sink to the bottom. With this system a concentrate is obtained, not a pure metal.

Electrolytic Methods. Electrolysis has been used to a greater extent in late years in refining metals, and has been more practical since extensive electric power has been made available through the many water-power projects. While copper refined by smelting is practical for some purposes, in electrical transmission a fraction of 1 per cent of impurities is objectionable, so that the electro-refining of copper has been highly developed. This is conducted in a cell of copper sulfate solution, with the coarse copper at the anode and pure copper being collected at the cathode. Electrolytic zinc is leached out in a bath of dilute sulfuric acid. Aluminum, as noted earlier, was made in a melt of cryolite.

Zinc. Zinc, tin, and nickel have found new uses in the age of metals. The use of zinc has been greatly accelerated for galvanizing. While it has been long used as an alloy with copper in making brass, the electrical industry has greatly increased this demand. The metallurgy of zinc has been improved by the introduction of the flotation process as a primary system, and the concentrate thus obtained has been further refined by electrolysis or by distillation. Zinc has been salvaged as a by-product in smelting other ores by a fusing process, embodying concentration of zinc fumes.

Tin and Nickel. There are no tin mines in the United States, and the shortage of tin has often created a crisis. America is the world's greatest consumer of tin, needed to satisfy the voracious appetite of the tin-can industry. The recent development of salvaging tin in detinning operations has been the source of a considerable supply. Nickel has risen to a top place in its role as an alloy in hardening steel. The newly opened nickel mines at Sudbury, Ontario, have assured a supply to this country. The flotation process has been effectively applied as a first step in refining nickel, by which a "matte" is produced. This is smelted with coke and "salt-cake" to produce the pure metal.

Alloys and the Newer Metals

Alloy Metals. Of the alloy metals—manganese, chromium, tungsten, vanadium, and molybdenum—90 per cent of the amount produced is used in steelmaking. When used as alloys, the oxides of the metal are employed. These are usually produced in an arc furnace. Where the pure metal is needed, it is produced by the Goldschmidt process in which the oxygen is removed from the oxide ores by a process involving aluminum, or it is produced by electrolysis. Although manganese was used in the early days of the Bessemer process, the other alloys have been more recently employed. Vanadium was not made commercially possible until the discovery of the Peru deposits after World War I. Wolframite, found in Colorado, is the best known of the tungsten ores outside of those from China. The Colorado ores are of such low grade they require concentration (or "beneficiation") before smelting. Another tungsten ore, sheelite, found in the Southwest, has been eagerly sought. Because it has the property of glowing in the dark, searchers go out at night with lanterns to discover deposits.

Molybdenum, the most recent addition to the alloy metals, is found in this country. When the supply of tungsten was cut off from China in World War II, molybdenum was used as a substitute. It takes 1 ton of American molybdenum ore to produce 6 pounds of metal. A nonalloy use of chromium is for plating metal. An American achievement in producing wrought tungsten made possible the tungsten filament for electric light bulbs.

New Metals. A metal-hungry world with a seemingly insatiable appetite has turned to the depths of the ocean and the sands of the sea, to the smoke from the furnaces and the tailings of the mines, to the coal beds and the gravel pits, and has found new metals which the metallurgist has discovered ways to process. He has also found new uses for them. Some of these have been known in the laboratory, and some have had limited uses; for the most part, the world had never heard of these new members of the metal family until the past decade. Included in the list are magnesium, an old friend used in fireworks manufacture, but now in a new guise; titanium, which has come into prominence since 1950, and its sister, zirconium, made available by the same process and used only since World War II; beryllium, a poor cousin to the emerald; uranium, searched for in discarded mines and dumps with the geiger counter; and germanium, a brand new metal found in smoke from the zinc furnace and in coal pits.

Magnesium from the Ocean. Magnesium has been called the "cinderella of metals" because it came into prominence overnight for use in structural purposes. Light, but sturdy, it appears destined to attain the rank of third greatest in our metallurgical economy. The supply of magnesium is unlimited, for it may be obtained from sea water. In this process, sea water is mixed with oyster shells, which provide lime, then filtered in baskets, neutralized with hydrochloric acid, and finally converted into metallic magnesium by the electrolytic process. Eight hundred tons of sea water produce one ton of magnesium.

Titanium, Zirconium. Titanium and zirconium were first made available by the discoveries of Dr. William Kroll, a German refugee to the United States. He gave his discovery to the United States Bureau of Mines. The ore from which titanium is obtained, ilmenite, is the fourth most abundant on the earth's crust. Zirconium is obtained from the sands of the beaches. It is the extraction of the metals that presented the hazards. This involved a dual process, first consolidating the elements by a reducing agent into a "sponge" of granular metallic particles. This, through a double-melting process, was transformed into ingots of pure metal. Improvements have been sought and a new sodium process evolved in England is being tried. Zirconium has been used as structural material for atomic reactors. Titanium is a superstrength light metal, with characteristics between those of aluminum and steel.

Beryllium. Beryllium, obtained from stones or beryl found in Argentina, has become a rival to aluminum as an alloy. It is lighter than aluminum, has great tensile strength which aluminum lacks, and, resists corrosion as well. Alloyed with copper it renders that metal hard enough to cut steel.

Germanium, Uranium. Germanium has been found to possess electronic potentialities and has been utilized in making transistors for hearing aids, replacing the bulky electronic vacuum tube. It has been applied to many other electronic uses. Uranium, which exists in more than a hundred ores, is being sought today more frantically than was gold by the "forty-niners." Until Professor Arthur Demster, while working with the metal in the laboratory of the University of Chicago in 1935, discovered its fissionable qualities, the metal was little known. With uranium, science has been able to split the atom and thus release the tremendous energy that holds the unit particles of the central core of the atoms together, their "binding" energy, thus producing atomic energy.

Powder Metallurgy. With alloys occupying an increasingly greater position in the nation's metal economy, a new method of compounding these alloys by means of powder metallurgy has been developed. In this process only pure metals are used. These are reduced to powder form, mixed according to specifications, and extruded through a die, casting the powder directly into the desired shape and form. The technology of the operation lies in the preparation of the powder, and in subjecting the mixed powders in the die to heat and pressure so that a uniform welding and fusion between the individual particles is effected. Carboloy, a cemented tungsten carbide and an alloy second only to diamonds in hardness, has been produced by this method. Parts for atomic reactors, and even stainless steel, have been produced by powder metallurgy.

Miracle of Modern Metallurgy

The history of metallurgy reveals that metals have been used during the past half-century in far greater quantity than ever before, and have been made to serve a number of new uses. The entire group of metals, from the old "stand-bys" to the strange new ones, have been in great demand. A recent government report predicts that by 1975 the mineral demand will exceed the present production by 90 per cent.[1]

While the quantity phase of development has been spectacular, the quality phase in metallurgical technology has been of even greater significance. New metals, or substitutes, have been sought to replace the exhaustible ones. Metallurgical technology has found ways to salvage metals from dumps, to collect them from gases, reclaim them from streams, and extract them from the ocean.

Metallurgical science has developed economies in smelting and processing metals by discovering ways for utilizing by-products, and by securing in one operation a variety of desired metals contained in ores that were once discarded. Metallurgical technology is creating new shifts in power and offers a new freedom of choice and action in this modern age of metals.

Notes for Chapter 39

1. *Paley Commission Report,* 1952.

See also *The Pirotechnia of Vannoccio Biringuccio,* translation by Smith and Gnudi (1942), American Institute for Mining and Metallurgical Engineers; Charles C. Carr, *Alcoa, An American Enterprise* (1952); Robert F. Mehl, *A Brief History of the Science of Metals,* American Institute for Mining and Metallurgical Engineers (1948).

Chapter 40

TECHNOLOGICAL ADVANCES IN NONMETALLIC INDUSTRIES SINCE 1920

While metals have played a prominent part in American technology, great progress has also been made in the nonmetallic industries since World War I. New products have been developed which in many cases rival the metals. These industries, like the metal industries, have been characterized by an increased application of science and more dependence on the findings of the laboratory.

The Glass Industry

The ancient industry of glassmaking has undergone a revolution since World War I. New processes have been introduced which have no resemblance to the age-old practices. New products have been developed that were never before conceived of as a function of glass. Machines have been designed to utilize the peculiar qualities of glass that the most highly skilled craftsman could not have developed. Due to applied science these hitherto little understood qualities of glass were ferreted out.

The Corning Ribbon Machine. One such example was the Corning ribbon machine for making electric light bulbs. This machine was fed by a continuous supply of molten glass, which flowed between two rollers fitted with a series of depressions and emerged as a ribbon of glass with biscuitlike protrusions suspended. Individual molds closed around each biscuit while tiny puffs of compressed air were injected into each. As the molds were released, the ribbon of glass continued and the tap of a hammer detached them as fully formed electric bulbs. The Corning ribbon machine was capable of producing six hundred to seven hundred bulbs a minute.

The Danner Machine. In making glass tubes and rods the Danner machine was made to utilize the viscous property of glass. In this machine the molten glass was allowed to flow continuously from an orifice in the melting furnace to a revolving mandrel, or core. Passing over this, it was seized by mechanical stretchers, which pulled it over a series of rollers, elongating it to the desired thickness. For making tubes rather than rods, a small pipe was inserted in the center of the mandrel, through which compressed air was constantly supplied as the glass was drawn off. A Danner machine could manufacture as much tubing in one day as ninety workers could produce by the old method.[1]

Vello Machine. The Vello machine, operating on a similar technological process but using an up-drawn principle, was capable of almost doubling the production of glass tubes. A tiny bore for thermometers was made, and tubing for such uses as neon lights, automobile fuses, and glass piping was provided.

Window Glass. The idea of utilizing the viscous property of glass in making window glass was introduced during World War I by I. W. Colburn of Franklin, Pennsylvania, and Emil Fourcault in Belgium, and evolved separately. Another ten years passed before the Colburn process had an appreciable effect on the commercial world. In the meantime, the Fourcault process was developed in this country, and these processes finally displaced the old cylinder machine.

Sheet Drawing Method. These methods operated on the principle known as "sheet drawing." Glass in its molten state was drawn from the tank and carried by means of continuous rollers during the hardening stage; then the hot ribbon was bent over another roller, and cut, and when cooled, it emerged a finished product ready for the packers. The operation was entirely mechanical from the time the ingredients entered the "batch" until the glass was ready for shipment. In each process the glass was drawn, not blown. It was made flat, not flattened as in the cylinder process. The sheet-drawing process produced a never-ending sheet of glass extending mile after mile for weeks at a time. Moreover, the quality of glass was superior. It had a harder surface, bore a more brilliant fire polish, and was of marked uniformity.

Continuous Plate Process. In the early 1920's a new technique was introduced in making plate glass. The glass was made to flow from the end of the melting tank through a gate, and over a wide spout in a continuous stream 4 to 6 feet wide, downward along an inclined plane carried between two water-cooled rollers. It emerged as a flat ribbon of uniform thickness and width, and was pulled by multiple rollers through the lehr where it was cut, inspected, and packaged.

Glass Ribbons. Grinding and polishing was also carried on in a continuous operation. The cut rough plates, laid on movable tables in a plaster bed, were propelled under the grinding and polishing machine. By this method the grinding and polishing operation was reduced from 24 hours to 4 hours. At the Ford Motor Company's plant at River Rouge, a ribbon of glass 51 inches wide was drawn without interruption for nearly two years. Moving at the rate of 3½ miles a day, this sheet had a length equivalent to 1,900 miles and a weight of 46,000 tons.

Semicontinuous Method. To produce larger, wider sheets than 4 to 6 feet, a semicontinuous method was introduced shortly after World War I. The basis for this new process was a new type of rolling machine. Pots of molten glass were carried by electric cranes which poured the glass on a moving table passing under iron rolls. As the glass emerged from these rolls, it was borne forward by moving tables under a rotating steel disc, where it was cut into desired lengths and moved on to the lehr. The semicontinuous method opened the way for many technological innovations in the glass industry. Larger and heavier pieces of glass were made, as there was no limit to the size, thickness, or quality of product attainable.

Improvements in Potmaking. Improvements in potmaking reduced the time from three years required in the old method to thirty days from the time the clay was mixed until the pot was ready for use. Special types of clay were devised for shortening the curing time, and pugging machines were made to knead the clay. With the elimination of the primitive methods in potmaking, the last of the handcraft techniques passed from the glass industry. It had become a mass production technology comparable to any in America.

Science in the Glass Industry. Behind the technological developments in glass production was the hand of science. Skilled engineering replaced the art of the master craftsman. Chemical formulas, universally accepted, replaced trade secrets. Scientific instruments took the place of earlier skills. With the entrance of science, glassmaking took on an exactness and precision techniques impossible under the old methods, and a great expansion in scope and extent of the enterprise. New products resulted and glass was put to new uses.[2]

New Uses for Glass. Lantern globes, made to withstand snow and cold, were used on the watchman's lantern or the railroad brakeman's signal. Laboratory battery jars were made that would resist corrosion. Red lamps were made that would not show orange at night, and telephone insulators, radio insulators, and countless other devices that an expanding civilization was calling into play were all made of glass.

With increased production and the introduction of machinery, the old problems in glassmaking became intensified. When a bad batch destroyed merely one pot, the loss was relatively small, but when a whole tank was ruined the loss was severe. Variations in viscosity due to heat or to composition of materials were sufficient to throw a finely regulated machine out of action. Such risks demanded the attention of highly skilled engineers and scientists.

Optical Glass. Once the American glass manufacturers took up optical glassmaking, a number of technological advances were introduced. Mechanical stirring was devised. Electricity was substituted in the annealing ovens. The plate-glass method of casting was introduced for the less important uses. Fewer ingredients were used in the batch, and a finer reflective quality resulted. Special glasses for motion pictures were made. Barium crown glass was made for bifocal lens. Colored optical glasses were fashioned. Tempered glass was used for spectacles. Polaroid lenses prevented glare, and polaroid discs were made to show three dimensions. Borate glasses were made for camera lenses. America assumed leadership in creating the finest optical glass in the world.

Fiber Glass. Many of the revolutionary developments in glass technology since World War II have produced such amazing products as fiber glass, foam glass, light sensitive glass, and building blocks of glass. Fiber glass was used in the manufacture of automobile bodies and in interlining and soundproofing cars. It is used in batteries as a new type of separator for battery plates. Fiberglass yarns have been introduced in draperies, and are used for interlining cloaks and top coats. Small airplanes built of fiber glass were placed on the market in the fall of 1954.

Foam Glass. Foam glass, used widely as insulating and water-resisting material, is especially adapted for blanketing large buildings and skyscrapers. The world's largest sphere, the atomic submarine engine-testing sphere near Schenectady, New York, large as a twenty-story building and more than 200 feet in diameter, is insulated with foam glass. This fireproof, waterproof product is used in the roof and walls of the National Archives Building to protect the nation's priceless documents. Here are housed the original copies of the federal constitution and the Declaration of Independence.

American Glass Exhibits. Three world expositions cover the period of America's technological advances in glassmaking and are significant of that progress. At the World's Columbian Exposition in 1893, the Libby Glass Company set up a typical glass factory and gave the two million visitors an opportunity to see how glass was made. The Chicago Exposition of 1933 and 1934 presented a marked contrast in exhibiting the new improvements. At the New York World's Fair, 1939–1940, the "Miracle of Glass, Past, Present and Future" proved the most comprehensive display in all history. There an "all-glass house" exemplified the part glass plays in modern living. The American people had become glass-conscious and were using more glass than ever before. While in 1920 the amount of glass in the average small home totaled forty-five dollars, in 1945 it approximated three hundred and forty-five dollars. Glass was used for automobiles. Seventy per cent of the total output of plate glass in 1935 went into passenger cars, buses, and trucks. Glass was used in institutional buildings, in factories, in plants, in business houses. The employment of glass ranged from ironing board covers to the walls of the United Nations Building, from dust filters to insulators for the atomic submarine engine-testing sphere, from Carrara glass for interior walls to the tiles for lining the Hudson tunnels. Glass leaders declared that the age of glass had arrived. In 1900 the value of glass production in America was valued at fifty-six and one-half million dollars, and in 1950 this value amounted to almost a billion dollars.

Glass in a Modern World. Of all the inventions that have contributed to man's comfort, convenience, and to his understanding of the world, none has helped more than glass. With the glass microscope he studies the minutest particles of matter and with the telescope he studies the stars millions of light-years away. With the eye of a camera he records objects and reproduces happenings which the spoken or written word can but partially describe. Glass bulbs, chimneys, and reflectors, all a product of glass technology, enable man to turn night into day and carry on his work throughout the four seasons of the year. Through panes of glass, light enters the home, offices, factories, cars, trains, planes—and the same panes of glass protect man against the elements which for thousands of years he has been striving to conquer.

The Age of Plastics

Need for Plastics. Plastics are the products of the laboratory and are a man-made material. They were devised, not especially to rival other products, but as substitutes for materials that were becoming increasingly scarce—the "ersatz" of the Germans. In the United States plastics were early designed to utilize the waste products from the petroleum industry, the coke ovens, and the cotton fields. But plastics were more than this: they were designed to improve on natural materials, to substitute for them where they could perform functions equally well, or offer a quality superior to that of the natural material.

The Fourth Kingdom. In addition to this substitute use, plastics have found countless uses in which other materials have never been employed. Plastics have created a new kind of material and constitute the fourth kingdom—a new man-made or chemical kingdom, adding to the animal, the vegetable, and the mineral kingdoms.

Growth of the Industry. The plastic industry has come to occupy a major position in our nation's economy. While the first plastic industry was introduced late in the nineteenth century, it was during the second quarter of the twentieth century that the greatest developments occurred. In 1922 production totaled 6,000,000 pounds, while in 1950, exclusive of the synthetic rubbers, the solvents, the synthetic fibers, and the plasticizers, the production had reached more than 2,000,000,000 pounds, with a prediction of this figure being doubled every few years. For non-industrial uses alone, it was estimated that a family of four was using 800 pounds of plastic products per year.

Two Major Plastics. Though plastics have been made from many different ingredients and are of varied chemical composition, two types have been differentiated: namely, the thermosetting (heat hardening) plastics, and the thermoplastics. The former are those in which a chemical change takes place when the ingredients are subjected to heat and pressure in the process of manufacture. These cannot be returned to their original form. Scrap from this type cannot be melted down and reformed. This type may be compared to concrete in which the mixture of sand and cement hardens in the presence of water and is said to have "set," forming a new solid body. In the thermoplastic type, no chemical change takes place in the process of molding and forming. It may be compared to sealing wax which becomes soft and moldable when heated, and becomes hard and rigid when cooled. Scrap from this type may be melted down and reformed, and the finished product is subject to the same chemical reactions after molding as were the ingredients before molding.

Growing List of Plastics. The industrial use of plastics began in the United States in 1869 when John Wesley Hyatt, a printer in Albany, New York, set out to capture the ten thousand dollar prize offered for a substitute for ivory for billiard balls. The new product experienced a great boom with the discovery of the cellulose acetate group of plastics in the late 1920's. When the patents on the new phenolic plastics called bakelite, discovered by Dr. Leo H. Baekeland, expired in 1920,[3] many innovations were introduced. The first of these was the cellulose acetate group. These later developments included styrene used in synthetic rubber, nylon, and tailor-made products produced by a laboratory process known as copolymerization. In this the hydrocarbon molecule is utilized.

Cellulose Acetate. The cellulose acetates were added to the earlier cellulose products and had the advantage over celluloid in that they were slow burning. Cellulose acetates were used for photographic films, for transparent wrapping material, for lacquers, and for all kinds of molded products.

Urea Plastics. Since phenolic plastics could be produced only in dark colors, a demand developed for white and pastel shades. To satisfy this need the urea plastics were developed. Urea, a white crystalline powder made from ammonia and formaldehyde, was first synthesized by Wohler in 1828. Little was done with it, however, even in the laboratory, until the late 1890's. It was commercially made in Germany during the early 1920's and was introduced into this country around 1930 when a product known as "Beetleware" was manufactured. It was used for dishes, lighting fixtures, and dental forms. Urea plastics were applied as surface coatings for refrigerators and ranges, for bathroom equipment and washing machines, for tabletops and wall tiles, and for countless other uses where the phenolics were less attractive.

Another new urea plastic with the trade name of Plaskon was developed at the Mellon Institute in Pittsburgh in 1930 to satisfy the demand of salesmen for the Toledo Scale Company, who found the 165-pound iron scale too heavy to carry around as a sample.

Alkyd Resins. The alkyd resins, hailed as bringing about a revolution in the paint industry, were thermosetting and were made from naphthalene and glycerin. First noted in 1901, they were not introduced into industry until the mid-1920's, when they became an ingredient in the nitrocellulose lacquers and paints. This quick-drying coating material was a major contribution to the automotive industry.

By the 1930's the vinyl resins, used for waterproofing fabrics, surface-coating metals, and for packaging foods, led to a big reduction in tin consumption. Impetus was given to its manufacture when it was used for safety glass for automobiles. By 1950 the vinyls ranked second in volume of production.

Coal Tar Products. It was during the 1920's that coal tar products, the cumarone-indene resins, came in to prominence. This product had been used in a limited way for coating shingles, and as a solvent in scrap rubber. Later it was used extensively as a binder for plastic tile and linoleum.

Styrene. The styrene plastics were not introduced extensively until the outbreak of World War II when they were used in making synthetic rubber. Styrene plastics, made from benzene and ethylene, the products of coal tar and gas, have many qualities not found in other plastics. They are crystal clear and have a resistance to chemicals that makes possible their use for acid bottles, and battery cases.

Silicones. Within the last decade new synthetic materials called "silicones" have been developed that bear the same relation to the silicon molecule that the plastics bear to the hydrocarbon molecule. Silicones, derived from sand, the most abundant compound in the crust of the earth, are produced in the form of fluids, oils, greases, and resins. Silicones, when compounded and mixed with fillers, can be fabricated.

The development of this new material was the result of almost a hundred years of research in the field of pure science. These silicones are thermosetting resins, with high heat resistance, enduring temperatures up to 1,000 degrees Fahrenheit. They are moisture-resistant, inert to chemicals, and are a nonconductor of electricity. Although the silicones are of relatively low mechanical strength, they can be reinforced with glass fiber, and have been used as vehicles in paints, as coatings for paper, glass, and fabrics. They are also used as permanent lubricants for bearings and as a calking material. The resins are excellent laminating materials, when used as impregnating and bonding agents. They are used with fiber glass as a flexible electrical insulation, and as a bonding agent in synthetic rubber. During the first decade of their adaption to industry they have already been put to more than a thousand uses, and are made under many trade names.[4]

Types of Fabrication. Equally important with the development of plastics themselves has been the development of techniques for fabricating them into desired products. The fact that the resins could be molded with fillers was one of the great attributes of the material, and improved molding methods were devised.

Plastic Processing of Films. The first use of plastic as a film-forming material was in the photographic field and in wrappings. In the 1930's the vinyl group began to be used in sheet and film form and has developed the most versatile applications of all the plastics in this field. Technical advances in processing films and sheetings made possible many new products. Films of less than 0.0002 inch in thickness were made on what was called a "wheel." Sheets were calendered to a few thousandths of an inch or more in thickness by passing through heated rolls. Sheets thus made were used to produce attractive upholstery, bookbinding materials, handbags and similar articles.

Laminating. Because of the adhesive qualities of plastics they were utilized both to impregnate materials and bind them together. Prior to World War II this process, termed "laminating," was applied to such simple structures as sheets, tubes, and a few special cross sections and contours molded under pressure. At first only the phenolic resins were laminated, but with the introduction of the urea resins, the attractive colorings available led to their utilization in laminated products. Structures of intricate contour were molded under pressure. New plastics were utilized in the laminating process, and a new sandwich press was developed. By the process of laminating, structural materials were reduced in weight. The most utilized materials included paper, wood, cotton fabrics, glass fiber, and asbestos.

Sandwiching. The sandwich construction in laminating, developed since World War II to reduce weight in airplane parts, involved utilizing a lightweight material as a core, a plastic adhesive, and outer surfaces of high-strength material. A honeycomb core was designed, made of cotton and paper sheeting impregnated with resin, with thin-gauge metal for the outer faces if structural properties were desired, or of wood veneer for decorative features. Sandwich construction was made in contour shapes as well as machineable pieces.

Casting. Another method of treating plastics is casting. In the middle 1930's it was discovered that certain of the thermoplastic resins could be formulated to produce liquid compositions that could be poured into molds, where they were allowed to harden without the use of pressure. Coating plastics in liquid form have been developed that are superior to shellacs, lacquers, and varnishes as coating materials. Plastic coatings have been developed for metals to render iron pipe acidproof, to make metals tarnishproof, for insulation for wires, and for leather, concrete, glass, paper, and ceramics.

Material of a Thousand Uses

By the mid-1950's plastic materials had found myriad uses. They were going into automobile bodies, aircraft parts, bathtubs, pipes, tubing, and hose. Reinforced plastics were applied to construction uses. They were molded, calendered, and formed and fabricated into products by every conceivable method. Advances were made in equipment, design, and operation. Plastics supply raw materials to virtually all man-made products, from unbreakable phonograph records to automobile parts. Polyethylene, of late years the most rapidly growing plastic, provides tough, translucent material, which in turn has stimulated the pre-packaging of foods, fresh fruits, and vegetables.

Plastics are generally thought of not as substitute material but as a new material designed to fill specific needs. Plastics are devised for those fields where they can do things better than any other materials can do them, and where they can improve on the natural limitations of other products. The plastic industry, within a few years, has become more than a billion dollar business.

Rubber and Synthetic Rubber

Synthetic rubber, really a plastic in a highly specialized field, had a dramatic development in the early days of World War II. Of great advantage in this development was the experience in the two previous decades with latex and with reclaimed rubber. All this culminated in the triumph of the synthetic tire.

Latex. Latex is the liquid rubber as it comes from the rubber tree. In the 1920's a business had developed (made possible by improvements in transportation and in refrigeration) of shipping from the rubber plantations in the Far East the rubber in its liquid form as well as in the creped or smoked rubber sheets, as had been done since the beginning of the industry. In the use of this product, latex, more chemistry and modern technology was required and less machinery. Dehydrated latex, which formed a powdery mass, was pressed into blocks by hydraulic pressure (a Dutch process used in roadbuilding) to form "Mealorub."

Uses of Latex. In the liquid form, latex was used as a cement and as a rubber coating to coat the cords of the carcass for rubber tires. It has been processed by mixing with compounding ingredients and vulcanized. It has been made into molded products, for when the latex sets it "jells." Sponge rubber used for mattresses, theater seats, and a wide range of products has been produced directly from latex. Shoe soles, leather substitutes, and many articles never before made of rubber have been made of latex. Latex thread has been made $\frac{1}{125}$-inch in diameter, so fine that a pound spool carries 10 miles of thread.

Reclaimed Rubber. Special processes for reclaiming rubber from used rubber articles developed during the 1920's likewise proved an advantage in the synthetic industry. Though this development was stimulated in order to reduce the cost of rubber products, it has been found that reclaimed rubber has a value in itself and is not merely a substitute for new rubber. Reclaimed rubber has proved of value in compounding much the same as scrap has in the steel industry.[5] Reclaimed rubber contains, besides the hydrocarbons found in rubber, the elements of the compounding ingredients as well.

Synthetic Rubber History. Ever since the usefulness of crude rubber was recognized, scientists have attempted to match nature by synthesizing it in the laboratory. At first the goal was to reproduce natural rubber, an object not yet achieved. Later, the aim was not to reproduce rubber, but to develop a superior product for special uses. Serious study was begun in 1860 when Grenville Williams, an Englishman, distilled rubber and discovered that the fundamental hydrocarbon in its composition was isoprene. This isoprene could be obtained from other substances such as turpentine, starch, petroleum, and coal tar. The next advance step was taken when science learned that it was not only the chemical composition of rubber that gave to it its qualities, but the structural arrangement as well, for the rubber molecule was made up of large aggregations of hydrocarbons. This discovery led to what was called the "polymerization" process, or building up these large hydrocarbon molecules.

German, Russian Experiments. German scientists led in the first commercial development of synthetic rubbers when in 1912 they produced tires for Kaiser Wilhelm's automobile from isoprene. German experiments continued and during World War I Germany produced several tons of methyl rubber. While this was not entirely satisfactory, it helped fill their need. Russia had a similar rubber product made from petroleum and alcohol which was produced in commercial quantities by the time of her entrance into World War II.

Neoprene. In this country some artificial rubber called "Thiokol" was produced commercially in 1930. The original purpose was not merely to produce artificial rubber, but to develop a rubberlike substance that would be superior to rubber, impervious to acid and oil, and a product that could be used for tank linings and hose connections. In 1931 the DuPont Company, under patents developed by Julius Arthur Nieuwland and Wallace Hume Carothers, was producing a product called "Neoprene," made from acetylene and salt and sulfuric acid. This product was used for wire and cable insulation, for printing presses, for oil hose, packing gaskets, and for power transmission.

Synthetic Rubber Tires. When the rubber lifeline to Singapore was menaced in World War II, scientists in the laboratories of all major tire companies worked feverishly to produce synthetic rubber. In June, 1940, while the Germans were storming across western Europe, the B. F. Goodrich Company announced it had produced the first synthetic rubber tires. They called their product "Ameripol." It was made from potatoes, grain sugar molasses, petroleum, and coal tar. Government Rubber Styrene (GRS) became the basis of the nation's wartime rubber industry.

From a total output of synthetic rubber in 1939 of 1,000 tons, the industry developed to 400,000-ton capacity in 1942, and 800,000 tons in 1952. Trade names in the several plants differed, but the products were all made on the same principle. Carbon-hydrogen molecules were made to polymerize to form an emulsion of synthetic rubber similar to the latex that exudes from nature's rubber trees.

Synthetic versus Natural Rubber. With the passing of the war emergency, synthetic rubber, instead of disappearing from the market, continued in great demand. New rubbers were produced, new rubberlike products to take the place of rubber and fill the specific need better than could be done by rubber were devised, and new uses developed. Many limitations in natural rubber such as deterioration by exposure to strong sunlight, by exposure to heat, and by the action of chemicals and petroleum products were overcome by the man-made product. Statistics for 1955 showed slightly less than 1,000,000 tons of synthetic rubber consumed, compared with 450,000 tons of natural rubber, and less than 300,000 tons of reclaimed rubber.

Whether rubber continues to be obtained from the gummy, sticky liquid that had been growing and rotting away for thousands of years before Charles Goodyear announced his vulcanization process, or whether it comes from the ingredients that scientists synthesize in their laboratories, rubber remains a determining factor in this machine-age civilization.

Petroleum Meets a Challenge

The petroleum industry in the period since World War I was challenged by the automobile and the airplane to meet the demands for both more and better fuel. It was also challenged by all machine operating industries to provide more and better lubricants. The wide diversification of the industry also presented new challenges in transportation. In all these, petroleum met the challenge, but did so only with the aid of science, with the improvements in new steels, and with the help of new power facilities.

A Slow-starting Industry. World War I left the petroleum industry a much larger one than it had been at the turn of the century, but there had been few technological improvements. Rotary drilling had been introduced, but was not at first considered applicable to deep wells or on other than soft formations. Refining was still carried on by the old "topping" direct distillation system, yielding about equal quantities of gasoline and kerosene. Western oils were still considered inferior because refining methods had not been developed sufficiently to utilize them. California oil found no better use in some cases than for road purposes, and Texas oil brought only one-eighth the price in the market which Pennsylvania crude commanded. Lubricants had not kept pace with the advance in machinery development that used high-speed steels. Foreign oil fields with different oils were presenting new problems. The petroleum industry did a great deal of growing up in the last thirty years.

Geologists Enter the Field. In locating new fields, the divining rod and the crooked stick gave way to the pursuits of science. The lone prospector was supplanted by the highly organized activity of groups of petroleum engineers, operating with co-ordinated efforts. The historical geologist continued to delineate the general area, while the subsurface geologist explored the field. The geophysical crews, equipped with the airborne magnetometer, the gravimeter, and seismographs, not only located new pools of oil, but also estimated their possible production.

With these advances in petroleum technology, new discoveries were made in the tidal area along the Gulf Coast; in Alberta, Western Canada; in the Williston Basin of North Dakota; in Mozambique, East Africa; in Tunisia, North Africa; and in the Kuweit field located near the ancient gold mines of King David and King

Solomon. The discoveries were made largely by American scientists.

Transportation. Large trucks, tank cars, barges, tankers, and pipe lines transport both the crude oil and the finished products. Hundreds of ocean tankers and thousands of barges loaded with oil travel the water routes. One hundred and fifty thousand miles of pipe lines stretch across the country; the largest is a 26-inch line with a capacity of three hundred thousand barrels a day.

Lubricating Oils. One of the greatest demands made on the petroleum industry after World War I was for improved lubricating oils. The new high-speed steels, making possible high-speed machinery, created a demand for better lubrication. Since practically the only oil accepted for lubricating purposes was Pennsylvania crude, production was not adequate to meet the new demands created by both the increased use of the automobile and increased use of machinery. Moreover, the oil industry was eager to utilize the naphthenic oils—those with an asphalt base (Pennsylvania crude has a paraffin base), and the varying grades of mixed types that were being pumped from wells all over the country.

Solvent Process. Instead of the old methods of filtration or of fractioning, a solvent process was introduced, applicable to all oils. It was employed to de-wax, de-asphalt, and remove any undesirable constituents in the oil. This was called "solvent fractionation." The chemical laboratory ascertained the required solvent for the specific crudes, and greases of high viscosity were produced without subjecting the crude to great heat, which lowers the viscosity. Propane was found to be a solvent for the asphalt-base oils, since petroleum fractions were soluble in propane and the undesired asphalt settled out.

Flash Distillation. A new distillation process was introduced that avoided overheating, and thus made the fractionation of lubricating oils possible. This was called "flash distillation." The oil was heated to the boiling point under pressure and the various fractions obtained by condensation. Under the regular distillation method fuel oils do not fractionate below a temperature of 375 degrees centigrade.

Additives. In the quest for better lubricants, it was discovered that lubricants could be improved by the addition of agents, as well as the subtraction of elements. There resulted the development of greasemaking as a diversified industry with special greases made to fit the individual use. In one industry more than eight hundred different kinds of greases were made, grease as soft as butter and grease as hard as wood. Greases were developed to lubricate machinery and to launch ships. The modern grease plant, with its kettles, some 20 feet high, its stills and its chilling apparatus, its pressure cookers and jacketed steam kettles, is evidence of the increased demand for different types of grease. Production since 1921 has nearly tripled.

Refining. Refining methods have produced better products and have brought vast increases in yield. In the old days gasoline was a waste product, and as late as World War I, demand was no greater for it than for kerosene. The automobile era and the high-compression engine that appeared in the early 1920's challenged the refiners. Thermal cracking, first introduced by William Burton in 1913, inaugurated the new era in refining. In 1920, tetraethyl lead was introduced, and then came the high-octane gases. Instead of distilling in batches, continuous distillation was introduced. Then came the addition of a catalyst to speed the operation and catalytic polymerization was introduced to convert more of the oil into gasoline. These processes have increased the yield more than 100 per cent. One of the newest catalytic reformers processed twenty-nine thousand barrels a day. A modern refinery, with its maze of towers, pipes, platforms and open steel stairways, stands as high as a fourteen-story building. It consists of an atmospheric and vacuum still with an atmospheric tower and a vacuum tower, a fluid catalytic unit, a catalytic polymerization unit, a hydrogen or platinum unit where more of the oil may be turned into gas, and the sweetening unit where the odors are removed. Some refineries process more than 60,000 gallons daily.

Petrochemicals. Modern refineries turn out a variety of products. Besides high-octane gasoline, they process lubricating oils and fuel oil. Petrochemicals, made from materials originally discarded, comprise a long list. These include ethylene glycol, from which permanent antifreeze is made, and ethylene oxide used in rubber, detergents, and plastics. Then, too, there are chemicals used in explosives, in resins, in cellophane, in processing textiles, and in the ingredients in brake fluids. In addition, there are fertilizers, insecticides, and countless other products, all by-products of the petroleum industry. These, with hundreds of other by-products resulting from petroleum technology, are all derived from what were once regarded as waste materials.

NOTES FOR CHAPTER 40

1. See article by Alexander Silverman, "Plate Glass," *Bulletin American Ceramic Society* (1947), p. 367.

2. See *Reports of Bureau of Domestic and Foreign Commerce,* Miscel. Series No. 60 (1917), p. 334.

3. Leo H. Baekeland, Belgian scientist, came to the United States as a lad in 1869

and won for himself fame and fortune for his invention of Velox developing paper. His invention of a thermosetting plastic, a resin which he patented in 1909, as "Bakelite," started the dramatic march of plastics on its way.

4. Rob Roy McGregor, *Silicones and Their Uses* (1953); a standard work on the history and uses of silicones.

5. See Walter Burton, *Engineering with Rubber.*

For standard books on glass, see N. Hudson Moore, *Old Glass, European and American* (1924); Helen and George L. McKearin, *Two Hundred Years of American Blown Glass* (1949); by the same authors, *American Glass* (1941) and (1948); Warren Scoville, *Revolution in Glass Making* (1948); Charles John Phillips, *Get Acquainted with Glass* (1950).

For developments in plastics, consult: "American Society For Testing Materials," *Symposium on Plastics* (March, 1938), (February, 1944); Stephen Bass, *Plastics and You* (1947); Leslie Bell and Mandell Thomas, *Making and Molding Plastics* (1938); D. M. Buttrey, *Plasticizers* (1950); Joseph B. Singer, *Plastics in Building* (1952); S.P.I. *Plastic Engineering Handbook* (1947).

The bibliography on rubber, including books, monographs, government reports, census reports, and proceedings of scientific societies, runs into hundreds of items. Most of the major tire and rubber companies have had company histories published. Numerous journals and magazines are devoted exclusively to the rubber industry. The American Society for Testing Materials has published many excellent reports of symposiums that have been held on the subject of rubber.

Chapter 41

TECHNOLOGY IN WORLD WAR II

The day following the Japanese attack on Pearl Harbor, (December 7, 1941) President Roosevelt asked Congress for a declaration of war. The next day, speaking to a nationwide radio audience, he predicted the character of the war into which we were entering, and pointed to the unprecedented demands that would be made on the American people. "This war," he declared,

> will not only be a long war, it will be a hard war. That is the basis on which we now lay our plans. . . . We shall need and demand money, materials, doubled and quadrupled production—ever increasing. The production must be not only for our own Army and Navy and Air Forces. It must reinforce the other armies and navies, and air forces fighting the Nazis and the war-lords of Japan, throughout America and the world.

The Mass-Production War Program

Two broad policies were decided upon: to speed up all existing production by going on a seven-day week in every war industry, including the production of all essential raw material; and, "rush additions to the capacity of production by building new plants, by adding to all plants, and by using many smaller plants for war needs."

The President explained that while a steady stream of tanks, planes, guns, ships, shells, and other war equipment had been going to Great Britain, France, and their allies in their fight against the Axis powers, this was only a beginning of things to come; that a steady stream of supplies had come from American industries which had only partially converted to war production. Now that the United States had entered the conflict, production of war materials had to reach a capacity never before contemplated.

American and British Scientists Co-operate. Meantime, another war potential was being developed. American scientists had been co-operating with those from Great Britain in developing new weapons. An agreement entered into in the summer of 1940 for the co-ordination of scientific efforts had brought a delegation of British scientists, headed by Sir Henry Tizard, Rector of the Imperial College of Science and Technology and advisor to the Ministry of Aircraft Production, to Washington. The Britishers brought numerous blueprints and scientific data which they revealed to American scientists in exchange for secret data which we possessed. From the British, the United States obtained valuable information relating to radar, chemical warfare, rockets, superchargers, underwater detection, and explosives. The United States in turn gave the British access to our arsenals, flying fields, army and navy experimental laboratories, and to the research programs that were carried on in producing the latest electrical equipment.[1]

Immediately following America's entrance into the war two major plans were undertaken: the nation's industrial potential was geared to mass production of war materials; and equally important, a project for developing new tools and new techniques by the nation's top scientists was launched. As Professor Baxter said, "Although the skill and valor of the fighting man remains the most important element in warfare, he has become increasingly dependent on the equipment needed to give him information, mobility, and fire power." [2]

NDRC, June, 1940. The equipment required to give the fighting man this latest war intelligence, mobility, and fighting power was speedily provided by mobilizing the forces of science and technology. President Roosevelt, like President Wilson in World War I, and like President Lincoln in the Civil War, turned to American scientists, inventors, and engineers for counsel. In June, 1940, a National Defense Research Committee (later to become the OSRD, Office of Scientific Research and Development) was organized and was headed by Dr. Vannevar Bush, then president of Carnegie Institution. This organization included many of the nation's top chemists, physicists, engineers, physicians, and psychologists. Within a year, some six thousand scientists were carrying on research in college laboratories and in industries throughout the nation, matching their scientific and technological know-how against the German and the Japanese scientists.

War Manpower Commission. Another important agency, the National Roster of Scientific and Specialized Personnel, was organized in June, 1942; it later merged with the War Manpower Commission. Its purpose was to provide for the most effective utilization of the nation's scientifically and professionally trained citizens. Its members assembled data on the needs of the army, navy, war research, and education; and estimated the probable needs for men in each of these fields. A special punchcard record system was developed which showed the qualifications and special skills of more than a half-million men.

Other Wartime Agencies. Other agencies were set up, such as the National Inventor's Council; the Office of Production Research and Development, which included a chemistry division; a committee on synthetic rubber; and scores of other groups organized for the specific purpose of carrying on scientific research aimed at winning the war. In May, 1942, the Joint Committee on New Weapons and Equipment was organized to co-ordinate the efforts of civilian research agencies and the armed services in developing and producing new weapons and equipment.

Radar Research and Development Committee. Working with the Joint Committee on New Weapons was an important subcommittee on Radar Research and Development, headed by Dr. Karl T. Compton. This top-planning group prepared the program for research and development in the field of radar, and early established close contacts with the British scientists working in the same fields. Close teamwork was also established between the civilian scientists and the Navy in developing antisubmarine weapons and the acoustic torpedo; and with the Army Air Force in developing the early use of radar.

Naval Technology

Technology in Prewar Years. In the midst of all these scientific preparations, the shooting war began. The United States Navy, despite the irreparable damages suffered at Pearl Harbor, was soon spearheading a number of attacks in various theaters of the war. Important advances in naval technology had occurred during the years preceding Pearl Harbor, notwithstanding international agreements and the Kellogg pact. The very restrictions which limited tonnage had proved a challenge to naval designers and technicians. Emphasis was shifted to quality in construction. Naval and metallurgical engineers produced stronger metals that weighed less; they substituted welding for riveting; they improved motive power by increasing temperatures and pressures; they improved fire-control apparatus, and made refinements in many other lines. Diesel engines were developed and installed on a number of ships. Steady advances had been made in re-fueling at sea. Ships were able to steam enormous distances without shipyard overhaul of machinery.

Naval Aviation. The building program had concentrated on destroyers and cruisers, bristling with antiaircraft guns; aircraft carriers, and mine sweepers, rather than battleships (although the battleships of the "Iowa" class had been authorized in 1935). Contracts let before Pearl Harbor called for long range antiaircraft guns, protection for personnel from fragments, special armor, and improved subdivisions in the holds of the vessels to offset underwater attack. Also, scientific experimental laboratories and testing stations had been established and staffed with technicians and scientists who laid the foundation for wartime success. In 1921 the Bureau of Aeronautics was established in the Navy, and aircraft and carriers were included in the war games. New and faster carriers had been developed, so that at the outset of World War II naval aviation was full-fledged.

Amphibious Preparation. Major technical developments had also been made in that ancient art of warfare, amphibian operations. This type of warfare had seemingly become obsolete, but modern naval technology had made it again possible. It was in the mid-1930's that the first marine climbed over the side of a battleship and down the cargo nets to a waiting motor launch in Chesapeake Bay to take part in the first amphibian practice. New landing craft were devised, and in 1938 the first manual on landing operations had been given to the fleet and Marine Corps.

Radar. Radar, which proved to be the most versatile instrument in modern warfare, owes its development to pre-World War II advances. Radar, so named by the Navy for *R*adio *D*etection *a*nd *R*anging, based on the invention of the cathode-ray oscilloscope, was used to measure time in such infinitesimal amounts that radio-echo ranges of objects miles away could be read with minute accuracy. Its application to naval operations was recognized in the early 1920's by the Naval Research Laboratory; and in the 1930's experiments in the use of radar were being made by the Navy and by the Signal Corps. A Microwave Committee of scientists and engineers from American colleges and industries, organized to further its development, found their work blocked by the lack of a vacuum tube which would generate microwave pulses sufficiently intense for practical pulse-detection equipment. This obstacle was solved when the Tizard mission in 1940, brought to America the British findings and sparked the whole development of microwave radar. No microwave sets had been delivered, however, to the armed forces prior to the outbreak of the war.

Rockets. The rocket was another ancient weapon that was revived by modern technology between World War I and World War II. A rocket, described as a projectile that could be fired without a recoil, was especially suited for firing from an airplane or light surface vessel. This weapon was a long tube, closed at one end and open at the other. It was filled with a propellent powder which, when fired, launched its warhead at the target. This projectile, devised in 1919, was intended for piercing armor, and for the projection of chemicals; it was used on tanks, submarines, small boats, and aircraft, and a special model was designed for the infantry. Again, it was the Tizard mission that added great impetus to its development. By 1941 the NDRC had under development antiaircraft rockets of both low- and high-altitude types: the 4.5-inch aircraft and artillery rocket; and the jet-accelerated armor-piercing bomb.

Explosives, RDX. Advances were also made in the manufacture of explosives, and in improvements on the TNT used in World War I. In the early 1920's, RDX, or cyclonite, was developed, but was too sensitive for operation in munitions. It was far more powerful than TNT, especially when used under water. EDNA, or Haleite, was a high explosive chemical, patented in the 1930's. It also proved too sensitive for large munitions except in mixture with other high explosives. The adaptation of RDX was one of the technological triumphs of the war. At the Eastman Works, Holston Valley, Tennessee, a mixture of RDX with TNT and aluminum developed Torpex, and was in production by 1943 at the rate of 340 tons a day.

Proximity Fuze. The proximity fuze, described as the most remarkable scientific achievement of the war, except the atom bomb, was not ready for mass production at the time of the attack on Pearl Harbor, but had survived the first hurdles of laboratory tests at that time and was soon ready for the assembly line. Other types of fuzes were the contact-fuze, which burst on coming in contact with an object, and the time-fuze, which operated by clock work. In both of these the percentage of inaccuracy was too high when it was estimated that an antiaircraft gun brought down only one plane for every twenty-eight hundred rounds. The answer to these mistakes was the proximity fuze, which exploded without actual contact as it neared the largest target. Early emphasis was on proximity fuzes for bombs and rockets, and work was done on photoelectric types, acoustic, and radio fuzes.

The major problem was that of constructing a power plant, together with a transmitter and receiver, that could be fitted into the nose of a Navy 5-inch, 38 caliber shell, in a space about the size of an ice cream cone. Moreover, such delicate instruments had to be constructed sufficiently rugged to withstand the force of more than 1,000 pounds of acceleration of the shell in the gun, and the high centrifugal force generated as it traveled through the air. In late 1940, contracts had been let for experimental tubes, rugged enough to withstand this impact.

Production Following Pearl Harbor. Another group of scientists, experimenting with radio fuzes for bombs and rockets, had carried the project beyond the laboratory tests early in 1941. The radio proximity fuze for shells had been successfully demonstrated. Next came the problem of devising accompanying batteries, condensers, resisters, and set-back switches; and finally of designing all these in sufficient simplicity to be capable of mass production. In November, 1941, orders were placed for pilot production of radio proximity fuzes, and for preparation for mass production. The events of the next month accelerated the fulfillment of this contract.

Progress had also been made in the prewar days in the development of smoke screens, incendiary bombs, mine detectors, and improved fire control; and in the electromagnetic method of separating the isotopes of uranium. All these scientific experiments bore fruit during the four years of combat. But it should be remembered, they had been undertaken before the United States entered World War II.

Navy Convoys. After the Japanese attack on our fleet in Pearl Harbor, December 7, 1941, and the declaration of war that followed, two immediate plans developed. One was the determination to meet and defeat the enemy in the Pacific and Far East (which will be discussed later), and to protect the huge flow of American troops and supplies that were being sent to aid our European Allies. Here, our first and most deadly weapon to be met was the submarine.

The battle against the submarine was fought with the convoy—a fleet of merchant vessels in orderly procession protected by destroyer escorts, battleships, cruisers, planes scouting the horizon, and submarines scouring the seas. The convoy of World War II bore little resemblance to the straggling parade of World War I. In the second world war there were five or six convoys at sea every day in the Atlantic: eight to ten daily in the coastal waters, and a dozen or more in the Gulf of Mexico. More than a thousand ships were engaged in convoy duty daily. Troop convoys often included an armada of a hundred vessels, covering a space on the ocean 20 by 30 miles.

Convoys Controlled by Scientific Instruments. Naval science as well as naval power was employed in the operation of convoys. Information to convoys as to the movements and whereabouts of enemy submarines came from shore-based high-frequency direction finders, whose bearings were plotted at London and were sent out on powerful low-frequency broadcasts. Fueling at sea was accomplished from merchant tankers proceeding at a speed of 8 knots. When a U-boat was sighted, newly designed weapons effected its demolition. Depth charges, filled with the more powerful Torpex, sent the U-boats to their watery graves. New methods of sighting the U-boat were employed, since radar gave long-range eyes to both ships and planes. Baxter quotes Admiral Doenitz as saying, late in 1943, that the United States and her Allies had rendered the U-boat ineffective. This had been achieved, "not through superior tactics or strategy, but through superiority in the field of science; this finds its expression in the modern battle weapon—detection." [3] The wolf packs had been thwarted. Not a single loaded troop ship was lost.

Role of Convoy Limited. Although the convoy, by keeping the supplies of strategic material moving to the aid of our Allies, prevented Germany from winning the war, it did not win the war for ourselves and the Allies. American industry produced material and American shipyards turned out vessels faster than the U-boats could sink them, yet Germany with her immense stockpiles of synthetic products continued to flout our sea power. During 1942, the monthly loss amounted to about 700,000 tons. (The United States participated in no fleet actions in the Atlantic.) The listening gear (ASDIC) on our ship bottoms could not detect the enemy submarine as it lay motionless, submerged, and hence our ships could not make contact. Also, the early aircraft lacked the means both of detecting the submerged boat or of damaging it underneath the water. The radar that was in use operated on long- or medium-length waves, and while this gave to both ships and planes long-range eyes, supplementing the "seaman's eye," it was not effective underwater. The early radar did not effect many kills, it merely drove the submarine underwater. The crying need was for better detection devices and for better weapons for the airplane.

MAD, Sonobuoy. In April, 1942, the first of several units of a new device, called MAD (magnetic airborne detector), was delivered to the Navy. With this, submerged submarines could be detected from an altitude of several hundred feet. The sonobuoy was another detector which had the advantage of being expendable. It was light enough to be carried by planes and could be dropped by either planes or ships. This was merely a radio-buoy which, as it floated on the water, picked up the sounds of any vessel approaching and radioed a warning signal. These buoys were used in patterns to scour large areas of water. Each had a different code signal. The sonobuoy became one of the outstanding developments in naval technology.

Microwaves. A much wider field for the application of radar was opened up with the development of the microwave. In 1940 when members of the Tizard mission brought Britain's development of the resonant cavity magnetron to the United States, they brought with them, "the most valuable cargo ever carried to our shores." [4] This device, the heart of every modern radar equipment, is a vacuum tube capable of producing power enough to operate in the super-high-frequency range, at wave lengths less than 50 centimeters. American physicists, after experimenting with the possibilities of the microwave, produced equipment operating on 10 centimeter wave-lengths. Microwave systems were applied to ground radars, to ships, and airborne radars. Early in 1942 the first airborne microwave interceptor was produced, and before the end of the year some fifty different systems were in operation. In the summer of 1943 a new search-set known as ASG, or "George," projected a radar map of the territory ahead and spotted convoys several miles away; it also could map coastlines at a distance of 100 miles.

Loran. Other radar systems that met with success in America's first year in the war included Loran (Long Range Aid to Navigation), which proved a navigational and blind bombing aid. This had been developed in 1941 and involved the establishment of shore stations from which the ship at sea could plot its position. In the operation of this system three stations were necessary, a master and two slaves. Pulsed radio waves were sent from the shore stations pictured on charts carried by the ship at sea. As the war progressed, Loran coverage was extended over more than half the globe.

Sonar. Sonar was a device that enabled a ship to evade the prowling wolf packs. Sonar gear was developed for the accurate measurement of sound under water. Depth indication, echo ranging, and the position of obstacles were determined by sonar. Sonar was so perfected that by the end of the war it was possible to measure sound under water as accurately as in the air. A Plan Position Indicator (PPI) screen became part of a ship's radar equipment. This was a place where the controller and his assistant on a ship marked on the plotting board the movements of planes and ships tracked by the various radars—Loran, Sonar, Huff-Duff (friendly voice). This enabled the controller to direct his own fighters in intercepting hostile airplanes.

Hedge-hog, Mouse-trap, Minnie-mouse. To combat the U-boat, a weapon designed by the British, called "hedge-hog," a spigot mortar used to project a forward-thrown depth charge, was adopted.

In American installations, however, the depth charge was made more deadly by using the new ammunition, Torpex. Depth charges could be used only on destroyer escorts, or equally large vessels. For smaller craft, America devised the "mouse-trap," a rocket weapon to create a "head-thrown" barrage of small depth charges. With this the rocket projectiles were thrown from a multiple rail-launcher that bore a resemblance to a mousetrap. The first deliveries of the mouse trap were under way by July, 1942. Another weapon used to combat the submarine was the "Minnie-mouse," which was launched backward from an airplane. Flight tests for these were made in July, 1942, marking the first firing of American rockets. Improvements were made in minesweepers, and in methods of laying mines to protect the harbors. Underwater flares were devised to show up submerged submarines at night.

Submarine Menace Passed. The steady onslaught of these new scientific devices and the mass production of ships and planes brought decisive results by March, 1943. The darkest days of the submarine warfare had passed. By May of that year, the tonnage lost was practically nil, and by the close of that year, the Allies, instead of concentrating on antisubmarine warfare, had developed the submarine as a war weapon and to fight submarines with submarines. The efforts paid off within the next two years.

Amphibious Operations, 1942. Amphibious operations, which Admiral King declared to be "the most difficult of all operations in modern warfare" were made possible through an amazing series of scientific and technological achievements. The two chief problems were that of directing traffic from ship to shore, and of protecting the troops in this passage. The trial reconnaissance invasion at Dieppe in August, 1942, to test the German defenses, was successful but costly. Its mission was accomplished. Withdrawal occurred on schedule with 9 hours occupation; this provided a proving ground for further amphibian operations. Self-propelled landing barges and improvised tank-landing craft, escorted by British destroyers and covered by an air umbrella, made a successful crossing of the channel and invaded the beachhead after minesweepers had cleared the path.

Operation "Torch," November, 1942. With Dieppe as a practice game, the first major amphibian operation was undertaken in November, 1942, in Operation Torch, when the great armada of one hundred two vessels steamed across the Atlantic and landed United States troops in French Morocco. Never before had such an undertaking been attempted. The fleet had traveled 4,000 miles, carrying seventy thousand troops, to effect a night surprise landing on a strange and hostile shore. The expedition included submarines, battleships, heavy cruisers, light cruisers, destroyers, thirty-five big transports, cargo vessels, tankers, and an air force, all of which covered a space on the ocean 20 by 30 miles. The landing craft, little more than motor launches, carried the troops up to the water's edge and from there they waded ashore.

The Army in Europe

Land Troops, 1942. While the American expeditionary forces won renown in Africa during 1942, the equipment of the infantryman at the beginning of the war contributed little to his fighting spirit. The Springfield rifle was outmoded, and there was no dependable defense against the modern tank or plane. American heavy 6-ton Sherman tanks were rolling in mass production in time for the African campaign. While these were not so powerful as the German "Tiger" and "Panther," it was deemed wiser to continue tank construction in quantity rather than to make readjustments for the improvement, and thereby delay the program.

Bazooka. A new weapon, the bazooka, became available for the infantry in 1942. Described as a rocket, which the foot soldier was able to carry, it was fired from a long tube by means of an electric battery. The missile was propelled with such force that it pierced the armor of a tank. It proved to be the infantryman's defense against the enemy's tank. It had the effectiveness of the armor-piercing shells of the artillery. The production of the bazooka was rushed. The first model was designed within a few days, and within a few weeks commercial production began. Production reached five thousand a month.

Mine Detectors. Land-mine detectors were developed early in the war for the aid of the infantry, but the first of these operated on the magnetic principle and were effective only where metallic mines were employed. Frequently, however, the mines were made of substances that failed to react to a magnet. In such cases the foot soldier was forced to risk the more hazardous task of searching for the hidden mines on his hands and knees. Eisenhower reports that in discussing the inefficiency of the systems of mine detection with Marshall Zhukov, he failed to strike a responsive chord. Zhukov said he found it more effective to explode the mines by marching the first contingent of his soldiers over them.[5]

1943 European Campaign. America's second year of the war was characterized by Churchill as "the end of the beginning." Though there were many months of hard fighting ahead, 1943 saw the crest of the U-boat campaign, the withdrawal of Italy from the conflict, the turning point in the German air raids over Britain, and the launching of the offensive in the Pacific.

The advance of American troops in Africa was marked by the utilization of masses of new equipment. Besides great quantities of tanks already in use by the ground forces, a supply of new and improved tanks, equipped with flamethrowers, leveled a sea of fire on the enemy, and supported the American troops in their gallant hold on Hill 609. This massive ground attack, accompanied by a concentration of air power which showered bombs against the enemy, brought the first break in General Rommel's line in Tunisia.

"Operation Husky," Sicily. Next came the campaigns across the Mediterranean in Sicily, Malta, and Italy. When the invasion of Sicily was undertaken in Operation Husky, rockets were fired from support boats off the shore. This resulted in a successful amphibious feat. Other victories, while costly, followed in rapid order due not only to the fighting spirit of the troops, but also to the new and more powerful mechanical equipment. Typical of the use of this power equipment was the construction of an air base off the island of Malta, which had stymied the British engineers who had been working with hand tools. When the American forces arrived they brought in every type of earth-moving machinery that had been shipped across the Atlantic, and within thirteen days after they stepped in, the first plane was launched. General Eisenhower described this achievement as a seeming miracle, which won from the British admiration little short of awe.[6]

Operation Avalanche. After the Mediterranean had fallen into the hands of the Allies and the threat to the Suez Canal was eliminated, a new device, a specialized plane equipped with search receivers, was able to plot the enemy radar stations from Spain to Crete. With this they were able to make maps and plot paths of approach, and to circumvent enemy radar coverage. The timing of the amphibious landing at Naples in Operation Avalanche, accompanied by equipment superior to that used a year earlier at Operation Torch, indicated the speed in which scientific aid was being supplied the fighting forces.

LAB, Radar in Aviation. Improved radar bombsights were effective aids in air battles over Germany. Until late 1943 the Eighth Air Force had been practically grounded in winter and during cloudy weather. A radar bombsight that was a pathfinder in aviation (Lower Altitude Bomb Sight) made possible a successful raid in November that year on Wilhelmshaven, a target missed in eight previous visual raids. Jamming the enemy's radar became as necessary as controls in our own. Radar installations were made effective for this, and for eliminating much of the enemy's flak, which had become highly disastrous. Other radar installations to locate the target and provide navigational aid made blind bombing possible, though less accurate than visual bombing. New and bigger planes, such as the P-51 fighter "Mustang" appeared, soundproofed and camouflaged and equipped with a newer and better system of fire control. (The European campaign will be resumed later.)

The Pacific Theater

Naval War in Pacific, 1942–1943. The Pacific naval campaigns which had begun in 1942 were launched under what appeared to be insurmountable obstacles. The Navy had to maintain itself for weeks at sea, remote from bases, and at the same time cope with land-based airpower and submarines. Naval history entered a new phase in May and June, 1942, when in two successive naval engagements—the battles of the Coral Sea and Midway—neither surface fleet came within sight or gun range of the other. They were carrier air battles. It was in the Pacific also that the submarine became a vital fighting force. Sonar gear on the submarines was effective in locating enemy ships, and in December, 1942, the "Salinoid" firing system (which operates electrically) improved the fire control.

Amphibian Operations, Dukws, Weasels. It was in the South Pacific that the greatest progress was made in amphibian operations. Landings on the many little islands proved the effectiveness of the new landing equipment. There were many new types of landing craft, each for a special purpose. The new amphibian craft, the Dukws which swam out of the LST's to the amazement of all, proved to be the most successful amphibious weapons in the war. The Dukw was a small seagoing truck with a waterproofed hull and a propeller mounted in a tunnel at the stern. It was equipped with a marine rudder interlocked with the steering wheel system. It could be operated on water or land, and could be manuevered at will. Dukws were used to expedite the discharge of cargoes and to carry preloaded artillery pieces. The Weasel was a smaller vehicle and was used in mud and swamps. Its chief advantage lay in its low ground pressure. Improved weasels were effective in later operations.

LCC, LSM. With improved equipment such a brilliant landing was made on Bougainville in the Pacific, November 1, 1943, that the marines hit the beach within 40 minutes after the ship anchored, and in less than 8 hours time, 14,000 men and 6,200 tons of supplies were ashore. Boats were led ashore by the LCC's (Landing Craft Control), which were equipped with navigational microfilm projectors. These were reflector-scopes for inshore piloting. They enabled the navigator to determine the position of his ship. The LSM's (Landing Ship Mechanized) were equipped with airborne rockets which discharged one hundred rockets a minute and had a range of 4,000 to 5,000 yards.

Carriers. Airplanes were also an important factor in these operations, and aircraft carriers played a part in all major engagements. An aircraft carrier could strike with its air group, suddenly and without warning, and within a short time be 100 miles away. Aircraft carriers were produced in huge quantities.

Bathy-thermograph. Submarines became an effective offensive weapon in the Pacific due to a number of technical devices. The Bathy-thermograph, a device developed at the Woods Hole Oceanographic Institute to measure the water temperature, was a factor in determining the acoustical character of the water through which a submarine traveled, and was thereby effective in detecting underwater sounds. A submarine could seek out a suitable temperature gradient and hide there with less risk of detection. With the new sound detection equipment, Allied submarines later made many daring raids, even into Tokyo Bay. New torpedo directors made the kill effective when enemy vessels were sighted.

Full-Blown War Production

1944 Campaign. In 1944 the culmination of all the nation's efforts was reached—a wealth of production, the genius of technology, and the marvels of science. That year saw the greatest of all naval actions in the Battle of the Philippines; the greatest of all amphibian events in Operation Overlord, in the invasion of Normandy; and the annihilation of the Japanese Navy in the Battle of Leyte Gulf.

"Operation Overlord." Operation Overlord, American in conception and largely American in execution was a demonstration of the nation's mighty war potential. Poised on the edge of the English Channel—for centuries never crossed by a hostile army—was a force of 2,876,000 men; 2,500,000 tons of supplies; and 11,000 planes. Every unit of the combined forces possessed the latest and most efficient equipment that science and technology could devise. Allied intelligence units, equipped with their secret instruments and fortified with strategic ruse and deception, were in constant communica-

tion with General Eisenhower, who was now ready to pierce the supposedly impregnable Atlantic wall.

Massing New Weapons. But there was more in that force than mass of equipment, gallant fighters, and brilliant command. For the amphibious operations, there were new and better landing craft: Amphtracks, armored as well as armed, designed to put concentrated fire on the beach immediately in advance of the first wave of troops; more than 2,000 Dukws carrying almost one-half of all the supplies. An army of scientists, engineers and technicians was also at hand. They were in command of microwave devices for guidance and control. There was "Mickey," H2x, used as a navigational aid and high-altitude bombsight on the B-17's, which sent its beams to the surface below and projected a maplike picture on its cathode-ray tube. There was Shoran, ground control of bombing; and "Cat's Eye," the new aircraft interception radar; and Mew, (Microwave Early Warning), the "Big Bertha" of radar. Mew guided the fighter and bomber sweeps into and around the invasion area, projecting on a large glass plotting screen a complete picture of operations at all times. It gave warnings of enemy planes 200 miles away and at 30,000-foot altitude. Radar aids were used for landing paratroopers; they operated as beacons and provided navigational information to the dropping points for night bombing. There were also guided missiles, uncanny in their remote control, as was Anzon used to bomb the roads leading to Paris prior to the invasion.

Rockets of Every Description. There were rockets that struck terror into the enemy ranks. As compared with artillery, the rocket lacked range, velocity, and point accuracy, and did not supplant artillery, but served a purpose artillery was unable to fill. Rockets, lacking recoil, could be launched from lightweight devices that could be easily transported, and when massed large numbers could be fired in a short time. Then, too, there were beach-barrage rockets, rockets on landing craft, and on airplanes. There was a gun called the "Holy Moses," 6 feet long, weighing 140 pounds, which had a velocity of 1,375 feet a second. It could pierce armor 1½ inches thick and reinforced concrete 4 feet thick. Bigger still, there was "Tiny Tim," which could hit a target with 1,250 pounds of explosives. Also there was the High Velocity Aircraft Rocket (HVAR) that hovered constantly over the advancing Third Army in attacking German heavy tanks, described as providing the Army Air Force "with the best anti-tank weapons of the war." And there were the rockets with multiple launching devices, the Whiz-Bangs mounted on Sherman tanks; and the Calliope mounted on a tank without exposing any of the tank crew. With these multiple launchers, a drenching fire of rockets could be spread so quickly that the enemy had no time to cover. Later, there were the Honeycombs, which fired seventy-one rounds in 15 seconds, used in pursuing the fleeing Germans.

Jamming. Technical deception devices were employed to disrupt the enemy's radar. These played a major part in Operation Overlord. Jammers were used to blind the eyes of the German radars and fill their "scopes" with an assortment of simulated echoes. Enemy radar stations, located every half mile or so along the coast had been mapped, and wherever possible, were bombed. Spoof raids were executed to clutter their radars. Technical jammers such as "Window" and "Chaff" were devised. Tons of aluminum strips, dropped with the rhythm of radio pulses, were scattered over large areas. Most efficient was electronic jamming, based on the discovery of a series of high-power magnetrons. Jamming devices were used not only to conceal our maneuvers from the enemy, but also to disrupt the radar control of his operations. These developments led to a search for jamproof controls or secret weapons.

Germany's V-1. As a reprisal to the Allied successful landing in Normandy, the V-1, a new German invention appeared, a "buzz" bomb over England. Rumors of a V-2 were widespread. "Had the Germans succeeded in perfecting and using these new weapons six months earlier," Eisenhower was quoted as saying, "our invasion of Europe would have proved exceedingly difficult, perhaps impossible." Until the atom bomb was developed, the V-2 was described as the finest achievement in the annals of military engineering. But the German scientists and engineers were held back in their development by the bureaucratic machinery of the German government, and their weapon was produced too late.

Anti-buzz Bomb Weapon. To combat the new "buzz bomb" which Germany had developed, the Allies came up with three new weapons that had been developed in the United States: the SCR-584 radar to give warning to antiaircraft of the approach of the enemy plane; the M-9 director, an electronic fire control; and the proximity fuze, hailed as "the most important innovation in artillery ammunition since the introduction of high explosive shells." In the first week of the V-1 attack, about 25 per cent of the targets engaged were destroyed; in the second, 46 per cent; in the third, 67 per cent; and in the fourth 79 per cent. By that date, United States intercepting equipment was in operation. On the last day on which a quantity of V-1's were launched, one hundred and four were detected by early warning radar, but only four reached London. The radar played a dual role: in the navigational control of fighter bombers, and in the control of antiaircraft fire.

Army's Use of Proximity Fuze. These new weapons were the determining factor in many operations. The proximity fuze had been kept as a guarded secret and used by the Navy only, but was released for the Army in time to be used by General Patton in the Battle of the Bulge, where it spread consternation among enemy forces.

Supply Lines. Supplying the huge AEF in France and Germany was also a great scientific, technological achievement. With thirty-eight divisions in action, it was necessary to deliver from the beaches to the front lines 20,000 tons of supplies daily. Other thousands of tons of equipment were delivered to airfields for construction and maintenance, and additional amounts were required for repair of bridges and roads. Gas and oil were pumped from the beaches to main distributing points through pipes laid on the surface of the ground and were brought to the Continent through flexible pipes laid under the English Channel, a system that was described as an astonishing engineering achievement.

The large fleets of motor transports necessary to keep the Army mobile were produced by mass-production methods. More than six hundred thirty thousand vehicles and almost one thousand locomotives, and more than one-third of the rolling stock used were shipped from America. They were shipped on the LST's, and stood on tracks laid in the bottoms of the vessels. In unloading the equipment, flexible connections between the ground tracks and those on the LST's enabled the cars and locomotives to be rolled to their destination, thereby eliminating the use of booms and cranes.

Operation 1945. Early in 1945 the effects of America's and the allied air offensive were becoming catastrophic to Germany's economy. By March, American forces had crossed the Rhine. Germany, however, was still frantically working to produce a secret weapon that would yet turn the tide in her favor. Disturbing information brought by our intelligence staff of a new invention by the Germans, a jet plane, kept the Allied attacking forces vigilant. Since manufacturing sites for jets required long runways and could be spotted, allied bombing planes equipped with all the latest scientific devices swept over these sites to delay the German production. Eisenhower saw his first jet in February of 1945, but allied bombers prevented the enemy from turning them out in sufficient numbers to inflict great damage.

The spring of 1945 saw the beginning of the final campaign in Europe. The huge airborne operation known as "Varsity," with more than one thousand six hundred planes and one thousand three hundred gliders carrying thousands of troops, co-ordinated their attacks with the ground forces to strike at any point along a front of a hundred miles. Strategic bombers, under the protection of accompanying fighters wreaked havoc behind enemy lines. They destroyed German oil reserves and prevented further production. Quantities of equipment were rushed to General Montgomery's forces. This enabled the crossing of the Rhine, March 24, 1945. The mighty fighting machine that had required three years to produce closed in on the enemy. General Eisenhower broadcast to his combined chiefs of staff, "The mission of this Allied Force was fulfilled at 1241, local time, May 7, 1945." The war in Europe was over.

Back to the Pacific. Turning to the Pacific, it should be noted that naval operations, together with the combined forces, reached proportions greater than were ever before witnessed in history. More than thirty thousand aircraft were employed, enabling fast carrier strikes. Submarines patroled the limitless seas. Amphibious operations were conducted over vast areas. New weapons and new techniques were brought into play. Perhaps the greatest feat of all was that this gigantic undertaking was accomplished at such remoteness from home bases. The year 1944 began with the amphibious operations at the Gilberts and Marshalls, and the close of the year saw MacArthur's expeditionary force—covered by battleships and cruisers, a flock of escort carriers, destroyers, and smaller craft—with the General himself splashing ashore from a landing barge to announce to the people of the Philippines, "I have returned."

Amassing New Weapons. But the Japs were not yet ready to give up. New strategy and new weapons of war were brought into play. Radar countermeasures had reached such perfection that at the Battle of Surigao Straits, the entire enemy fleet was lost when the eyes of its own radar were completely blinded by jamming techniques. The air bristled with rockets. Guided missiles were used by General Claire Chennault's Flying Tigers in their attacks on the Burma Road to cut the Japanese lifeline. Flame throwers were used, and incendiary bombs filled with a new combustible, Napalm, wrecked vast areas of flimsy structures in Tokyo. Superbazookas were used to pierce the Japanese rock enclosures, as were the radar navigation and bombing techniques. By the close of the year 1944, and early in 1945, the combined attacking forces were closing in on the enemy.

Iwo Jima, March, 1945. Iwo Jima, which had been pounded for many weeks with some 6,000 tons of shells and explosives in addition to several raids by cruisers and destroyers, finally was taken in March, 1945, by men who went ashore amid a murderous fire. The entrenchments had to be dug out with pick, sledge, shovel and crowbars. With Okinawa captured, the Philippine campaign ended, and Tokyo being bombarded by the fleet, the enemy still refused to acknowledge that the end had come. There yet remained the employment of the major scientific and technological achievement of American scientists, the atom bomb.

Coming of Atomic Weapons

The Atom Bomb. The scope of scientific research that produced the atom bomb, the most devastating of all man-made weapons, has not yet and probably never will be recorded in one single chapter of any one book. It is only the result of the researches that began in Europe and were carried to completion in the laboratories of America which can be noted here.

American scientists had originally been interested in discovering a way to use uranium fission as a source of peacetime energy, rather than for destructive war purposes. As early as 1939, President Roosevelt, acting upon the recommendation of a group of scientists, appointed an Advisory Committee on Uranium, headed by Dr. Lyman J. Briggs of the National Bureau of Standards "to explore the possibilities of developing this new source of energy." The Committee was first placed under the NDRC. As the months passed and as the United States seemed to be headed toward war in the summer and fall of 1941, American and British scientists began to study the possibility of developing a more deadly weapon of destruction—an atomic bomb. Following the attack on Pearl Harbor, December 7, 1941, steps were taken to rush the production of such a weapon at the greatest possible speed.

Never in history did a group of scientists accomplish such a revolutionary achievement in such a short time. Proceeding on the theory that, "if a sufficient mass of element U-235 is brought quickly together, a fission bomb of superlatively destructive power will result," and that "the explosive energy liberated per kilogram of U-235 would be equivalent to approximately 300 tons of TNT," frantic steps were taken to bring this about.

Dr. Vannevar Bush, director of OSRD, secured the endorsement of President Roosevelt to proceed with all haste. Pilot plants were erected, and many of the nation's top physicists went to work in an effort to discover a method for manufacturing a bomb. Sufficient quantities of U-235 had to be separated from the more prevalent isotope U-238. Plans were undertaken to separate the isotopes by both the gaseous diffusion method and the centrifuge method. Scientists at a number of universities went to work on both methods.

Atomic Reactor, December, 1942. When Professor E. O. Lawrence of the University of California had transmuted uranium into two entirely new elements (neptunium and plutonium) by using a huge cyclotron, and when Professor Arthur H. Compton at the University of Chicago announced in December, 1942, the world's first atomic reactor to sustain chain reaction (the atomic pile), the halfway mark toward achieving the atom bomb had been reached.

New Elements Discovered. Next came the problem of producing a bomb by using an element that had just been discovered, when only a few millionths of a gram of the new element were available. Transmutation was the basis for the atomic bomb. Uranium had been transformed into plutonium, something that never before existed. Scientists then went to work in an attempt to discover one or more new elements to be added to the list of ninety-two; and they succeeded. Number ninety-three was called Neptunium; ninety-four, Plutonium; ninety-five, Americinum; ninety-six, Curium; and ninety-seven, Berkelium. Modern science had achieved what the ancient alchemists had always dreamed of—the creation of a new element.

Professor H. D. Smyth, who has written extensively on this pioneer development of the atomic bomb, described this achievement as follows: "The technological gap between producing a chain reaction and using it as a large scale power source or an explosive is comparable to the gap between the discovery of fire and the manufacture of a steam locomotive."

Speeding Up the Research. Scientists were then faced with the task of speeding up the next process and closing a time gap of several thousand years. Large-scale production of a new element of which "only a few millionths of a gram were available" offered a challenge such as the scientific world had never known. Scientists knew that if a large mass of U-235 or plutonium was brought together, a chain reaction would be started and an explosion would result. What they did not know was the efficiency to be expected.

The Manhattan Project. Work on the production of plutonium and the project of separating uranium isotopes was speeded up by awarding a number of contracts. In the spring of 1943, the OSRD scientists joined forces with the Army Corps of Engineers. Brigadier General Leslie R. Groves, working with a Military Policy Committee composed of American, Canadian, and British scientists, decided on a site near Knoxville, Tennessee, on the Clinch and Tennessee Rivers, for building a plutonium pilot plant and electromagnetic separation plant, a huge gaseous diffusion plant, and a thermal diffusion plant.

Hanford, Washington. A large plutonium plant was erected at Hanford, Washington, on the Columbia River, where the atomic pile was cooled by water from the river. Uranium was fed into the pile in aluminum cans; and small quantities of the new chemical element plutonium were produced in the pile and then stockpiled. It was to be used for the proposed atom bomb.

Oak Ridge, Tennessee. Meantime, hundreds of scientists from colleges, universities, and corporations, and tens of thousands of workers began flocking to the site called Oak Ridge, Tennessee, where large funds had been voted for building the Clinton pile. A modern city of nearly eighty thousand population sprang up within a few months. There, a huge gaseous diffusion plant and a large scale electro-magnetic plant were built.

Los Alamos Laboratory. To ensure greater safety, as well as secrecy, a separate laboratory was erected at Los Alamos, New Mexico, for those scientists assigned to work on the bomb itself. There, in the spring of 1943, was constructed "the best equipped physics research laboratory in the world," staffed by world-renowned scientists. While the details of the actual construction of the atom bomb were veiled in secrecy, it was known that a different design for creating atomic piles had been followed. After two years of frantic research, plans were made for the world's first atomic experiment. Two billion dollars had been spent on the project.

In the early morning of July 16, 1945, the first atom bomb in history was exploded from the top of a steel tower erected in a remote site 120 miles southeast of Albuquerque, where the whole countryside "was lighted by a searing light with an intensity many times that of the mid-day sun."

Bombs on Hiroshima, Nagasaki. Man's most devastating weapon was now ready to be used against the one enemy, Japan, with which we were still at war. On August 6, 1945, an atomic bomb was dropped on Hiroshima, and three days later another bomb was dropped on Nagasaki. Both cities were wiped out. The following day the Japanese Emperor sued for peace. World War II was over.

American Technology Wins the War

In conclusion it may be said that the outcome of World War II, which was called a total war, depended upon both the scientist and the fighting man. While the men at the front, in the air, and on the sea were out-maneuvering and out-slugging the enemy, back home thousands of others in the laboratories and on the assembly lines were turning out the finest fighting equipment ever carried into battle. Produced in limitless quantities, and unleashed with devastating accuracy against the enemy, there was no recourse left for him but to surrender. Sheer manpower could no longer compete against the forces of applied science and technology.

Notes for Chapter 41

1. See James Phinney Baxter, III, *Scientists Against Time, With A Foreword* by Vannevar Bush (1946). (By permission of Little, Brown, & Company.)
2. *Ibid.*, p. 4.
3. *Ibid.*, p. 46.
4. *Ibid.*, p. 142.
5. Dwight D. Eisenhower, *Crusade in Europe,* (1948), p. 468. (By permission of Doubleday & Company.)
6. *Ibid.*, p. 171.

Among the many books, other than the two mentioned above, which contain some data on the scientific and technological developments in World War II, see: George C. Marshall, *Biennial Report to Secretary of War, Our Weapons* (July, 1943), (June, 1945); Edgar McInnis, *The War (1940–1946),* 6 vol.; Andrian Van Sinderen, *Four Years, A Chronicle of World War II* (1945); Walter Millis, *The Last Phase* (1946); *Supreme Command, European Theater of Operation, U.S. Army in World War II,* Government Printing Office (1954).

On the United States Navy, see the nine-volume work of Samuel Eliot Morison, *History of United States Naval Operations,* (1947–1954); Archibald D. Turnbull, *History of United States Naval Aviation* (1949); Com. Edward Beach, *Submarine* (1952); John E. Burchard, *Rockets, Guns, and Targets* (1948). Harold G. Bowen, *Ships, Machinery and Mossbacks: The Autobiography of a Naval Engineer* (1955); *United States Army in World War II,* a series of volumes currently edited by Kent R. Greenfield.

Chapter 42

TODAY'S TECHNOLOGIES

In this concluding chapter the reader is reminded once again of the basic theme of this study; namely, that American civilization is fundamentally a technological civilization. It is a civilization in which man at long last has succeeded in conquering his environment. He has harnessed the forces of nature and made them his servant. He can prevent famines and pestilences. Hunger and starvation need no longer exist. A small percentage of our population produces all the food needed by the entire nation, with a surplus sufficient to feed millions in other lands: the same holds true in many other commodities. The word surplus, formerly used only by bankers and financiers, has today, because of the advances in applied science and technology, taken on a new meaning, and has become a common word in our daily vocabulary.

Food Supplies Become Abundant

The raising and processing of foods continues to be the largest single industry in America (measured in dollars), comprising one-sixth of the total manufacturing production of the nation. The American farmers' gross income in the mid-1950's was approximately thirty billion dollars yearly.

Corn. Corn production alone amounted to more than three million bushels. Scientific research over the past quarter-century has resulted in the development of a disease-resisting grain. The introduction of hybrid corn, also the result of scientific research, has greatly increased the nation's yield. With gas-powered tractors to operate plows, harrows, planters, seeders, cultivators, and corn pickers, a complete system of mechanization has resulted.

Wheat. Wheat production in the mid-1950's was averaging more than one billion bushels annually, half again as much as the nation consumed. Again, mechanical seeders and huge combines for harvesting grain, operated by crews that moved from farm to farm across the Midwest, were replacing the traditional harvest hand. These mechanized outfits, ranging from a one-man unit to the larger operators equipped with dozens of machines, cut the average farmer's wheat crop, threshed it, and trucked it to elevators and warehouses in a matter of hours.

Chemistry on the Farm. The chemical industry, besides finding new uses for agricultural products, has performed myriads of miracles for the farmer. To meet the increasing battle between man and insects, new chemicals have been developed to make plants internally resistant to attacking pests. Commercial fertilizers, amounting to several million tons annually, are being used to increase the yield. The introduction of these chemical compounds has had an effect upon American agricultural production comparable to the increase that followed the shift from horse power to tractor power a generation earlier.

The Farm as a Factory. Operating with machines much like those used in industry, the American farm has become a factory. Plants and raw materials, with the energy provided by the sun, are processed into products of food and fiber for man. Mechanical pickers gather up truckloads of beets, carrots, and turnips every hour. Mechanized potato diggers dig and sack tons of potatoes daily. Ladder machines speed up the operations of pruning and picking in the production of deciduous fruits. Irrigation systems have made the farmer virtually independent of rainfall. The middle of the twentieth century saw agriculture restored to the high position conceded to it by Jefferson—not subservient to industry, but through science and technology co-operating with industry as an equal factor in the nation's economy.

Chemurgy. Promoting this co-operation, a new science called "chemurgy" has been developed. By scientific methods, waste materials are being converted into new and useful products. New plants have been introduced, and crops of grain are being turned to new uses. Among such have been the cultivation of bamboo for wood pulp, sunflower and sesame for oil, castor-bean oil for aviation, yams for cortisone for the pharmaceutical industry, and canigre for tannin.

The advances in agricultural technology have produced such a surplus of foods that their distribution has become a major problem in American politics. It has even come to plague the Food and Agricultural Organization (FAO) of the United Nations, and scientists and statesmen have had to join forces in dealing with the world's food supply.

Nutrition. Nutrition, the newest science in food technology, has been championed by the food industry which established in 1942 the Nutrition Foundation for research and education. Its studies have revealed the food requirements of the human system, and food technologists have devised means for their preparation. Americans are assured of being not only the best fed nation in the world, but also the most scientifically fed.

Mass-Producing Buildings and Shelter

Next to food comes shelter. The mid-twentieth century will be remembered as the age of mass production in building technology, and the age of prefabricated houses. "Packaged" buildings of every description—homes, office buildings, stores, warehouses, and even school buildings—were not just temporary makeshift structures, but were built to endure. Even though mass-produced, the finished buildings were of every conceivable design.

Aluminum, Glass, Steel Walls. Metal-clad buildings have been introduced. The ease and speed of this type of construction marked another advance in building technology. Architects and builders began to abandon the accustomed concrete and steel designs in favor of aluminum, glass, and steel. In the early 1950's, a large office building with aluminum slabs was erected in New York City with the speed and efficiency of an automobile assembly line. The eighteen hundred panels, each two stories high, were swung into place and the curtain metal walls were hung without the use of exterior scaffolding. Both glass and steel-clad walls were also introduced in building technology during the same period.

The far-reaching effects of the new technologies on the furnishings and equipment of the American home have already been discussed in Chapter 37.

New Clothing Materials

New Man-made Fibers. Man's third basic need following food and shelter is clothing. The overworked word *revolution* has entered into the discussion of clothing and textile technology. Man-made fibers are responsible for this revolution in textiles. While for thousands of years man had made his clothes from such products as wool, cotton, silk, linen, and leather, recent researches in textile technology have produced numerous man-made synthetic fibers which rival the age-old materials. The introduction of rayon, acetate, nylon, dacron, orlon, and acrilon have brought major changes in the clothing customs of the American people.

Growth of Synthetic Fibers. While the present production of man-made fibers amounts to almost 6,000,000 pounds annually, the industry is described as only in the beginning stage, both in quantity of production and quality of material. The first acetate fiber, developed from the surplus "dope" used for airplane wings in World War I, bears little resemblance to the new synthetics that rival the fleece of the Cashmere goat in luxury. Although the earlier fibers, the rayons and the acetates, still comprise the bulk of today's production, noncellulose fibers have been developed and are taking the place of the fine woolen fabrics used in both men's and women's clothing.

These newly developed fibers permit a greater variety in designs and patterns, and thus offer greater individuality, the very opposite of standardization.

Processing Natural Resources

Production of Raw Materials. The United States continues to lead the world in producing most of the basic materials that have brought about a machine-age civilization. More than half of the steel produced in the world is manufactured in this country. By the mid-1950's the steel industry had reached a production capacity of more than 125,000,000 tons. At the opening of this century, the amount of steel produced per capita was approximately 300 pounds; in the mid-1950's it amounted to almost 1,500 pounds per capita.

Petroleum. Petroleum production which furnishes a large part of the nation's power and heat (in 1953 more than 40 per cent of the total as compared with less than 5 per cent in 1900) had reached a daily average of more than six million barrels by mid-1950's. Advances in petroleum technology include scientific methods for discovering new fields, new techniques in drilling and processing, improvements in distribution, and the manufacture and utilization of thousands of new products. The nation's annual production equals approximately one-half of the world supply.

Coal. Mid-twentieth century America saw a power-hungry nation regarding with alarm the rapid consumption of its exhaustible supplies of coal and gas. Science and technology found new methods for utilizing hitherto discarded low-grade coal and discovered hundreds of new uses for by-products derived from coal. Bituminous coal had by mid-century reached an all-time-high peak of production.

Natural Gas. The production and use of natural gas reflects major changes in energy sources. In 1900 gas was utilized for about 3 per cent of the nation's power. In 1955 it provided more than 25 per cent. Natural gas used for fuel and power was being produced at the rate of approximately 10 trillion cubic feet annually in the mid-1950's. With new discoveries being made periodically, the proved recoverable natural gas reserves were reported in 1955 to be in excess of 200 trillion cubic feet.

Transportation and Communication

Trains and Automobiles. Advances in transportation technology during the first half of the nineteenth century had reduced the size of the American continent to a fractional part of its vast extent during the era of the covered wagon. Diesel-powered trains have almost driven the steam locomotives, once the pride of all railroad transportation, to the scrap pile. More recently, jet engines have taken to the rails. One of the nation's major railroads has introduced gas turbine electric locomotives, which pack the greatest horsepower per foot-length of any internally powered locomotive ever built. Another road has pioneered a coal-burning steam turbine, a 4,500-horsepower locomotive, measuring 161 feet in length, and described as the longest single unit locomotive in the world. It is a steam-electric plant on wheels.

The automobile has become the most universally used vehicle of transportation in the nation's history, and all within the last half century. With the number of motor vehicles approaching sixty million (with a predicted eighty million by 1965) the United States has become a nation on wheels. President Eisenhower, early in 1955, outlined a program for local and national highways requiring more than one hundred billion dollars to build, the greatest peacetime project ever proposed in American history.

The Airplane. Of all the advances in transportation technology the airplane has made the most marked progress. Within a fifty-year period, a method of transportation has been developed that enables people to travel across the American continent in less than 5 hours. An air trail has been established across the North Pole from Los Angeles to Copenhagen. Jet planes had been developed by the mid-

1950's. A truck-mounted launching system had been instituted, similar to those first used for launching guided missiles, a first step in the elimination of runways.

Communication Technology. Advances in communication technology have helped to break down the barriers of regionalism and nationalism. Neither geographical differences nor distances can any longer prevent the inhabitants of one area from knowing what is going on in other areas of their own country, or in lands thousands of miles away. By means of the radio, telephone, and television a system of instant communication has finally been established between peoples in every area of the globe.

Science in the Atomic Age

The National Science Foundation, 1950. Fundamental scientific research, independent of applied science and technology, had reached such proportions by the mid-century to cause Congress to create The National Science Foundation. It was founded upon the creed that "The security of our nation depends upon the maintenance of our present leadership in scientific research and development." The avowed purpose of the Foundation is "to promote the progress of science, to advance national health, to secure national defense, and for other purposes." In its first annual report, November 1, 1951, it was pointed out that "the leading agricultural, economic, industrial, and military position of the United States springs, in a large part, from the technological talent of the American people. Our real genius as a Nation has been the power to convert scientific knowledge into practical utility." The report declared that evidence of the people's genius "is found on every hand, in farming, in industry, in business, in public health, and during the two World Wars, in our military power." Federal support of scientific research was declared to be fundamental, not only to America's high standard of living, but also to national security.

Scientific Research. Scientific research, supported by both federal and nongovernmental agencies, continues at the mid-twentieth century in increasing proportions. An estimated five billion dollars are being spent annually on thousands of different research projects. Considerably more than half of the amount is being allotted to projects that scientists have developed within the past quarter-century. Included among these are the synthetic-fiber industries, radar and television, the development of atomic energy, jet aircraft, the practical use of the helicopter, electronic digital computers, diesel electric and gas turbines, silicones, the germanium diode and transistor, magnesium and titanium for structural uses, cloud-seeding and artificial rainmaking—plus an endless number of other projects. They all contribute to exploring what Dr. Vannevar Bush has described as "an endless frontier of science."

Power from Atoms. America, and the world at large, had moved into a new power age by the mid-1950's—that of atomic power. Anticipating the day when existing sources of power will not be sufficient for a power-hungry nation, scientists began the work of harnessing atomic energy.

Construction of the first commercial electric plant in the nation to be powered by atomic energy (begun in 1954 at Shippingsport, Pennsylvania, near Pittsburgh), and a similar project by the Consolidated Edison Company on the Hudson River, marked the beginning of a new power age. A 3,000-ton submarine, the "Nautilus," made history in January, 1955, when it moved about, on and under the sea, propelled by atomic power. By mid-summer of 1955, nearly one thousand companies had built "hot" laboratories, while some two hundred companies had become regular customers of the Atomic Energy Commission's radioisotopes. Confident of the ability of scientists to bring about the universal use of atomic power, President Eisenhower, in June 1955, proposed that America help the free countries of the world build atomic reactors, "and to supply them with the nuclear material needed to fuel the reactor." Further, he proposed that, "within prudent security considerations," the United States offer to instruct scientists in other countries in the technological process of constructing and operating power reactors for peaceful purposes.

World Conference on Atomic Energy. Two months later, in August 1955, the world's First International Conference on The Peaceful Uses of Atomic Energy met in Geneva, Switzerland, attended by some twelve hundred top scientists, representing seventy-two countries. Again, the President of the United States sent a special message in which he proposed the creation of "A pool of atomic knowledge for all nations to tap." This was followed by the surprise announcement that the United States would offer 200 kilograms (a total of 440 pounds) of fissionable material to serve as "fuel" in experimental reactors, an amount in excess of that offered by all other nations combined. The aim for pooling such scientific knowledge was to make available to backward nations as well as to the most advanced, the uses of the greatest discovery made since man learned to build a fire a half-million years ago.

Atomic Power Uses in United States. The United States delegates were able to report that atomic power had been put to a number of industrial uses in this country. More than a dozen firms in the United States were prepared to build nuclear-power reactors. Radioisotope devices were being used to spot defects in virtually all manufactured products, ranging from the thickness of rolled steel and tin cans to such items as paints, papers, lubricants, and floor wax.

In the field of medicine and surgery, radioisotopes were being used in treatment of cancer, as tracers in blood streams to detect hidden tumors and abscesses, and to diagnose liver ailments. Radioisotopes have been used in studying nature's greatest reaction, photosynthesis—the process by which vegetation builds itself, using only water and carbon dioxide of the air. Efforts are being made to discover how sunlight, by falling on green leaves, produces cellulose, sugars, and starch. Atomic radiation is used successfully to influence chemical reactions to eradicate rust fungus in wheat and oats, to keep potatoes from spoiling, to keep food products fresh, and to alter the qualities of plastics.

The peacetime uses of atomic energy point the way toward making possible the highest standard of living ever enjoyed by the human race.[1]

Automation: New Concept of Mechanization

Accompanying the mid-twentieth century advances in providing food, shelter, and clothing; in the processing of basic raw materials into finished products; in the creation of new materials; in communication and transportation technology; and in the introduction of new sources of energy came another development—automation.

Automation has been defined in simple terms as "the use of machines to run machines." A fuller definition of automation describes it as "an advanced mechanism through which production is increased, through which quality standards are raised and controlled, and through which production costs are cut by means of automatic machines, conveyor systems, electronic controls, feedback systems, and brain-like computers." Automation not only replaces the muscle of man, "but also replaces a part of his mind." [2]

Beginnings of Automation. Although the word "automation" is comparatively new, it describes a process that has been going on at least as far back as the days of James Watt, who in 1788 devised a fly-ball "governor" to control the speed of his steam engine. The device relieved Watt, and his helper, from having to regulate by hand the speed of the engine—the "governor" did the job. During the first decade of the twentieth century, the terminology of automatic controls was employed when it meant nothing more than dropping a coin in a machine (an automat), and out came a sandwich, a piece of pie, and a cup of coffee. The system was later extended to player pianos, juke boxes, and slot machines.

Myriad Uses. With the coming of electric power, a number of other automatic mechanisms developed. During World War II air warfare demanded rapid and precise automatic control of searchlights and antiaircraft guns. Also, during that war, servo-controlled turrets, servo-steered torpedoes, radar eyes, and closed-loop missiles like V-1, carrying boxes of automatic guiding instruments, were used extensively. These devices, plus a number of simple voltage regulators, comprised some of the principal achievements in automatic controls.

Thousands of automatic controls have been introduced in industry, in military establishments, in offices, and in homes. A "governor" in a central power station makes an electric clock keep correct time. Electronic governors stabilize the performance of radio and television sets. Our whole communication system is one vast automatic control circuit. An intricate network of automatic control instruments runs many of the large petroleum refineries. Chemical plants introduced automatic mechanisms to regulate temperature and flow. Large food-processing plants and bottling works have introduced automatic control in virtually all operations. These "robots," as they were first called, have been a part of the American industrial system for years.

Meantime, as various automatic controls were being introduced, there seemed to be no understanding or connection between the different developments. It was only within the last few years that scientists reached the conclusion that all control and regulation systems depend on common principles. When, for example, it was discovered that, "the design of controls for a modern boiler or a guided missile is based primarily on principles first developed in the design of radio amplifiers," rapid technological progress was made in automation. Underneath all these various phenomena lies a common pattern, an interdependence, called the "feedback" principle.

Feedback Principle. A simple illustration of the difference in function of two small well-known instruments, the thermometer and the thermostat, indicates the point at which the feedback system steps in. First consider the action of a room thermometer. When the room becomes warmer, the thermometer reads higher; when it becomes cooler, it shows a lower reading. The room is not made warmer by action of the thermometer. Rather, the rise and fall of the mercury in the thermometer merely registers the different degrees of temperature. But note the thermostat. It is so contrived that as the temperature readings go below a selected setting, it agitates a valve which "kicks on" the fuel (the furnace). Conversely, when the temperature rises to a fixed setting, it directs the furnace to be "shut off." The thermostat, with its contrivance, not only registers the temperature in the room, but the room temperature is regulated by the thermostat. The two conditions are independent, yet each is a cause, and each is an effect of the other. Few systems of automatic feedback are this simple. Rather, they become more intricate when applied to other and larger quantity mechanisms and manufactured products, and by the large demand for goods of uniform high quality.[3]

Need for Automation. In the present-day system of complicated technologies, many products are processed under conditions of speed, temperature, pressure, and chemical exchange that make human controls impossible, or at least impracticable. Then, too, modern machines and instruments themselves must often satisfy standards of quality and precision in operation never before attained. Beyond certain limits the control of qualitative differences eludes human capacity, and as the tempo and complexity increases, human operators simply cannot keep up with the demands of the machines. Hence, scientists found it necessary to devise instruments and give them control functions. Greater power was given to the machines to replace human muscles, but more than that, mechanisms were devised that enabled them to assume some of the functions of the human brain—anticipate, proportion, and integrate.

Continuous Petroleum Refining. A number of industries have introduced automatic controls and the feedback system. The petroleum industry offers a good illustration. While the oil refineries are not yet entirely automatic, still their operation points the way toward the controlled plants of the future. In a modern refinery crude petroleum is fed in at one end of the plant, flows continuously through a series of treatment chambers and pipes, and pours out a variety of finished products at the other end. All the successive operations are amenable to automatic controls.

A refinery in which these processes occur represents a bewildering kind of a factory, with metallic towers rising twenty stories high, and equipped with hundreds of miles of pipe. The plant is next to noiseless. Only a few lonely men wander about to perform supervisory or maintenance tasks. Despite its seeming inertness, the plant is throbbing with internal heat and motion. Every day a quarter of a million barrels of oil flow through its pipes, and come out in the form of dozens of finished petroleum products. In the process, a great volume of different chemicals are used, and an equally large number of new chemicals are manufactured.

The nerve center of the huge mechanical plant is the control room with its control panel. A few men sitting at the controls perform the few manual operations that are necessary to keep the oil flowing. Complete as are all these operations, yet they are not wholly automatic. Automation in oil refineries, as in other industries, is still a relative term. There are as yet few, if any, factories operated completely by a "push-button" control.

The Computer System. In order to establish fully automatic controls, machines must be designed that simulate human abilities. In an effort to approach this stage, a device called the "computer," has been designed. While the chief function of this special mechanism is computation, yet it does more. It aids the human operator in performing many specialized functions, and also aids in the over-all operation of the control mechanisms. For these reasons, another name, "information center, or information machine," has been applied to computers. The term "information machine" distinguishes its function from a power machine, such as a loom. The loom does the physical work of weaving a fabric, whereas the information machine controls the pattern that is being woven by the loom. Its purpose is not to perform the work, but to order and supervise the way the work is done.

The information center is called the "lifeblood" of automatic controls. To receive and act on information (note the lowly thermostat again, signaling the temperature and indicating whether the furnace is to be turned "on" or "off") is the essential function of every control system, from the simplest to the most complex.

Industries and business organizations of all kinds are approaching the era of automation. The many advantages offered by this newest development in technology are impelling. One of these is the need for fewer employees, without reducing the total output. The number of man-hours required for "the average unit of output" has declined steadily in the last seventy-five years. But, in the past quarter-century it has declined at an astounding rate. In 1940 the unit of output consumed only about half as many man-hours as were required in 1920. By 1955 the time consumed was much less. Wassily Leontief, writing on "Machines and Man" (*Scientific American,*

September, 1952, p. 150 *ff*) said: "Man has almost ceased to be a lifter and mover, and [has] become primarily a starter and stopper, a setter and an assembler, and repairer." Leontief predicts that with the further introduction of controlled machines, man's role will be even less; the starter and stopper will disappear first, the setter and assembler will go next. Troubleshooters and repair men will of course always be needed. So, too, will inventors and designers. Able scientists and technicians will always be in demand to fill highly skilled positions, design and install new machines, prepare precise information and feed it into the machine. Engineers, mathematicians, and logicians will play a more prominent role in determining our nation's future.

Potential Effects of Automation. As to the effects of automation, automatic controls, computers, and mechanical information centers on American society, the economist, sociologist, and the philosopher will be arguing for years to come. Meantime, science will move into new fields not yet dreamed of, into developments, described by General David Sarnoff as those that will be born from "the numberless embryos in the womb of science."

As man masters his physical environment and succeeds in producing surpluses in all needed commodities, he will move toward that greater goal, a spiritual revolution.

Arnold J. Toynbee, distinguished British historian, wrote (New York *Times Magazine,* December 2, 1951) that "This twentieth century will be remembered as the era when the benefits of civilization will have become available to the whole human race." The idea of the common welfare for all people, he noted, had come initially from early America. It was founded upon the belief that the vast American continent could provide sufficient food and material possessions for everyone. Mankind's hope of better things, according to Toynbee, "lies in a permanent industrial revolution."

The industrial-technological revolution has been a continuing force in American history for more than three hundred years. Instead of reducing man to a cog in a vast machine-age civilization, as many feared, it has done just the opposite. It has given him greater freedom and more leisure time: time to be creative. Never in any civilization have so many people had so much time to be themselves—thanks to today's technologies.

Notes for Chapter 42

1. Three Ford brothers, Henry Ford II, Benson, and William Clay announced during the conference the establishment of a one million dollar fund, "to provide annual atoms-for-peace awards for the next ten years." Each year an international jury of awards selects the individual or group of individuals who does most to develop atoms-for-peace, "without regard for nationality or political belief." The total annual award amounts to $75,000.00, and a "suitable medal."

2. See article by E. E. Evanson, "You and Automation," *Industry And Power* (July, 1952), p. 47 *ff*.

3. See *Scientific American* (September, 1952); the entire issue of eight articles in all, each by a specialist, deals with "Automatic Controls."

INDEX

A

D

F

H

M

N

Q

R

S

Y

Z